Water Science and Technology Library

Volume 98

The aim of the *Water Science and Technology Library* is to provide a forum for dissemination of the state-of-the-art of topics of current interest in the area of water science and technology. This is accomplished through publication of reference books and monographs, authored or edited. Occasionally also proceedings volumes are accepted for publication in the series. *Water Science and Technology Library* encompasses a wide range of topics dealing with science as well as socio-economic aspects of water, environment, and ecology. Both the water quantity and quality issues are relevant and are embraced by *Water Science and Technology Library*. The emphasis may be on either the scientific content, or techniques of solution, or both. There is increasing emphasis these days on processes and *Water Science and Technology Library* is committed to promoting this emphasis by publishing books emphasizing scientific discussions of physical, chemical, and/or biological aspects of water resources. Likewise, current or emerging solution techniques receive high priority. Interdisciplinary coverage is encouraged. Case studies contributing to our knowledge of water science and technology are also embraced by the series. Innovative ideas and novel techniques are of particular interest.

Comments or suggestions for future volumes are welcomed.

Vijay P. Singh, Department of Biological and Agricultural Engineering & Zachry Department of Civil and Environment Engineering, Texas A &M University, USA
Email: vsingh@tamu.edu

More information about this series at http://www.springer.com/series/6689

Ramakar Jha · Vijay P. Singh · Vivekanand Singh ·
L. B. Roy · Roshni Thendiyath
Editors

Climate Change Impacts on Water Resources

Hydraulics, Water Resources and Coastal Engineering

Springer

Editors
Ramakar Jha
Department of Civil Engineering
National Institute of Technology Patna
Patna, India

Vivekanand Singh
Department of Civil Engineering
National Institute of Technology Patna
Patna, India

Roshni Thendiyath
Department of Civil Engineering
National Institute of Technology Patna
Patna, India

Vijay P. Singh
Department of Biological and Agricultural Engineering
Texas A&M University
College Station, TX, USA

L. B. Roy
Department of Civil Engineering
National Institute of Technology Patna
Patna, India

ISSN 0921-092X ISSN 1872-4663 (electronic)
Water Science and Technology Library
ISBN 978-3-030-64201-3 ISBN 978-3-030-64202-0 (eBook)
https://doi.org/10.1007/978-3-030-64202-0

This Springer imprint is published by the registered company Springer Nature Switzerland AG
The registered company address is: Gewerbestrasse 11, 6330 Cham, Switzerland

Preface

Climate change has been emerging as one of the major challenges in the world. Changes in climate may lead to adverse negative impacts on both natural and human systems. Continued emissions of greenhouse gases would further amplify the existing risks and create new complications for people and ecosystems. Climate change impact on water resources is an important aspect of effective water resources management. Increasing temperature, intense precipitation, and development of heat islands are causing hydrologically extreme events in the Indian subcontinent. To analyze the possible impacts of climate change on a river basin, it is required to predict the impact of future climate changes, which in turn, will help in the planning and management of water resources in the basin. General Circulations Models (GCMs) and statistical downscaling of GCM output are one of the most credible tools presently available for modeling climate change. Downscaling is one of the approaches where GCM outputs are interpolated to the scale of hydrological modeling or local scale requirements.

Over the years, a number of books on climate change impact have been brought out. Most of the books published so far are theoretical, with a limited number of examples and case studies. This book is an amalgamation of available resources and is divided into various sub-themes.

In the present book, different climate scenarios have been evolved using different soft computing techniques and statistical downscaling models. The seasonal and inter-annual variability of sea surface temperature and its correlation with maximum sustained wind speed; trend analyses of temperature series; trend analysis of rainfall series and gridded rainfall; analysis of intensity-duration-frequency and depth-duration-frequency curve projections under climate variability; and trend analysis of evapotranspiration and reference crop evapotranspiration are discussed for many river basins of India, including Tapi basin, Godavari river basin, Eastern Ganga Canal command, Naini river basin, western Maharashtra, Kerela, Chattisgarh, Ganga basin and other basins.

Observed spatiotemporal trends of precipitation and temperature over Afghanistan is an interesting study in the book. The climate change impacts on water resources in Ethiopia is also an interesting inclusion in the book.

Different case studies indicating the impact of climate change in water resources at varying climatic conditions are demonstrated effectively. Assessment of climate change on crop water requirement and on hydrological regime and other studies are described for Narmada basin and other river systems of India.

Some of the improved algorithms and mathematical formulations include statistical downscaling of GCM output and simulation of rainfall scenarios, identification of rainfall trends using singular spectral analysis, use of EM algorithm for missing data on monthly maximum and minimum temperatures, characteristics of GLDAS evapotranspiration and its response to climate variability, fuzzy maximum likelihood segmentation and trend analyses, adaptive neuro-fuzzy inference system, comparison of CMIP5 wind speed from global climate models, comparison of selection of predictors for statistical downscaling of precipitation using different statistical techniques, neural network techniques, and neuro-fuzzy techniques, forecasting reference evapotranspiration using artificial neural networks, and error analysis of TMPA near-real-time precipitation will be of interest.

The book will help understand the climate change phenomenon in the basins of India, Afghanistan, and Ethiopia. Also, the book highlights the impact of climate changes on rivers and flood plains with respect to floods and droughts.

Patna, India — Ramakar Jha
College Station, TX, USA — Vijay P. Singh

Acknowledgements

The editors are grateful to Prof. P. K. Jain, Director, NIT Patna, for his constructive support and encouragement for completing the book. They also thank reviewers for reviewing the papers included in the book.

This book wouldn't have been possible without the support of the Indian Society for Hydraulics, CWPRS Pune, and sponsors who supported the annual conference. This book constitutes a portion of the proceedings of the conference.

The editors thank all the experts who delivered the Keynote address, all the participants, faculties, staff, and the students who contributed to the completion of the book. Some of the students are Kamakshi Singh, Saba Khurshid, Ratnesh Kumar, Shalini, and Ravi Ganga.

Finally, the editors acknowledge all those who helped with bringing the book to fruition, especially the authors of all the papers.

Contents

1 **Trend Analyses of Seasonal Mean Temperature Series Pertaining to the Tapi River Basin Using Monthly Data** 1
Ganesh D. Kale

2 **Dry Spell and Wet Spell Characterisation of Nandani River Basin, Western Maharashtra, India** 9
Abhijit Mohanrao Zende and Prashant Basavaraj Bhagawati

3 **Assessment of Climate Change on Crop Water Requirement in Tandula Command of Chhattisgarh (India)** 19
Rahul Kumar Jaiswal, H. L. Tiwari, and Anil Kumar Lohani

4 **Impact of Climate Change on Hydrological Regime of Narmada River Basin** 35
Deepak Kumar Tiwari, H. L. Tiwari, Raman Nateriya, and Satanand Mishra

5 **Climate Change Impacts on Water Resources in Ethiopia** 47
Abiot Ketema and G. S. Dwarakish

6 **Spatio-Temporal Trend Analysis of Long-Term IMD-Gridded Precipitation in Godavari River Basin, India** 59
C. H. Praveenkumar and V. Jothiprakash

7 **Forecasting Reference Evapotranspiration Using Artificial Neural Network for Nagpur Region** 67
Nikhil Band, Aniruddha Ghare, and Avinash Vasudeo

8 **Time-Varying Downscaling Model (TVDM) and its Benefit to Capture Extreme Rainfall** 79
Subbarao Pichuka and Rajib Maity

9 **An Assessment of Impact of Land Use–Land Cover and Climate Change on Quality of River Using Water Quality Index** 91
Manisha Jamgade and Shrikant Charhate

10 **Assessment of Tail Behavior of Probability Distributions of Daily Precipitation Data Over India** ... 101
Neha Gupta and Sagar Rohidas Chavan

11 **Benefit of Time-Varying Models Developed Using Graphical Modeling Approach for Probabilistic Prediction of Monthly Streamflow** ... 115
Riya Dutta and Rajib Maity

12 **Determination of Effective Discharge Responsible for Sediment Transport in Cauvery River Basin** ... 127
Shobhit Maheshwari and Sagar Rohidas Chavan

13 **A Comparative Study of Potential Evapotranspiration in an Agroforestry Region of Western Ghats, India** ... 139
Pandu Narayana and K. Varija

14 **Influence of Air Temperature on Local Precipitation Extremes Across India** ... 149
Sachidanand Kumar, Kironmala Chanda, and Srinivas Pasupuleti

15 **Effect of Spatial and Temporal Land Use-Land Cover Change on the Rainfall Trend: A Case Study in Kerala** ... 161
Lini R. Chandran and P. G. Jairaj

16 **Innovations and Application of Operational Ocean Data Products for Security of Marine Environment** ... 173
Madhulika Sinha and Shrikant Charhate

17 **Statistical Downscaling of Sea Level by Support Vector Machine and Regression Tree Approaches** ... 183
S. Sithara, S. K. Pramada, and Santosh G. Thampi

18 **Assessing the Impacts of Climate Change on Crop Yield in Upper Godavari River Sub-basin Using H08 Hydrological Model** ... 193
Pushpendra Raghav and T. I. Eldho

19 **Evaluation of Time Discretization of Daily Rainfall From the Literature for a Specific Site** ... 207
R. Harshanth, Saha Dauji, and P. K. Srivastava

20 **Quality Checks on Continuous Rainfall Records: A Case Study** ... 223
R. Harshanth, Saha Dauji, and P. K. Srivastava

21 **Assuring Water Intake Sustainability Under Changing Climate** ... 233
Gaurav Talukdar and Arup Kumar Sarma

22 **Characteristics of Gldas Evapotranspiration and Its Response to Climate Variability Across Ganga Basin, India** 241
Lalit Pal, C. S. P. Ojha, and Amit Kumar

23 **Seasonal and Inter-Annual Variability of Sea Surface Temperature and Its Correlation with Maximum Sustained Wind Speed in Bay of Bengal** 253
Jiya Albert and Prasad K. Bhaskaran

24 **Comparison of CMIP5 Wind Speed from Global Climate Models with In-Situ Observations for the Bay of Bengal** 267
Athira Krishnan and Prasad K. Bhaskaran

25 **Trend Analysis of Temperature for Eastern Ganga Canal Command** 279
Radha Krishan, Bhaskar R. Nikam, and Deepak Khare

26 **Analysis of Long-Term Rainfall Trends in Rajasthan, India** 293
Darshan Mehta and S. M. Yadav

27 **Statistical Downscaling of GCM Output and Simulation of Rainfall Scenarios for Brahmani Basin** 307
Lasyamayee Lopamudra Sahoo and Kanhu Charan Patra

28 **Impact of Land Use–Land Cover Changes on the Streamflow of the Kolab River Basin Using SWAT Model** 319
Partha Sarathi Bhunia and Kanhu Charan Patra

29 **Statistical Downscaling of Climatic Variables in Indo-Gangatic Alluvial Plain** 333
Prabhakar Shukla and Raj Mohan Singh

30 **Comparing Global High-Resolution Precipitation Data with Rain Gauge Data in Assam, India** 341
Pulendra Dutta, Dipsikha Devi, and Arup Kr. Sarma

31 **Variability of Rainfall, Temperature and Potential Evapotranspiration at Annual Time Scale Over Tapi to Tadri River Basin, India** 349
Prem Mahyavanshi, V. D. Loliyana, and Priyank J. Sharma

32 **Climate Change and Water Resources: Emerging Challenges, Vulnerability and Adaptation in Indian Scenario** 365
Y. Shiva Shankar, Abhishek Kumar, and Devendra Mohan

33 **Observed Spatio-Temporal Trends of Precipitation and Temperature Over Afghanistan** 377
S. Rehana, P. Krishna Reddy, N. Sai Bhaskar Reddy, Abdul Raheem Daud, Shoaib Saboory, Shoaib Khaksari, S. K. Tomer, and U. Sowjanya

34 Identification of Historical Shift, Dispersion, and Trend of the Monsoon Season for Guwahati City Using Fuzzy Segmentation and Trend Analyses ... 393
Amrutha Suresh and Pekkat Sreeja

35 Analysis of Intensity–Duration– Frequency and Depth–Duration– Frequency Curve Projections Under Climate Variability ... 407
Manish Kumar Sinha, Klaus Baier, Rafig Azzam, M. K. Verma, and Ramakar Jha

36 Changes in Monthly Hydro-Climatic Indices for Middle Tapi Basin, India ... 423
Priyank J. Sharma, P. L. Patel, and V. Jothiprakash

37 Multiobjective Automatic Calibration of a Physically Based Hydrologic Model Using Multiobjective Self-Adaptive Differential Evolution Algorithm ... 435
Saswata Nandi and M. Janga Reddy

38 Adaptive Neuro-Fuzzy Inference System-Based Yield Forecast Using Climatic Variables ... 449
Kalpesh Borse and P. G. Agnihotri

39 Impact of Climate Change on Hydrological Parameters ... 457
Arunima Priyadarsini Patnaik and Bandita Naik

40 Morphometric Analysis of Kosi River Basin, Bihar, India Using Remote Sensing and GIS Techniques ... 469
Niraj Kumar and Ramakar Jha

41 Simulation of Impact of Climate Change on the Performance of a Reservoir System in Eastern India ... 483
Satabdi Saha, Debasri Roy, and Rajib Das

42 Assessing the Impact of Spatial Resolution on Land Surface Model Based on Hydrologic Simulations ... 493
Aiendrila Dey and Renji Remesan

43 Infilling Missing Monthly Maximum and Minimum Temperature Dataset by EM Algorithm Followed by Distribution Based Statistical Assessment Using Eight Absolute Homogeneity Tests ... 503
P. Kabbilawsh, D. Sathish Kumar, and N. R. Chithra

44 Multisite Monthly to Daily Naturalised Streamflow Disaggregation Using Daily Flow Pattern Hydrograph ... 519
Vivek Verma

45 Error Analysis of TMPA Near Real-Time Precipitation Estimates for an Indian Monsoon Region 529
Ashish Kumar and RAAJ Ramsankaran

46 Comparison of Selection of Predictors for Statistical Downscaling of Precipitation Using Different Statistical Techniques 541
Kumar Keshav, Vivekanand Singh, and Roshni Thendiyath

About the Editors

Ramakar Jha is a Chair Professor at the Department of Civil Engineering and has 30 years of experience in the field of hydrology and water resources engineering. Dr. Jha is presently working as Chair Professor in the Department of Civil Engineering, National Institute of Technology (NIT) Patna-INDIA, which is a Premier Institute in India under the Ministry of Human Resource Development, Government of India. Dr. Jha has served at various levels from Scientist-B to Scientist-E1 at National Institute of Hydrology (NIH), Roorkee, India, and as a Professor in the Department of Civil Engineering, NIT Rourkela. He has worked and is working as Country Co-ordinator of UNESCO-GWADI and Principal Investigator for many International (EU-FP7, DAAD, ADB, AUS-Aid) and National research and consultancy projects (ISRO, DST, MoWR, MHRD). Moreover, he served as Chair for many administrative positions and received a couple of international and national awards for research papers. Presently, he is working as Dr. Rajendra Prasad Chair for Water Resources under the Ministry of Water Resources, Government of India in the Department of Civil Engineering, NIT Patna, Bihar.

Vijay P. Singh is a University Distinguished Professor, a Regents Professor, and Caroline and William N. Lehrer Distinguished Chair in Water Engineering at Texas A&M University. He received his B.S., M.S., Ph.D., and D.Sc. in engineering. He is a registered professional engineer, a registered professional hydrologist, and an Honorary Diplomate of ASCE-AAWRE. He is a Distinguished Member of ASCE, a Distinguished Fellow of AGGS, and an Honorary Member of AWRA, and a Fellow of EWRI-ASCE, IAH, ISAE, IWRS, and IASWC. He has published extensively in the areas of hydrology, irrigation engineering, hydraulics, groundwater, water quality, and water resources (more than 1320 journal articles, 31 textbooks, 75 edited reference books, 110 book chapters, and 315 conference papers). He has received more than 95 national and international awards, including 3 honorary doctorates. He is a member of 11 international science/engineering academies. He has served as President of the American Institute of Hydrology (AIH), Chair of Watershed Council of American Society of Civil Engineers, and is currently President of the American Academy of Water Resources Engineers. He has served/serves as Editor-in-Chief of

3 journals and 2 book series and serves on editorial boards of more than 25 journals and 3 book series. His Google Scholar citations include 65120; h-index: 113, and I10-index: 993.

Vivekanand Singh is a professor at the Department of Civil Engineering and has 28 years of experience in the field of river hydraulics, groundwater, and water resources engineering. Dr. Singh has published research papers in international journals including ASCE Journal. He has done national projects and consultancy during his work at NIT Patna. Prior to this, Dr. Singh was a Scientist at the National Institute of Hydrology Roorkee. He organized several summer courses, conferences, and workshops. He has been working as an Editor in some Indian Journals.

L. B. Roy is a Professor at the Department of Civil Engineering and has 33 years of experience in the field of water resources engineering as well as geotechnical engineering. Dr. Roy has been serving the Water Resources Engineering Department for a long time and carried out various field-based research activities for different river systems of the region with special emphasis on floods. Dr. Roy has been involved in many research projects and published several research papers in peer-reviewed journals. He has been working for Intellectual Property Rights (IPR) and received awards for various activities.

Roshni Thendiyath is an Assistant Professor at the Department of Civil Engineering and has 9 years of experience in the field of water resources engineering. Dr. Roshni earned her Ph.D. in Civil Engineering at the University of Pisa, Italy. After obtaining Ph.D., Dr. Roshni had joined the National Institute of Technology Calicut, India as an Ad hoc faculty and later joined the National Institute of Technology Patna as an Assistant Professor. She has supervised 2 Ph.Ds., more than 30 Master's theses. Furthermore, she has been awarded the DST-SERB project for the topic Two Phase flows and Water Quality in Rivers, funded by the Department of Science and Technology, MHRD India. Results obtained from her research have been published in nearly 30 papers in international journals and more than 15 papers in international conferences and 3 book chapters. Dr. Roshni is active in a variety of professional bodies and she has organized numerous workshops and conferences in her academic career.

Chapter 1
Trend Analyses of Seasonal Mean Temperature Series Pertaining to the Tapi River Basin Using Monthly Data

Ganesh D. Kale

Abstract Information about climate change on the scale of basin is very crucial for development, planning and utilization of water. The Tapi basin is climatically responsive. One of the most crucial parameters of the climate is temperature. Alterations in climate at regional scales influence rudimentary aspects of our life. It is important for policymakers that individual seasonal alterations are illuminated. Therefore, in the current work, trend detection analysis of regional seasonal (four seasons) mean temperature (T_{mean}) series (1971–2004) appertaining to the Tapi basin is conducted. Evaluation of trend magnitude is performed by using Sen's slope (SS) test. Identification of dependence of data is performed by using correlogram. Assessment of trend significance for independent and dependent data is performed by using Mann–Kendall (MK) test and block bootstrapping approach with MK test (MKBBS test), respectively. The results showed the presence of statistically significant increasing trend only in regional winter T_{mean} series of the basin.

Keywords Regional series · Seasonal series · Mean temperature · Tapi basin · Trend detection

1.1 Introduction

Warming of climate system is definite, which is now recognizable through observations of incrementing global mean temperatures of ocean and air, extensive defrosting of ice and snow and ascending global mean sea level (IPCC 2007). Thus, it is clear that, there is impact of climate change on temperature. Information about climate change on the scale of basin is very crucial for development, planning and utilization of water (Singh et al. 2008a). Significant increase is noticed in maximum and mean temperatures of the Narmada–Tapi basin for the post-monsoon season (Deshpande et al. 2016). For the Tapi basin, which is located in the Central region of India, 2 °C

G. D. Kale (✉)
Civil Engineering Department, SV NIT, Surat 395007, Gujarat, India
e-mail: gdk@ced.svnit.ac.in

R. Jha et al. (eds.), *Climate Change Impacts on Water Resources*, Water Science and Technology Library 98, https://doi.org/10.1007/978-3-030-64202-0_1

rise in surface temperature is observed over the period 1990–2010 (Bhamare and Agone 2011). Thus, it can be inferred that Tapi basin is climatically sensitive.

One of the most crucial parameters of the climate is temperature. Temperature affects hydrologic processes occurring in a basin (Singh et al. 2008b). Alterations in climate at regional scales influence rudimentary aspects of our life, containing welfare and health, natural ecosystems and economy (Fahad et al. 2017). It is important for policymakers that individual season alterations are illuminated, as various seasons are vital for different aspects of management of water (Hannaford and Harvey 2010).

In the preparation of regional seasonal data of T_{mean}, if average of monthly values of months in given season is considered as value of seasonal T_{mean} for given season, then there may be possibility of loss of climate change signal because of averaging. But, if T_{mean} values of individual months in respective season are contemplated then aforementioned possibility may be averted. Thus, in the current work, trend analysis of regional seasonal T_{mean} series appertaining to the Tapi basin is conducted by contemplating individual month T_{mean} values in the given season.

1.2 Data

Trend analysis of regional T_{mean} series appertaining to the Tapi basin is carried out in the current study. For the aforesaid analysis, Climate Research Unit Timeseries (CRU TS) 3.22 data set of T_{mean} (1901–2013) is downloaded. The aforesaid data is monthly gridded data having lat/long resolution of 0.5° × 0.5° which is available at "http://www.cru.uea.ac.uk/cru/data/hrg/cru_ts_3.22/" ("last accessed on January 14, 2018"). The analysis period used in this study is 1971–2004 (34 years). For each month of every year of analysis period, mean of monthly T_{mean} values of all grids over the Tapi basin is taken as regional monthly T_{mean} value for given month to get the regional monthly T_{mean} series. Regional seasonal T_{mean} series are prepared from regional monthly T_{mean} series. For example, regional pre-monsoon T_{mean} series consists of monthly T_{mean} values of March to May months of every year in analysis period. Sonali and Nagesh Kumar (2013) stated that majority of the trend both in minimum and maximum temperature commence after 1970 either in seasonal or annual levels. Thus, 1971 is the beginning year of analysis period. In Sonali and Nagesh Kumar (2013), it was also stated that, for trend detection, minimal tally of years needed is 34. In the current work, analysis period is also 34 years.

1.3 Study Area

The Tapi basin is extended over the states of Gujarat, Maharashtra and Madhya Pradesh (MP) covering 65,145 km^2 area. The basin has maximum width and length of 196 and 534 km, respectively. Basin is situated between north latitudes 20°9′ to 21°50′ and east longitudes of 72°33′ to 78°17′. Tapi river is the second largest

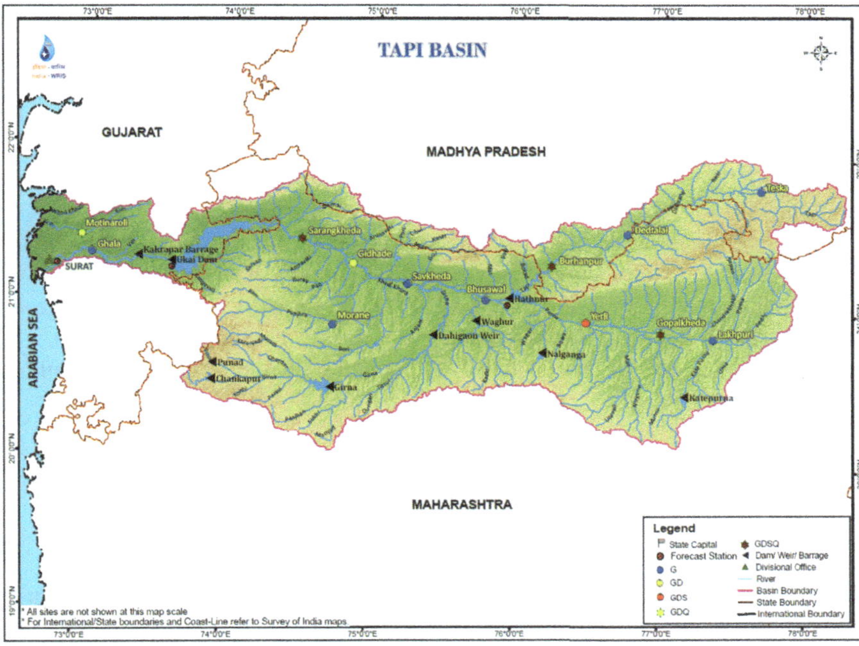

Fig. 1.1 Tapi Basin. (*Source* http://india-wris.nrsc.gov.in/wrpinfo/index.php?title=Tapi)

river of the peninsula, draining toward the west. It emerges at 752 m elevation, near Multai reserve forest located in Betul district, MP. Important tributaries of the Tapi river joining from the left are Amravati, Vaghur, Panjhra, Buray, Girna, Bori, Purna, Sipna and Mona, while those joining from the right are Gamai, Suki, Aner and Arunavati. Plain areas of the basin are fertile, broad and proper for cultivation, while hilly area of the basin is well timbered. The total river length is 724 km, from start to way out into the Arabian Sea. The dominant part of the basin is covered by agriculture which accounts for 66.19% of the basin area. Water bodies covered 2.99% area of the basin (http://india-wris.nrsc.gov.in/wrpinfo/index.php?title=Tapi). The Tapi basin is shown in Fig. 1.1.

1.4 Methodology

In the current work, SS test (Sen 1968) is used for evaluation of trend magnitude, while MK test (Mann 1945) or MKBBS test is applied for evaluation of trend's statistical significance for independent or dependent data, respectively. Sonali and Nagesh Kumar (2013) and Khaliq et al. (2009) among others have also used the aforesaid three tests.

Existence of serial correlation at one or more than one lags shows that data is dependent (Kundzewicz and Robson 2000). Kundzewicz and Robson (2004) proposed the use of autocorrelation plots as one amongst the usual types of graph which can be useful for hydrologic data. So, autocorrelation plot/correlogram is used in this study for identifying dependent or independent data. Innovative trend analysis (ITA) (Sen 2012) is utilized for assessment of nature of trend (non-monotonous or monotonous).

When smoothing curve is plotted for seasonal data, the purpose is generally to find changes in long term and not only seasonal variation (Kundzewicz and Robson 2000). Thus, smoothing curve is also used in the current work to identify changes in long term. Window 12 is utilized in the current work for smoothening of data. The detailed information about aforesaid tests and graphical methods (ITA and smoothing curve) is available in papers cited above.

1.5 Results

In the current work, trend analysis of regional seasonal T_{mean} series appertaining to the Tapi basin is conducted. For MKBBS test, data resampling is conducted for 2000 times by using 'k+1' block size, where 'k' is the number of significant conterminous lag correlations as given in Khaliq et al. (2009).

1.5.1 Trend Analysis of Regional Winter T_{mean} Series Corresponding to the Period 1971–2004

Significant increasing trend is found in regional winter T_{mean} series by application of MKBBS test. The trend magnitude as evaluated by SS test is 0.0122 °C/month. The correlogram, smoothing curve and ITA plot corresponding to regional winter T_{mean} series are shown in Fig. 1.2a, b and c, respectively. Correlogram shown in Fig. 1.2a shows that data is not independent, thus MKBBS test is applied on given data. Statistically significant increasing trend present in regional winter T_{mean} series is aided by increasing pattern of data present in smoothing curve and ITA plot of given series as shown in Fig. 1.2b and c, respectively.

1.5.2 Trend Analysis of Regional Pre-monsoon T_{mean} Series Corresponding to the Period 1971–2004

Significant trend is not found in regional pre-monsoon T_{mean} series on application of MKBBS test. Trend magnitude as evaluated by SS test is 0.0037 °C/month. The

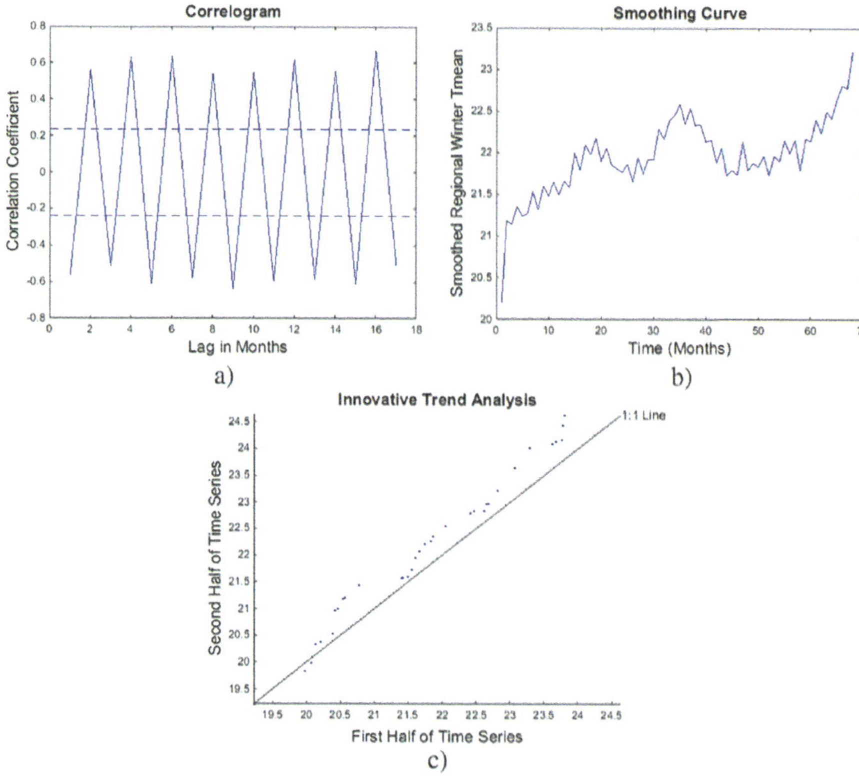

Fig. 1.2 Correlogram, smoothing curve and ITA plot of regional winter T_{mean} series appertaining to the Tapi basin are shown in (**a**), (**b**) and (**c**), respectively

small magnitude of SS supports the non-existence of statistically significant trend in the given series.

1.5.3 Trend Analysis of Regional Monsoon T_{mean} Series Corresponding to the Period 1971–2004

Significant trend is not detected in regional monsoon T_{mean} series on application of MKBBS test. Trend magnitude as evaluated by SS test is 0.0013 °C/month. The small magnitude of SS supports the non-existence of statistically significant trend in the given series.

1.5.4 Trend Analysis of Regional Post-monsoon T_{mean} Series Corresponding to the Period 1971–2004

Significant trend is not found in regional post-monsoon T_{mean} series on application of MKBBS test. Trend magnitude as assessed by SS test is 0.0053 °C/month. The small magnitude of SS supports the absence of significant trend in the given series.

1.6 Conclusion

Significant increasing trend is found in regional winter T_{mean} series (1971–2004) of the Tapi basin. So, if the same trend is followed in future, it will lead to more evaporation and consequently less availability of water during the winter season, which will increase the burden on management of water resources in given area in corresponding season.

Acknowledgement The author is grateful to the Climate Research Unit, UK for making available the required data for the current work, on their website.

References

Bhamare SM, Agone V (2011) Change detection of surface temperature and its consequence using multi-temporal remote sensing data and GIS application to Tapi Basin of India. In: Global conference on global warming. Lisbon, Portugal, pp 1–12

Deshpande NR, Kothawale DR, Kulkarni A (2016) Changes in climate extremes over major river basins of India. Int J Climatol. https://doi.org/10.1002/joc.4651. Accessed 05 Feb 2018

Fahad GR, Nazari R, Daraio J, Lundberg DJ (2017) Regional study of future temperature and precipitation changes using bias corrected multi-model ensemble projections considering high emission pathways. J Earth Sci Clim Change 8(9):1–10. https://doi.org/10.4172/2157-7617.1000409

Hannaford J, Harvey CL (2010) UK seasonal river flow variability in near-natural catchments, regional outflows and long hydrometric records. In: BHS Third international symposium managing consequences of a changing global environment. Newcastle, pp 1–7

IPCC (2007) Summary for policymakers. In: Solomon S, Qin D, Manning M, Chen Z, Marquis M, Averyt KB, Tignor M, Miller HL (eds.) Climate change 2007: the physical science basis. Contribution of working group I to the fourth assessment report of the intergovernmental panel on climate change. Cambridge University Press, Cambridge, United Kingdom and New York, NY, USA

Khaliq MN, Ouarda TBMJ, Gachon P, Sushama L, St-Hilaire A (2009) Identification of hydrological trends in the presence of serial and cross correlations: a review of selected methods and their application to annual flow regimes of Canadian Rivers. J Hydrol 368:117–130

Kundzewicz ZW, Robson A, (ed) (2000) Detecting trend and other changes in hydrological data. World climate programme–water, world climate programme data and monitoring, WCDMP-45, WMO/TD –No. 1013. World Meteorological Organization, Geneva, Switzerland

Kundzewicz ZW, Robson AJ (2004) Change detection in hydrological records-a review of methodology. Hydrol Sci J des Sci Hydrol 49(1):7–19
Mann HB (1945) Nonparametric test against trend. Econometrica 13(3):245–259
Singh P, Kumar V, Thomas T, Arora M (2008a) Changes in rainfall and relative humidity in river basins in northwest and central India. Hydrol Processes 22:2982–2992
Singh P, Kumar V, Thomas T, Arora M (2008b) Basin-wide assessment of temperature records in northeast and central India. Hydrol Sci J des Sci Hydrol 53(2):421–433
Sen PK (1968) Estimates of regression coefficient based on Kendall's Tau. J Am Stat Assoc 63(324):1379–1389
Sen Z (2012) Innovative trend analysis methodology. J Hydrol Eng ASCE 17(9):1042–1046
Sonali P, Nagesh Kumar D (2013) Review of trend detection methods and their application to detect temperature changes in India. J Hydrol 476:212–227

Chapter 2
Dry Spell and Wet Spell Characterisation of Nandani River Basin, Western Maharashtra, India

Abhijit Mohanrao Zende and Prashant Basavaraj Bhagawati

Abstract Soil and water conservation measures are necessary to know the sequence of dry and wet periods along with the onset and withdrawal of rainy season for successful agricultural management and planning. Daily observed rainfall (1998–2017) are analysed to compare and contrast the large-scale duration characteristics of rainfall over semi-arid region. This paper analyses the trend of rain spell frequency in terms of duration by using standard statistical methods. The analysis has been carried out at four locations, namely Kadegaon, Karad, Vaduj and Vita, in and nearby the Nandani river basin. In the Nandani river basin, the duration of dry spell varies from 69 to 119 days, and wet spell varies from 34 to 87 days. In this study, the rain spells were classified into low, medium, high, very high and extreme rain spells. The important results have been found through analysis of this study. These results are: the spells were examined only for the monsoon season (June–October) because all the above categories of rain spells occur only in this monsoon season. The rain spells help to coordinate various activities, like water effect on crop growth, supplementary irrigation, water release schedule and so on. The maximum dry spell (DS) was 119 days at Vaduj in 2003 and minimum 69 days at Kadegaon in 2006. The maximum wet spell (WS) was 87 days at Karad in 2005 and minimum 34 days at Vaduj in 2003.

Keywords Crop growth · Dry and wet spell · Supplementary irrigation · Water release schedule

---

A. M. Zende (✉)
Department of Civil Engineering, Dr. Daulatrao Aher College of Engineering, Karad, Vidyanagar, Banawadi, Karad, Dist-Satara 415124, Maharshtra, India
e-mail: zenabhi31@yahoo.co.in; hodcivil@dacoe.ac.in

P. B. Bhagawati
Department of Civil Engineering, Annasaheb Dange College of Engineering and Technology, Ashta Tal-Walwa, Dist-Sangli 416301, Maharashtra, India
e-mail: prashantbhagawati@gmail.com

R. Jha et al. (eds.), *Climate Change Impacts on Water Resources*, Water Science and Technology Library 98, https://doi.org/10.1007/978-3-030-64202-0_2

2.1 Introduction

Global water resources are highly sensitive to both climate change and climate variation. Rainfall, the main input to the global hydrological cycle and an important indicator of water resources availability, has shown significant change and variations over the years both globally and regionally (Zende et al. 2012). Drought is the interruption of the rainy season by a so-called dry spell (DS). Dry spell (DS) can be defined as a sequence of dry days including days with less than a threshold value of rainfall. The information of spell is extremely useful for planning and design applications in agriculture and environment and many other sectors (Atal and Zende 2015). Rainfall is a seasonal phenomenon in tropical monsoonal climate. The start and end of rainy season, rainfall amount, rainfall intensity and duration of wet spells (WS) and duration and severity of intervening dry spells (DS) are characterised by large spatial and temporal variations. Climatology and unevenness of the parameters of rainy season and wet and dry spells are valuable information for planners, scientists, engineers and managers working in water-related sectors (water resources, agriculture, ecology, hydrology etc.). Crucial issue of rainfall spell is determination of year-wise starting and ending dates of the rainy season and identification of wet and dry spells in the rainfall and its time distribution. In this study, we attempt to demonstrate that demarcation of start and end of rainy season, and identification of wet/dry spells are fundamentally one problem, and single-objective criterion can serve both the purposes (Atal and Zende 2015).

Recently, Ranade et al. (2008) applied this criterion to determine year-wise start and end of the hydrological wet season (HWS) over 11 major and 36 minor river basins as well as the West Coast Drainage System (WCDS) and studied the variability of HWS parameters (starting and ending dates and duration, seasonal rainfall/rainwater and surplus rainfall/rainwater potential) over the period of longest available instrumental records (1813–2006). Dash et al. (2009) applied the IMD criteria of rainy day (rainfall equal to or greater than 2.5 mm per day) to identify wet and dry events across India. Parts of India experience considerably longer duration of rainy season than the summer monsoon period as they get considerable rainfall from other systems also in the pre- and post-monsoon periods. To identify wet spells across China, Bai et al. (2007) used the following criteria: (i) In a wet spell (WS), the number of rainy days (rainfall ≥ 0.10 mm/day) should not be less than 4 days, and the first three days should not include a dry day. (ii) A wet spell (WS) is over whenever there are two consecutive dry days. The duration of a wet spell (WS) is defined as the number of days between the two consecutive dry days. They have also documented different definitions attempted earlier in different geographical locations of China. A sophisticated hydrological scheme, the available water resources index (AWRI), has been developed by Byun and Wilhite (1999) for detecting the onset, end and accumulated stress of drought. In the AWRI, the accumulated precipitation, the duration of accumulation and the daily reduction of water by runoff, evapo-transpiration, infiltration and so on are taken into account quantitatively. Using this scheme, the onset and ending dates, and intensities of the three rainy seasons in Korea were determined for each year and each station, and also climatologically (Byun and Lee 2002).

For Peninsular Malaysia, Deni et al. (2008) defined a wet day with rainfall of at least 0.1 mm. A wet (dry) spell is defined as a period of consecutive days of exactly, say x, wet (dry) days immediately preceded and followed by a dry (wet) day. Aviad et al. (2004) studied the variation in beginning, end and length of the rainy season along a Mediterranean-arid climate transect with the change in rainfall threshold. They changed the thresholds from 0.1 to 80 mm with interval of 2.5 mm, and from 80 to 200 mm with intervals of 5 mm, and found that as the threshold increased the rainy season began later, ended earlier, and consequently its length was shortened. Cook and Heerdegen (2001) defined rainy season as that period when the probability of 10-day dry spells was less than 0.5, and the wet season (monsoonal influence) as that period within the rainy season when the probability of dry spells was less than 0.1. A dry-day was defined with rainfall less than 5 mm. This definition was essentially meant for ecological purposes.

Understanding climatic and hydroclimatic features of wet and dry spell is essential for effective agricultural and hydrological operations. On the face of global climate change background and scenario projections, the problem assumed greater importance; therefore attempts have been made in recent decades world-wide to understand the problem on regional/local scales. This comprehensive study of the wet and dry spells (including their extremes) in a semi-arid region is undertaken with the following main objective.

2.2 Objective

The objective of this study is analysis of dry spell (DS) and wet spell (WS) to minimise unexpected damage due to long dry spells and to have effective and efficient planning for various stakeholders.

2.3 Study Area and Methodology

The study area is Nandani river basin located at upper Yerala river basin, a tributary of Krishna river basin India. Nandani river originates at Jaigoan and flows from Aundh, Khatav tahsil, Satara district and confluence to Yerala river at Shivani, Kadegaon tahsil, Sangli district. Nandani river flows through plateau region. There are high agricultural lands, including approximately 70% natives of Nandani river basin as cultivators. Moreover, the crop practice at Nandani river basin includes rainy season crops (Kharif) which includes soyabean, ginger, potato and groundnut. Geology of the area is covered by basaltic rocks. In this region temperature ranges from 10–42 °C. The coordinates are latitude 17°13′28″N to 17°33′53″N longitude 74°14′16″E to 74°25′19″E (Fig. 2.1). Basin is appx. 20 km away from main town

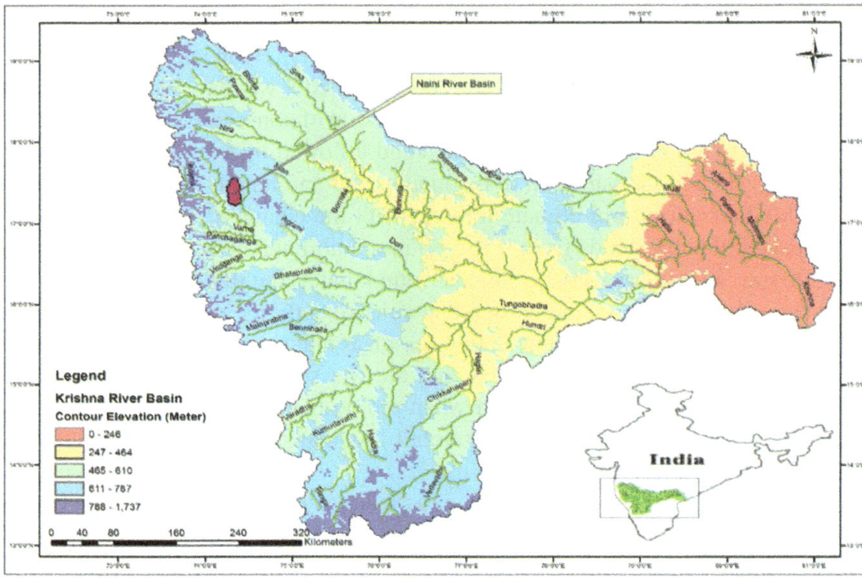

Fig. 2.1 Location of Nandani river basin

Karad. Average elevation of study area is 660–840 m. June to October is the monsoon season in the study area. The river basin experiences tropical monsoon weather with normal temperature, humidity and evaporation throughout a year. The study area falling under Survey of India (SOI) topographical sheet nos. 47 K/6/SE, 47 K/6/NW, 47 K/6/SW, 47 K/6/NE, 47 K/7/NE, 47 K/7/SE, 47 K/7/NW, 47 K/7/SW, 47 K/8/SE, 47 K/8/NW, 47 K/8/SW, 47 K/8/NE on a scale of 1:25000. With the help of Arcgis 10 GIS software, first all toposheets were georeferenced, mosaic, and marked particular region river basin boundary (by cropping study area). Also, the following features get achieved like drainage pattern of basin, location of settlement/roads/railway, location of existing water bodies and contours lines and so on.

2.3.1 Data Used

Nandani river is one among the tributaries of the Yerala river basin. The total length of the river is 39 km and its area is 420 km^2. It starts from Jaigaon located at 914 m above MSL and confluence Yerala river at 551 m above MSL. 60% rainfall occurs during June to September due to South-West monsoon, 18% during September to December due to North-East monsoon and 12% after December. Most of the areas of this basin receive rainfall less than 700 mm. The highest annual of rainfall of 1038.4 mm was recorded during 2006 at Kadegaon and the lowest of 175.3 mm during 2003 at Khatav. It is expected that the characteristics of wet and dry spells on basin/watershed scales would provide useful information for science (to understand physics and dynamics of rain-producing weather systems) and society (hydrology and water resources, ecology and agriculture).

Rainfall data of four stations operated by the meteorological organisation of Western Maharashtra were used to evaluate rainfall variability in semi-arid zones (Table 2.1).

These stations were selected because they have rainfall records for 20 years (1998–2017) (Fig. 2.2). The data set has included annual rainfall, number of rainy days per year and monthly rainfall. The number of rainy days per year was defined as the number of days in a year with a rainfall amount greater than 1 mm. Following physico-climatological factors are also qualitatively considered for detailed analysis

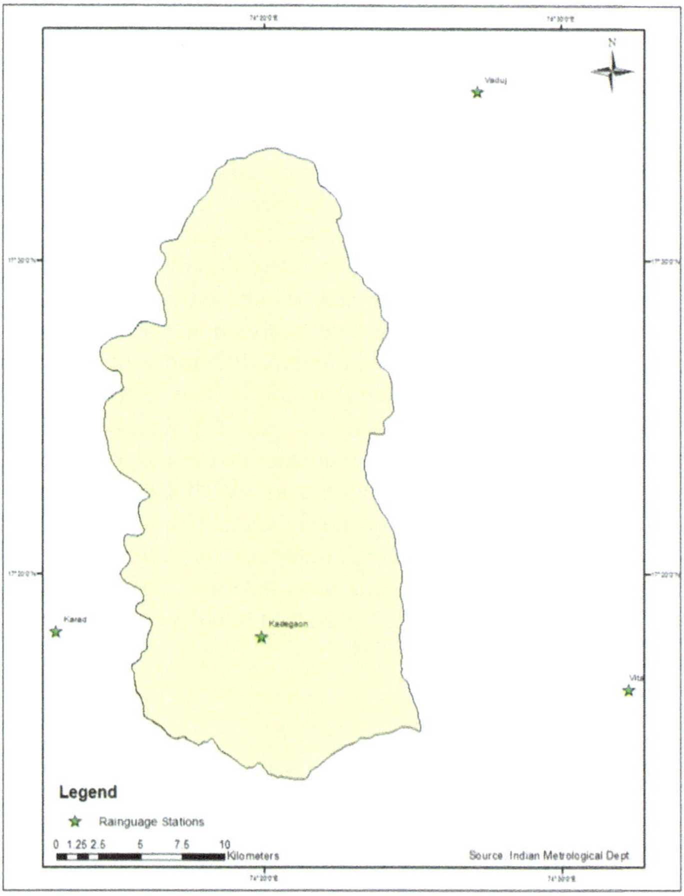

Fig. 2.2 Raingauge stations location in Nandani river basin

Table 2.1 Details of Rainguage stations

S. No.	Rainguage station	Latitude	Longitude	Sample size (Year)	Altitude (m)
01	Kadegaon	74°19′56.40″E	17°18′7.58″N	20	675
02	Karad	74°13′33.01″E	17°18′11.73″N	20	580
03	Khatav	74°27′14.37″E	17°35′24.82″N	20	720
04	Khanapur	74°32′16.04″E	17°16′27.70″N	20	680

of rain spell. (a) Topographic features; (b) Spatial pattern of the mean annual/South-West monsoon rainfall; (c) Drainage pattern; (d) Pre and post-monsoon ground water level across the river basin; and (e) Daily rainfall data.

2.3.2 Method

In the wet/dry spell approach, the time-axis is split up into intervals called wet periods and dry periods. A rainfall event is an interval in which it rains continuously (it is an uninterrupted sequence of wet periods). The definition of event is associated with a rainfall threshold value which defines wet. Rainfall information from four raingauge stations was collected. Four stations are within or on the periphery of the basin. Yearly, monthly and daily rainfall information were collected. Maximum evaporation is 236 mm as of May and minimum is 102 mm as of December. Though average rainfall is not more, total annual evaporation is also not more in comparison with other low rainfall areas. But since less rainfall is spread over 7 months in overall, this area does not receive rainfall more than natural evaporation in any month. Generally, annual evaporation is thrice the annual rainfall. During any of the months, evaporation is more than the naturally occurring rainfall. The scenario lasts for all the four monsoon months. Especially during September, the deficit is quite appreciable from this count. The annual total evaporation is 1833 mm. It varies between 108 and 134 mm during June to September, 126 and 120 mm during October to February, and 219 and 234 mm during March to May.

2.3.3 Identification of Wet and Dry Spell

Rainfall threshold used in the criterion is derived from local rainfall climatology that is ‘daily mean rainfall’ (DMR) of the climatological (long-term mean or normal) summer monsoon period over the area of interest.

Computational steps of the schemes are as follows:

1. Calculation of daily mean rainfall (DMR) of the normal summer monsoon period (mm/day).

2. Identification of continuous period with normalised, smoothed daily rainfall values equal to or greater than 1.0 mm as wet spell (WS) and less than 1.0 mm as dry spell (DS).

Climatological and fluctuation features of the following parameters of actual wet and dry spells have been studied.

i. Number of wet/dry spells; ii. Total rainfall of the wet/dry spells; iii. Total duration of the wet/dry spell; iv. Starting date of first wet/dry spell; v. Ending date of last wet/dry spell; vi. Rainfall amount of individual wet/dry spell; vii. Rainfall intensity of individual wet/day spell; viii. Duration of individual wet/dry spell.

Farming in most parts of Nandani basin depends on rainfall. With other environmental factors assumed to be fair and good, dry spells are considered to be the major cause of poor crop production, hence a study on dry spell analysis is important. The success and failure in crop production during rainy seasons are dependent on the frequency and length of dry spells. Knowledge on dry spell distribution within a rainy season plays a vital role in trying to maximise benefits in rainfed agricultural regions.

2.4 Results and Discussion

To comprehend broad spatial climatological features of the dry and wet spells across the country, a generalised description of the parameters in five categories is provided.

a. Number and total duration of wet/dry spells and duration of actual and different extreme wet/dry spells.
b. Total rainfall of wet/dry spells and rainfall amount and rainfall intensity of actual and extreme wet/dry spells.
c. Start of the first wet spell.
d. End of the last wet spell.
e. Occurrence of different extreme wet/dry spells.

If the series of successive precipitations do not form independent events, the waiting time follows a gamma distribution with two parameters instead of an exponential distribution. For planning purposes, the longest dry spells associated with different return periods are of fundamental importance.

Weekly rainfall analysis was carried out for the months of June to October, in understanding the adequacy rainfall during the 24th to 46th meteorological weeks for rainfed cultivation (Sudhishri, 2004). 25 mm of rainfall in a week will be able to meet 0.5–0.75 times the evaporation demand. During the early stages of growth, the crop water requirement will be about half of the evaporation demand and would increase during the reproductive stage of crop growth. Therefore, a week with rainfall less than 25 mm was considered as dry and additional water is (other than rainfall) demanded for this week. However, during dry week, the crops may meet their insufficient water requirements through the moisture (stored) available in the soil. If the rainfall is less than 25 mm per week for two or more consecutive weeks, the crops will be subjected to moisture stress due to lack of adequate stored soil moisture. Supplementary irrigation is to be provided for those weeks. It was observed that the rainfall starts in the 23rd week and continues up to 41th meteorological week. The dry spell is the period between two rainy events, in which no rainfall is reported. Dry spells in all the raingauge station are shown in Fig. 2.3.

Dry and wet spell duration of raingauge stations is given in Table 2.2.

The number of dry spell ranges between 2 and 5. Hence, 2–5 or more than 5 supplementary irrigation iterations are required. The details of crop calendar are shown in Table 2.3. Hence, the crops that totally depend on rainfall could not yield.

Sowing Vegetation Growth + Flowering + grain formation Harvest.

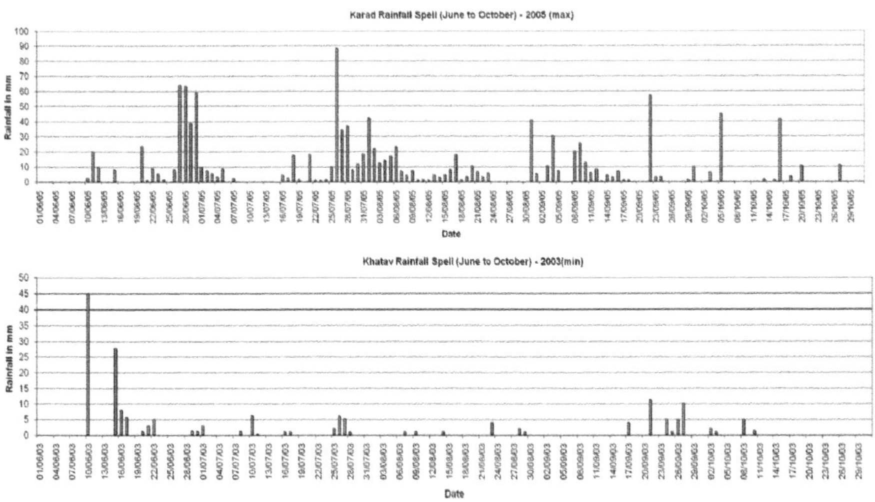

Fig. 2.3 Duration of dry spell within monsoon event

Table 2.2 Duration of dry and wet spell days at various raingauge stations

Raingauge station	Dry spell (Days)	Wet spell (Days)	No. of dry spell	Year
Kadegaon	114	40	29	2003
	69	84	16	2006
Karad	96	55	23	2015
	102	87	12	2005
Vaduj	119	34	28	2003
	94	57	21	2009
Vita	99	48	20	2003
	85	68	14	1998

Table 2.3 Crop calendar

Sr. No.	Crops	Month											
		Jan	Feb	Mar	Apr	May	Jun	Jul	Aug	Sep	Oct	Nov	Dec
01	Sorghum (Kharif)							Sowing	Vegetation Growth +Flowering + grain formation	Vegetation Growth +Flowering + grain formation	Harvest		
02	Sorghum (Rabbi)	Vegetation Growth +Flowering + grain formation										Sowing	Vegetation Growth +Flowering + grain formation
03	Groundnut (Kharif)							Sowing	Vegetation Growth +Flowering + grain formation	Vegetation Growth +Flowering + grain formation	Harvest	Harvest	
04	Groundnut (HW)			Sowing			Harvest						
05	Wheat	Vegetation Growth +Flowering + grain formation	Harvest									Sowing	Vegetation Growth +Flowering + grain formation
06	Pearl Millet								Vegetation Growth +Flowering + grain formation	Vegetation Growth +Flowering + grain formation	Harvest		
07	Soya bean								Vegetation Growth +Flowering + grain formation	Vegetation Growth +Flowering + grain formation	Harvest		
08	Gram	Vegetation Growth +Flowering + grain formation	Harvest									Sowing	Vegetation Growth +Flowering + grain formation
09	Sunflower								Vegetation Growth +Flowering + grain formation	Vegetation Growth +Flowering + grain formation	Harvest		
10	Turmeric Or Ginger	Harvest							Vegetation Growth +Flowering + grain formation	Vegetation Growth +Flowering + grain formation	Vegetation Growth +Flowering + grain formation	Vegetation Growth +Flowering + grain formation	Vegetation Growth +Flowering + grain formation
11	Sugarcane	Sowing	Vegetation Growth +Flowering + grain formation	Vegetation Growth +Flowering + grain formation	Vegetation Growth +Flowering + grain formation	Vegetation Growth +Flowering + grain formation	Vegetation Growth +Flowering + grain formation	Vegetation Growth +Flowering + grain formation	Vegetation Growth +Flowering + grain formation	Vegetation Growth +Flowering + grain formation	Vegetation Growth +Flowering + grain formation	Vegetation Growth +Flowering + grain formation	Harvest

Sowing Vegetation Growth +Flowering + grain formation Harvest

2.5 Conclusion

Spatial and temporal variability of rainfall for 1998–2017 indicates no systematic trend and correlation between the stations. Dry spell (DS) of the monsoon was used in the identification of ground water supplementary irrigation. Weeks receiving rainfall about 25 mm are not able to meet 0.5–0.75 times the evaporation demand. Irrespective of the increase or decrease in rainfall and days, the number of dry spells ranges between 2 and 5. Hence, 2–5 or more than 5 supplementary irrigation iterations are required to prevent the crop failure due to water stress. These results also indicated that for the analysed time period, there was no significant climate change in the semi-arid region of Western Maharashtra. The results also suggest the need for further investigation on local anthropogenic intervention in the environment, which could be one of the major causes of climate change in semi-arid regions.

References

Atal KR, Zende AM (2015) Wet and dry spell characteristics of Semi-arid region, Western Maharashtra, India. E-proceedings of the 36th IAHR World Congress, 28 June–3 July, 2015. The Hague, the Netherlands, pp 1–7

Aviad Y, Kutiel H, Lavee H (2004) Analysis of beginning, end, and length of the rainy season along a Mediterranean-arid climate transect for geomorphic purposes. J. Arid Environ 59:189–204

Bai A, Zhai P, Liu X (2007) Climatology and trends of wet spells in China. Theor Appl Climatol 88:139–148

Byun HR, Wilhite DA (1999) Objective Quantification of Drought Severity and Duration. J Climatol 12:2747–2756

Byun HR, Lee DK (2002) Defining three rainy seasons and the hydrological summer monsoon in Korea using available water resources index. J Meteor Soc Japan 80:33–44

Cook GD, Heerdegen RG (2001) Spatial variation in the duration of the rainy season in monsoonal Australia. Int J Climatol 21:1723–1732

Dash SK, Kulkarni MA, Mohanty UC, Prasad K (2009) Changes in the characteristics of rain events in India. J Geophy Res 114:101–109

Deni SM, Jemain AA, Ibrahim K (2008) The spatial distribution of wet and dry spells over Peninsular Malaysia. Theor Appl Climatol 94(3–4):163–173

Ranade A, Singh N, Singh HN, Sontakke NA (2008) On variability of hydrological wet season, seasonal rainfall and rainwater potential of the river basins of India (1813–2006). J Hydrol Res Devpt 23:79–108

Sudhishri S, Panda RK, Patnaik US (2004) Models for agricultural drought investigation at Koraput (Orissa). J Soil and Water Conserv India 3:157–168

Zende AM, Nagarajan R, Atal KR (2012) Rainfall Trend in Semi arid region–Yerala river basin of Western Maharashtra. India. Int J Adv Technol 3(3):137–145

Chapter 3
Assessment of Climate Change on Crop Water Requirement in Tandula Command of Chhattisgarh (India)

Rahul Kumar Jaiswal, H. L. Tiwari, and Anil Kumar Lohani

Abstract India is an agrarian country where more than 70% of population depend largely on agriculture and agro-related business. The projected scenarios of precipitation from climate models predict decrease in some region and increase in others, albeit with large uncertainty in most of the places. The Chhattisgarh state which is called rice bowl of India has number of water resources projects where climate change can change crop water requirement. The present study has been carried out in the command of Tandula reservoir using statistical downscaling of climatic parameters for computation of crop water requirement using RCP2.6, RCP4.5 and RCP8.5 scenarios of coupled model inter-comparison project 5 (CMIP5). For calibration and validation, 26 NCEP rescaled climatic variables from 1971 to 2003 have been used with minimum and maximum temperature and rainfall for concurrent period. The percentage reduction and *k-fold* cross-validation techniques have been used for selection of best-suited climatic parameters for statistical downscaling and used to generate multiple ensembles of temperature and rainfall for three future assessment periods, namely near century period as FP-1 (2020–2035), mid-century period as FP-2 (2046–2064) and far century period as FP-3 (2081–2099). The projected multiple series of climatic variables were further used to compute evapotranspiration using CLIMWAT and then crop water requirement in the command and compared with corresponding requirement during the base period (BP: 1971–2014). The results of analysis suggested that the mean monthly maximum temperature showed a rising trend in all the months, while significant increase of minimum temperature during winter and rainy season. The average crop water requirement for designed cropping pattern of 82089 ha of kharif paddy during base period under present overall efficiency of 51% during base period may be about 473.7 Mm^3 and will increase to 479.0 Mm^3 during near century period (2020–2035), 492.7 Mm^3 during mid-century

R. K. Jaiswal (✉)
National Institute of Hydrology, Central India Hydrology Regional Centre, Bhopal, MP, India
e-mail: rkjaiswal_sagar@yahoo.co.in

H. L. Tiwari
Maulana Azad National Institute of Technology, Bhopal, MP, India

A. K. Lohani
National Institute of Hydrology, Jal Vigyan Bhavan, Roorkee, UT, India

R. Jha et al. (eds.), *Climate Change Impacts on Water Resources*, Water Science and Technology Library 98, https://doi.org/10.1007/978-3-030-64202-0_3

period and reduce to 387.9 Mm^3 during far century period (2018–2099). The mid-century period may be the most critical among all and it is recommended to develop adaptation measures to combat climate change especially in mid-century period.

Keywords Climate change · Global circulation model · Representative concentration pathways · Downscaling · Crop water requirement · Evapotranspiration

3.1 Introduction

One of the major challenges for scientific community is to develop appropriate adaptation strategies for water resources management which is adversely affected by changing demand and supply scenarios due to plausible climate change. The different reports of Intergovernmental Panel on Climate Changes (IPCC 2012, 2013) and other independent researches has confirmed that climate is changing on global and regional scale, which is likely to affect precipitation, runoff availability and supply of water (Tabari et al. 2016); in addition, irrigation water demand (Rosenzweig et al. 2004) must be evaluated to assess demand–supply scenarios due to future climate change. For simulation of future climate on the earth, General Circulation Models (GCMs) are the most advanced tools used in different hydrologic and other studies (Anandhi et al. 2008). These models can predict climate for future hundreds of years based on probable greenhouse gas (GHG) concentrations in the atmosphere under different development conditions. However, the GCMs are the most credible tools to simulate time series of climate variables (Ghosh and Mujumdar 2008), but unable to resolve significant sub-grid scale features including topography and land use, as needed in hydrologic modelling and impact assessment. This problem of coarse grid data can be solved by downscaling GCMs to local and basin scale with the help of dynamic or statistical downscaling techniques that bridge the large-scale atmospheric conditions with local-scale climatic data (Tisseuil et al. 2010). Both these techniques have their own advantages or disadvantages in future projection due to climate change (Tukimat and Harun 2013). The dynamic downscaling techniques use physically based model run in time-slice mode and limited area having major drawback of complexity and high computation cost (Anandhi et al. 2008) and propagation of systematic bias from GCM to RCM (Salathe 2003).

The statistical downscaling techniques reasonably accurate in developing relationships between GCM predictors and regional/station climatic data are simple, flexible in adjustment and movement to different regions, less costly and computationally undemanding in comparison to dynamic downscaling and proved its reliability and compatibility in future projections (Fowler et al. 2007; Lopes 2009; Sharma et al. 2011; Ethan et al. 2011). Wilby et al. (2002) explained application of statistical downscaling model (SDSM) for prediction of future climatic variables using GCMs data. Mahmood and Babel 2014 applied SDSM for prediction of future scenarios of extreme temperature in trans-boundary region of India and Pakistan. In the study, bias correction has been applied to make generated data de-biased, and then eight intensity and four frequency indices were determined to examine the change in climate in three future periods of 2011–2040, 2041–2070 and 2071–2099 s at the Jhelum basin. The results of the study confirmed increase of hot days and hot nights and decrease of cold days and cold nights in all three future periods in comparison to base period (1961–1990). Similarly, several other researchers (Wilby et al. 2002; Jaiswal and Tiwari 2015 etc.) applied user-friendly and convenient SDSM for downscaling of climatic parameters.

3.2 Study Area and Data Used

The study area selected for the study is Tandula command that receives water from Tandula and Gondli reservoirs situated in Chhattisgarh state of India. The location map of Tandula command is presented in Fig. 3.1. The designed cropping pattern of Tandula command consists of 82095 ha, kharif paddy. As complete cropping area cannot be shown at once, the total cropping area of paddy has been divided into two equal parts grown on second and third 10-daily period of June each year. The cropping pattern of Tandula command is given below.

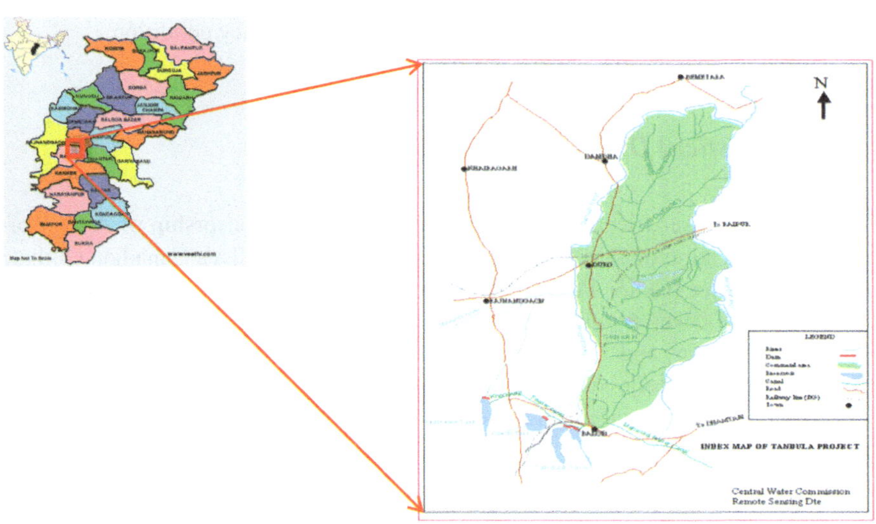

Fig. 3.1 Location map of Tandula command in Chhattisgarh (India)

Crop	June			July			August			September			October			November	
	D-1	D-2	D-3	D-1	D-2	D-3	D-1	D-2	D-3	D-1	D-2	D-3	D-1	D-2	D-3	D-1	D-2
Paddy-1 41048 ha		S														H	
Paddy-2 41047 ha			S														H

The daily time series of minimum and maximum temperature of Raipur from 1971 to 2014 and rainfall of Balod R.G. station from 1981 to 2014 have been used. Twenty-six rescaled NCEP predictors and projected predictors from three different scenarios RCP2.6, RCP4.5 and RCP8.5 from Canadian GCM CANESM were used in the analysis for generation of multiple ensembles of climatic parameters and rainfall.

3.3 Methodology

The methodology for computation of crop water requirement under climate change proposed under the study consists of statistical downscaling of maximum and minimum temperature and precipitation for Tandula command for three future periods, namely near century period from 2020–2035 as FP-1, mid-century period from 2046–2064 as FP-2 and far century period from 2081–2099. The multiple projected climatic series were used for computation of evapotranspiration using ET_0 calculator and crop water requirement under present efficiency condition.

3.3.1 Statistical Downscaling

The SDSM is a user-friendly software developed under sponsorship of A Consortium for Application of Climate Impact Assessments (ACACIA), Canadian Climate Impacts Scenarios (CCIS) Project and Environment Agency of England and Wales. The SDSM can develop multiple, low-cost scenarios of daily surface weather variables under present and future climatic forcing. Presently, SDSM version 5.2 is available where the following seven key functions are required to perform the task of daily weather downscaling and forecasting (Fig. 3.2).

1. Quality control and data transformation
2. Selection of downscaling predictor variables
3. Model calibration
4. Weather generator
5. Data analysis
6. Graphical analysis
7. Scenarios generation.

The steps involved in statistical downscaling consist of quality control, selection of set of predictors, calibration and validation of model, uncertainty analysis, and

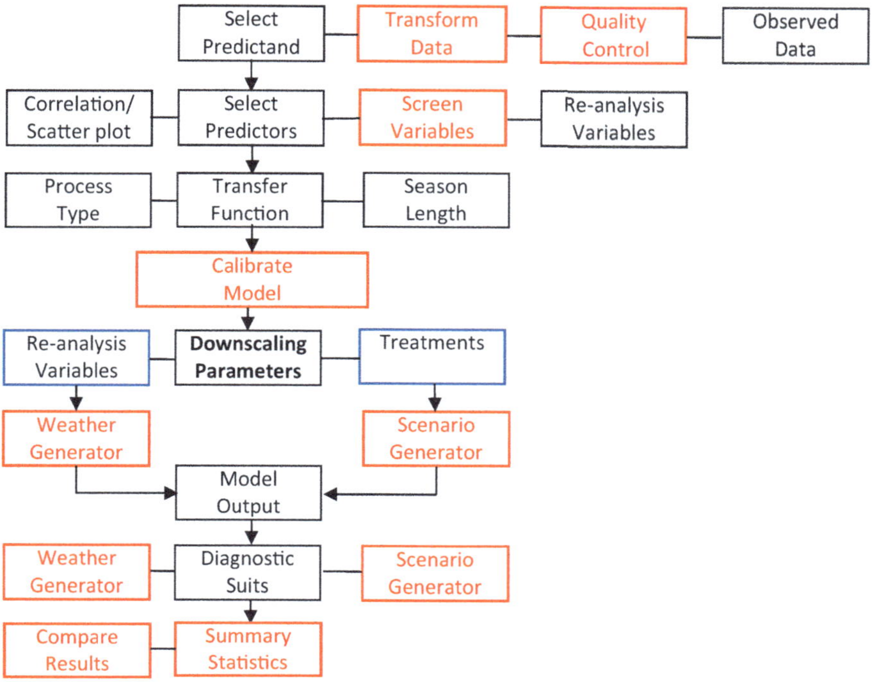

Fig. 3.2 Workflow in SDSM-DC (Reproduced from Wilby et al. 2014)

generation of series and statistical analysis/comparison. The selection of a set of appropriate predictors is an important task in downscaling process and proper understanding of physical process and knowledge of physically sensible predictors that can represent atmospheric process are essential (Huang et al. 2011). Several methods were propagated by different researchers for selection of appropriate set of predictors (Benestad et al. 2007; Shongwe et al. 2006; Tripathi et al. 2006 etc.) and in the present study, percentage reduction method (Mahmood and Babel 2014; Jaiswal et al. 2018) along with scatter diagram were used where the correlation coefficient between predictand (precipitation for present study) and 26 NCEP rescaled predictors on monthly, seasonal and annual basis have been computed using unconditional approach for temperature and conditional for rainfall. The predictor ranked first was termed as super predictor (SP) and using it, absolute correlation coefficient, absolute partial correlation and percentage reduction (PR) can be computed for remaining predictors using the following equation.

$$PR = \frac{P_r - R}{R} \tag{3.1}$$

where P_r and R are the partial and absolute correlation coefficient, respectively. The predictor having lowest PR value has been recognised as second super predictor. The second super predictor along with remaining predictors was used again to get third super predictors. In general, one to three predictors are enough to model climatic variability (Chu et al. 2010). After selecting a set of appropriate predictors, *k-fold* cross-validation (Casanueva et al. 2014) has been used. In this technique, the complete data series is divided in two parts, where first $((K-1)/K)$ part is taken for calibration to develop statistical relationships with appropriate transformation and model type while remaining $(1/K)$ part of series is used for validation purpose. The series generated during calibration and validation was de-biased using appropriate correction and used to compute efficiency and other goodness-of-fit statistics of the proposed model. After successful calibration and validation, the developed models were used to generate multiple ensembles of climatic parameters and rainfall using predictors of RCP2.6, RCP4.5 and RCP8.5 scenarios from CANESM GCM for three future assessment periods as near century period as FP-1 (2020–2035), mid-century period as FP-2 (2046–2064) and far century period as FP-3 (2081–2099).

3.3.2 Reference Crop Evapotranspiration ($\mathbf{ET_0}$)

The evapotranspiration for reference crop is computed using ET_0 calculator software which uses Penman Monteith equation for computation of long-term reference evapotranspiration. The ET_0 calculator is a software developed by the Land and Water Division of Food and Agriculture Organization to calculate reference evapotranspiration (ET_o) according to FAO standards. The ET_0 calculator uses minimum, maximum temperature, sunshine hours, wind speed and relative humidity for computation of evapotranspiration. Alternatively, only temperature data along with location can be used to compute evapotranspiration. A screenshot of ET_0 calculator is presented in Fig. 3.3.

3.3.3 Computation of Irrigation Water Requirement

The crop water requirement can be defined as the total amount of water needed to be supplied from source to meet all the requirements of a crop, including evapotranspiration, field preparation, nursery, leaching and losses during conveyance and application. The paddy in kharif season is the only crop in design cropping pattern where water is supplied from Tandula reservoir. The crop water requirement at dam head has been computed using present overall efficiency as 51% (conveyance efficiency as 75% and application efficiency as 68%). The results of crop water requirements for three future assessment periods were compared with base period crop water requirement during as BP (1971–2014).

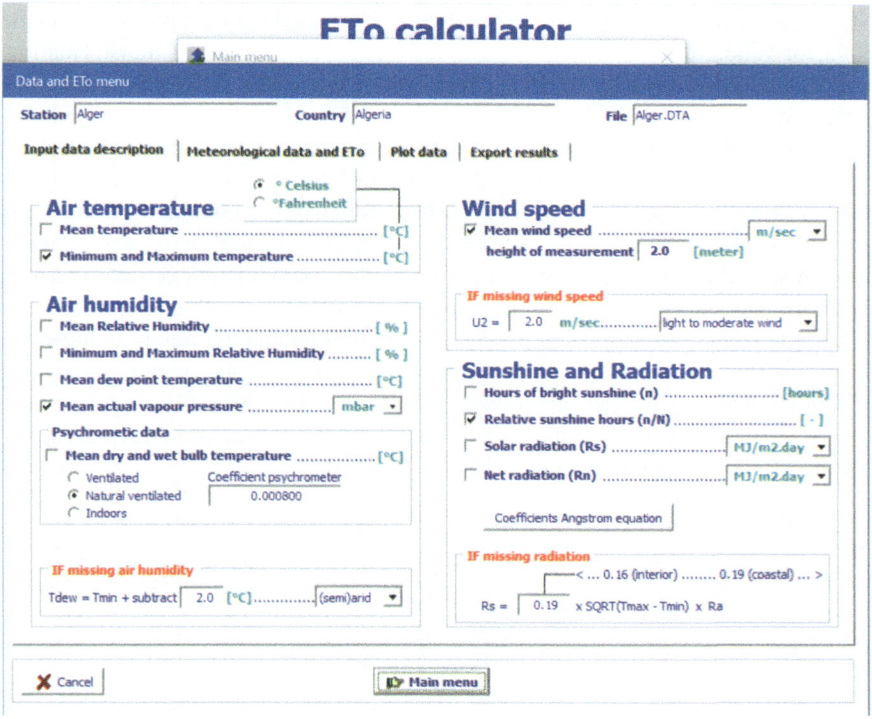

Fig. 3.3 Screenshot of ET_0 calculator

3.4 Analysis of Results

The assessment of crop water requirement in a command is an essential step for development of adaptation measures to cope up the harmful effects of climate change. For computation of crop water requirement in Tandula command, climate and rainfall have been projected for three future assessment periods and used for computation of crop water requirement with the help of ET_0 calculator, prevailing norms and present efficiencies.

3.4.1 Future Projection of Climate Parameters and Rainfall

The statistical downscaling technique has been used in k-fold cross-validation technique where minimum and maximum temperatures were modelled with unconditional, while rainfall in conditional approach. In the analysis, three-fold cross-validation technique has been used where depending on availability of at-site and

NCEP rescaled data, the period 1971–1990 was selected for calibration and 1991–2003 for validation. During calibration, different combinations of transformations and model types were tested and a series of climatic parameters for calibration period was generated and de-biased using linear difference and scaling techniques. The generated and observed climatic data were taken outside of SDSM and the coefficient of correlation (R), adjusted R^2 and Nash-Sutcliffe efficiency (η) for calibration period were computed on monthly basis for all the combinations. Few best-fit combinations of predictors were selected during calibration and were used for generation of series with independent series of predictors for validation periods, and after de-biasing goodness-of-fit measures were determined. The graphical representation of computed and observed monthly mean of minimum and maximum temperature during calibration and validation has been presented in Fig. 3.4a and monthly rainfall of Balod during similar periods in Fig. 3.4b. The list of selected predictors, model type, transformation and goodness-of-fit measures for climatic variables and rainfall at Balod R.G. station has been presented in Table 3.1.

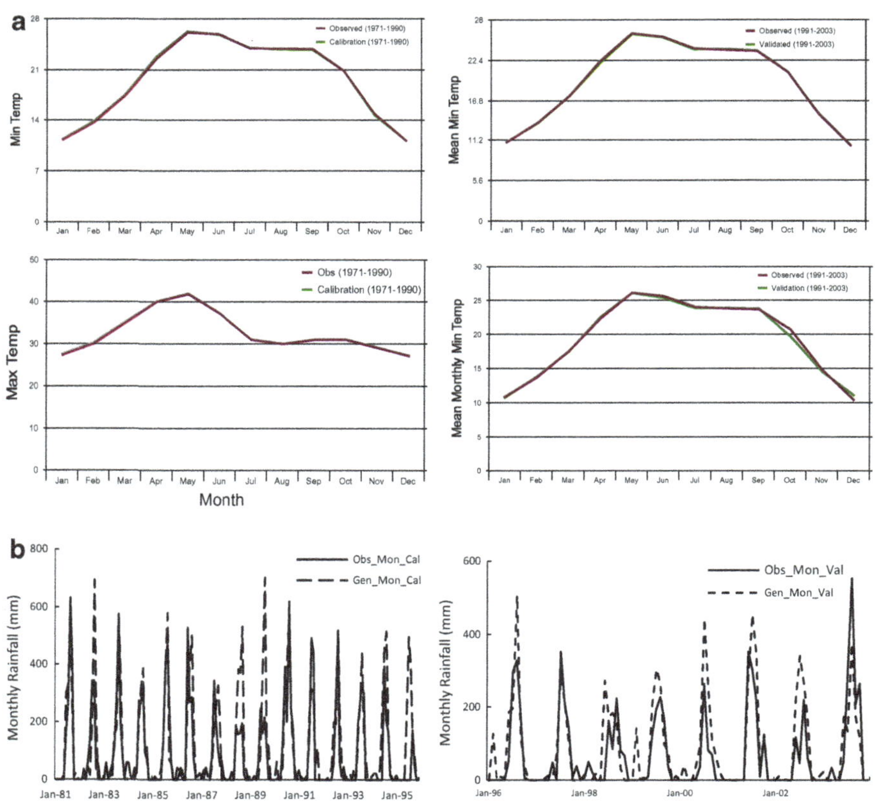

Fig. 3.4 **a** Mean monthly minimum and maximum temperature during calibration and validation. **b** Monthly rainfall at Balod R.G. station during calibration and validation

Table 3.1 Goodness-of-fit measures for statistical model for climatic variables and rainfall

S. No.	Climatic variables	Selected predictors	Model type/transformation/process	Calibration			Validation		
				C_c	$Adj\ R^2$	η	C_c	$Adj\ R^2$	η
1.	Minimum temperature	ncepp_fgl, ncepp500gl, nceps850gl	Monthly/none/unconditional	0.96	0.91	91.3	0.96	0.92	91.5
2.	Maximum temperature	ncepp5_zgl, ncepp500gl, ncepp8zhgl	Monthly/none/unconditional	0.73	0.73	98.5	0.84	0.70	99.8
3.	Rainfall at Balod	ncepp5_ugl, ncepp8_ugl, ncepp850gl	Monthly/none/conditional	0.69	0.55	85.9	0.66	0.66	75.3

C_c = Coefficient of correlation, $Adj\ R^2$ = Adjusted coefficient of determination, η = Nash-Sutcliffe efficiency

From the analysis, it has been found that the Nash-Sutcliffe efficiencies are 98.5 and 99.8% for minimum temperature, 91.3 and 91.5% for maximum temperature and 85.9 and 75.3% for rainfall at Balod R.G. station during calibration and validation, respectively. From the analysis of different goodness-of-fit tests and graphical representation, it may be concluded that the suggested models perform satisfactory during calibration and validation periods and can be used for further projection. Multiple series of climate and rainfall were generated and de-biased for three future assessment periods and used for computation of crop water requirement. The mean monthly values and ranges of projections of minimum and maximum temperature have been presented in Fig. 3.5a, while the same for rainfall at Balod in Fig. 3.5b.

The results from multiple projected series from RCP2.6, 4.5 and 8.5 scenarios confirmed the increase of monthly minimum temperature in the months of March, April and October to December. The annual minimum temperature for the base period (BP: 1971–2105) is 19.8 °C, which may increase to 20.1, 20 and 20.3 °C during early, mid and far century periods, respectively. The monthly maximum temperature showed an increasing trend in all the months except December in all three assessment periods. The annual maximum temperature for the base period (1971–2105) is 32.7 °C which may increase to 34.3, 35.1 and 35.6 °C, respectively, during near, mid and far century periods. The more intense heat wave in summer months may create several health as well as water-related problems and request more water for crops in rabi and kharif seasons in the region. The increase of temperature, in general, may intense the water requirement in the region, and appropriate management and adaptation option should be planned to cope up the adverse impact of rising temperature due to global warming. The temporal and spatial distribution of rainfall plays an important role for irrigation supply and projected reduced rainfall on an average of 943, 874 and 836 mm, respectively, during near, mid and far century periods with respect to 936 mm seasonal rainfall during base period. The June and July month may receive lower rainfall which will have adverse impact on water supply for planting and nursery operation.

3.4.2 Evapotranspiration and Irrigation Water Requirement

The observed and multiple future projected climate data (3 series) for future assessment period have been used to compute crop evapotranspiration. The evapotranspiration and rainfall series were further used in excel programming considering requirement of nursery, plantation, leaching along with conveyance and application losses to determine gross water required at the head of reservoir. As per the prevailing norms in the region, 100 mm each has been used in first two 10-daily periods for nursery and 150 mm for plantation during third 10-daily period. The present conveyance efficiency as 75% and application efficiency as 68% have been used to compute gross water requirement of the command. The average yearly future requirements computed in this way were compared with average base period requirement (Fig. 3.6). The computed average yearly gross water requirement (GWR) for the base period is

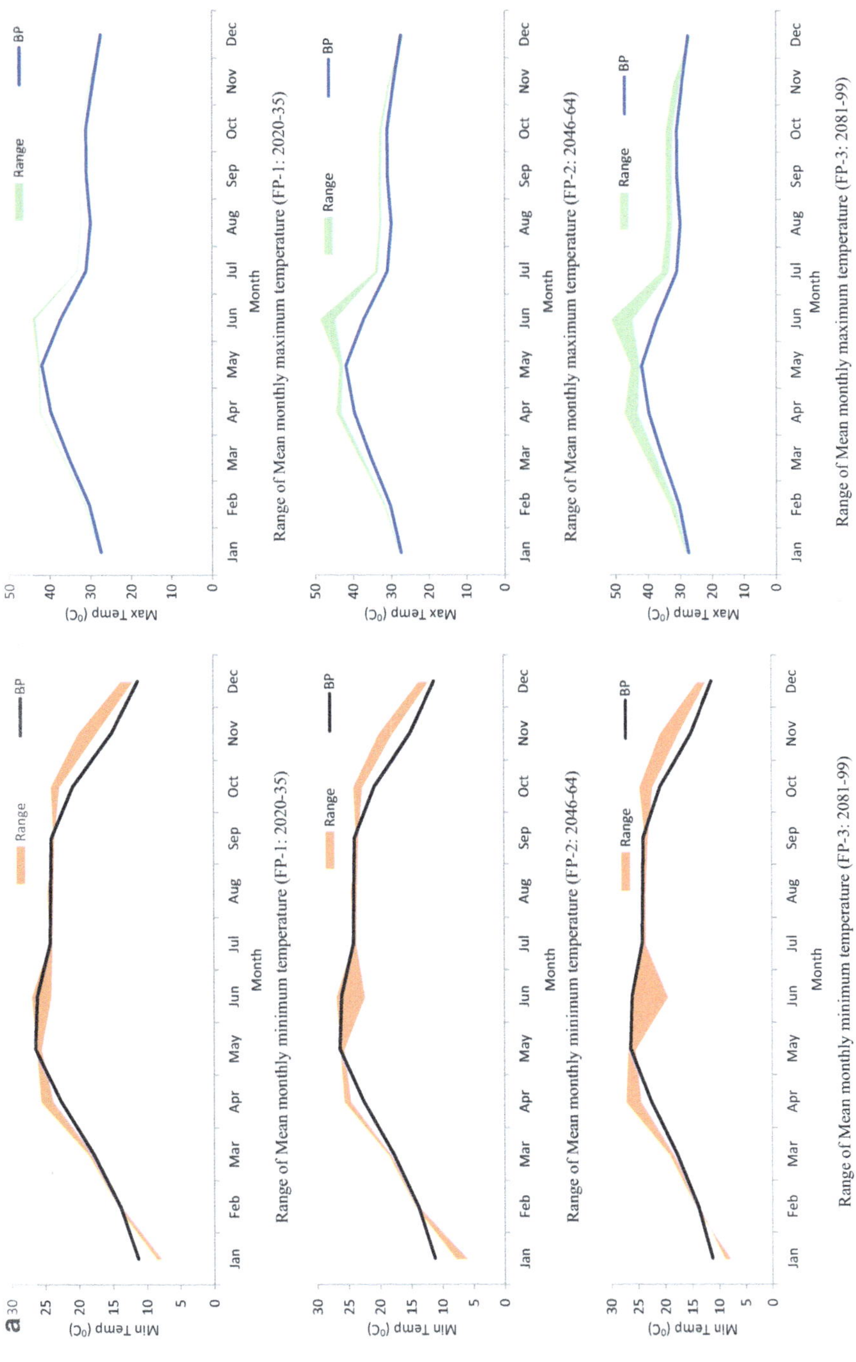

Fig. 3.5 **a** Projected range of mean monthly minimum and maximum temperature for future assessment periods. **b** Projected monthly rainfall at Balod R.G. station

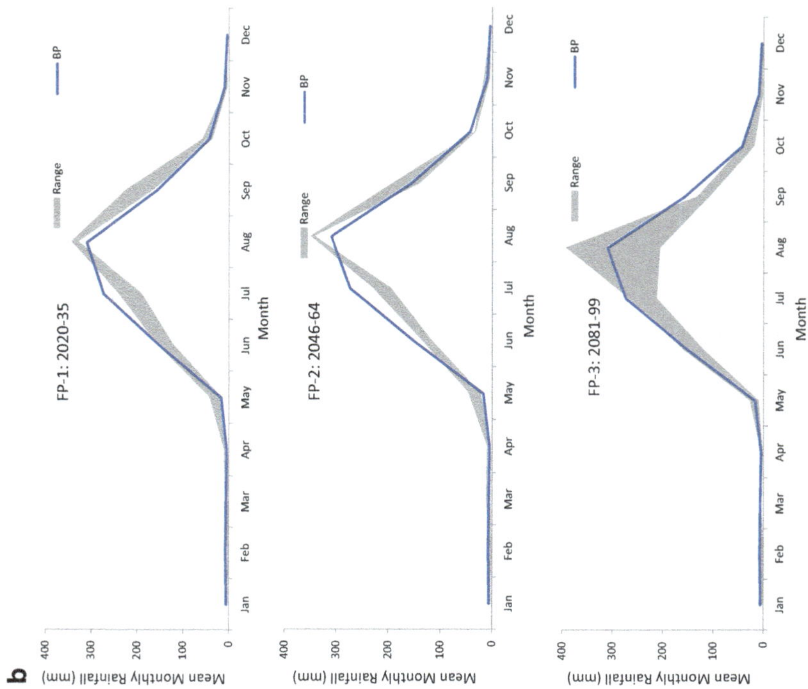

Fig. 3.5 (continued)

437.3 MCM. The computed average yearly crop water requirement for near century period may be in the range of 424–587.4 MCM where RCP 8.5 projected higher water demand. During mid-century period, all projections except RCP 2.5 predicted more requirement of water than present-day requirement. The far century period may be safe for irrigation water requirement point of view, where multiple series from all three scenarios projected comparatively less water (336.4–407.9 MCM) demand.

3.5 Conclusions

The impacts of climate change on water resource are inevitable and can be solved by future assessment of demand and supply analysis using future projection of climate data. In the present study, crop water requirement for design cropping pattern of Tandula command using RCP2.5, 4.5 and 8.5 scenarios data of CNESM GCM downscaled with the help of statistical downscaling. The analysis has been carried out for near (FP-1: 2020–2035), mid (FP-2: 2046–2064) and far century (FP-3: 2081–2099) periods and found that minimum and maximum temperature in the region may increase, and rainfall at Balod R. G. station filled in command area may get reduced rainfall during mid and far century periods. The crop water requirement under present

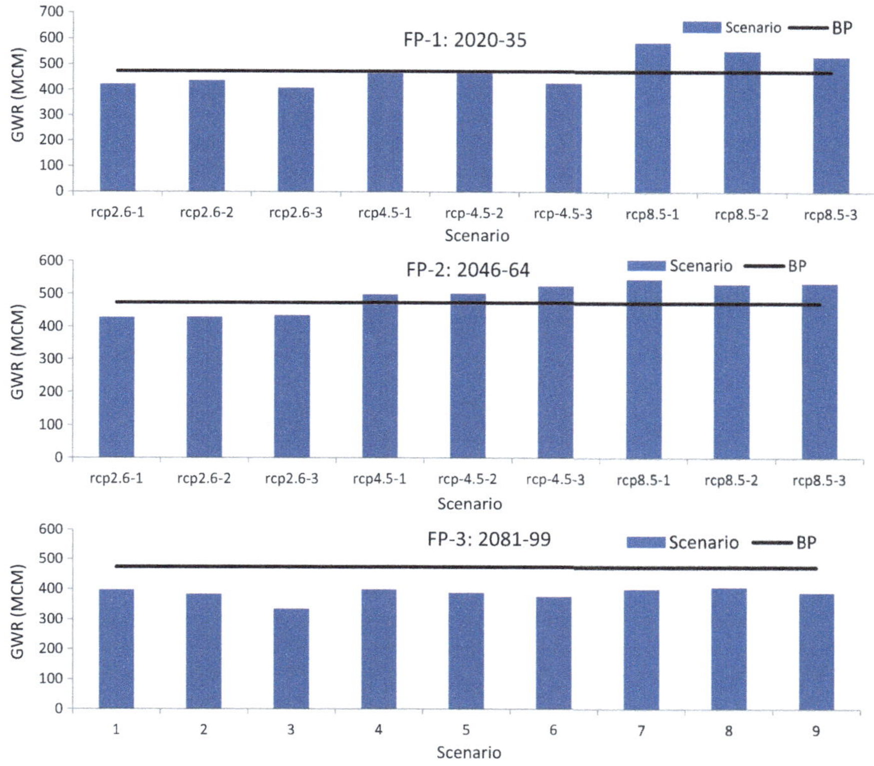

Fig. 3.6 Projection of average crop water requirement for Tandula command

51% efficiency may be more or less similar during near century, higher in mid-century and less during far century period for Tandula command. The mid-century period may be more critical where demand may be higher than present water need, and it is necessary to take adaptive management options along with precision irrigation and efficient operation of reservoir.

Acknowledgements The authors are grateful to Dr. Sharad K. Jain, Director NIH, Dr. N. C. Ghosh Coordinator & Scientist-G, National Institute of Hydrology, Roorkee and Dr. N. S. Raghuvansi, Director, MANIT, Bhopal for their constant support and valuable suggestion during the study. They are thankful to Water Resources Department, Raipur for providing the data and other information for the study.

References

Anandhi A, Srinvas VV, Nanjundiah RS, Nagesh DK (2008) Downscaling precipitation to river basin in India for IPCC SRES scenarios using support vector machine. Int J Climatol 28:401–420

Benestad RE, Hanssen-Bauer I, Forland EJ (2007) An evaluation of statistical models for downscaling precipitation and their ability to capture long-term trends. Int J Climatol 27(5):649–655

Casanueva A, Frias MD, Herrera S, San-Martin D, Kamonovic K, Gutierrez JM (2014) Statistical downscaling of climate impact indices. Clim Change 127:547–556

Chu JT, Xia J, Xu CY, Singh VP (2010) Statistical downscaling of daily mean temperature, pan evaporation and precipitation for climate change scenarios in Haihe River, China. Theor Appl Climatol 99(1-2):149–161

Ethan DG, Roy MR, Changhai L, Kyoko I, David JG, Martyn PC, Jimy D, Gregory TA (2011) Comparison of statistical and dynamical downscaling of winter precipitation over complex terrain. J Clim Am Meteorol Soc 25:262–281

Fowler HJ, Blenkinsop S, Tebaldi C (2007) Linking climate change modelling to impacts studies: recent advances in downscaling techniques for hydrological modelling. Int J Climatol 27(12):1547–1578

Ghosh S, Mujumdar PP (2008) Statistical downscaling of GCM simulations to streamflow using relevance vector machine. Adv Water Resour 31(1):132–146

Huang J, Zhang J, Zhang Z, Xu C, Wang B, Yao J (2011) Estimation of future precipitation change in the Yangtze river basin by using statistical downscaling method. Stoch Environ Res Risk A 25(6):781–792. https://doi.org/10.1007/s00477-010-0441-9

Intergovernmental Panel on Climate Change (IPCC) (2012) Managing the risks of extreme events and disasters to advance climate change adaptation. In: Field CB, Barros V, Stocker TF, Qin D, Dokken DJ, Mastrandrea MD, Mach KJ, Plattner GK, Allen SK, Tignor M, Midgley PM (eds) a special report of working groups I and II of the intergovernmental panel on climate change. Cambridge University Press, New York

Intergovernmental Panel on Climate Change (IPCC) (2013). "Summary for policymakers: In climate change 2013. In: Stocker TF, Qin D, Plattner GK, Tignor M, Allen SK, Boschung J, Nauels A, Xia Y, Bex V, Midgley PM (eds) The physical science basis, contribution of working group I to the fifth assessment report of the intergovernmental panel on climate change. IPCC, Geneva, Switzerland, pp 1–27

Jaiswal RK, Tiwari HL (2015) Downscaling and generation of future scenarios for maximum temperature in upper Mahanadi basin. Int J Frontier Technol 2(2):40–46

Jaiswal RK, Tiwari HL, Lohani AK (2018) Climate change assessment of precipitation in Tandula reservoir System. J Inst Eng India Ser A 99(1):17–27

Lopes P (2009) Assessment of statistical downscaling methods-application and comparison of two statistical methods to a single site in Lisbon. IOP Conf Sereis Earth Environ Sci 6, 1. https://doi.org/10.1088/1755-1307/6/2/022015022015

Mahmood R, Babel M (2014) Future changes in extreme temperature events using the statistical downscaling model (SDSM) in the trans-boundary region of the Jhelum river basin. Weather Clim Extremes 5–6:56–66

Rosenzweig C, Strzepek KM, Major DC, Iglesias A, Yates DN, McCluskey A, Hillel D (2004) Water resources for agriculture in a changing climate: international case studies. Global Environ Ch 14:345–360

Salathe EP Jr (2003) Comparison of various precipitation downscaling methods for the simulation of streamflow in a rainshadow river basin. Int J Climatol 23:887–901

Sharma M, Coulibaly P, Dibike YB (2011) Assessing the need for downscaling RCM data for hydrologic impact study. J Hydrol Eng 16(6):534–539. https://doi.org/10.1061/(ASCE)HE.1943-5584.0000349

Shongwe EM, Landman WA, Mason SJ (2006) Performance of recalibration systems for GCM forecasts for southern Africa. Int J Climatol 26:1567–1585

Tabari H, Hosseinzadehtalaei P, Willems P, Saeed S, Brisson E, Van Lipzig N (2016) How will be future rainfall IDF curves in the context of climate change?" In Proc. 4th IAHR Europe Cong., Liege, Belgium

Tisseuil C, Vrac M, Lek S, Wade AJ (2010) Statistical downscaling of river flow. J Hydrol 385(1):279–291

Tripathi MP, Raghuwansi NS, Rao GP (2006) Effect of watershed subdivision on simulation of water balance components Hydrol Processes 20:1137–1156

Tukimat NNA, Harun S (2013) Multi-correlation matrix (M-CM) for the screening complexity in he statistical downscaling model (SDSM). Int J Eng Sci Innov Technol 2(6):331–342

Wilby RL, Dawson CW, Barrow EM (2002) SDSM—a decision support tool for the assessment of regional climate change impacts. Environ Model Softw 17(2):145–157. https://doi.org/10.1016/s1364-8152(01)00060-3

Wilby RL, Dawson CW, Murphy CO, Connor P, Hawkins E (2014) The statistical downscaling model-Decision centric (SDSM-DC): conceptual basis and applications. Clim Res 61:251–268

Chapter 4
Impact of Climate Change on Hydrological Regime of Narmada River Basin

Deepak Kumar Tiwari, H. L. Tiwari, Raman Nateriya, and Satanand Mishra

Abstract The predicted climate change may introduce more stress and need various adaptation strategies to be implemented. Climate change in the Indian context has caused an increase in surface temperature at the rate of 0.2 °C per decade from 1971 to 2007. The present work is done to assess the impact of climate change through review of literatures pertaining to it.

4.1 Introduction

The per capita annual water resource (AWR) has been used to classify countries with respect to the water scarcity. Countries with an AWR per capita of 1700 cum and above have been termed as countries where shortage will be rare; those with an AWR per capita of less than 1000 cum as water-stressed countries; and those with an AWR per capita of 500 cum and below as countries where availability of water is a primary constraint to life. In 1955, only seven countries were found to be with water-stressed conditions. In 1990, this number rose to 20 and it is expected that by the year 2025 another 10–15 countries shall be added to this list. It is further predicted that by 2050, two-thirds of the world population may face water-stressed conditions. It is worth noting that this assessment has been made without taking into account the possible impact due to predicted changes in global climate. Such consideration may aggravate the situation of AWR further (Ministry of Water Resources of India 1999).

D. K. Tiwari (✉) · H. L. Tiwari · R. Nateriya
Maulana Azad National Institute of Technology, MANIT, Bhopal, Madhya Pradesh, India
e-mail: deepaktiwaricivil7sept@gmail.com

H. L. Tiwari
e-mail: hltiwari@rediffmail.com; hltiwari@manit.ac.in

R. Nateriya
e-mail: raman_nateriya@rediffmail.com

S. Mishra
Advanced Materials & Processes Research Institute (AMPRI), Bhopal, Madhya Pradesh, India
e-mail: snmishra07@gmail.com

R. Jha et al. (eds.), *Climate Change Impacts on Water Resources*, Water Science and Technology Library 98, https://doi.org/10.1007/978-3-030-64202-0_4

The predicted climate change may introduce more stress and need various adaptation strategies to be implemented. The strategies may range from land use, cropping pattern, flood warning system, and so on (Gosain et al. 2006). The rise in surface air temperature and subsurface ocean temperature, which are perceived to be the primary cause of sea level rise resulting from a continuously warming climate and increased moisture content associated with expected higher rainfall amounts, are considered to be strongly affecting the temporal and spatial pattern of Indian Summer Monsoon Rainfall (ISMR) (Goswami et al. 2006; Dash et al. 2011). Climate change in the Indian context has caused an increase in surface temperature at the rate of 0.2 °C per decade from 1971 to 2007 (Kothawale et al. 2010). Such an increase in temperature is expected to bring out a change in the hydrological cycle mainly due to an increase in average evaporation, water vapour and precipitation. Recently, an increase in the intensity and frequency of extreme rainfall events has been reported over Central India (where the basin is located), which makes these areas susceptible to flash floods and drought at the same time. So, this raises the most important question on whether the climate change is a treat to human water security and increased severity to hydrological disasters. Previous studies have shown that simulated climate change impacts vary substantially depending upon climate model and emission scenario of greenhouse gases used (Arnell 1999; Bronstert 2004; Hurkmans et al. 2008).

The objective of the study is to assess the impact of climate change through review of literatures pertaining to it. A detailed literature review is conducted followed by methodology involving study area and description of methodologies used previously, data details and followed by conclusion.

4.2 Literature Review

Hosseinizadeh et al. (2015) tudied the impact of climate change on the severity, duration and frequency of drought in a semi-arid agricultural basin in Khuzestan, Iran. The study applied the Standardized Precipitation Index (SPI) along with a combination of GCM scenarios to create the severity–duration–frequency (SDF) curves of drought for the period 2020–2044. An average period of six months (ending in May) was used for the SPI, corresponding to the agricultural growing season of the region, to assess drought conditions fewer than five plausible climate scenarios. The selected GCM scenarios were GISS-ER A1B (warmer and drier), CSIROMk3.5 B1 (cooler and drier), INGV-SXG A1B (median conditions), ECHO-G A2 (warmer and wetter) and ECHAM5 B1 (cooler and wetter) and they were downscaled with an artificial neural network (ANN) approach. Results reveal that most scenarios exhibit an increase in the duration of extreme drought while the duration of moderate drought decreases under all scenarios (Hosseinizadeh et al. 2015).

Shamir et al. (2015) studied the stream flow events in Upper Santa Cruz River in Arizona. They analysed climate change projections of precipitation for the Upper Santa Cruz River from eight dynamically downscaled global circulation models

(GCMs). Our analysis indicates an increase (decrease) in the frequency of occurrence of dry (wet) summers. The winter rainfall projections indicate an increased frequency of both dry and wet winter seasons, which implies lower chance for medium-precipitation winters. The climate analysis results were also compared with resampled coarse GCMs, bias adjusted and statistically downscaled CMIP3 and CMIP5 projections readily available for the contiguous U.S. The impact of the projected climatic change was assessed through a water resources management case study (Shamir 2015).

Dan (2015) conducted study on Pearl river located in South of China which has second largest stream flow in China. They projected that due to global warming emission of greenhouse gases increased, which affects the precipitation, run-off processes and water resources, and can also alter the hydrological extremes. In the end, it was concluded that the effect of climate change can be assessed by model simulations. There are substantial differences in results between the different climate models, for example, all the models point to the same direction in terms of flow changes with the exception of HadGEM2-ES. This model shows much higher precipitation and thus runoff compared to the other four GCMs. The main conclusion of their work is that dry seasons are projected to become drier throughout the basin. Wet seasons projected to become drier in the upper reach and wetter in the middle and lower reaches of the Pearl river basin (Dan 2015).

Devkota and Gyawali (2015) studied the middle hilly region of the Koshi river basin in Nepal. Hydrological impact simulations conducted using two models (PRECIS-HADCM3Q0 and PRECIS-ECHAM05) based on IPCC-SRES A1B scenario and also using SWAT model. Design flood estimation was done after extreme value analysis based on annual flow maxima. New hydrological insights for the region: The study found that climate change does not pose major threat on average water availability. However, temporal flow variations are expected to increase in the future (Devkota and Gyawali 2015).

Pandey et al. (2016) studied hydrology Godavari river basin. A GIS-based semi-distributed hydrological model, soil and water assessment tool (SWAT), has been used to estimate the water balance components on the basis of unique combinations of slope, soil and land cover classes for the baseline (1961–1990) and future climate scenarios (2071–2100). Sensitivity analysis of the model has been performed to identify the most critical parameters of the watershed. Coefficient of determination (R^2), Nash–Sutcliffe efficiency (ENS) and root mean square error (RMSE) were used to evaluate the model performance. Calibrated SWAT setup has been used to evaluate the changes in water balance components of future projection over the study area. HadRM3, a regional climatic data, has been used as input of the hydrological model for climate change impact studies. The study is useful in decision-making for the application of best management practices (Pandey et al. 2016).

Nepal (2015) studied the climate trends in Koshi river basin using statistical analysis. The study had objectives of identifying the present climatic trends from historic data and to assess the projected impact of climate change on hydrological regime. The hydrological system dynamics were analysed in the Dudh Koshi (sometimes written as Kosi) sub-basin using the process-oriented distributed J2000 hydrological model, which includes both hydrological and cryospheric processes (snow and glacier melt). Finally, the projected meteorological data from the PRECIS RCM A1B scenario (research period 1995–2096; baseline 2000–2010) were used as an input in the hydrological model to assess the impact on the hydrological regime. This type of systematic approach contributes to improved understanding of the effect of projected climate change on different watershed components, and the potential influence on water distribution in the basin, including future water availability downstream (Nepal 2015). Thomas et al. (2015) studied the Narmada basin in Central India for the trend analysis of 1-day maximum rainfall series. Drought duration estimated by the standardized precipitation index for the periods 1951–1970 and 1989–2008 indicated that the entire basin has experienced frequent droughts during the recent two decades, with the middle zone of the basin being more prone to droughts. The analysis also suggested that appropriate measures may be proposed for better management of the water resources in the basin, and also for mitigation of floods and droughts, considering the increased risk of the high-intensity storms as well as the increased frequency of drought occurrence during the recent two decades (Thomas et al. 2015).

4.3 Methodology

4.3.1 Study Area

Narmada starts from Amarkantak in Annupur district in Madhya Pradesh and flows towards Arabian Sea in Gulf of Khambhat. Its total length is of 1312 km, out of which first 1079 km in MP, rest in Maharashtra and Gujarat. It has a stretch of 159 km. The basin spreads over 20 parliamentary constituencies (2009) comprising 15 of Madhya Pradesh, 3 of Gujarat, and 1 each of Chhattisgarh and Maharashtra. Table 4.1 describes the salient features of Narmada river at a glance. It has average water resource potential of 45639 MCm, out of which 75% was used by Madhya Pradesh. Also, Central Water Commission has made 26 hydrological observation stations for its monitoring. Total catchment area is approximately one lac km^2 and 4 flood-forecasting stations available presently.

Thick blue indicates the main course of Narmada river basin. Also, major barrages, Sardar Sarovar dam and sites of hydrological importance have been shown in Fig. 4.1. The boundary of basin is lined in red colour. Salient features of Narmada river have been tabulated in Table 4.1.

Narmada river basin has huge potential for water storage. The Master Plan on the utilization of Narmada waters was submitted to Khosla Committee by the Madhya Pradesh Government. It proposed that by constructing 19 dams on Narmada and its tributaries, irrigation can be provided to 3.1 million ha area by utilizing 29,295 MCM of water. Out of these 31 projects only five are completed. So, it shows that a lot of potential exploitation can be achieved by proper modelling of its hydrological characteristics.

Table 4.1 Salient features of Narmada basin

Basin extent	
Longitude	72°38′ to 81°43′ E
Latitude	21°27′ to 23°37′ N
Length of Narmada river (km)	1312
Catchment area (sq. km)	98796
Average water resource potential (million cubic metre-MCM)	45639
Utilizable surface water resource (MCM)	34500
Live storage capacity of completed projects (MCM)	17806.0
Live storage capacity of projects under construction (MCM)	6835.00
Total live storage capacity of projects (MCM)	24641.0
No. of hydrological observation stations of CWC	26 (including 8 Gauge site)
No. of flood forecasting stations of CWC	4

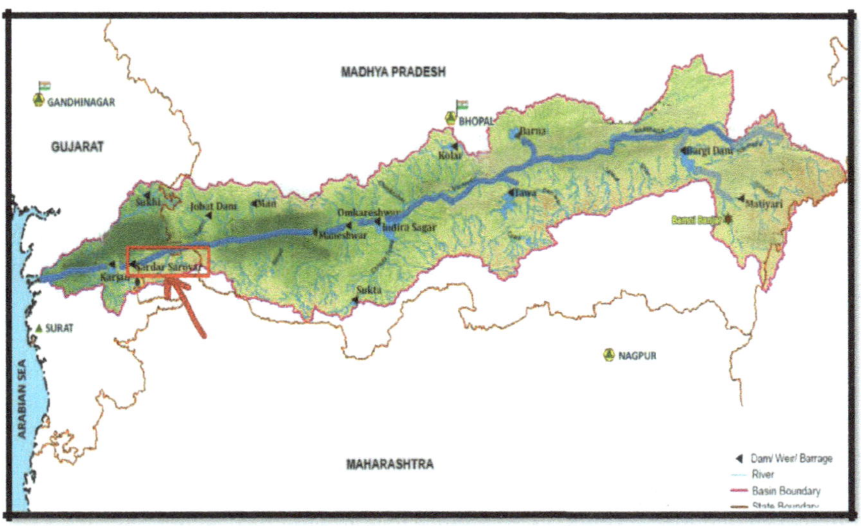

Fig. 4.1 Narmada basin along with main CWC stations

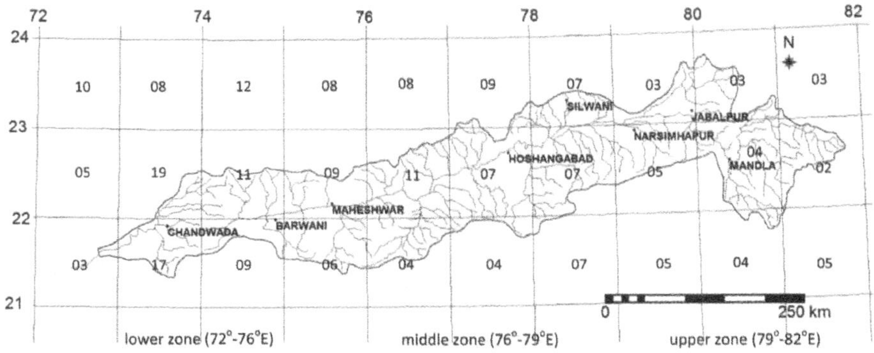

Fig. 4.2 Grid and rain gauge station information of the Narmada basin

4.3.2 Data and Methodology

The 1 W × 1 W daily gridded rainfall data prepared by the India Meteorological Department (IMD) for the Indian land mass (6.5WN–38.5WN and 66.5WE–100.5WE) will be used in the proposed study (Rajeevan et al. 2011). All the rain gauge stations concerned with the project along with 90% minimum data availability will be used for the period of 50 years at least. The data interpolation method is based on Shepherd method (Shepard 1968) and the weighted sum of the observations will be considered at surrounding rain gauge stations that lie within predetermined radius of influence. Also, to validate the possibilities and usefulness of data used in present studies, daily gridded interpolated rainfall data will be compared with direct observations for grids where rainfall data is available (in this case Hosangabad). The Narmada basin is covered by 23 grid cells (1 W × 1 W) as per the IMD interpolated information (Maxey et al. 2012), of which the lower zone (72WE–76WE and 21WN–24WN) constitutes seven grid cells, the middle zone (76WN–79WN and 21WN–24WN) eight grid cells and the upper zone (79WE–82WE and 21WN–24WN) eight grid cells as shown in Fig. 4.2 (Smith et al. 2012).

4.3.3 Model Working Flowchart

Digital elevation model (DEM) represents a topographic surface in terms of a set of elevation values derived at a finite number of points. DEM has been generated using contours taken from 1:250,000 scale topographic maps (Mekonnen and Jayawardena 2009). After getting DEM automatic delineation of the river basins is done by using the DEM as input and the final outflow point on the drainage of the river basin as the final pour/drainage point (self-delineating with the help of GIS). The total area of the river basin as obtained from the automatic delineation has also been used for forecasting. Classified land cover data (13 categories) produced by the University

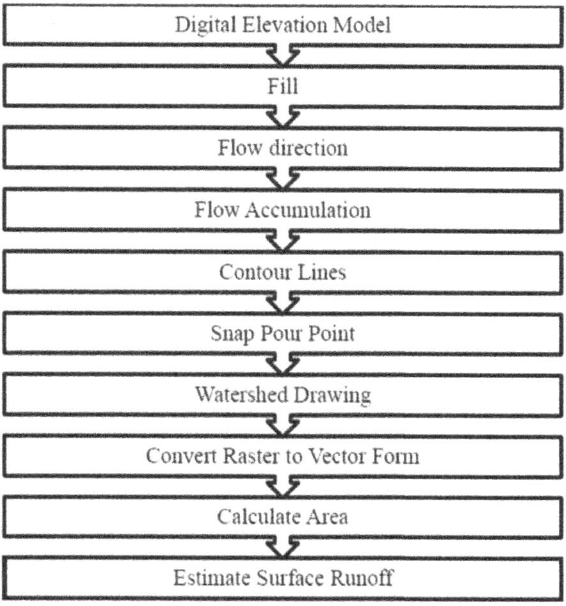

Fig. 4.3 Steps involved in calculation of surface runoff

of Maryland Global and cover facility using remote sensing with resolution of 1 km grid cell has been used. Soil map adapted from FAO digital soil map of the world and derived soil properties with a resolution of 1:5,000,000 have been used. Here, a flowchart for the model construction is given for better understanding. It represents steps involved in estimation of runoff of the river. Artificial neural network will be applied for data correction and forecasting of the discharge and precipitation (Smith et al. 2010) (Fig. 4.3).

4.4 ANN Used in Runoff Modelling

Fundamentally, stream-flow forecasting is mainly classified into two main temporal categories, namely short-term (real-time) forecasting (hourly and daily) is critical for reliable operation of flood and mitigation systems and secondly long-term forecasting (weekly, monthly and yearly) which is important for planning and operations of irrigation management decisions, scheduling releases of reservoir water, hydropower generation, sediment transport and many other applications. Time series models such as auto regressive (AR) moving average (MA), auto regressive integrated moving average (ARIMA), auto regressive integrated moving average with exogenous input (ARIMAX), linear regression (LR) and multiple linear regression (MLR) are used vastly for stream flow forecasting since 1970s. Early artificial neural networks were inspired by perceptions of how the human brain operates. In the recent years, ANN technological developments have made it more of a technique of applied mathematics

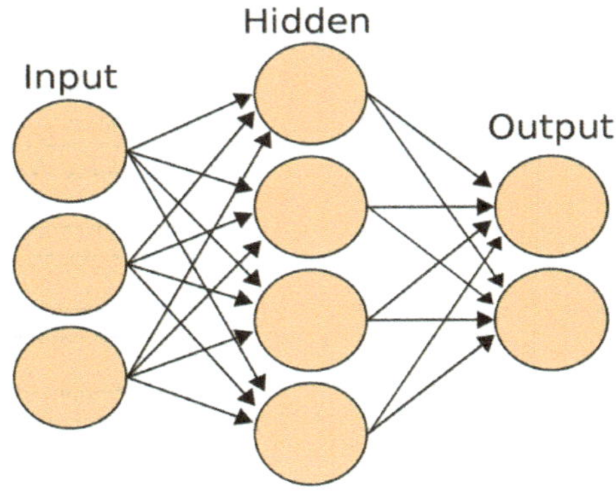

Fig. 4.4 Simple three-layered artificial neural network

with some similarities to the human brain. ANNs retain two characteristics of the brain as primary features: the ability to (1) 'learn' and (2) generalize from limited information. A general neural network is depicted in Fig. 4.4.

The hidden layer receives signals from input and uses one activation function for output generation. Once a network is modelled, for a specific application, it is ready to be trained. To start this, initial weights and training parameters are chosen at random. After that training or learning begins. There are two approaches to learning: supervised and unsupervised. In supervised training mechanism, network is provided with desired output either by manually or by providing desired out with input. In case of unsupervised one, network has to itself predict the input without being helped by outside.

4.5 Wavelet Transform: A Data Processing Technique

Data pre-processing technique is an essential alteration to be smeared before giving the data input to ANN networks. It has been found to increase the efficiency of the network amazingly. Mostly methods used for data gathering are casually controlled, results in outliers, impossible data combination, missing values, human errors, improper calibration of recording instruments, and so on. Interpreting the data which has not been carefully separated may result in confusing results. Thus, depiction and quality of data is of prime importance to run any analysis of dataset. The performance of wavelet-AI models was found to be superior to the standard-AI models. This is due to the advantages of wavelet transform, which are de-noising the time series to improve the AI-based modelling performance and extracting dynamic and multi-scale features of non-stationary time series.

One-dimensional discrete wavelet function is best for stream flow forecasting mathematically represented as

$$W_{\varphi}(j_0, k) = \frac{1}{\sqrt{M}} \sum_{x} f(x)\varphi_{jo,k}(x) \quad (4.1)$$

where $W_{\varphi}(j_0, k)$ = discrete wavelet transforms functions
$f(x)$ and $\varphi_{jo,k}(x)$ = functions of discrete variables.
M = number of vales.

4.6 Efficiency Parameters

The author is intended to use coefficient of determination (R^2), Nash–Sutcliff efficiency (ENS) and root mean square error as efficiency parameters for validation and testing of model. Coefficient of determination (R^2) measures the dispersion between observed value and simulated value from model (Mellor et al. 2000). It is given as:

$$\textbf{Coefficient of determination } \mathbf{R^2} = \frac{n(\sum Q_{obs} * Q_{sim}) - (\sum Q_{obs}) * (\sum Q_{sim})}{{}^{1}/\sqrt[2]{[n(\sum Q_{obs}^2) - (\sum Q_{obs})^2]X[n(\sum Q_{sim}^2) - (\sum Q_{sim})^2]}} \quad (4.2)$$

where
Q_{obs} = observed value of discharge
Q_{sim} = simulated value of discharge
n = number of observations
The range of R^2 lies between 0 and 1, which represents no correlation and perfect correlation between observed and simulated value.

Nash–Sutcliffe efficiency (Castro et al. 2014): It is defined as the one minus the sum of absolute squared differences between the predicted and observed value normalized by the variance of the observed values during the period under investigation. It is calculated as follows:

$$\textbf{Nash–Sutcliffe efficiency}, \ \mathbf{N} = \frac{\sum_{t=1}^{T}\left(Q_m^t - Q_o^t\right)^2}{\sum_{t=1}^{T}\left(Q_o^t - Q_{mo}^t\right)^2} \quad (4.3)$$

where Q_m^t = modelled discharge at time t
Q_o^t = observed discharge at time t
Q_{mo}^t = mean of observed discharges.

Root mean square error: RMSE is very commonly used and makes for an excellent general-purpose error metric for numerical predictions. Compared to the similar mean absolute error, RMSE amplifies and severely punishes large errors. The RMSE represents the sample standard deviation of the differences between predicted values and observed values. These individual differences are called residuals when the calculations are performed over the data sample that was used for estimation, and are called prediction errors when computed out-of-sample. The RMSD serves to aggregate the magnitudes of the errors in predictions for various times into a single measure of predictive power. RMSE is a measure of accuracy, to compare forecasting errors of different models for a particular data and not between datasets, as it is scale-dependent (Seo et al. 2014). It is given by:

$$\text{RMSE} = \sqrt{\frac{\sum_{n=1}^{N} (\boldsymbol{P}_i - \boldsymbol{A}_i)^2}{N}} \tag{4.4}$$

4.7 Conclusion

At present, the performance of many flood-forecasting system in an operational context is sub-optimal and often below expectation. The information they designed to provide fail to reach much of the target audience. Existing flood-warning systems even with their manifest deficiencies can be effective in the mitigation of flood damage. It is very likely that if they are recast in terms of completeness, carefully planned and kept alive, their effectiveness can be augmented considerably. In India most of the techniques for formulating the real-time flood forecast are based on statistical approach. For some projects, network model and multi-parameter hydrological model are used. Conventional systems of communication are normally used for transmitting the data in real time. Flash floods are usually experienced. As such there is no system for formulating the flash flood forecast. It results in heavy losses of lives and properties. There is a need for improvement of the real-time flood-forecasting system in India. Efficient automatic communication systems are required to be established for transmitting the data in real time. The forecasting techniques such as deterministic models, stochastic models, ANN and fuzzy logic techniques are required to be studied, and a suitable method may be recommended for field applications based on the performance evaluation criteria and considering the data availability. The information about the flood have to be disseminated well in advance to the people likely to be affected so that an emergency evacuation plan may be developed for the river basin.

References

Arnell NW (1999) The effect of climate change on hydrological regimes in Europe: a continental perspective. Global Environ Chang 9(1):5–23

Bronstert A (2004) Rainfall-runoff modelling for assessing impacts of climate and land-use change. Hydrol Process 18(3):567–570

Castro LM, Giron J, Fern_andez B (2014) Spatial estimation of daily precipitation in regions with complex relief and scarce data using terrain orientation. J Hydrol 517:481–492
Dash SK, Archana AN, Makarand AK, Mohanty UC (2011) Characteristic changes in the long and short spells of different rain intensities in India. Theor Appl Climatol 105:563–570
Dan Y (2015) Hydrological response to climate change: the Pearl River, China under different RCP scenarios. J Hydrol Reg Stud 4:228–245
Devkota LP, Gyawali DR (2015) Impacts of climate change on hydrological regime and water resources management of the Koshi River Basin, Nepal. J Hydrol Reg Stud 4:502–515
Gosain AK, Rao S, Basuray D (2006) Climate change impact assessment on hydrology of Indian river basins. Curr Sci 90(3
Goswami BN, Venugopal V, Dengupta D, Madhusoosanan MS, Xavier KP (2006) Increasing trend of extreme rain events over India in a warming environment. Curr Sci 314:1442–1445
Hosseinizadeh A, SeyedKaboli H, Zareie H (2015) Impact of climate change on the severity, duration, and frequency of drought in a semi-arid agricultural basin" Geo environmental Disasters. Springer Open J. https://doi.org/10.1186/s40677-015-031-8
Hurkmans RTWL, Moel H, Aerts JCJH, Troch PA (2008) Water balance versus land surface model in the simulation of Rhine River discharges. Water Resour 44(1)
Kothawale DR, Munota A, Krishna KK (2010) Recent trends in pre-monsoon daily temperature extremes over India. J Earth Syst Sci 119:51–65
Maxey R, Cranston M, Tavendale A, Buchanan P (2012) The use of deterministic and probabilistic forecasting in countrywide flood guidance in Scotland. BHS Eleventh National Symposium, Hydrology for a changing world, Dundee 2012. ISBN: 1903741181, British Hydrological Society. https://doi.org/10.7558/bhs.2012.ns33
Mekonnen ZZ, Jayawardena AW (2009) Development of flood forecasting model in middle awash river basin of Ethiopia. Hydrology Department, Ministry of Water Resources, Ethiopia
Mellor D, Sheffeld J, O'Connell PE, Metcalve AV (2000) A stochastic rainfall forecasting system for real time flow forecasting I: Development of MTB conditional rainfall generator. Hydrol Earth Syst Sci 4(4):603–615
Ministry of Water Resources of India (1999) Integrated water resource development–a plan for action. Report of the National Commission for Integrated Water Resources Development. Volume–I, Government of India, New Delhi
Narmada Basin (2012). http://nihroorkee.gov.in/rbis/basin%20maps/narmada_about.html
Nepal S (2015) Impacts of climate change on the hydrological regime of the Koshi river basin in the Himalayan region. J Hydro-Environ Res 10(2016):76–89
Pandey BK, Gosain AK, Paul G, Khare D (2016) Climate change impact assessment on hydrology of a small watershed using semi-distributed model. Springer, Appl Water Sci J (2017) 7:2029–2041
Rajeevan M, Bhate J, Kale JD, Lal B (2011) High resolution daily gridded rainfall data for the Indian region: analysis of break and active monsoon spells. Curr Sci 91:296–306
Seo D-J, Siddique R, Zhang Y, Kim D (2014) Improving real-time estimation of heavy-to-extreme precipitation using rain gauge data via conditional bias-penalized optimal estimation. J Hydrol 519:1824–1835
Shamir E, Megdal SB, Carrillo C, Castro CL, Chang H, Chief K, Corkhill FE, Eden S, Georgakakos KP, Nelson KM, Prietto J (2015) Climate change and water resources management in the Upper Santa Cruz River , Arizona. J Hydrol Elsevier B.V. 521:18–33
Shepard D (1968) A two-dimensional interpolation function for irregularly spaced data. Proceedings of the ACM National Conference. New York, USA, pp 517–524
Smith JA, Villarini G, Baeck ML (2010) Mixture distributions and the hydro climatology of extreme rainfall and flooding in the Eastern United States. J Hydrometeorol 12:295–309. https://doi.org/10.1175/2010JHM1242.1
Smith PJ, Beven KJ, Weerts AH, Leedal D (2012) Adaptive correction of deterministic models to produce probabilistic forecasts. Hydrol Earth Syst Sci. www.hydrol-earth-syst-sci.net/16/2783/2012, https://doi.org/10.5194/hess-16-2783-2012

Thomas T, Gunthe S, Ghosh NC, Sudheer KP (2015) Analysis of monsoon rainfall variability over Narmada basin in central India: Implication of climate change. J Water Clim Change, 615–627. https://doi.org/10.2166/wcc.2014.041
Water Resources Information System of India (2017) India-WRIS. http://india-wris.nrsc.gov.in/wrpinfo/index.php?title=Narmada#Description

Chapter 5
Climate Change Impacts on Water Resources in Ethiopia

Abiot Ketema and G. S. Dwarakish

Abstract The impact of climate change on water resource is a primary concern. The focus of this review was to recognize how climate change affects the water resource of Ethiopia and to identify gaps that can be addressed through research. The result revealed that the peoples living in the country already suffered from the extreme events of climate change, like drought and flood. The nature and magnitude of how global change will affect the Ethiopian water resource is not yet adequately understood. Besides, the studies of climate change impacts on water resources in Ethiopia are concentrated in northern part of the country, based on old climate change emission scenarios and limited number of global climate models (GCM). Making a conclusion based on a single GCM output may not give a clear representation of the future changes, and it is most likely will mislead the decision makers and policy developers. A comprehensive study is highly needed, and the issue has to be addressed at a scale relevant to decision-making based on multiple GCMs under new climate change emission scenarios for better understanding of the potential impacts, informed decision-making, and effectively respond and adapt to projected changes, otherwise the consequences becoming awful.

Keywords Climate change · Ethiopia · Potential impacts · Water resource

5.1 Introduction

The relationship between water and climate change is a primary subject. The most important outcome of climate change is the alteration of the available water. A warmer climate will speed up the hydrological cycle, varying the amount and timing of both rainfall and runoff. Increased rainfall variability and intensity are projected to enhance the possibility of drought and flooding in many areas (Bates et al. 2008). Climate change impacts are location-specific. Its magnitude is not the same throughout the world and the extent of the undesirable outcomes of climate change will vary notably

A. Ketema (✉) · G. S. Dwarakish
Department of Water Resources and Ocean Engineering, National Institute of Technology Karnataka, Surathkal, 575 025 Mangalore, Karnataka, India
e-mail: ketemaabiot80@gmail.com

R. Jha et al. (eds.), *Climate Change Impacts on Water Resources*, Water Science and Technology Library 98, https://doi.org/10.1007/978-3-030-64202-0_5

across regions, nations and socioeconomic groups. Usually, developing nations like Ethiopia are more likely vulnerable for the climate extremes than developed countries because of their competence to execute mitigation and adaptation techniques to alleviate its harmful impacts (Intergovernmental Panel on Climate Change [IPCC] 2014).

Climate-associated natural hazards such as drought and flood happen everywhere in the planet, however, its consequences are very serious in Africa and particularly in Ethiopia (National Meteorological Agency [NMA] 2007). Ethiopia is a developing country in Africa and its development is extremely reliant on rain-fed farming (NMA 2007; World Bank 2007; Ethiopian Panel on Climate Change [EPCC] 2015a). In a country with a predominantly rain-fed agricultural system like Ethiopia, variation in a rainfall may unfavourably have an effect on crop yield as the rainfall is very critical climatic determinants of crop production (NMA 2007). Ethiopia has implemented some water development programs, watershed management activities, and mitigation and adaptation measures to safeguard the country from the undesirable repercussions of global change and enhance the level of water security. However, water shortage is one of the greatest problems of the country. The nature and magnitude of how global change affects the Ethiopian water resource is poorly understood. This review assessed climate change impacts on water resources in Ethiopia and identifies gaps that can be addressed through research.

5.2 Water Resources of Ethiopia

Ethiopia is located in East Africa between 3°–15° north and 33°–48° east, just north of the equator. Its area is around 1.13×10^6 km^2 (Berhanu et al. 2014). The climate of Ethiopia ranges from temperate in highlands to tropical in the lowlands (Fazzini et al. 2015). The yearly average rainfall of the country is about 744 mm (Awulachew et al. 2007). Lowlands are vulnerable to increasing temperatures and prolonged droughts, whereas highlands are prone to intense and irregular rainfall (USAID 2016). Bega, Belg, and Kiremt are the three seasons in Ethiopia. The Bega is the period from October to January. It is a dry and harvesting period for various parts of Ethiopia. The Belg is a squat rainy period from February to May, and in this season rainfalls are very erratic in nature. The Kiremt is the major rainy period from June to September; 50–80% of the country's annual rainfall is measured in this period (NMA 2013; Fazzini et al. 2015; USAID 2016).

Ethiopia has copious surface water resources with estimated yearly average flows from the country's 12 river basins at 122 billion cubic meters per annum (MoWR 2002) and has lake water from 12 main lakes at 70 billion cubic meters (Berhanu et al. 2014). The accurate potential of the country's groundwater is not yet known; however, it is reported that Ethiopia has a groundwater potential of approximately up to 30 billion cubic meters (MoWR and GW-MATE 2011). Despite the fact that Ethiopia has considerable water resource potential, the spatio-temporal distribution of water is highly uneven (EPCC 2015a), and Ethiopians face water-related problems.

In spite of plenty in some parts, Ethiopia is highly water scarce due to poor water management and shortage of storage infrastructure.

The availability of surface water and rainfall in Ethiopia is erratic and extremely varies with space and time (Robinson et al. 2013; EPCC 2015a). Spatially, 80–90% of the water resources are found in western parts of Ethiopia where no more than 30–40% of Ethiopia's population lives. In contrast, more than 60% of the populations were living in the eastern and central part of Ethiopia where the available surface water resource is only 10–20%. Temporally, the majority of the rivers in Ethiopia become full in June, July and August within the duration of the Kiremt season (Berhanu et al. 2014). The amount of rainfall varies spatially within the country and within different regions in the country. As an example, within Amhara national regional state, yearly precipitation ranges from 770 mm in the eastern part to greater than 1660 mm western part of the region (Bewket and Conway 2007). An increment in rainfall from the northeast to the western portion of Tigray region was also recorded. A study by Gebrehiwot and Van der Veen (2013) reported that the yearly rainfall ranges from 300 mm in the northeast to 1260 mm in the west of Tigray.

Too little has been done to date in utilizing the water resources in Ethiopia as engines to boost country's socioeconomic development. Even though the water resource of the country has a capacity to develop an irrigation area of 3.8×10^6 ha (Berhanu et al. 2014), only less than 5% is developed (World Bank 2006). Hydropower potential has been estimated to 45,000 MW per annum (Berhanu et al. 2014), of which the country developed only 3813 MW (International Hydropower Association [IHA] 2017). Currently, the country pays more attention to generate electric energy from hydropower. As an example, 6000 MW hydropower project is under construction from a single project called Great Ethiopian Renaissance Dam (GERD). Inadequate finance for investment to control water, technical challenge like having less number of hydrological gauging stations, hydro-politics, and lack of good authority and skilled experts in the water sector are some of the reasons for not fully using the water resource potential in Ethiopia (Berhanu et al. 2014; Mosello et al. 2015).

5.3 Ethiopian Contribution to Global Greenhouse Emission

Ethiopia's contribution to global annual greenhouse gas (GHG) emission is negligible. According to EPCCb report, GHG emission from the country was 48 Mt CO2e (0.9 tone CO2e per capita) in 1994, 150 MtCO2e in 2010 (1.8 tone CO2e per capita) and will increase to 400 MtCO2e (3 tone CO2e per capita) by 2030. Agriculture is the main GHG emission sector in Ethiopia. In 2010, land use change and agriculture account for 87% of the national emissions. The rest 13% is from industry, transport and energy sectors (EPCC 2015b). Although the country contributes little to GHG emission, Ethiopians already suffered from the extreme events of climate change like drought and flood (NMA 2007).

5.4 Climate Change Impacts on Water Resources in Ethiopia

5.4.1 *Observed (Historical) Impacts on Water Resource*

In Ethiopia floods and droughts occur repetitively as an outcome of climate change (Robinson et al. 2013; EPCC 2015a). These events can lead to famine and disruption of socioeconomic well-being. Yearly the country loses up to 6% of the entire crop yield as a result of rainfall variability (EPCC 2015b). Too much surface water runoff in some parts of Ethiopia results in flood, and at the same time, other parts of the country suffer from drought.

The principal and direct consequence of climate change would be the changes in water availability. A decline of soil moisture, reduction of freshwater availability, repeated droughts and floods are some of the negative impacts observed in Ethiopia due to rise in temperature, and changing the amount and pattern of rainfall (USAID 2016). According to the USAID report, the occurrence of flood and drought increased since 1960 in Ethiopia (USAID 2016). As the Ethiopian socioeconomic development is extremely dependent on rain-fed agriculture, Ethiopia suffered a lot from climate variability and weather extremes. The global warming resulted in repeated droughts and heavy rainfall in different parts of Ethiopia, and decreasing crop productivity. Yearly many Ethiopians are exposed to famine, serious health problem, flood hazard and drought due to rainfall variability in response to ongoing climate change (NMA 2007; EPCC 2015a). Drought is very critical climate-associated disaster in Ethiopia. For example, as a result of drought, 7.75 million people in 1983, 2.1 million in 1996, 13.2 million in 2003 and 7.1 million in 2004 were exposed to famine (Dorosh and Rashid 2013; EPCC 2015b). Few years back, in 2015/16 more than 10,000,000 people were affected by drought in Ethiopia (Cochrane and Singh 2017). Like drought, flood has a considerable impact in different parts of Ethiopia. For instance, the 2006 flood occurred in Gambella regional state that damaged 1650 ha of maize and decreased agricultural production by 20% in the region (Gambella Region Disaster Prevention and Preparedness Agency 2007). In addition, 364 people were killed and 6000–10,000 people were displaced, in South nation nationality people region (South Omo zone) (World Food Program [WFP] 2006). In the same year, as a result of the flash flood, 256 people died, more than 9956 people displaced and more than 1000 traders' property were damaged in Dire Dawa city (Alemu 2015). The spatio-temporal variability of rainfall is more important than total rainfall deficits for water-related problems in Ethiopia.

5.4.2 *Potential Impacts on Water Resource*

Long-term climate change impacts in Ethiopia is primarily linked to the rise in temperature and rainfall variability (EPCC 2015a). So, recognition and analysis of their pattern in the country are highly needed. Future prediction of temperature and rainfall patterns in Ethiopia shows a high degree of uncertainty. Most GCMs and regional projections of climate models agreed that yearly average temperature is estimated to raise by between 1 and 2 °C (USAID 2016) and between 1.4 and 2.9 °C (Cochrane and Singh 2017) by 2050. It is uncertain whether rainfall will rise/drop in Ethiopia, and its projections varies between − 25% and + 30% by 2050s. The percentage of extreme total rainfall rises up to 18%. Studies suggest a 30% decrease in the runoff on several Nile tributaries by 2050 (USAID 2016). The increase in temperature results in an increasing water stress, whereas an increment of extreme rainfall results in increasing flood. As an outcome, climate change affects the development of the country by reducing Ethiopia's gross domestic product (GDP) nearly 10% in the upcoming (2045) (USAID 2016). The followings are the summaries of the studies in Ethiopia related to future climate change impact on water resources.

Serur and Sarma (2016) studied the climate change effects on water resource of Weyib watershed, Ethiopia. They used CanESM2 of CMIP5 climate change model with three RCPs emission scenarios to predict the future rainfall and temperature, and soil and water assessment tool (SWAT) to estimate the availability water until 2100. They detected that both precipitation and temperature showed an increasing trend in the upcoming periods. As well, they identified that, a rising trend of yearly average daily water availability in the ranges of 11.82–12.68% (RCP 2.6), 3.98–20.40% (RCP 4.5), and 9.18–24.49% (RCP 8.5), and a decreasing trend of potential evapotranspiration (PET) in all the three RCPs climate change scenarios compared to reference period (1984–2004). They also estimated the seasonal variation of water availability in the basin. The results showed an increase of water availability in the rainy and transitional season, whereas a decline in the dry season. Gebre and Ludwing (2015) used five GCM outputs of CMIP5 based on high and stabilization scenarios to appraise climate change impacts on water resource of Giligel Abay, Ribb, Gummer, and Megech catchments of Lake Tana watershed in Ethiopia. The result revealed that, in all months and seasons, and in all five GCMs, both temperature and PET increases. Although it varies in both magnitude and direction, precipitation shows an increasing trend in the future. The average yearly runoff may raise about + 55.7% (RCP 4.5) and + 74.8% (RCP 8.5) for the period of 2035–2064, and by + 73.5% (RCP 4.5) and + 127.4% (RCP 8.5) for the period (2071–2100) as compared to baseline period (1960–2005). They concluded that the increase in runoff may have a positive contribution to existing development projects and crop productivity based on rain-fed agriculture in the basin on the conditions that appropriate mitigation and adaptation measures are applied to diminish harmful possible impacts. The effect of climate change on the water availability in Lake Tana watershed in Ethiopia for early (2010–2039), mid (2040–2069) and late (2070–2099) twenty-first century was studied by Nigatu et al. (2016) using HadCM3 GCM output of A2 and B2 emission scenarios under SRES. Their result revealed that for both emission scenarios and all the three future periods, the water storage of Lake Tana increases constantly relative to the baseline period (1981–2010) due to the increase in rainfall noticeably from 9 to 11%. A study by Kim et al. (2008) showed that water resource of UBNRB in northern Ethiopia may not be adversely affected by climate change. They used the ensemble mean of six

GCMs under A2 of SRES. The result revealed that both rainfall and runoff showed an increasing trend within the period (2040–2069) compared to (1961–1990). The increasing trend of rainfall in the UBNRB implies the water availability for crop production, the reduction of severe drought events and a slight increase in flood risk by 2050s.

According to Dile et al. (2013), both rainfall and discharge of Gilel Abay River basin in Ethiopia are decreased for the period 2010–2039, and increases for the period 2070–2100. They used HadCM3 GCM output based on A2 and B2 scenarios of SRES to project the future temperature and precipitation, and SWAT for simulation of its effect on discharge. They concluded that the rise in rainfall in the wet period (Belg and Kiremt) in the future enhances crop production in the basin, which is rainfall dependent, even though the rise in ET is expected due to an increase in temperature. Depending on adaptation measures implemented in the area, the increase in flow may have positive impacts (on water development projects) or negative impacts (recurrent flooding problems) in the area. Mekonnen and Disse (2016) examined the effect of climate change on water resource of Blue Nile River basin in Ethiopia based on both CMIP3 and CMIP5 GCMs. The ensemble mean result showed an increasing pattern in rainfall ranging from 1.0 to 14.4%, maximum temperature from 0.4 to 4.3°, and minimum temperature from 0.3 to 4.1 °C. Wagesho et al. (2013) examined the climate change impacts on runoff of Blate and Hare watersheds in Ethiopia by using two GCM outputs under CMIP3 based on A2 and A1B emission scenarios as an input to SWAT hydrological model. Results revealed that extreme events of daily rainfall and temperature will be increasing in the future compared to observed events. The stream flow varies from −4 to 18% and − 4 to 14% for Hare and Blate river basins, respectively, during the simulation period (2081–2090) compared to baseline period (1990–1999).

Setegn et al. (2011) analysed future temperature and precipitation in Lake Tana basin of Ethiopia by using A1B, A2 and B1 emission scenarios. Then, they applied the temperature and precipitation generated from the 15 GCMs as an input to the SWAT model to predict stream flow and other hydrological components in the two future periods. They found that, for the period of 2070–2100 and for the three emission scenarios, the temperature increases 2–4.4 °C and 2.2–4.9 °C in the wet and dry period, respectively. For the same period and scenarios, the precipitation varies from −13 to + 12% and −14 to + 16%, respectively. Also, they observed that statistically significant decline in the magnitude of stream flow following the same direction as rainfall compared to the baseline period (1980–2000). Both soil moisture and groundwater also show a reduction in the future. They concluded that, unless ample irrigation water is available, the basin may be exposed to a great agricultural drought. A study was carried out by Chaemiso et al. (2016) in Omo-Gibe basin in southern Ethiopia to spot the impact of global change on water resource. They used ArcSWAT model and A1B emission scenario under SRES. The result indicated that both temperature and the annual PET will increase in the future (the 2030s and 2090s) and rainfall varies considerably relative to the base period (1980–2005). The surface water availability is declining within the dry period and rising within the wet period in the basin.

As revealed by Abraham et al. (2018) there is a high reduction of water availability in the future in the Katar and Meki sub-watersheds of Lake Ziway, Ethiopia. They applied the outputs of multiple/three GCMs under stabilization (RCP4.5) and high (RCP8.5) climate change emission scenarios and HBV hydrological model. They reported that temperature (maximum and minimum) will rise for the above scenarios. Rainfall shows a decreasing trend by up to 51.19% within the 2050s (2041–2070). Annual runoff depth reduced by up to 19.45% and 20.28% in Katar and Meki sub-watersheds, respectively, by 2080s (2071–2099) compared to base period (1980–2005). Another study in the same watershed by Zeray et al. (2006) also indicates that the water resources in the watershed are greatly affected by climate change in the future. They used B2A and A2A scenarios and SWAT hydrological model to analyse the future interaction between climate change and water resource in the Lake Ziway watershed. They reported that average inflow volume to Lake Ziway decreases notably by about 19.47 and 27.43% for A2A and B2A scenarios, respectively, for the time of 2001–2099 relative to 1981–2000, even though the average monthly and annual rainfall will increase in Lake Zeway watershed by up to 29 and 9.4%, respectively. The rise in rainfall is concealed by the rise in temperature (minimum and maximum) in the region.

Previously, most of the climate change studies in Ethiopia were based on CMIP3 climate change models (such as Zeray et al. 2006; Setegn et al. 2011; Dile et al. 2013; Wagesho et al. 2013; Chaemiso et al. 2016; Nigatu et al. 2016), and had been relied on a particular/finite number of climate models (such asDile et al. 2013; Serur and Sarma 2016; Nigatu et al. 2016). If the simulation is based on a particular/finite number of GCMs, the results will be highly uncertain. It is also most likely to mislead the decision makers and policy developers (IPCC 2007; Taye et al. 2011). Thus, predicting climate change impacts based on multiple GCMs under CMIP5 is quite important in Ethiopia for well perceptive of the potential climate change effects, informed decision-making for proper water resource management, and effectively respond and adapt to projected changes, otherwise the consequences becoming awful.

5.5 State Reaction for Threats of Climate Change in Ethiopia

In Ethiopia, a number of strategies were implemented to reduce the detrimental outcomes of climate change and to maximize the adaptive capability of the country. The country has signed some of global and regional environmental agreements associated with global warming. Ethiopia also approved the UNFCCC (United Nations Framework Convention on Climate Change) and the Kyoto Protocol on 31 May 1994 and on 21 February 2005, respectively. The 1992 Rio Convention on Environment and Development guide to design conservation strategy of the country in 1993 (EPCC 2015b). The 1995 Constitution of Ethiopia provides for environmental rights and a policy of promoting sustainable development. Environmental right is stipulated in the constitution under Article 44 (1), which states the citizens have the right to a clean and healthy environment.

Ethiopia has prepared the National Adaptation Programme of Action (NAPA) in June 2007 (Tadege 2007). The NAPA includes the following 11 projects: Encouraging crop/drought coverage package, reinforcement flood and drought timely caution structures, implementation of small-scale irrigation techniques and wise utilization of water, pasture management measures, sustainable usage of wetlands, capacity building program for climate change adaptation, multipurpose large-scale water development project, community-based carbon sequestration project, establishing of state research centre for climate change, strengthening malaria control package and upgrade of agroforestry practices in selected parts of the country. In addition to NAPA, based on the Copenhagen's agreement Ethiopia submitted Nationally Appropriate Mitigation Actions (NAMAs) to UNFCCC. The document lists about 88 projects (EPCC 2015b).

For long, emergency food aid was the principal response to challenges of food insecurity in the country. However, the delivery of emergency support repeatedly as food for work had not solved the hitch lastingly. It is this observation that influenced the launching of a productive safety net program (PSNP) in 2005. PSNP is credited for improvements in natural resource management and environment through the conservation measures (watershed management techniques) executed by the beneficiary households. PSNP builds households resilience to climate variability and weather extremes (EPCC 2015b). On the other hand, Weldegebriel and Prows (2013) reported that despite the fact that the PSNP has been effective in protecting farmers from malnutrition and safeguard farmers for squat period, it is not developing resilience to hazards lastingly (Cochrane and Singh 2017). Presently, it is supported by climate smart initiative project to enhance its contribution to climate resilience.

Ethiopia launched a Sustainable Land Management Program in 2008. The second phase of the program aims to introduce methods to tackle climate variability/change interrelated hazards and to take advantage of the reduction of GHG emission with a view to meeting the growth and transformation plan (GTP) and climate resilience green economy (CRGE) targets. The current GTP has a section entitled "Environment and Climate Change". The CRGE, which predates GTP II, provides a blueprint for achieving middle-income status by 2025 with no net increases in greenhouse gas emissions relative to 2010 levels (USAID 2016). The CRGE approach concentrates on both climate change adaptation and mitigation objectives. Having a policy framework is important, but it is also equally noteworthy to have a structure for monitoring the implementation of policies. In this regard, it is reported that there is a problem of implementation and enforcement capacity. The country has already started implementation of some of the key CRGE components. The policy framework is not yet complete. Literature and studies on the country's policy and institutional response for threats of ongoing climate change are limited.

5.6 Adaptation and Mitigation Techniques in Ethiopia

There are two general approaches to control the undesirable impacts of ongoing climate change: adaptation and mitigation; to slow of GHGs emissions and to reduce their consequences, respectively. Concerning to water resource and crop production, adaptation measures consist of using varieties and drought-resistant crops; using water-efficient irrigation techniques like drip irrigation; implementing water harvesting technology, and adjustment of crop calendar (Bates et al. 2008). Hedgu et al. (2015) point out that alteration in crop type/variety, watershed management measures, crop diversifications, adjustment in planting calendar and water-efficient irrigation practices are adaptation measures that were implemented in Tigray regional state in Northern Ethiopia. Another report by Ahmed (2016) showed that the above-listed strategies of adaptations are commonly implemented by farmers in central Rift valley of Ethiopia.

5.7 Conclusions and Recommendations

The followings are the main conclusions and recommendations based on the review:

- Ethiopia's contribution to global annual greenhouse gas (GHG) emission is negligible.
- Historically drought, flood and rainfall variability have a great impact on Ethiopian agricultural productivity and livelihood.
- The spatio-temporal variability of rainfall is more important than total rainfall deficits for water-related problems in Ethiopia.
- Land use change and agriculture are the main drivers of GHG emission in Ethiopia.
- The rise in temperature and variability of rainfall are the major features of climate change; those climate elements have an effect on almost all other hydrological processes, and the long-term climate change impacts in Ethiopia is mainly linked to these two climate elements. So, recognition and analysis of their pattern in the watershed are highly needed.
- Climate change impacts are location-specific and adaptation measures are implemented at watershed level, recognizing the effects of climate change at a watershed level that is sensibly imperative. However, the exact impact at watershed level in Ethiopia is poorly understood. The potential impacts because of these changes on water resources at applicable level are not examined in most parts of Ethiopia and particularly in southern region. Studies on climate change in Ethiopia are concentrated on UBNRB or Lake Tana Watershed and others areas were neglected.
- Various previous climate change studies carried out in the nation are based on old emission scenarios under SRES and single or finite number of GCMs. Forming an inference regarding the impacts of climate change on the water resources based on a single GCM may not provide a graspable picture of the coming changes.

- SWAT model is widely used in climate change impact assessments in Ethiopia and has good performance to predict an interaction of climate change and water.
- The nature and magnitude of how global change will affect the Ethiopian water resource is not yet adequately understood and there are a lot of knowledge gaps related to climate and water interface even though the water sector is listed under the most vulnerable sectors in Ethiopia. This is a current challenge for planners and policy makers to develop long-term water resource management and climate change adaptation strategies in the water sector. Therefore, comprehensive study is highly needed and the issue has to be addressed at a scale relevant to decision-making to improve the level of water security, water management and the planning for mitigation and adaptation practices of climate change.

References

Abraham T, Woldemicheala A, Muluneha A, Abateb B (2018) Hydrological responses of climate change on Lake Ziway Catchment, Central Rift Valley of Ethiopia. J Earth Sci Clim Chang 9(6):1–16

Ahmed MH (2016) Climate change adaptation strategies of maize producers of the Central Rift Valley of Ethiopia. J Agric Rural Dev Trop Subtrop 117(1):175–186

Alemu YT (2015) Flash flood hazard in Dire Dawa, Ethiopia. J Soc Sci Humanit 1(4):400–414

Awulachew SB, Yilma AD, Loulseged M, Loiskandl W, Ayana M, Alamirew T (2007) Water resources and irrigation development in Ethiopia. In: International Water Management Institute (Iwmi), vol 123, pp 78.

Bates BC, Kundzewicz ZW, Wu S, Palutikof JP (eds) (2008). Climate change and water. In: Technical paper of the intergovernmental panel on climate change, IPCC Secretariat, Geneva, pp 210

Berhanu B, Seleshi Y, Melesse AM (2014) Surface water and groundwater resources of Ethiopia: potentials and challenges of water resources development. Nile River Basin. Springer, Cham, pp 97–117

Bewket W, Conway D (2007) A note on the temporal and spatial variability of rainfall in the drought-prone Amhara region of Ethiopia. Int J Climatoly J R Meteorol Soc 27(11):1467–1477

Chaemiso SE, Abebe A, Pingale SM (2016) Assessment of the impact of climate change on surface hydrological processes using SWAT: a case study of Omo-Gibe River basin, Ethiopia. Model Earth Syst Environ 2(205):1–15

Cochrane L, Singh R (2017) Climate services for resilience: the changing roles of NGOs in Ethiopia. In: Fighting climate change: human solidarity in a divided world, Palgrave Macmillan UK, pp 1–18

Dile YT, Berndtsson R, Setegn SG (2013) Hydrological response to climate change for gilgel Abay River, in the Lake Tana basin-upper blue Nile basin of Ethiopia. PLoS ONE 8(10):1–13

Dorosh P, Rashid S (eds) (2013) Food and agriculture in Ethiopia: progress and policy challenges. University of Pennsylvania Press, pp 345

Ethiopian Panel on Climate Change (EPCC) (2015a) First assessment report, summary of reports for policy makers. The Ethiopian Academy of Sciences, pp 1–92

Ethiopian Panel on Climate Change (EPCC) (2015b) First assessment report, an assessment of Ethiopia's policy and institutional frameworks for addressing climate change. Published by the Ethiopian Academy of Sciences, pp 1–88

Fazzini M, Bisci C, Billi P (2015) The climate of Ethiopia. In: Landscapes and landforms of Ethiopia. Springer, Dordrecht, pp 65–87

Gambella Region Disaster Prevention and Preparedenes Agency (2007) Report on emergency disaster study. Gambella, Ethiopia
Gebre SL, Ludwig F (2015) Hydrological Response to climate change of the Upper Blue Nile River Basin: based on IPCC fifth assessment report (AR5). J Climatol Weather Forecast 3(1):1–15
Gebrehiwot T, van der Veen A (2013) Assessing the evidence of climate variability in the northern part of Ethiopia. J Dev Agric Econ 5(3):104–119
Hadgu G, Tesfaye K, Mamo G, Kassa B (2015) Farmers climate change adaptation options and their determinants in Tigray Region, Northern Ethiopia. Afr J Agric Res 10(9):956–964
IPCC (2007) Climate Change (2007). Impacts, adaptation, and vulnerability contribution of working group II to the third assessment report of the intergovernmental panel on climate change, Parry ML et al (eds). Cambridge University Press, Cambridge, UK
IPCC (2014) Annex II: Glossary. In: Mach KJ, Planton S, von Stechow C (eds) Climate change 2014: Synthesis report. contribution of working groups I, II and III to the fifth assessment report of the intergovernmental panel on climate change. Core WritingTeam, Pachauri RK, Meyer LA (eds). IPCC, Geneva, Switzerland, pp 117–130
International Hydropower Association (IHA) (2017) Hydropower status report.
Kim U, Kaluarachchi JJ, Smakhtin VU (2008) Climate change impacts on hydrology and water resources of the Upper Blue Nile River Basin, Ethiopia. International Water Management Institute (Iwmi), vol 126
Mekonnen DF, Disse M (2016) Analyzing the future climate change of Upper Blue Nile River Basin (UBNRB) using statistical down scaling techniques. J Hydrol Earth Syst Sci 1–27
Mosello B, Calow R, Tucker J, Parker H, Alamirew T, Kebede S, Alemseged T, Gudina A (2015) Building adaptive water resources management in Ethiopia. Overseas Development Institute, London, pp 1–100
MoWR (2002) Water sector development program main report volume II. Ministry of Water Resources Federal Democratic Republic of Ethiopia, Addis Ababa
MoWR, GW-MATE (2011) Ethiopia: strategic framework for managed groundwater development. Addis Ababa
Nigatu ZM, Rientjes T, Haile AT (2016) Hydrological impact assessment of climate change on Lake Tana's Water Balance, Ethiopia'. Am J Clim Chang 5:27–37
National Meteorological Agency (NMA) (2007) Climate change national adaptation programme of action (NAPA) of Ethiopia. National Meteorological Services Agency, Ministry of Water Resources, Federal Democratic Republic of Ethiopia, Addis Ababa, pp 1–99
NMA (2013) National meteorological agency. Meteorological data and climatology directorate annual climate bulletin for the year 2013, pp 1–15
Robinson S, Strzepek K, Cervigni R (2013) The cost of adapting to climate change in Ethiopia: sector-wise and macro-economic estimates, Ethiopian strategy support program II, vol 53, pp 1–26
Serur AB, Sarma AK (2016) Impact of climate change on water availability in the Weyib River Basin, Southeastren Ethiopia. Int J Innov Res Sci Eng Technol 5(6):1–9
Setegn SG, Rayner D, Melesse AM, Dargahi B, Srinivasan R (2011) Impact of climate change on the hydroclimatology of Lake Tana Basin, Ethiopia. Water Resour Res 47(4):1–13
Tadege A (2007) Climate change national adaptation programme of action (NAPA) of Ethiopia. National Meteorological Services Agency, Ministry of Water Resources, Federal Democratic Republic of Ethiopia, Addis Ababa, pp 1–96
Taye MT, Ntegeka V, Ogiramoi NP, Willems P (2011) Assessment of climate change impact on hydrological extremes in two source regions of the Nile River Basin. Hydrol Earth Syst Sci 15(1):209–222
USAID (2016) Climate change risk in Ethiopia. Country fact sheet.
Wagesho N, Jain MK, Goel NK (2013) Effect of climate change on runoff generation: application to Rift Valley Lakes Basin of Ethiopia. J Hydrol Eng 18(8):1048–1063
Weldegebriel ZB, Prowse M (2013) Climate-change adaptation in Ethiopia: To what extent does social protection influence livelihood diversification? Dev Policy Rev 31:35–56

World Bank (2006) Ethiopia: managing water resources to maximize sustainable growth, pp 1–119
World Bank (2007) The little green data book 2007 world development indicators. Washington, DC
World Food Program (WFP) (2006) Extensive flooding in Ethiopia: WFP update. United Nations WFP, Fighting Hunger Worldwide, pp 1–5
Zeray L, Roehrig J, Chekol DA (2006) Climate change impact on Lake Ziway watershed water availability, Ethiopia. In: Conference on international agricultural research for development, pp 1–4

Chapter 6
Spatio-Temporal Trend Analysis of Long-Term IMD-Gridded Precipitation in Godavari River Basin, India

C. H. Praveenkumar and V. Jothiprakash

Abstract The study aims to analyze spatio-temporal variations in India Meteorological Department (IMD) gauge-based gridded rainfall data over Godavari river basin, India. A total of 502 rainfall grid stations which covers the entire Godavari basin and 53 years long-term daily rainfall data from 1961 to 2013 is used in the present study. The study focused on trend analysis of individual rainfall grid stations in Godavari basin rather than averaged at sub-basin level to check the internal characteristics of IMD-gridded rainfall. Trend analysis is carried out for three rainfall time scales, namely daily, monthly and annual. As this region receives major contribution of rainfall during South-west monsoon period from June to September, the trend analysis is separately carried out for monsoon time series. To overcome the assumption of normality of input rainfall data, a widely used nonparametric trend analysis test, Mann–Kendall (MK), is adopted in the present study. From the results of trend analysis, a decreasing trend is observed at majority of the stations in all sub-basins of Godavari both for daily and monsoon daily time series data. However, monthly rainfall is observed to be free from trend except for very few stations scattered all over the basin. Further, the annual rainfall, monsoon monthly rainfall and monsoon total rainfall showed an increasing trend of the rainfall stations at sub-basins of Godavari Upper, eastern portion of the Indravati and Godavari Lower. It is worth to mention that all increasing rainfall stations observed at three time scales discussed above are located either at foothills of Western Ghats or foothills of Eastern Ghats along the Coast of Bay of Bengal. Therefore, the present study analysis reveals that except in the hill shadow regions and coastal areas, IMD grid rainfall is free from trend in Godavari river basin during the study period.

Keywords Spatio-temporal variations · Trend analysis · India Meteorological Department (IMD) grid rainfall · Mann–kendall test

---

C. H. Praveenkumar · V. Jothiprakash (✉)
Indian Institute of Technology (IIT) Bombay, Mumbai 400076, India
e-mail: vprakash@iitb.ac.in

R. Jha et al. (eds.), *Climate Change Impacts on Water Resources*, Water Science and Technology Library 98, https://doi.org/10.1007/978-3-030-64202-0_6

6.1 Introduction

Climate change refers to the drastic variations in the weather phenomenon that leads to an extreme weather events over the long duration of time. Rainfall, being a key meteorological variable, changes in the local weather process can be addressed by analyzing its internal characteristics both spatially and temporally within the river catchment. Trends of rainfall data pertaining to large river catchments may directly indicate the changes in the climate; therefore, it is possible to take necessary actions while making new policies and further efficient management of available natural resources. Many works in the past literature also reported that rainfall is the promising meteorological variable, and changes in its internal characteristics such as trend directly indicate the climate change (Tomozeiu et al. 2000; Jhajharia et al. 2012; Minaei and Irannezhad 2018).

There exist various methods to assess the trend in a given time series data (Pingale et al. 2014; Bisht et al. 2018). Among all the available methods, Mann–Kendall (MK) test by Mann (1945) and Kendall (1975) is the widely used method for analyzing trends of hydrological time series data. Several studies have been carried out to find the trends of rainfall data in various river catchments of India (Jhajharia et al. 2012; Pingale et al. 2014; Bisht et al. 2018; Praveenkumar and Jothiprakash 2018). The advantage of adopting MK test in the present study is to overcome the normality of input rainfall data, since nonparametric methods can be applied without passing though the normality of input data. This is the major obstacle in other methods because daily rainfall time series data never follows this normality assumption, especially countries having complex weather processes like India due to most of the river catchments receive major quantity of rainfall only during monsoon season from June to September in recent times. Therefore, MK test is one of the best methods to quantify the trend in the rainfall time series data and has been used in the present study.

The daily gauge-based gridded rainfall time series data available from 1961 to 2013 (53 years), obtained from India Meteorological Department (IMD), is used in the present study and the trend is examined for different time scales, namely daily, monthly and annual. Bisht et al. (2018) also carried out the trend analysis of all Indian river basins using IMD grid rainfall from 1901 to 2015. But their study focused on trend analysis of rainfall which is average at sub-basin level in all catchment rather than each grid station. Thus, the present study is aimed to analyze the long-term trend of IMD-gridded precipitation data in Godavari river basin, India at each individual grid station rather than averaged over a sub-basin level. The detailed explanation of study area, data analysis and results are given in the following sections.

6.2 Study Area and Datasets Used

Godavari river basin, the second longest river in India and first largest river catchment in peninsular India, is considered as study area in the present study. The catchment lies within the global coordinates of 73° 28′ 4″ to 83° 9′ 33″ E and 16° 18′ 10″ to 22° 44′ 27″ N. The Godavari catchment covers a geographical area of 312,812 km^2. The highest elevation is 1067 m (Source: SRTM Digital Elevation Model with 30 m resolution) and the lowest elevation where it joins with Bay of Bengal is 8 m below MSL. The location map of Godavari river basin along with IMD grid stations is shown in Fig. 6.1. The Godavari river catchment, being a largest in Peninsular India, flows through eight different states of India, namely Maharashtra, Telangana, Andhra Pradesh, Chhattisgarh, Madhya Pradesh, Odisha, Karnataka and Puducherry.

This basin is divided into eight sub-catchments, namely Godavari Upper and Godavari Middle located in the west side, Wardha and Weinganga in the north side, Indravati and Godavari Lower in the east side, and Manjira and Pranahita with others in the south side of the river catchment. The Godavari main river originates in the Western Ghats and flows toward east. The other tributaries of river Godavari are Purna, Manjira and Manyar which flow toward east, Weinganga and Wardha flow toward south, and Indravati and Sabari flow toward west direction.

The daily 0.25° x 0.25° gauge-based gridded rainfall data provided by IMD (Pai et al. 2014) is used in the present study. There exist 502 grid stations which cover the entire river basin as shown in Fig. 6.1. The study period considered is from 1961 to 2013. This basin receives the majority of its rainfall during the South-west monsoon period (June–Sept). The average annual rainfall in this river basin as per IMD grid rainfall during the study period is 2900 mm highest at Western Ghats

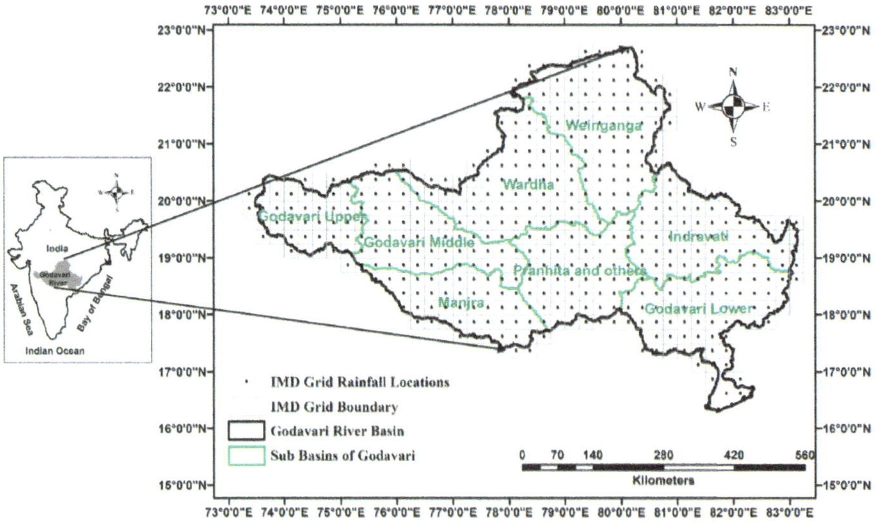

Fig. 6.1 Location map of Godavari river basin along with IMD grid rainfall stations

region and 520 mm lowest in the downstream area of Upper Godavari sub-basin. The methodology adopted in the present study is discussed in the following section.

6.3 Methodology

The methodology applied in the present study includes trend analysis of IMD grid rainfall from 1961 to 2013. All the analysis of trend was carried out for daily, monthly and annual time scales. As Godavari river basin receives its major contribution of rainfall during the monsoon period from June to September, trend analysis is carried out separately for this monsoon period also. The MK test proposed by Mann (1945) and Kendall (1975) is a non-parametric test, which is a widely used method for trend analysis of rainfall, and hence considered in the present study. The results of trend analysis are discussed below.

6.4 Results and Discussion

The trend analysis of the IMD grid rainfall data at all three time scales, namely daily, monthly and annual, is analyzed using the MK test and the results are shown in Table 6.1. From Table 6.1, it can be clearly observed that at daily time scale, sub-basins Wardha and Weinganga located in the north side of the basin showed decrease in rainfall trend at 43 and 41 grid stations, respectively. Similarly, sub-basins namely Indravati and Godavari Lower also showed decrease in rainfall trend at 36 and 15 stations, respectively (Table 6.1). However, the number of stations with an increase in rainfall trend at daily time scale is observed at very few stations in all sub-basins except Godavari Lower, which has 17 stations with increasing trend in daily grid rainfall. It is worth to mention from Fig. 6.2a that all the increasing rainfall grid stations at daily time scale is located either at foothills of Eastern Ghats, along Coast of Bay of Bengal or at the foothills of Western Ghats. Overall, at daily time scale, there exist 200 decreasing trend stations, 44 increasing trend stations and 258 no trend stations of IMD grid rainfall in Godavari river basin.

Further, it is observed from Fig. 6.2(b) that the monthly IMD grid rainfall is free from trend in entire river catchment. Very few significant stations are observed with increasing trend (three stations in Weinganga) and decreasing trend (four stations in Godavari Lower) at monthly time scale analyzed with MK test (Table 6.1). However, surprisingly, annual rainfall time series data showed more stations of increasing trend (51 stations) in Godavari basin. It is also observed again from Fig. 6.3a that all the increasing rainfall grid stations located either at foothills of Eastern Ghats, along the coast of Bay of Bengal, or at the foothills of Western Ghats, which is also observed in case of daily rainfall time series data.

As the Godavari river basin receives major quantity of rainfall during monsoon season, trend analysis is separately carried out for monsoon alone time series data.

Table 6.1 Number of stations showing increasing, decreasing and no trend in Godavari basin at different time scales

S. no	Location	Sub-basin of Godavari	Number of trend stations for time scales analyzed									Total no. of stations in each sub-basin
			Daily			Monthly			Annual			
			DT	*IT*	*NT*	*DT*	*IT*	*NT*	*DT*	*IT*	*NT*	
1.	North	Wardha	43	2	30	0	0	75	0	2	73	75
		Weinganga	41	1	42	3	0	81	1	5	78	84
2.	East	Indravati	36	2	26	0	0	64	0	9	55	64
		Godavari Lower	15	17	42	0	4	70	0	24	50	74
3.	South	Manjira	14	5	31	0	1	49	1	0	49	50
		Pranahita and Others	5	6	44	0	0	55	0	2	53	55
4.	West	Godavari Upper	11	6	26	0	0	43	1	9	33	43
		Godavari Middle	35	5	17	0	0	57	0	0	57	57
Total no. of stations in Godavari river basin with trend			200	44	258	3	5	494	3	51	448	502

Note ***DT*** indicates decreasing trend, ***IT*** indicates increasing trend and ***NT*** indicates no trend

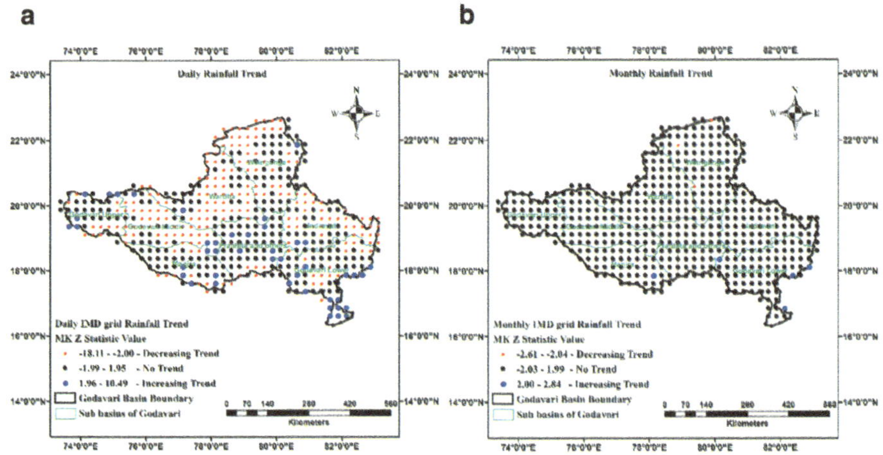

Fig. 6.2 Spatial distribution maps of trend for IMD grid rainfall in Godavari basin at 95% confidence level. **a** Daily rainfall and **b** Monthly rainfall

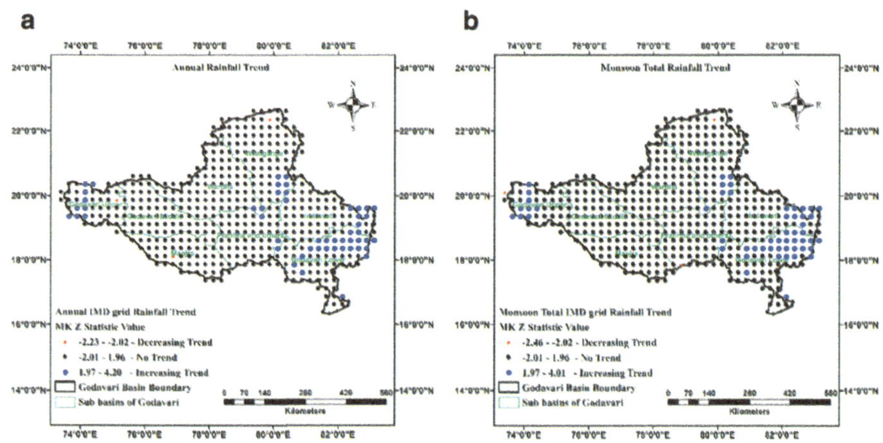

Fig. 6.3 Spatial distribution maps of trend for IMD grid rainfall in Godavari basin at 95% confidence level. **a** Annual rainfall and **b** Monsoon total rainfall

From Fig. 3(b), it can be observed that monsoon total rainfall showed exactly similar behavior of annual full time series data. It is also observed from our analysis that monsoon monthly rainfall, monsoon total rainfall and annual full time series rainfall showed exactly similar rainfall trends in Godavari river basin.

Therefore, it can be concluded that the IMD grid rainfall in Godavari river basin is free from trend at majority of the grid stations (Figs. 6.2b, 6.3a and b). Stations located at the foothills of Eastern and Western Ghats are showed an increase in rainfall trend. The reason behind receiving more rainfall at above-mentioned areas may be due to high altitude and presence of hill shadows. Generally, hill shadow regions with higher altitudes receive more rainfall magnitudes compared to low-level areas due to orographic precipitation.

6.5 Conclusions

The long-term trend analysis of IMD gauge-based gridded rainfall is carried out using MK test. The results of daily rainfall trend indicate that stations located at the foothills of both Western and Eastern Ghats have an increasing trend, whereas majority of stations located in the Godavari basin is observed with decreasing trend for daily time scale analyzed with MK test. The monthly rainfall time series data is observed to be free from trend in Godavari basin except at very few increasing trend stations located in Weinganga and Godavari Lower sub-basins. Further, the annual, monsoon monthly and monsoon total rainfall time series show an increasing trend stations at sub-basins Upper Godavari, Indravati and Godavari Lower. Surprisingly, all the increasing trend stations observed in above-mentioned three time scales are exactly located either at foothills of Western Ghats in Upper Godavari sub-basin or foothills of Eastern Ghats in Indravati and Godavari Lower sub-basins as similar trend is observed in case of daily time scale also. It is also observed from annual, monsoon monthly and monsoon total rainfall trend analysis that other than hill shadow regions, no trend is observed in entire basin. Therefore, it can be concluded that IMD grid rainfall in Godavari river basin has increasing trend nature only at hill shadow regions and it is the trend free from other regions.

Acknowledgements The authors would like to thank the India Meteorological Department (IMD) for providing gridded rainfall data for carrying the present study.

References

Bisht DS, Chatterjee C, Raghuwanshi NS, Sridhar V (2018) Spatio-temporal trends of rainfall across Indian river basins. Theoret Appl Climatol 132(1–2):419–436

Jhajharia D, Yadav BK, Maske S, Chattopadhyay S, Kar AK (2012) Identification of trends in rainfall, rainy days and 24 h maximum rainfall over subtropical Assam in Northeast India. CR Geosci 344(1):1–13

Kendall MG (1975) Rank Correlation Methods. Charles Griffin and Co., Ltd., London, UK

Mann HB (1945) Nonparametric tests against trend. Econometrica 13:245–259

Minaei M, Irannezhad M (2018) Spatio-temporal trend analysis of precipitation, temperature, and river discharge in the northeast of Iran in recent decades. Theoret Appl Climatol 131(1–2):167–179

Pai DS, Sridhar L, Rajeevan M, Sreejith OP, Satbhai NS, Mukhopadhyay B (2014) Development of a new high spatial resolution (0.25° × 0.25°) long period (1901–2010) daily gridded rainfall data set over India and its comparison with existing data sets over the region. MAUSAM 65(1):1–18 (January 2014)

Pingale SM, Khare D, Jat MK, Adamowski J (2014) Spatial and temporal trends of mean and extreme rainfall and temperature for the 33 urban centers of the arid and semi-arid state of Rajasthan, India. Atmos Res 138:73–90

Praveenkumar C, Jothiprakash V (2018) Spatio-temporal trend and homogeneity analysis of gridded and gauge precipitation in Indravati river basin, India. J Water Clim Chang jwc2018183. https://doi.org/10.2166/wcc.2018.183

Tomozeiu R, Busuioc A, Marletto V, Zinoni F, Cacciamani C (2000) Detection of changes in the summer precipitation time series of the region Emilia-Romagna, Italy. Theor Appl Climatol 67(3–4):193–200

Chapter 7
Forecasting Reference Evapotranspiration Using Artificial Neural Network for Nagpur Region

Nikhil Band, Aniruddha Ghare, and Avinash Vasudeo

Abstract There are different methods of forecasting, such as time series analysis, genetic programming regression analysis, wavelet transform, support vector machines, etc. In the present study, forecasting of reference evapotranspiration has been attempted for Nagpur region, Maharashtra State, India so that it can be used to take decisions for reservoir planning. The past meteorological information, such as relative humidity, temperature, wind velocity, solar rays were used for the calculation of reference evapotranspiration (ET_0) value by using the Penman-Monteith method, which is recommended as the standard method for determining ET_0 by Food and Agricultural Organization (FAO). The data has been trained in artificial neural network (ANN) and the two learning techniques, namely Levenberg–Marquardt algorithm and quasi-Newton algorithm, were compared with the untrained data. Sum square error (SSE), mean square error (MSE) and regression value (R^2) have been used as criteria for the performance assessment. Levenberg–Marquardt algorithm is found to be better among the two, for the study area under consideration.

Keywords Evapotranspiration · Artificial neural network · Forecasting · Metrological data

N. Band (✉)
Department of Municipal Administration, Bhandara 441912, Maharashtra, India
e-mail: nikhil.band060893@gmail.com

A. Ghare · A. Vasudeo
Department of Civil Engineering, Visvesvaraya National Institute of Technology, Nagpur 440010, India
e-mail: adghare@yahoo.co.in

A. Vasudeo
e-mail: avasudeo@yahoo.com

R. Jha et al. (eds.), *Climate Change Impacts on Water Resources*, Water Science and Technology Library 98, https://doi.org/10.1007/978-3-030-64202-0_7

7.1 Introduction

Management and planning of water resource system demands the investigation of different hydrological entities like rainfall, evaporation, infiltration/seepage from the reservoir, requirement for generation of electricity, domestic water supply and irrigation, etc. Operational research of reservoir systems might also need diffusion of partial historical facts to calculate the feasible future performance. For the convenient operation of irrigation system, prediction of reference evapotranspiration is very important. For forecasting the evapotranspiration (ET) there are different techniques. Traey et al. (1992) developed time series model to forecast irregularity of ET obtained by yearly difference or month-to-month average models. In most of the existing techniques in time series analysis, linear connection is considered among variables. Since temporal/spatial variation in the data shows irregularities, therefore it is hard to analyse the forecasting accurately. Since this world has drastic temporal problem, the artificial neural network (ANN), a nonlinear model, can be used for complicated nonlinear system. Neural networks have so many applications in water resources. Prediction is considered as the challenging application area for artificial neural network. ANN had been employed for ET_0 prediction; a sequentially adaptive radial basis function network was applied (Trajkovic et al. 2003). Landeras et al. (2009) compared the forecasted weekly ET using ARIMA and ANN model for Alava region situated in the northern Spain. Chauhan and Shrivastava (2009) applied different ANN algorithms for the prediction of ET_0. Perera et al. (2014) forecasted daily ET_0 in Australia with the help of climate prediction outputs. Kisi and Ozturk (2007) estimated ET_0 with the help of Levenberg–Marquardt (LM) algorithm for different input combinations of existing meteorological data. Zanetti et al. (2007) employed artificial neural network for evapotranspiration estimation using minimum and maximum temperature, and Levenberg–Marquardt algorithm had used it for training the network.

This study represents the capability of a feed-forward NN for ET_0 prediction. Two distinct algorithms of ANN based on various key patterns have been considered. In this study, quasi-Newton (QN) and Levenberg–Marquardt (LM) algorithms have been employed for training algorithms. This has been carried out to determine the algorithm that gives good outcome for the study area.

7.2 Study Area

For the current study, monthly average meteorological information of maximum and minimum temperature, wind velocity, relative humidity and solar radiation in Nagpur region were collected for the duration January 1990 to July 2014. This climatic parameter information is available on Global Weather Data for SWAT website, which is under Texas University. Average monthly ET_0 values were calculated by FAO Penman-Monteith (PM) method, which is considered as the standard method for

ET_0 estimation (Allen 1986). The computed ET_0 values were considered for training and testing of distinct ANN algorithms. The location of the study area is Nagpur region (21.1458° N, 79.0882° E) in the State of Maharashtra, India.

7.3 Methodology

Revised form of FAO 56 PM equation (Allen 1986) was used for estimation of reference evapotranspiration values for the study area, which is shown in Eq. (7.1)

$$ET_0 = \frac{0.408\Delta(R_n - G) + \gamma \frac{900}{T+273} U_2(e_s - e_a)}{\Delta + \gamma(1 + 0.34\ U_2)} \tag{7.1}$$

where ET_0 represents reference evapotranspiration [mm/day], T represents mean daily air temperature measured at elevation of 2 m [°C], G represents soil heat flux density [MJ/m^2/day] (generally very small and assumed to be zero), U_2 represents wind velocity at elevation of 2 m [m/s], R_n represents net radiation at the crop surface [MJ/m^2/day], e_s represents saturation vapour pressure [kPa], e_a represents actual vapour pressure [kPa], e_s-e_a represents saturation vapour pressure deficit [kPa], γ is the psychrometric constant [kPa°C^{-1}], and Δ represents slope vapour pressure curve [kPa °C^{-1}].

The calculated ET_0 values were employed for the training of the neural network. After the completion of training stage, testing of data was done using ANN. To obtain the greatest ANN configuration, a variety of possibilities were considered. Feed-forward single hidden layer ANNs has been carried out in this study. Different models were developed from the combination of input variables which are shown in Table 7.1.

There are five input variables, and six combinations were made by taking these five climatic parameters. For each input parameter, two training algorithms (QN algorithm and LM algorithm) were used to train ANN models. This was practiced to obtain an optimum number of hidden neurons for every model. Usually number of trials are performed to determine the hidden layer nodes inside the neural networks. For

Table 7.1 Input variable combinations considered for the development of ANN models

Model	Input Parameters
Model 1	Relative humidity, maximum temperature
Model 2	Relative humidity, maximum temperature, wind velocity
Model 3	Minimum temperature, relative humidity
Model 4	Relative humidity, wind velocity, solar radiation
Model 5	Minimum temperature, relative humidity, wind velocity
Model 6	Relative humidity, solar radiation

training a network, the overall performance indices considered were mean squared error, coefficient of regression and sum square error. Utilizing the input test data, the performance of ET_0 prediction for each network had been assessed. With the help of test dataset, linear regression analysis has been carried out with estimated ET_0 as a dependent variable and artificial neural networks forecasted values as independent variables.

In this study, linear regression (LR) was employed to fit the relationship between the inputs (x), viz., solar radiation, wind speed, relative humidity and minimum–maximum air temperature and the targeted outputs (y), as:

$$y_i = w_i x_i + a + E_i \tag{7.2}$$

where

y_i ET_0 fitted value
x_i weather variable
w_i slope
a intercept
E_i instantaneous error

Figure 7.1 explains the linear regression analysis used in this study. If the value of coefficient of regression is near to 1, it shows very well correlation between the two considered variables. After linear regression, the input values of reference evapotranspiration have been fed to neural network and forecasted using Levenberg–Marquardt algorithm. For forecasting, the values from 1990 to 2010 are used for predictive model and it is validated for the years 2011–2014.

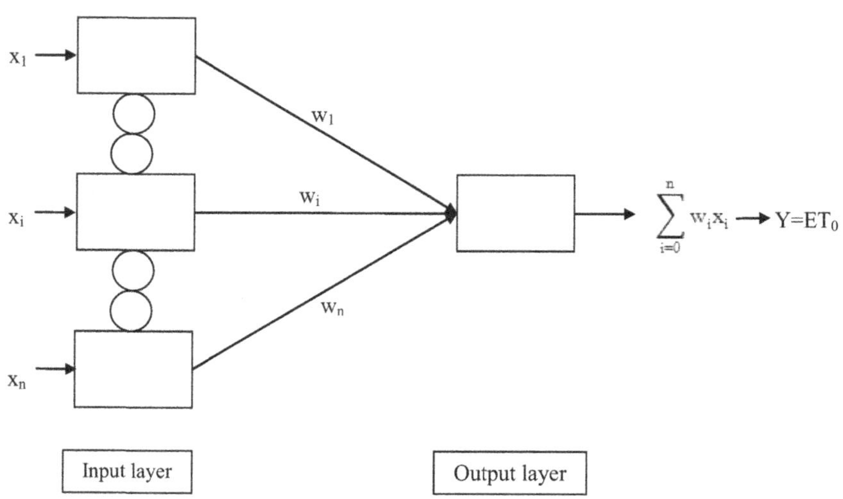

Fig. 7.1 Diagram representing a linear regression network

7.4 Results and Discussions

The meteorological data utilized for the validation and training of the ANNs were used for calibration of estimated ET_0 equations. QN and LM algorithms of ANN models have been trained for 100 epochs (training iterations) with three consecutive runs. On abscissas ET_0 values by PM are plotted, while ordinate consists of ET_0 values by ANN. The training and testing errors of all considered ANN configurations are presented in the Figs. 7.2, 7.3, 7.4, 7.5, 7.6 and 7.7 for each model. It was observed that MSE reduces with increase in number of training iterations only up to a certain value (depending upon ANN models), thereafter remains constant with further increase in iterations.

Figure 7.2 shows scattered plots between estimated ET_0 by PM method and predicted model 1 ET_0 for a trial epoch with the network configuration of 3-10-1 using the LM algorithm.

Figure 7.3 shows scattered plots between estimated ET_0 by PM method and predicted model 2 ET_0 for a trial epoch with the network configuration of 3-10-1 using the LM algorithm.

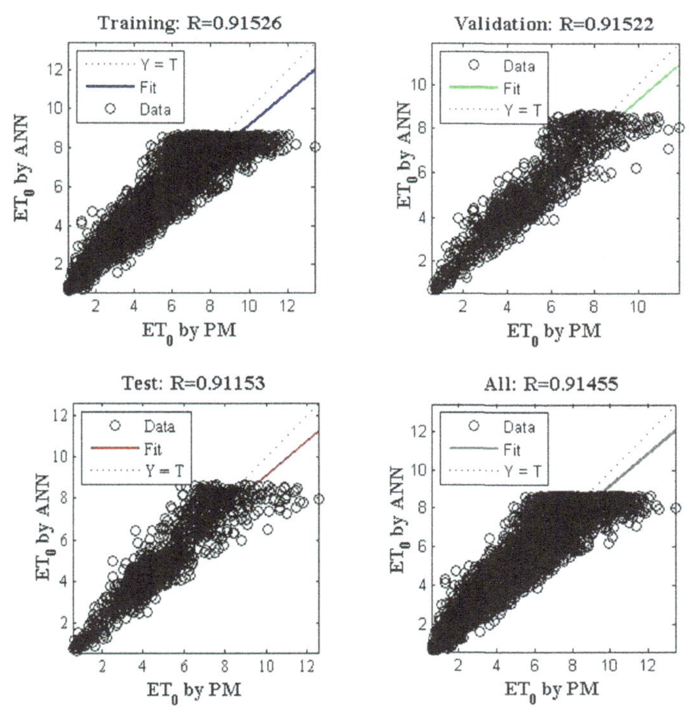

Fig. 7.2 Plots of predicted PM values with ANN Model 1

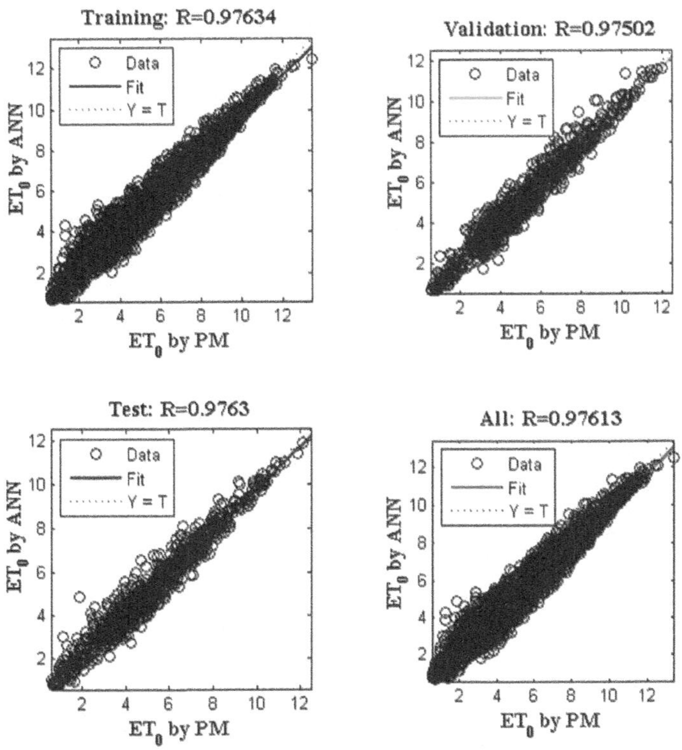

Fig. 7.3 Plots of predicted PM values with ANN Model 2

Figure 7.4 shows scattered plots between estimated ET_0 by PM method and predicted model 3 ET_0 for a trial epoch with the network configuration of 3-10-1 using the LM algorithm.

Figure 7.5 shows scattered plots between estimated ET_0 by PM method and predicted model 4 ET_0 for a trial epoch with the network configuration of 3-10-1 using the LM algorithm.

Figure 7.6 shows scattered plots between estimated ET_0 by PM method and predicted model 5 ET_0 for a trial epoch with the network configuration of 3-10-1 using the LM algorithm.

Figure 7.7 shows scattered plots between estimated ET_0 by PM method and predicted model 6 ET_0 for a trial epoch with the network configuration of 3-10-1 using the LM algorithm.

During the period of training, the objective was to attain minimum MSE value and maximum regression value. When the objective was achieved, training stopped. ET_0 values forecasted by LM algorithm were close to the actual one. The validation graph is shown in Fig. 7.8. The observed MSE values for all models are shown in Table 7.2 using LM algorithm for ANN model 1 to model 6, respectively.

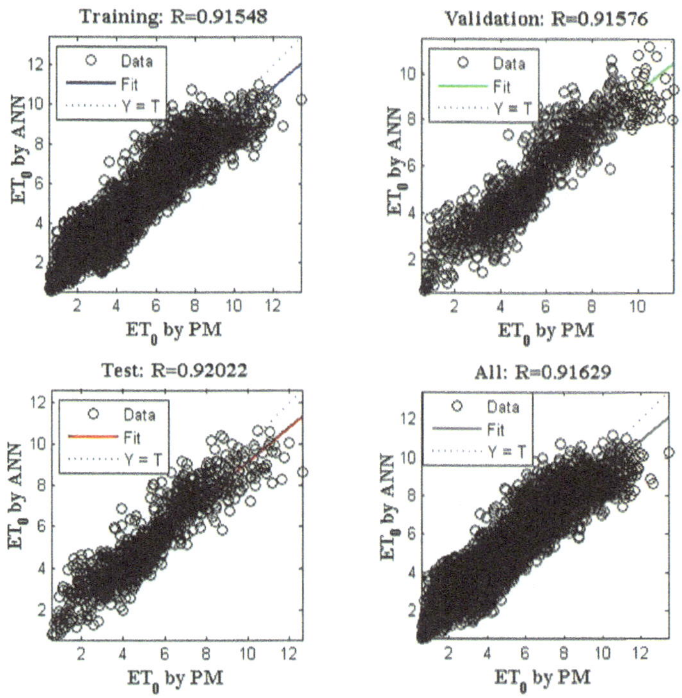

Fig. 7.4 Plots of predicted PM values with ANN Model 3

With the help of testing dataset, ET_0 values calculated by PM and forecasted ET_0 values using LM algorithm were compared for year January 2011 to July 2014. All configurations of ANN models 1, 3, 4, 5 and 6 showed poor performance as compared to model 2, so these have been neglected for the remaining analysis. The entire network configuration of model 2 of ANN displayed superior results. Owing to this, it was selected for further analysis of forecasting. From two training algorithms, each model was chosen based on SSE, MSE and coefficient of regression (R^2). The ANN forecasted ET_0 values were compared with PM estimated values and imitated the same pattern. Figure 9 shows the comparison of forecasted ET_0 values and ET_0 values calculated by Penman-Monteith equation. The forecast was prepared for a period of three and half years, from January 2011 to July 2014.

As the natural phenomenon is complicated, evapotranspiration prediction can be a difficult and extensive work. In this operation, the usage of ANN as a tool allows in the betterment of consistency and simplicity of this work as it is capable of carrying huge quantities of data for processing. Figures 7.2 and 7.7 also show that ANN model 1 and model 6 evinced the poorest performance, with an MSE of 0.631 and 0.673, respectively, while regression coefficient of 0.914 and 0.909, respectively. These results demonstrated the potential of the ANN to detect a nonlinear input and output

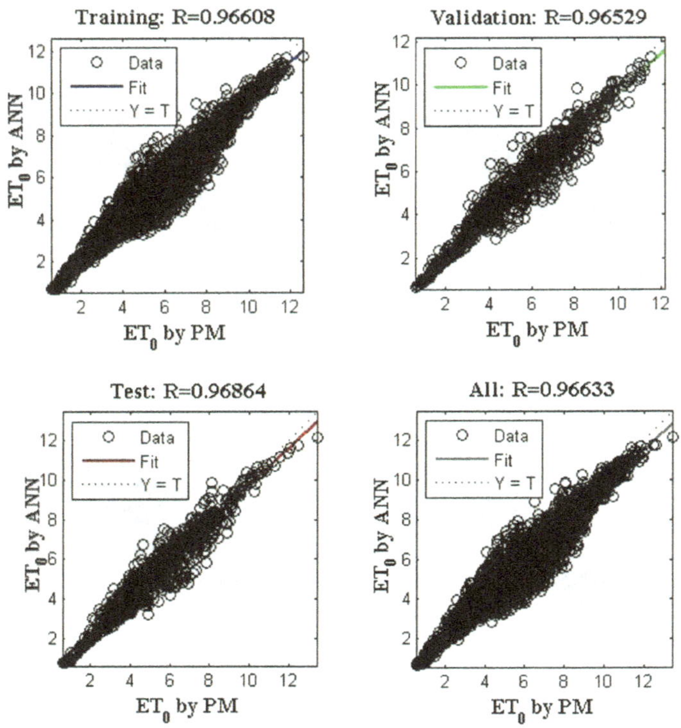

Fig. 7.5 Plots of predicted PM values with ANN Model 4

connection. For the perfect estimation of ET_0 using an artificial neural network, wind speed, humidity and temperature are considered to be the most important input variables.

7.5 Conclusions

This study shows the capability of an artificial neural network for forecasting ET_0. For the prediction of reference evapotranspiration, a good quality of meteorological data is required. For accurate evaluation of ET_0 using ANN, solar radiations, humidity and temperature are the most important input variables. Based on the techniques and concept of ANNs, different types of model have been utilized for forecasting ET_0 for Nagpur region, India. Two training algorithms have been used for the training of each model. The obtained results from ANN model with less number of input data used for testing and training are very encouraging, since it gave fairly accurate results as compared with PM values of reference evapotranspiration. The model 2 of

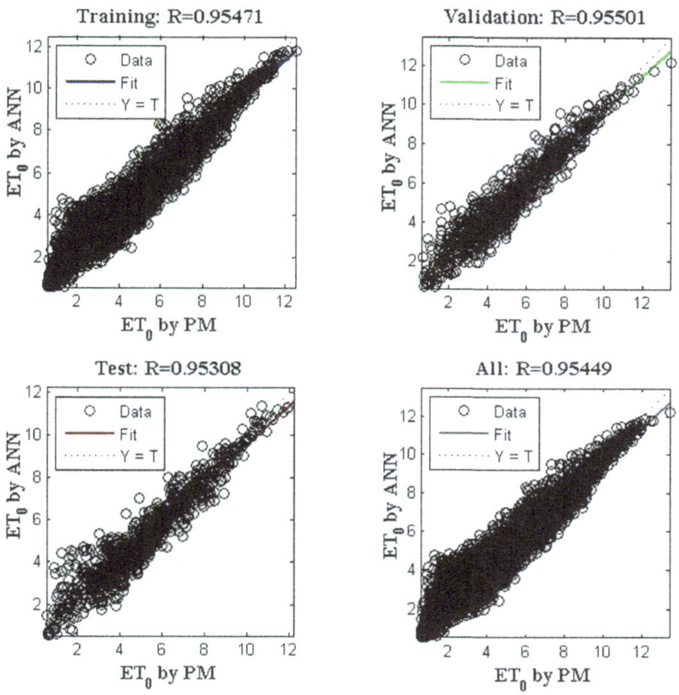

Fig. 7.6 Plots of predicted PM values with ANN Model 5

ANN with training configuration of 3-10-1 using Levenberg–Marquardt algorithm was found better based on the performance criteria. It is observed that the learning performance generally increases by increasing number of hidden neurons. Trained architecture using the Levenberg–Marquardt algorithm is found to be better, based on the least mean square error (0.191) and highest coefficient of regression (0.975) among the compared models. Also, it is found that the forecasted reference evapotranspiration values are approximately equal to PM reference evapotranspiration values. The results indicate that ANN algorithms such as Levenberg–Marquardt and quasi-Newton can be applied to evaluate the forecasting of reference evapotranspiration values in different study areas.

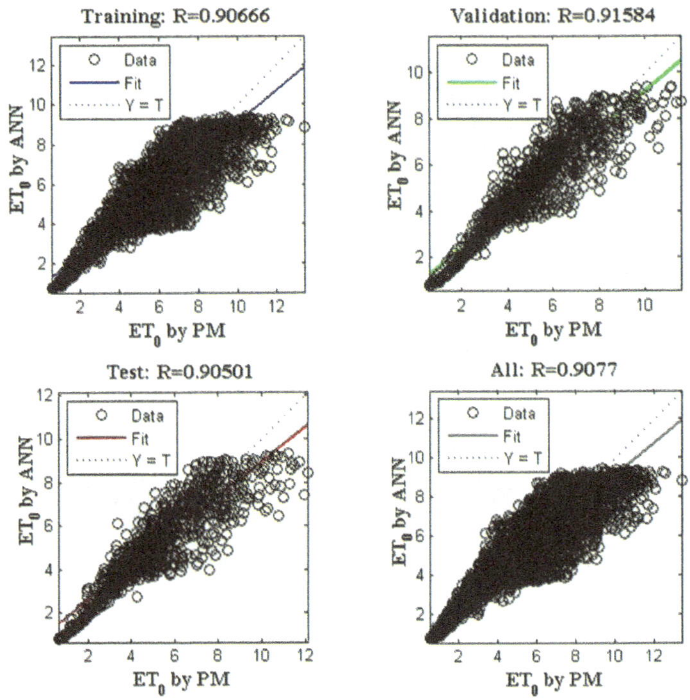

Fig. 7.7 Plots of predicted PM values with ANN Model 6

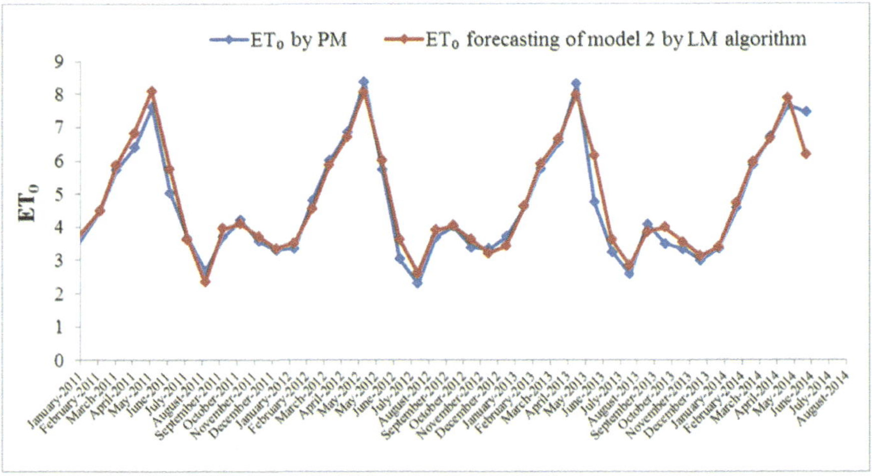

Fig. 7.8 Comparison of forecasted ET_0 values and calculated PM values for the year 2011 to July 2014

Table 7.2 Mean square error, regression coefficient and sum square error of all ANN models for testing data using LM

Model no.	Mean square error (MSE)	Regression coefficient (R^2)	Sum square error (SSE)
1	0.631	0.914	878.174
2	0.191	0.976	274.886
3	0.683	0.915	869.55
4	0.262	0.967	375.72
5	0.415	0.953	501.237
6	0.673	0.909	846.155

References

Allen RG (1986) A Penman for all season. J Irrig Drain Eng 12(4):348–368. https://doi.org/10.1061/(ASCE)0733-9437(1986)112:4(348)

Araújo GL, Oliveira Jr MM, Pinto FAC, Mantovani EC (2014) Forecasting reference evapotranspiration with artificial neural network. In: II INOVAGRI international meeting, Fortaleza, Brazil, pp 4287–4293 (January 2014). https://doi.org/10.12702/ii.inovagri.2014-a577

Chauhan S, Shrivastava RK (2009) Reference evapotranspiration forecasting using different artificial neural networks algorithms. Can J Civ Eng 36(9):1491–1505

Kisi O, Ozturk O (2007) Adaptive neuro-fuzzy computing technique for evapotranspiration estimation. J Irrig Drain Eng 133(4):368–379. https://doi.org/10.1061/(ASCE)0733-9437(2007)133:4(368)

Landeras G, Ortiz-Barredo A, López JJ (2009) Forecasting weekly evapotranspiration with ARIMA and artificial neural network models. J Irrig Drain Eng 135(3):323–334

Marino MA, Dandy GC, Taghavi SA (1993) Forecasting of reference crop evapotranspiration. Agric Water Manag 24:163–187. https://doi.org/10.1016/0378-3774(93)90022-3

Perera KC, Western AW, Nawarathna B, George B (2014) Forecasting daily reference evapotranspiration for Australia using numerical weather prediction outputs. Agric For Meteorol 194:50–63

Trajkovic S, Todorovic B, Stankovic M (2003) Forecasting of reference evapotranspiration by artificial neural networks. J Irrig Drain Eng 129(6):454–457

Traey JC, Marino MA, Taghavi SA (1992) Predicting water demand in agriculture regions using time series forecasts of reference crop evapotranspiration. In: Karamouz M (ed) Water resources planning and management: saving a threatened resources-in search of solution. ASCE, New York, pp 50–55

Zanetti SS, Sousa EF, Oliveira VPS, Almeida FT, Bernardo S (2007) Estimating evapotranspiration using artificial neural network and minimum climatological data. J Irrig Drain Eng 133(2):83–89

Chapter 8
Time-Varying Downscaling Model (TVDM) and its Benefit to Capture Extreme Rainfall

Subbarao Pichuka and Rajib Maity

Abstract The focus of this chapter is to explore the benefits of time-varying downscaling model (TVDM) to preserve the characteristics of extreme rainfall in the downscaled products. The TVDM was developed based on the hypothesis that the relationship between causal and target variables is non-stationary in the context of climate change. The TVDM utilizes the skill of the Bayesian approach in updating its parameters and thus, incorporates the time-varying relationship between the causal variables and the target variable to be downscaled. In this study, the potential of the TVDM to replicate the extremes in the downscaled precipitation field is explored. The entire Indian landmass is considered as the study area. The output of TVDM is compared with the existing statistical downscaling model (SDSM) and regional climate model (RCM) outputs procured from the coordinated regional climate downscaling experiment (CORDEX). It may be noted that the SDSM assumes the causal–target relationship is time-invariant, whereas the TVDM considers the time-varying relationship since it is based on the non-stationarity assumption. Causal variables from the coupled Hadley centre Global Environmental Model (version-2)—Earth System model (HadGEM2-ES) general circulation model are considered as input to the TVDM and SDSM. Downscaled precipitation field is compared with (i) $0.25^\circ \times 0.25^\circ$ gridded observed precipitation data, procured from the India Meteorological Department (IMD), (ii) output of SDSM at selected locations, and (iii) RCM downscaled precipitation field obtained from CORDEX. Observed precipitation field is found to best correspond to the TVDM downscaled precipitation field as compared to CORDEX and SDSM outputs. This analysis reveals that the TVDM is a promising downscaling technique for the assessment of extreme values. Overall, the TVDM

S. Pichuka
Department of Civil Engineering, National Institute of Technology Andhra Pradesh, Tadepalligudem 534101, Andhra Pradesh, India
e-mail: srpichuka@gmail.com

R. Maity (✉)
Department of Civil Engineering, Indian Institute of Technology Kharagpur, Kharagpur 721302, West Bengal, India
e-mail: rajib@civil.iitkgp.ac.in; rajibmaity@gmail.com

R. Jha et al. (eds.), *Climate Change Impacts on Water Resources*, Water Science and Technology Library 98, https://doi.org/10.1007/978-3-030-64202-0_8

is found to be potential in a changing climate due to its time-varying characteristics considering the non-stationarity issue that exists in the relationship between the causal–target variables.

Keywords Climate change · Extreme events · TVDM · GCM · CORDEX

8.1 Introduction

The extreme rainfall values are the natural disasters and affect large spatial extent. Recent studies concluded that the climate change has a significant impact on the hydrological extremes (Arnell, 1999, Haddeland et al. 2014; Pichuka and Maity 2020a). Moreover, the impact varies spatio-temporally because of global hydroclimatic teleconnection (Maity et al. 2007) and local physical features like topography and climatology of the location (Maity and Kashid 2011; Pichuka and Maity 2016; Silverman and Maneta 2016). There are many studies around the world that assessed different hydrological extreme events. For instance, Mishra et al. (2011) have investigated the changes in extreme precipitation in Texas. Lindström and Bergström (2004) explore the annual and seasonal flood peaks in Sweden. Tofiq and Guven (2014) attempted to identify the peak monthly discharge through statistical downscaling approach at a dam site in Iraq. Pichuka et al. (2017) assessed the extreme stream flow events over the Bhadra basin using downscaled rainfall as input to rainfall-runoff (RR) model. Jiang et al. (2007) concluded that the change in amount of average water availability causes change in the extreme rainfall values. Pichuka and Maity (2020a) assessed the future variation of extreme rainfall events over the Bhadra basin. The occurrence of extremes often causes loss of lives, agriculture and can also change the fate of country's economy. Therefore, assessment of extreme values owing to climate change is of utmost importance.

The outputs of the general circulation models (GCMs) have a too poor spatial resolution (~250 km × 250 km) to use for local-scale hydrologic assessment. The technique to improve the spatial resolution of GCM outputs to finer scale is known as *downscaling*. It is accomplished either by establishing the statistical relationship, i.e., statistical downscaling (Hessami et al. 2008; Wilby and Wigley 1997) between causal variables and target variable to be downscaled or by regional climate model (RCM), i.e., dynamical downscaling. The GCM outputs are used as boundary conditions of the RCMs in the dynamical downscaling (Chirivella et al. 2016; Giorgi et al. 2015). However, the usage of dynamic downscaling technique is restricted because of the requirement of high computational facilities. On the other hand, the statistical downscaling method assumes the relationship between causal–target variable as time-invariant, which is not true. Therefore, the applicability of statistical downscaling is debatable as it is a stationary-based approach (Merkenschlager et al. 2017; Sachindra and Perera 2016; Pichuka and Maity 2016).

Recently, Pichuka and Maity (2018) have developed a non-stationary-based model for statistical downscaling of various hydroclimatic variables and named it as time-varying downscaling model (TVDM). The TVDM was found to perform better in downscaling the monthly and seasonal precipitations as compared to existing statistical downscaling model (SDSM) and regional climate model (RCM). The objective of this chapter is to explore the ability of TVDM to represent stochastic features of observed rainfall including extreme values and its comparison with SDSM (stationary approach) and RCM downscaled data.

8.2 Study Area

The entire Indian landmass is considered as the study area (Fig. 8.1). A wide range of climate is noticed across India. For instance, dry region (low rainfall) in the West side to wet region (high rainfall) in the East. The South-west monsoon contributes to the principal portion (~80%) of total annual rainfall. The North-west part of India experiences drought whereas North-east part is a flood-prone region. Hence, India consists of a variety of climatology that offers an ideal study region to compare the

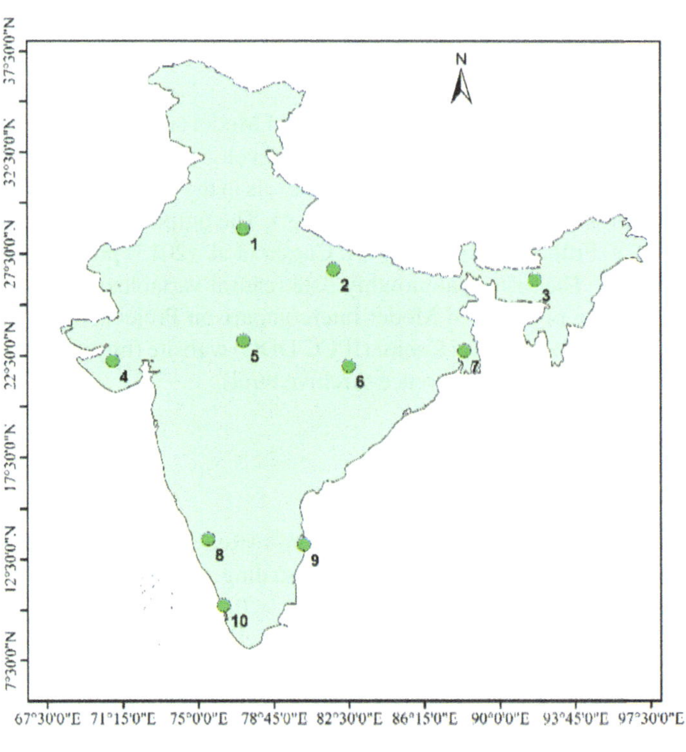

Fig. 8.1 Study area map showing 10 key locations

downscaled products from different models. In this study, ten key locations (Fig. 8.1) are selected, which are well-spread and represent the variation of Indian climate. The location-specific assessment is carried out at these locations and the results are explored.

8.3 Materials and Methodology

8.3.1 Data Used

8.3.1.1 Observed Data

The daily observed rainfall data is procured from the India Meteorological Department (IMD) at a horizontal resolution of $0.25° \times 0.25°$ for the study region. Further, the monthly precipitation values are calculated at all the grid intersections. Part of the observed data is used for calibrating the models and named as calibration period (1951–1990) and other part of data is used for validating the downscaled results and named as validation period (1991–2005).

8.3.1.2 GCM Data

The coupled Hadley centre Global Environmental Model (version-2)—Earth System model (HadGEM2-ES) is used in this study. It is developed at the Hadley centre in the UK meteorological office and has 38 pressure levels in the atmosphere (vertical). The spatial resolution is 1.25° (lat.) by 1.875° (long.). The outputs of HadGEM2-ES are used in the IPCC Fifth Assessment report. Caesar et al. (2013) presented the details of HadGEM2-ES. The historical monthly data (causal variables) were downloaded from the fifth phase of Coupled Model Intercomparison Project (CMIP5) supplied through IPCC Data Distribution Centre (IPCC DDC) website (http://www.ipcc-data.org/sim/gcm_monthly/AR5/Reference-Archive.html).

8.3.1.3 RCM Data

The outputs of regional climate model, version-4 (RegCM4) monthly precipitation are obtained from coordinated regional downscaling experiment (CORDEX) data portal. The spatial resolution of RegCM4 is $0.5° \times 0.5°$ (Li et al. 2015). The outputs of RegCM4 were utilized in many projects to assess the impact of climate change on extreme values and other purposes. The Indian Institute of Tropical Meteorology (IITM), Pune provides the outputs from RegCM4. This data for the South Asia region is downloaded from IITM for the historical period (1951–2005) (URL: http://cccr.tropmet.res.in/home/ftp_data.jsp).

8.3.2 Methodology

In this study, two models, i.e., TVDM and SDSM version-5.2 (SDSM-5.2) are adopted to downscale the monthly precipitation using GCM data as input to the model. SDSM is developed by considering causal–target relationship as time-invariant (stationary). The methodological details of SDSM are presented in Wilby et al. (2002).

8.3.2.1 Outline of TVDM Methodology

TVDM was developed by Pichuka and Maity (2018), which is based on the Bayesian approach (West and Harrison 1997). It considers the relationship between causal–target variables as time-varying (non-stationary). In this chapter, the methodology of TVDM is explained using key equations.

The first step is to standardize all the causal variables by subtracting the mean (μ) and dividing the difference with standard deviation (σ). It ensures all the variables are in similar range. Then the target variable is downscaled at time step t using the following equation.

$$\left(Y_t / D_{t-1}\right) \sim T_n[F_t, Q_t] \tag{8.1}$$

where Y_t is the target variable to be downscaled at time step t, D_{t-1} is the information of model parameters at time step t-1 and n is the degree of freedom. The value of n for tth time step is given as

$$n = t - 1 \tag{8.2}$$

The equations for F_t and Q_t are given as

$$F_t = \overline{Y_t} + x_t^1 \times m_{t-1}^1 + x_t^2 \times m_{t-1}^2 + x_t^3 \times m_{t-1}^3 + \cdots + x_t^z \times m_{t-1}^z \tag{8.3}$$

$$Q_t = \left(x_t^1\right)^2 \times R_t^1 + \left(x_t^2\right)^2 \times R_t^2 + \left(x_t^3\right)^2 \times R_t^3 + \cdots + \left(x_t^z\right)^2 \times R_t^z + S_{t-1} \tag{8.4}$$

where $\overline{Y_t}$ is the climatological mean value of the target variable at tth time step. $x_t^1, x_t^2, x_t^3, \ldots x_t^z$ are the standardized causal variables at the tth time step. $m_{t-1}^1, m_{t-1}^2, \ldots m_{t-1}^z$ are the model parameters supplied by the modeler at initial time step. R_t is expressed as

$$R_t = C_{t-1} / \delta \tag{8.5}$$

where δ is known as discount factor which ranges between 0 and 1. The value of δ indicates the presence of uncertainty in the future information. The lower value of δ denotes the higher rate of decay in past information and vice versa (West and Harrison 1997). Therefore, the optimum value of δ is estimated on the basis of model performance.

S_{t-1} is expressed as

$$S_{t-1} = \frac{d_{t-1}}{n_{t-1}} \tag{8.6}$$

The parameters m_0, C_0, n_0 and d_0 are to be supplied to the TVDM as the initial information for all the causal variables. Next, the parameter values (*m-values*) are updated using a set of equations until the model calibration. Then the target variable is downscaled at every time step using these updated *m-values* and causals data in Eq. 8.3. However, the parameter updation is possible only when observed data is available, and for future period it cannot be updated using set of equations. The target variable during future period is downscaled by considering the temporal evolution of model parameters and its regeneration. It is achieved by projecting the deterministic and stochastic components of historical *m-values*. The trend and periodic components are the deterministic part of historical time series which are removed from the *m-values* and the residual left is purely random in nature which can be modeled using auto-regressive (AR) model. Using these deterministic (trend and periodicity) and stochastic (AR1) components, ensemble of realizations (any number) can be generated. Each time one realization (set of regenerated *m-values*) is considered for downscaling the target variable during future period. It can be achieved by substituting the regenerated *m-values* and the causal variables obtained from the CMIP-5 portal in Eq. 8.3. The detailed procedure of regenerating *m-values*, set of equations to update these *m-values* and their regeneration during the future period are presented in Pichuka and Maity (2018). The 75th, 90th and 95th percentile values are considered as extreme values to compare the observed and downscaled products.

8.4 Results and Discussion

8.4.1 Calibration Period

The monthly precipitation is downscaled using TVDM and SDSM-5.2 and it is also obtained from RegCM4 RCM (CORDEX). The TVDM is calibrated at all the grid (0.25° × 0.25°) intersections and the SDSM is calibrated at each of the 10 key locations using the observed data for 40 years (1951–1990). Then the extreme values of downscaled precipitation are modeled at 75th, 90th and 95th percentiles. These are presented in Fig. 8.2. The correspondence between TVDM and the observed data is found better as compared to that between TVDM and the RCM (CORDEX) data. The RCM seems to be overestimating the 75th percentile values and underestimating the 95th percentiles in the North-east region, whereas the association between observed

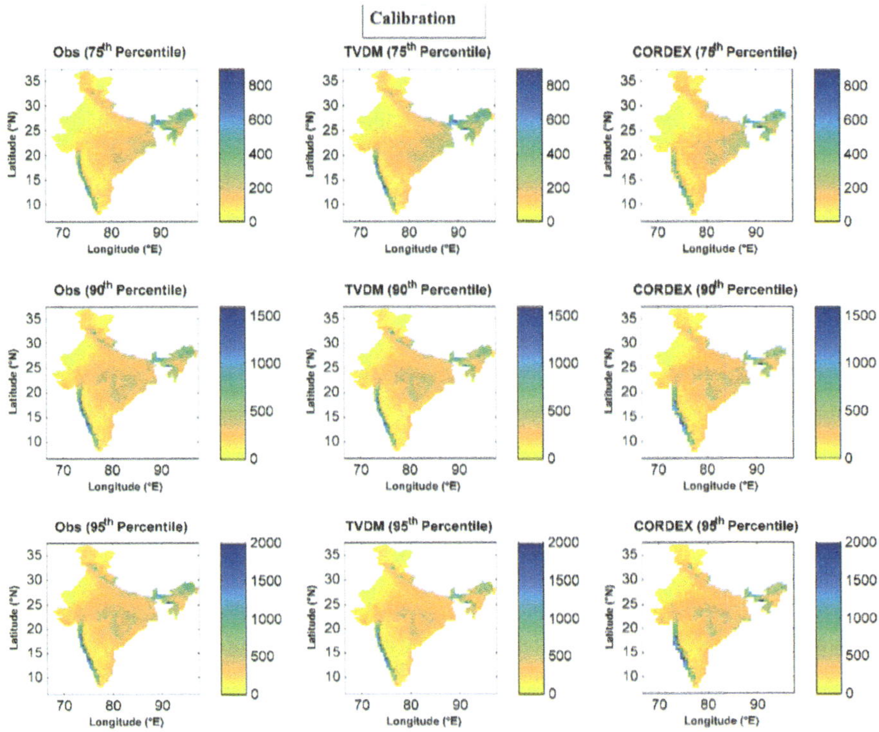

Fig. 8.2 Comparison of extreme events obtained using TVDM and CORDEX outputs with respect to observed data during calibration period

and TVDM downscaled precipitation is found better in these regions. In fact, the observed and downscaled precipitation maps look almost similar in case of TVDM, whereas discrepancies at places are noticed in case of maps developed with the RCM data (Fig. 8.2). It is worth mentioning that Pichuka and Maity (2020b) carried out the analysis on the applicability of CORDEX data in representing the observed extreme rainfall events (75th, 90th and 95th percentiles). They concluded that the CORDEX data is unable to model the extreme events in some parts (especially high rainfall regions) of the country. Similar result is noticed from the analysis carried out in this chapter and shown in Table 8.1. The performance measures during calibration period at ten key locations are presented in Table 8.1 for all three models. The differences between observed and TVDM downscaled extreme values are found to be minimum as compared to the other cases, i.e., observed-SDSM and observed-RCM. These observations indicate the best performance of TVDM as compared to other two models.

Table 8.1 Extreme values at different quantiles for different models during calibration period (1951–1990)

Loc no.	Observed			TVDM			SDSM			CORDEX		
	75th	90th	95th	75th	90th	95th	75th	90th	95th	75th	90th	95th
1	60.7	202.4	300.2	62.2	205.4	284.6	72.9	150.6	188.2	73.2	187.3	239.2
2	143.9	330.2	438.8	156.0	332.2	389.3	87.7	233.7	306.7	171.0	299.4	352.9
3	256.3	377.0	429.4	262.5	361.6	407.8	240.4	346.7	389.1	278.4	392.0	461.8
4	61.0	196.7	279.7	71.6	193.2	275.3	57.0	126.3	170.6	48.9	192.6	245.5
5	121.7	340.6	465.3	133.5	348.5	456.5	87.1	329.4	503.2	110.5	342.6	428.5
6	169.2	409.0	497.9	184.5	416.8	475.9	112.5	261.6	397.8	202.6	384.1	435.0
7	235.1	363.7	465.0	231.8	357.5	408.2	247.4	324.5	385.0	241.6	358.0	393.2
8	214.6	479.2	721.5	229.0	481.9	738.0	174.7	473.4	794.4	181.5	402.6	696.2
9	142.4	257.0	367.0	136.3	252.6	360.0	109.6	268.6	423.1	137.6	213.6	351.6
10	420.9	631.6	812.3	386.6	630.0	794.4	304.3	621.3	915.5	377.8	650.9	782.4

8.4.2 Validation Period

The calibrated TVDM and SDSM models are used for validation. It can be noted from Pichuka and Maity (2018) that in case of TVDM, the time series of model parameters during the calibration period are utilized to regenerate by considering deterministic (trend and periodicity) and stochastic (auto-regressive component) components of the time series. The regenerated parameters are utilized along with the causals data (GCM variables downloaded from IPCC data portal) to obtain the monthly downscaled precipitation. Then the extreme values are computed at different percentile for all three cases, i.e., observed data, TVDM downscaled data and RCM (CORDEX) data. The magnitudes of these extreme values are presented in Fig. 8.3.

It can be noticed from Fig. 8.3 that the TVDM outputs are more or less matching alike with the observed data. The RCM outputs seem to be underestimated in the North-east and Western Ghat regions. The specific values at ten key locations are presented in Table 8.2 for observed data and downscaled values using TVDM, SDSM and RCM.

The best performance of TVDM can be noticed from the values shown in Table 8.2. Almost for all the cases, the difference between observed and TVDM outputs are minimum. It can be worth mentioning that the performance of RCM is found to be better than that of SDSM at majority of the key locations. These results reflect a fact that the time-varying or dynamic approaches (TVDM and RCM) are performing better when compared to the time-invariant (SDSM) downscaling approach. However, the superiority of TVDM is found to perform even better than RCM in assessing the extreme values. It is achieved due to the incorporation of time-varying component between the causal–target relationships which is essential to consider in a changing climate.

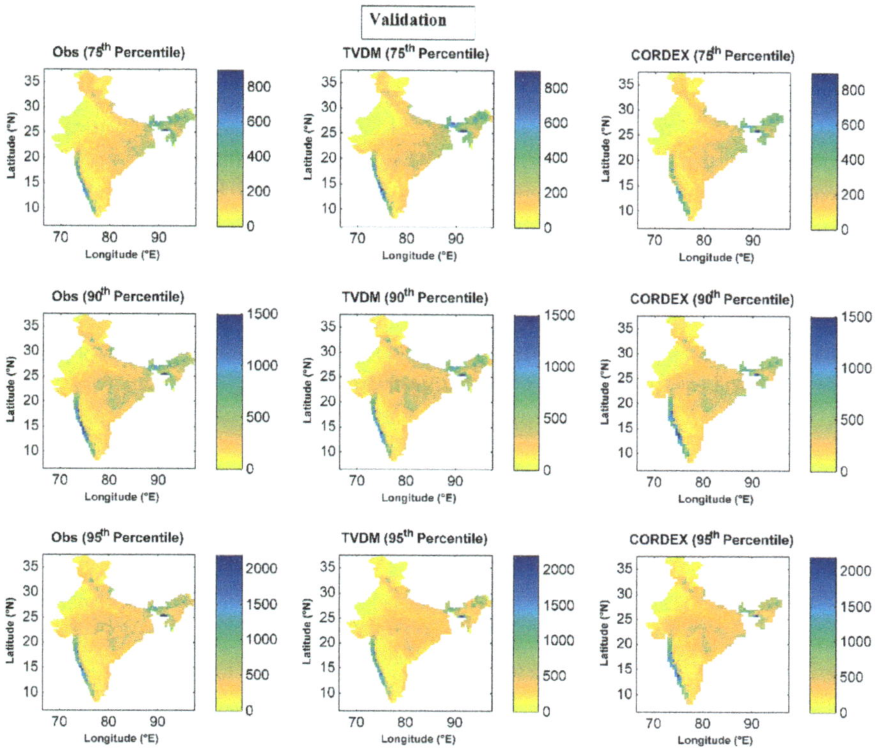

Fig. 8.3 Comparison of extreme events obtained using TVDM and CORDEX outputs with respect to observed data during validation period

Table 8.2 Extreme values at different quantiles for different models during validation period (1991–2005)

Loc no.	Observed			TVDM			SDSM			CORDEX		
	75th	90th	95th	75th	90th	95th	75th	90th	95th	75th	90th	95th
1	64.9	192.8	271.9	62.2	199.9	269.9	69.5	153.6	261.0	73.2	187.3	259.2
2	143.6	248.0	317.6	163.2	255.1	323.1	111.5	230.7	365.9	171.0	239.4	352.9
3	256.3	411.4	467.3	277.1	378.4	463.7	230.1	343.3	426.9	278.4	392.0	461.8
4	48.8	184.3	264.9	64.1	181.7	259.1	71.3	165.5	214.0	48.9	192.6	275.5
5	120.1	284.6	456.0	124.4	299.0	453.2	126.6	366.5	520.1	110.5	342.6	456.5
6	166.0	369.4	439.6	179.2	377.4	428.3	124.1	308.2	414.6	202.6	384.1	474.0
7	256.1	362.7	407.5	252.0	368.9	410.6	300.6	342.8	373.2	261.6	358.0	393.2
8	332.6	518.6	616.3	272.1	475.5	609.8	229.9	588.0	982.5	261.5	402.6	586.2
9	112.2	284.8	359.2	114.8	261.9	356.2	115.6	256.3	422.5	117.6	213.6	351.6
10	458.5	672.6	783.8	387.2	661.1	758.0	288.9	655.7	841.2	376.8	650.9	812.4

8.5 Conclusions

This study establishes the ability of a newly developed time-varying downscaling model, i.e., TVDM to capture extreme rainfall values. Considering the entire India as study region and ten key locations for comparison, the outputs of TVDM are compared with the SDSM (a time-invariant approach) and RCM data (obtained from CORDEX portal). Then the correspondence between observed and downscaled (using TVDM, SDSM and RCM) extreme rainfall values is explored during the calibration (1951–1990) and validation (1991–2005) periods. The TVDM performs much better in representing the extreme rainfall values during both calibration and validation periods as compared to the other models. The difference between observed and downscaled extreme values is found minimum in case of TVDM.

In brief, TVDM is a promising downscaling technique in a changing climate due to its time-varying characteristics considering the non-stationarity issue that exists in the relationship between the causal–target variables. It is effective for the assessment of extreme values of any climatic region in the world and computationally less intensive as compared to RCM.

Acknowledgements The work was partially supported by the sponsored projects supported by Department of Science and Technology, Climate Change Programme (SPLICE), Government of India (Ref No. DST/CCP/CoE/79/2017(G)) through a sponsored project.

References

Arnell NW (1999) Climate change and global water resources. Glob Environ Change 9:S31–S49

Caesar J, Palin E, Liddicoat S, Lowe J, Burke E, Pardaens A, Sanderson M et al (2013) Response of the HadGEM2 earth system model to future greenhouse gas emissions pathways to the year 2300. J Clim 2009:3275–3284. https://doi.org/10.1175/jcli-d-12-00577.1

Chirivella V, Capilla JE, Pérez-martín MA (2016) Dynamical versus statistical downscaling for the generation of regional climate change scenarios at a Western Mediterranean basin: the Júcar River District. J Water Clim Chang 7(June):379–392. https://doi.org/10.2166/wcc.2015.207

Giorgi F, Gutowski J, William J (2015) Regional dynamical downscaling and the CORDEX initiative. Annu Rev Environ Resour 40(1):150724171620008. https://doi.org/10.1146/annurev-environ-102014-021217

Haddeland I, Heinke J, Biemans H, Eisner S, Flörke M, Hanasaki N, Markus K et al (2014) Global water resources affected by human interventions and climate change. Proc Natl Acad Sci 111:3251–3256. https://doi.org/10.1073/pnas.1222475110

Hessami M, Gachon P, Ouarda TBMJ, St-Hilaire A (2008) Automated regression-based statistical downscaling tool. Environ Model Softw 23(6):813–834. https://doi.org/10.1016/j.envsoft.2007.10.004

Jiang T, Chen DY, Xu C, Chen X, Chen X, Singh VP (2007) Comparison of hydrological impacts of climate change simulated by six hydrological models in the Dongjiang Basin, South China. J Hydrol 336:316–333. https://doi.org/10.1016/j.jhydrol.2007.01.010

Li L, Diallo I, Xu C, Stordal F (2015) Hydrological projections under climate change in the near future by RegCM4 in Southern Africa using a large-scale hydrological model. J Hydrol 528(May):1–16. https://doi.org/10.1016/j.jhydrol.2015.05.028

Lindström G, Bergström S (2004) Runoff trends in Sweden 1807–2002. Hydrol Sci J. https://doi.org/10.1623/hysj.49.1.69.54000.maity

Maity R, Kashid SS (2011) Importance analysis of local and global climate inputs for basin-scale streamflow prediction. Water Resour Res 47(11):1–17. https://doi.org/10.1029/2010wr009742

Maity R, Nagesh Kumar D, Nanjundiah RS (2007) Review of hydroclimatic teleconnection between hydrologic variables and large-scale atmospheric circulation indices with Indian perspective. ISH J Hydraul Eng 13(1): 77-92. Taylor & Francis. https://doi.org/10.1080/09715010.2007.10514859

Merkenschlager C, Hertig E, Jacobeit J (2017) Non-stationarities in the relationships of heavy precipitation events in the Mediterranean area and the large-scale circulation in the second half of the 20th century. Global Planet Chang 151:108–121. Elsevier B.V. https://doi.org/10.1016/j.gloplacha.2016.10.009

Mishra AK, Singh VP, Özger M (2011) Seasonal streamflow extremes in Texas river basins : uncertainty, trends, and teleconnections. J Geophys Res 116:1–28. (June 2010). https://doi.org/10.1029/2010jd014597

Pichuka S, Maity R (2016) Spatio-temporal downscaling of projected precipitation in the 21st century : indication of a wetter monsoon over the Upper Mahanadi Basin, India. Hydrol Sci J 62(3):467–482. Taylor & Francis. https://doi.org/10.1080/02626667.2016.1241882

Pichuka S, Rajendra Prasad R, Maity R, Kunstmann H (2017) Development of a method to identify change in the pattern of extreme streamflow events in future climate: application on the Bhadra reservoir inflow in India. J Hydrol Reg Stud 9:236–246. https://doi.org/10.1016/j.ejrh.2016.12.084

Pichuka S, Maity R (2018) Development of a time-varying downscaling model considering non-stationarity using a Bayesian approach. Int J Climatol 1–20. https://doi.org/10.1002/joc.5491

Pichuka S, Maity R (2020a) Assessment of extreme precipitation in future through time-invariant and time-varying downscaling approaches. Water Resour Manag 34:1809–1826. https://doi.org/10.1007/s11269-020-02531-6

Pichuka S, Maity R (2020b) How far the CORDEX high-resolution data represents observed precipitation for the Indian region: an analysis across Indian mainland. Theoret Appl Climatol. https://doi.org/10.1007/s00704-020-03355-5

Sachindra DA, Perera BJC (2016) Statistical downscaling of general circulation model outputs to precipitation accounting for non-stationarities in predictor-predictand relationships. PLOSone 11:1–21. https://doi.org/10.1371/journal.pone.0168701

Silverman NL, Maneta MP (2016) Detectability of change in winter precipitation within mountain landscapes: spatial patterns and uncertainty. Water Resour Res 52:4301–4320. https://doi.org/10.1002/2014wr016493

Tofiq FA, Guven A (2014) Prediction of design flood discharge by statistical downscaling and General Circulation Models. J Hydrol 517:1145–1153. Elsevier B.V. https://doi.org/10.1016/j.jhydrol.2014.06.028

West M, Harrison PJ (1997) Bayesian forecasting and dynamic models, 2nd edn. Springer, New York

Wilby RL, Dawson CW, Barrow EM (2002) SDSM—a decision support tool for the assessment of regional climate change impacts. Environ Model Softw 17(2):145–157. https://doi.org/10.1016/s1364-8152(01)00060-3

Wilby RL, Wigley TML (1997) Downscaling general circulation model output: a review of methods and limitations. Prog Phys Geogr 21(4):530–548. https://doi.org/10.1177/030913339702100403

Chapter 9
An Assessment of Impact of Land Use–Land Cover and Climate Change on Quality of River Using Water Quality Index

Manisha Jamgade and Shrikant Charhate

Abstract Globally rapid population, urbanization and industrialization are posing problems such as changes in the land use pattern and climate and contamination of surface waters. Land use changes have potentially large impacts on river water quality, e.g., adaptation of cropland to urban area, crop rotations, establishment of new industries, commercial and residential areas etc. Unplanned developments associated with land use change have notably influenced the river water quality. Further due to climatic changeability such as increased temperature, rainfall change affects surface water quality and aquatic environment. The purpose of this study is to analyze river water quality related to spatial land use in river basin using GIS analysis. To manage river water quality in the most effective and efficient way, the cause and effect relationship of the river basin must be identified. Water quality indices are an effective and easy assessment method to analyze the effect of land use and climate change on stream water quality. The present study focused on development of improved surface water quality indices considering various water quality parameters affected by land use–land cover and climate change scenarios. These indices will help to classify the river and will provide simple interpretation of the monitoring data to aid citizens and managers for decision making to improve river water quality.

Keywords Land use–land cover · Climate change · Water quality index · River water quality

---

M. Jamgade (✉)
Pillai HOC College of Engineering & Technology, Rasayani 410207, India
e-mail: manisha21jamgade@gmail.com

S. Charhate
Amity School of Engineering & Technology, Panvel 410206, India
e-mail: sbcharate@yahoo.co.in

R. Jha et al. (eds.), *Climate Change Impacts on Water Resources*, Water Science and Technology Library 98, https://doi.org/10.1007/978-3-030-64202-0_9

9.1 Introduction

9.1.1 Land Use–Land Cover

Land use–land cover has become an environmental issue worldwide due to changes in forestland, agricultural land, waterways and air. It is well known that there is a close relationship between the land use type and water quality (Pullanikkatil et al. 2015). Some researches focus on this relationship from the perspective of the spatial configuration of land use in the recent years (Gazzaz et al. 2012). Several research studies have been conducted related to the change in investigation of watersheds, which are important in developing effective management strategies to protect water resources.

Wan et al. (2014) showed the influence of land use and land cover (LULC) changes on soil properties, geomorphology, hydrological processes and quality of water at local, regional and global scales. Land use and land cover changes are known to be main drivers of water quality deterioration in watercourses (Yong and Chen (2002). Knowledge of the spatial and temporal variability of land use change characteristics and its linkage to water quality parameters in basins is limited. Land use changes have the capability to affect the land cover and vice versa.

Ahearn et al. (2005) correlated the impact of land use and land cover and climate changes on stream water quality through mixed effects model considering the parameters like nitrate and total suspended solids loading between dry and average water years 1999 and 2001. In their study they contrast to the TSS model, the nitrate loading model was more complex with agriculture, grassland, and the presence or absence of waste water treatment plants. Juan Huang et al. (2013) indicated that there was considerable negative correlation between grassland, forest land and built-up area on the water quality, while the influence of the cultivated land on the water quality was very intricate. Moreover, the impact of the landscape variety on the indicators of water quality within the basin was also analyzed, which indicated the significant negative relationship between them. Land use and land cover (LULC) changes affect geomorphology, soil properties, hydrological processes and water quality at global, regional and local scales (Wan et al. 2014).

Many research studies investigating the linkages between LULC and stream water quality have concluded that a significant relationship exists between land use and water quality parameters at a catchment level (du Plessis et al. 2014; Kibena et al. 2014; Teixeira et al. 2014), while others have shown complex relationships, varying from one region in the world to another and which are dependent on the depth of analysis (Miserendino et al. 2011; Dabrowski and de Klerk 2013). Wan et al. (2014) emphasized the importance of temporal and spatial variations and distinguished that upstream river reaches are subject to fewer impacts, compared to those downstream. Geographic information system (GIS) has the capability for capture, storage, manipulation, analysis and retrieval of multiple layer resource information occurring both in spatial and non-spatial forms (Mishra and Patel 2001). du Plessis et al. (2014) found that LULC changes have various negative impacts on the water quality of a watercourse, as they lead to both increases and declines in the concentration of water quality variables. Fewer studies have considered the effects of spatial patterns of land use on the water quality, which could provide suggestion on landscape planning and land use management. Therefore, this paper outlines a study aimed to assess the long-term association between land cover/land use change, climate change and water quality changes that occurred in the Indian river basin. The water quality and quantity data available for the analysis in this study belong to the river gauge station located at the outlet of the catchment.

9.1.2 Climate Change

Climate change is one important factor that is known to affect ecosystems. The main impact of climate change on water quality is attributed to changing air temperature and hydrology (Alam et al. 2013). Water temperature is directly affected by ambient air temperature (Akomeah et al. 2015) and is expected to increase as a result of global warming. Variations in water temperature govern physico-chemical equilibriums (e.g., nitrification, mineralization of organic matter, etc.) in rivers and hence change transport and concentration of contaminants. Increases in water temperature result in reduced oxygen solubility, thus reducing dissolved oxygen (DO) concentrations at which saturation occurs. Reduced DO concentration of contaminants increases in water temperature and result in reduced oxygen solubility, thus reducing dissolved oxygen (DO) concentrations at which saturation occurs. Temperature alone can decrease the volume of surface waters through increase in ET. The south-eastern United States has little seasonal variability in precipitation, but significant variability in streamflow because of seasonal changes in rates of ET (Murdoch et al. 2000). Increased water temperature is anticipated to result in increased anoxia of already eutrophied waters of the south-eastern United States if global warming occurs. The duration and intensity of stratification in surface waters is a major factor in determining seasonal changes in surface water quality. The interaction of several factors includes the initial temperature of surface and ground water inputs to a lake, lake trophic state and the physical geometry (Whitehead et al. 2009).

9.1.3 Water Quality Index

Water quality index is an important way to assess the quality of water in recent years. WQI is a superior way of understanding water quality issues by integrating complex data and generating a score, which ultimately describes the water quality status (Mishra and Patel 2001). GIS is a computer system for capturing, storing, querying, analyzing and displaying all types of geographical data. It permits observing, understanding, enquiring, interpreting and visualizing data in several ways which reveal relationships, patterns and trends in the form of maps, globes, reports and charts. GIS is not only an effective tool for collection, storage, management and retrieval of an assembly of spatial and non-spatial data but also for spatial analysis and integration of the data to arise useful outputs and modeling (Gupta and Srivastava 2010).

A water quality index offers a single number that represents overall water quality of a particular water sample (time and location) for several water quality parameters (Singh et al. 2013). It is an important factor to judge environment changes, which are strongly associated with social and economic development. The index allows a general analysis of water quality on many levels that affect a stream's ability of hosting life. WQI is the rating which reflects the composite influence of different water quality parameters. The major objective of the study is to provide an overall picture of the river water quality using WQI.

9.2 Study Area

The Patalganga River is situated between the Western Ghats and the Arabian Sea. The effluent of the Khopoli power project is discharged into this river near Khopoli in Maharashtra, India. It comes at foremost in terms of pollution. It is known to be one of the most polluted rivers of Maharashtra. It is the significant source of intake water and raw water for the nearby industries and drinking water for neighboring villages. The sewage from the towns and villages along the river is directly disposed into the river without any treatment. Mainly textile, pharmaceuticals and dye intermediate manufacturing industries are located in the catchment of the Patalganga River. It is, therefore, of vital importance to monitor the water quality parameters of the Patalganga River to ascertain whether the water quality is still suitable for various purposes. So far no systematic study has been undertaken to assess the water quality of Patalganga River.

Nowadays, local villagers nearby Patalganga River complain regarding the poisoning caused by consuming fishes. Biodiversity of this river basin is facing serious threat due to the harmful chemicals from the fertilizer, dyeing, insecticides, pesticides and alkyl amines industries. The stretch of the study area of the river is shown in Fig. 9.1.

9.3 Methodology

The land use data was extracted from the Landsat TM images. The land types are classified as the cultivated (agricultural) land, forest land, grassland, water area, built-up area and unused (barren) land. Proportion or fraction of each land type is found out.

Organizing and presentation of the data sets: The data sets for water quality parameters like temperature, dissolved oxygen, pH, conductivity, BOD, COD, and faecal coliform etc. are some of the data obtained for sampling and experimentation.

Fig. 9.1 Stretch of Patalganga River understudy

The data sets were organized in spread sheets for further analysis and illustrative presentation. The standard deviation values were calculated and have been presented along with the data sets in the spread sheets.

9.3.1 Estimation of Water Quality Index (WQI)

Water quality parameters, i.e., pH, dissolved oxygen (DO), biochemical oxygen demand (BOD), turbidity, faecal coliform (FC), temperature, total carbon (TC), nitrate, total solids were selected for calculating the index. The WQI is calculated for the river stretch near Patalganga MIDC area using the formula provided by National Sanitation Foundation (NSF) and the relative weights modified by Central Pollution Control Board. To present the data in a spatial format, geographical information system (GIS) maps were generated. The expression for calculating the NSFWQI is expressed as Eq. (9.1). Table 9.1 is used as a reference for assessing the river water

Table 9.1 Relation between water quality index and surface water pollution

S. no.	Water quality status	WQI	Remark
1	Bad to very bad	38 and less	Heavily polluted
2	Bad	39–50	Polluted
3	Medium to good	51–63	Not polluted
4	Good to excellent	64–100	Not polluted

Table 9.2 Analysis of water quality data from the experimentation

S. no.	Particulars	pH	DO	BOD	Ammonia	TC	Nitrate	FC
1	Mean	7.5	5.8	24	0.56	485	0.75	144
2	Std. deviation	0.35	0.55	0.80	0.65	450.2	0.87	155.4
3	No. of samples	25	20	20	25	20	20	25

quality status.

$$\mathrm{WQI} = \sum\nolimits_{i=1}^{P} Wi * Ii \tag{9.1}$$

where I_i = sub-index for ith water quality parameter; W_i = weight (in terms of importance) associated with water quality parameter; P = number of water quality parameters.

9.3.2 Analysis of WQI and Data

The WQI evaluation of the available data of Patalganga River indicated that the water quality was good to excellent at all the stations except at Turade where it varied from moderately good to good and then to excellent. The reason for such variation could be release of some domestic waste from Turade village. Mean and standard deviation are shown for different water quality parameters in Table 9.2.

The descriptive statistics of Patalganga River showed that the pH value and dissolved oxygen complied with the MPCB standards at some of the stations near MIDC waterworks intakes. The BOD value exceeded the standards in the range of 7–20% at most of the stations. Percent exceedance for ammonical nitrogen was 7, 7 and 31% at Shilphata Bridge, near intake of MIDC waterworks and at D/s of Kharpada Bridge, respectively. The reason for exceedance could be the effluent discharged from MIDC, HOC Ltd, HIL, domestic waste and fishing.

9.3.3 Relationship Between Land Use, Climate Change and Water Quality

There was a positive relationship between the cultivated land area (%) and the concentration of ammonia and DO. This is mainly due to the developed agriculture in area and emission of ammonia from the exposure of soil surface resulting from the agricultural practices and the application of chemical fertilizers. The fertilizers used in the agricultural land will be dissolved into the runoff of river ultimately polluting the river water. Also, the vegetation in the surface soil of agricultural land absorbs and retain the various pollutants. As a result, the cultivated land plays a complicated role in influencing the water quality. The forest land and grassland both have significant positive influence on the water quality. The forest land and grassland played a key role in reducing the nitrogen pollutants and phosphorus pollutants and played a controlling role in re-grading the water quality. The forest vegetation and forest grass land effectively reduce the nutrient salts brought to the river by the surface runoff because they play a significant role in reducing the surface runoff and its concentration helping in the conservation of water and soil. Therefore, the increase of the forest land and grassland area will reduce the concentration of ammonia and oxygen-consuming substances, increase the concentration of dissolved oxygen and consequently improve the water quality. The built-up land has negatively influenced the water quality as a whole. The built-up area was positively related to TDS, nitrate, ammonia, and BOD and was negatively related to DO, indicating that the increase of the built-up area tends to degrade the water quality. It is due to the increase of the nutrient concentration. The population density and economic events both are focused in the built-up area which causes serious pollution. Furthermore, increase in paved road will contribute to increase of surface runoff and may increase the concentration of nutrient. More runoff will reach the river and consequently de-grading in the water quality happens within the basin. There was a positive relationship between the built-up area and total dissolved solids and BOD. Since the vegetation can protect the soil from raindrops and tend to slow down the movement of runoff and allow the excessive surface water to infiltrate into the soil, the conversion of land with vegetation into built-up area will aggravate the soil erosion and consequently increase the amount of TDS into the runoff.

9.4 Discussion

The analysis revealed that with the exception of pH (no trend) and sulfate (negative trend), all other water quality variables including total dissolved solids and hardness show significant increased value (positive trends). The study indicated that there was a significant negative correlation between forest land and grassland and the water pollution, and the built-up area had negative impacts on the water quality, while the influence of the agricultural land on the water quality was very complex.

According to the result mentioned previously and the current conditions of the local water quality, it is necessary to increase the area of forest land, grassland and water area in the local land use planning. Since the forest land proportion influences the local river water quality, it is particularly important to increase the area of forest land. Landscape diversity should be increased.

9.5 Conclusion

1. A decline of the natural vegetation occurs, relative to large increases of urban/built-up, cultivation and other land uses. These changes in LULC have direct negative or positive impacts on runoff and water quality of surface water, due to alteration of hydrological system.
2. The pledge between increasing trends of developed areas (infrastructural development such as residential, roads etc.) and increasing trends of most of the water quality variables reveals that land use change influences the river water quality.
3. Increases in water temperature decreases the oxygen-holding capacity of surface waters, which can decrease productivity in surface waters already stressed by biological oxygen demand (BOD).
4. Changes in water quality anticipated from increased global temperature, global increases in air temperature and the associated increases in water temperature can cause measurable changes in water quality that are independent of changes in moisture.
5. Significant changes in water quality have occurred as a direct result of short-term changes in climate.

References

Akomeah E, Chun KP, Lindenschmidt KE (2015) Dynamic water quality modelling and uncertainty analysis of phytoplankton and nutrient cycles for the upper South Saskatchewan River. Environ Sci Pollut 22:18239–18251

Alam A, Badruzzaman ABM, Ali MA (2013) Assessing effect of climate change on the water quality of the Sitalakhya river using WASP model. J Civil Eng 41:21–30

Boori MS, Vozenilek V (2014) Land cover disturbance due to tourism in Jeseniky mountain region: a remote sensing and GIS based approach. In: Earth resources and environmental remote sensing/GIS applications, proceedings of SPIE, international society for optics and photonics, Amsterdam, The Netherlands, vol 9245

Dabrowski JM, de Klerk LP (2013) An assessment of the impact of different land use activities on water quality in the upper Olifants River catchment. Water SA 39(2):231–244

Ahearn DS, Sheibley RW, Dahlgren RA, Anderson M, Johnson J, Tate KW (2005) Land use and land cover influence on water quality in the last free-flowing river draining the western Sierra Nevada, California. J Hydrol 313(3–4):234–247

Du Plessis A, Harmse T, Ahmed F (2014) Quantifying and predicting the water quality associated with land cover change: a case study of the Blesbok Spruit Catchment, South Africa. Water 6(10):2946–2968

Gazzaz NM, Yusoff MK, Ramli MF, Aris AZ, Juahir H (2012) Characterization of spatial patterns in river water quality using chemometric pattern recognition techniques. Mar Pollut Bull 64(4):688–698

Gupta M, Srivastava PK (2010) Integrating GIS and remote sensing for identification of groundwater potential zones in the hilly terrain of Pavagarh, Gujarat, India. Water Int 35:233–245

Huang J, Zhan J, Yan H, Wu F, Deng X (2013) Evaluation of the impacts of land use on water quality: a case study in the Chaohu Lake Basin. Hindawi Publ Corp Sci World J Article ID 329187

Kibena J, Nhapi I, Gumindoga W (2014) Assessing the relationship between water quality parameters and changes in land use patterns in the Upper Manyame River, Zimbabwe. Phys Chem Earth Parts A/B/C 67–69 (0):153–163

Miserendino ML, Casaux R, Archangelsky M, Di Prinzio CY, Brand C, Kutschker AM (2011) Assessing land-use effects on water quality, in-stream habitat, riparian ecosystems and biodiversity in Patagonian northwest streams. Sci Total Environ 409(3):612–624

Mishra PC, Patel RK (2001) Study of the pollution load in the drinking water of Rairangpur, a small tribal dominated town of North Orissa. Indian J Environ Ecoplanning 5(2):293–298

Murdoch PS, Baron JS, Miller TL (2000) Potential effects of climate change on surface-water quality in North America. J Am Water Resour Assoc 36(2)

Pullanikkatil D, Palamuleni LG, Ruhiiga TM (2015) Impact of land use on water quality in the Likangala catchment, southern Malawi. Taylor and Francis. Afr J Aquat Sci 40(3):277–286

Singh PK, Tiwari AK, Mahato M K (2013) Qualitative assessment of surface water of West Bokaro Coalfield, Jharkhand by using water quality index method. Int J Chem Tech Res 5(5)

Teixeira Z, Teixeira H, Marques JC (2014) Systematic processes of land use/land cover change to identify relevant driving forces: implications on water quality. Sci Total Environ 470–471:1320–1335

Wan R, Cai S, Li H, Yang G, Li Z, Nie X (2014) Inferring land use and land cover impact on stream water quality using a Bayesian hierarchical modeling approach in the Xitiaoxi River Watershed, China. J Environ Manag 133:1–11

Whitehead PG, Wilby RL, Battarbee RW, Kernan M, Wade AJ (2009) A review of the potential impacts of climate change on surface water quality. Hydrol Sci J 54:101–123

Yong TY, Chen W (2002) Modeling the relationship between land use and surface water quality. J Environ Manag 66:377–393

Chapter 10
Assessment of Tail Behavior of Probability Distributions of Daily Precipitation Data Over India

Neha Gupta and Sagar Rohidas Chavan

Abstract Reliable estimation of extreme precipitation is of utmost importance to ensure the structural safety of major civil engineering infrastructures (e.g., urban storm drainage network, spillways of major water control structures, etc.). Conventional practice for design of those infrastructures includes determination of magnitude and frequency of extreme precipitation events. The extreme events usually lie in the tail part of the probability distribution of daily precipitation data. Based on the nature of the tails, different distributions are classified into two categories: heavy-tailed and light-tailed distributions. The tails of heavy-tailed distributions tend to approach zero less rapidly when compared to those of light-tailed distribution (which are having exponential tails). Heavier tails imply more frequency of extreme precipitation events as compared to lighter tails. There is a dearth of attempts to assess the tail behavior of probability distributions of daily precipitation data over India. In this paper, we compare the tails of empirical distributions of daily precipitation data and tails of fitted theoretical distributions (e.g., pareto, lognormal, Weibull and gamma distributions). Gridded precipitation data prepared by the India Meteorological Department (IMD) having a resolution of 0.25° was used for the analysis. The results indicated that heavy-tailed distributions describe the observed precipitation extremes more effectively than the light-tailed distributions. This result shows that the light-tailed distributions, which are widely adopted for the determination of extreme precipitation events, are inadequate to capture the tail behavior of precipitation data accurately. This study reveals the importance of considering heavy-tailed distributions for reliable estimation of extreme precipitation events required for the design of major civil engineering infrastructure. The results from the study can find use in the design of urban storm drainage network, spillways of major water control structures.

Keywords Heavy-tailed distribution · Light-tailed distribution · Pareto · Lognormal · Weibull · Gamma distribution · Mean square error · Genetic algorithm

---

N. Gupta · S. R. Chavan (✉)
Department of Civil Engineering, Indian Institute of Technology, Ropar 140001, India
e-mail: ng0907@gmail.com; sagar@iitrpr.ac.in

R. Jha et al. (eds.), *Climate Change Impacts on Water Resources*, Water Science and Technology Library 98, https://doi.org/10.1007/978-3-030-64202-0_10

10.1 Introduction

Catastrophic rainfall events can lead to flooding which results in harm to human life, damage to buildings and infrastructure, loss of crops and livestock. Such rainfall events having less frequency of occurrence are characterized by adjectives like "abnormal", "rare" or "extreme". Generally, an event is pronounced as "extreme" event based on the necessity of infrastructure design and human losses. The infrastructure design like urban storm drainage network and spillways of major water control structures are based on long-term rainfall prediction as against to the short-term prediction for their efficient operation. It is infeasible to determine the long-term rainfall prediction accurately. Hence the rainfall is treated as a random variable (RV) which has some associated probability and follows some distribution laws. Those laws enable us to find the magnitudes of rainfall event that could be viewed as extreme based on the return period, e.g., rainfall event with return period 500 or 1000 years or more is indeed an extreme (Koutsoyiannis 2004a; b). Often, the question which is yet debatable is, which and how the distribution law should be chosen?

The selection of distribution law for rainfall could be made either by trying one out of the many distributions from parametric families of distributions or by parameters estimation according to one of many existing fitting methods (e.g., rank regression, maximum likelihood estimation, Bayesian estimation methods, etc.). Since the selection of the distribution requires curve fitting, we can also choose best-fitted one according to some metric or fitting test. The estimated fitting parameters always point toward the distribution that describes the most significant portion of the data, by definition which does not belong to the tail part. The tail of the distribution, in general, refers to the upper part of the distribution that controls both the magnitude and frequency of extreme events (Papalexiou and Koutsoyiannis 2013). Unless a very large sample of data is available, a very small portion of the empirical dataset belongs to the tail. Hence these procedures do not guarantee the modeling of the tail accurately, as all fitting methods are "biased" against the tail. An ill-fitted tail results in severe errors for modeling the extreme event having severe consequences in hydrological design. The distributions based on the asymptotic behavior of their tail can be classified into two general classes: (a) the subexponential class with tails decreasing more slowly than any exponential tail (it is a special case of heavy-tailed distribution) (Klüppelberg 1988, Goldie and Klüppelberg 1998), and (b) the hyper-exponential or the superexponential class, with tails approaching zero more rapidly than an exponential tail (Teugels 1975).

Mathematically, let F be a distribution function on $(0, \infty)$ for which $F(0+) = 0$, $F(x) < 1$ for all $x > 0$, $F(\infty) = 1$. The distribution F is said to belong to subexponential class if:

$$\lim_{x \to \infty} \frac{1 - F^{(2)}(x)}{1 - F(x)} = 2 \tag{10.1}$$

Furthermore, different terms like "heavy tails", "fat tails", "thick tails" or "long tails" have been used in the literature for tails "heavier" than the exponential. El Adouni et al. (2008) deduced the classification for subexponential distribution, class I contains distributions for which the tail is a power function of the return period. Class II contains the lognormal distribution. The tail of the distributions in class III is a power function of the logarithm of the return period. The last class (class IV) contains distributions with an upper bounded support. Terms like "light tail" and "thin tail" are used in the literature for superexponential distributions having thinner tails than an exponential distribution. A heavy tail implies that large values can occur in a sample with nonnegligible probabilities making them more suitable for modeling extreme events as compared to light tails. Hence, the usual approach of adopting light-tailed models (e.g., the gamma distribution) and fitting them on the whole sample of empirical data would result in a significant underestimation of risk with potential implications for human lives. Availability of large datasets of rainfall has allowed us to check which distribution fits data better and also investigate the appropriateness of light or heavy tails for modeling extreme events. "India is highly vulnerable to floods, out of the total geographical area of 329 million hectares (Mha), more than (40 Mha) is flood prone" according to the National Disaster Management Authority, a government body. Witnessing such situations, it becomes more important to have a reliable estimation of extreme events over India. This paper contributes toward a better understanding of extreme events by assessing the tail behavior of probability distributions of daily precipitation data over India.

10.2 The Dataset

The data used in the study are $0.25° \times 0.25°$ resolution daily rainfall data product procured from India Meteorological Department (IMD) which was prepared based on extensive coverage of 6955 actual observation sites. The data is archived at the National Data Centre, IMD, Pune. The gridded data was prepared using interpolation based on Shepard's method (Pai et al. 2014). There are 4949 grids/stations with a uniform record length of 109 years (1901–2010) with no suspicious value even for a single day over entire India. The locations of all stations can be seen in Fig. 10.1.

10.3 Methodology

10.3.1 Defining the Tail

The marginal distribution of daily rainfall has a discrete part describing the probability of zero rainfall (no rain, dry period), or the probability dry, and a continuous part expressing the magnitude of the nonzero (wet/day) rainfall. We focus on the

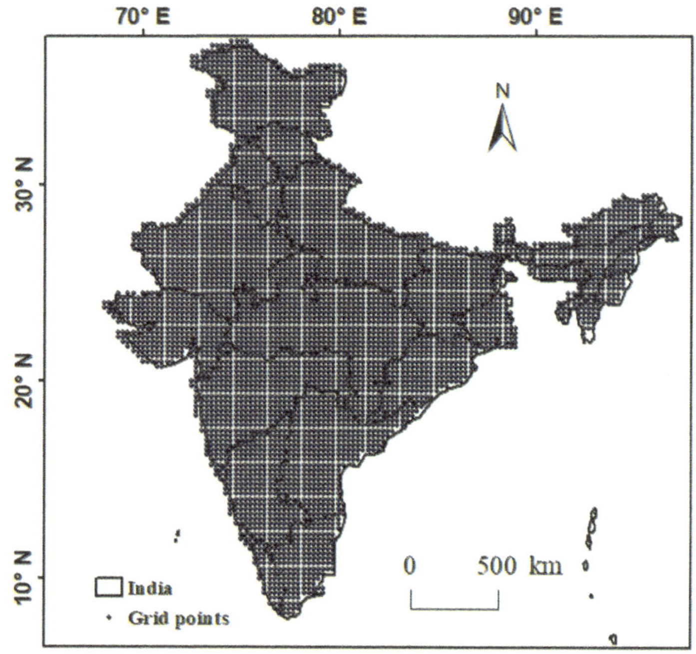

Fig. 10.1 Locations of the grids/stations over India (a total of 4949 daily rainfall records with time series length of 109 years)

behavior of the distribution's right tail for studying extreme rainfall as it governs the frequency and the magnitude of extremes.

If we denote the rainfall with X, and the nonzero rainfall with X|X > 0, then the exceedance probability function (EPF; also known as survival function, complementary distribution function, or tail function) of the nonzero rainfall is defined as:

$$P\ (X > x|X > 0)\ = \overline{F}_{X|X>0}\ (x) = 1 - F_{X|X>0}(x) \tag{10.2}$$

where $F_{X|X>0}(x)$ is any valid probability distribution function chosen to describe nonzero rainfall.

We focus only on the upper part of the EPF, as it contains "right tail" which describes the extremes. Generally, a lower threshold value x_L is set (Cunnane 1973; Ben-Zvi 2009) and the values above it are considered as the "upper part". No universal method is yet accepted to choose this threshold. In this study we use the standard method of studying extremes in hydrology (known as partial duration series method), determining the threshold indirectly based on the empirical distribution, in such a way that the number of values above the threshold equals the number of years of record N (Cunnane 1973). The resulting series, defined in this way, is known as annual exceedance series (Gupta 2011).

We prefer the N largest daily values of the record over the maximum value of each year. The method of annual maxima may distort the tail behavior by taking into account only one of the n-largest daily values that occur within a single year. Thus, the approach adopted by us in the study has the advantage of better representing the exact tail of the parent distribution (Papalexiou and Koutsoyiannis 2013).

Given that each station has an N-year record of daily values and a total number n of nonzero values, the probability of exceedance EPF $\overline{F}_N(x_i)$ (according to the Weibull plotting position) is defined as:

$$\overline{F}_N(x_i) = 1 - \frac{r(x_i)}{n+1} \tag{10.3}$$

where $r(x_i)$ is the rank of the rainfall data x_i in the ordered sample as $x(1) \leq \ldots \leq x(\text{n})$ of the nonzero values. Thus, the empirical tail is determined by the N largest nonzero rainfall values of $\overline{F}_N(x_i)$ with $n - N + 1 \leq i \leq n$ (note that $x_L = x_{(n-N+1)}$). The number of nonzero daily rainfall values is $n = (1 - p_0)n_d N$, where $n_d = 365.25$ is the average number of days in a year. According to the Weibull plotting position given in Eq. (10.3), the exceedance probability $\overline{p}(x_L)$ x_L will be

$$\overline{p}(x_L) = 1 - \frac{n-N+1}{n+1} = \frac{N}{(1-p_0)n_d N + 1} \approx \frac{1}{(I - p_0)n_d} \tag{10.4}$$

This equation shows that the dependency of the exceedance probability of the threshold x_L on the probability dry p_0. The average dry probability p_0 of the record comes out to be approximately 0.6961, which implies that the exceedance probability of is on average as low as 0.0090. This value set a threshold and the values above this can be assumed to belong to the tail of the distribution. Beguería et al. 2009 chose the threshold value corresponding to the 90th percentile, a value much smaller than our choice of threshold.

10.3.2 Fitting Method

The fitting method followed in the study directly fits and compares the tails of empirical distributions and fitted theoretical distributions estimated from the daily rainfall records already described. It provides the best possible description of the tail and is not affected by lower values. The theoretical tails are fitted to the empirical ones by minimizing a modified mean square error (MSE) norm N1 using the genetic algorithm (GA).

$$N1 = \frac{1}{N} \sum_{i\,=\,n-N+1}^{n} \left(\frac{\overline{F}(x_{(i)})}{\overline{F_N}(x_{(i)})} - 1 \right)^2 \tag{10.5}$$

As shown in Eq. (10.5), the modified MSE considers the relative error between the values of the theoretical and empirical distribution rather than using the daily precipitation values themselves. An advantage of not considering daily precipitation values is that the error measure MSE would not get affected by the large deviations in daily precipitation values.

Goldberg (1989) and Michalewicz (1992) developed a heuristic, stochastic, combinatorial, optimization technique based on the biological process of natural evolution. Three heuristic processes of reproduction, crossover and mutation are applied probabilistically to discrete decision variables that are coded into binary or real numbers strings. In this study, GA has been used effectively by minimizing the norm given in Eq. (10.5) in two ways: (a) by fitting all the points of the empirical distribution, and (b) in only the largest N points. It is also used to estimate the parameters of candidate distributions (Papalexiou and Koutsoyiannis 2013). Figure 10.2 depicts (a) Weibull, (b) gamma, (c) lognormal and (d) pareto distribution of a station, randomly selected (stations code given in the figure). It is clear that the first approach where distribution is fitted to the whole empirical distribution points (black dashed

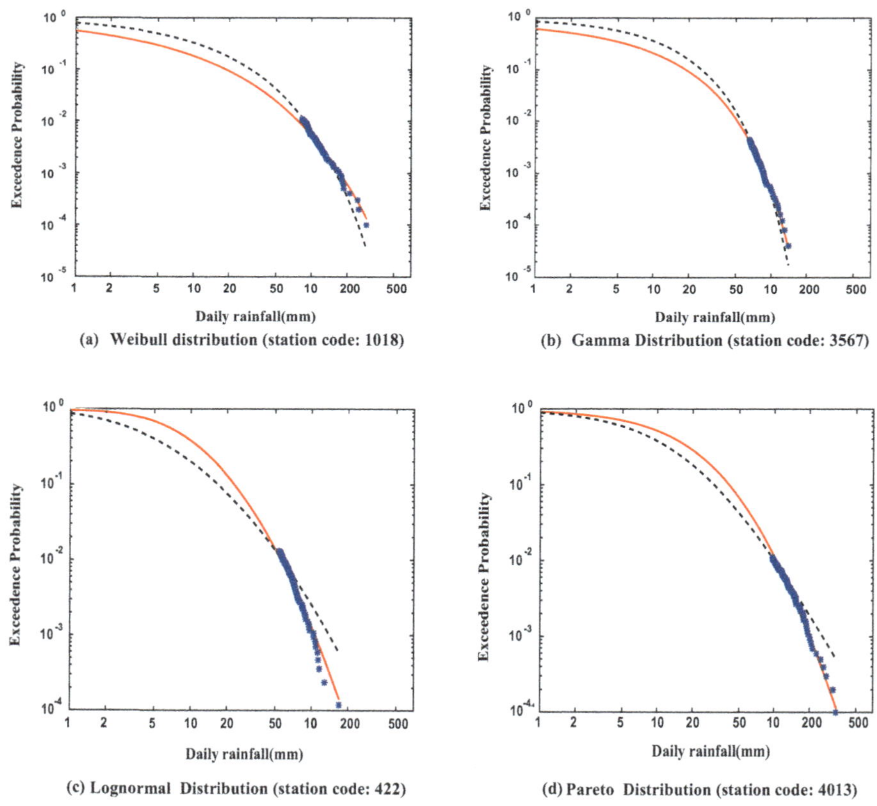

Fig. 10.2 Explanatory diagram of the fitting approach followed

line) does not adequately describe the tail, while solid red line depicts the distribution fitted only to the tail points shows the tail behavior efficiently. Least MSE value of the proposed method indicated the best fit. However, other fitting methods like lognormal maximum likelihood and the log-probability plot regression (Kroll and Stedinger 1996), and the partial L-moments (Wang 1996) require estimation of an additional measure to compare the performance of the fitted distributions.

10.3.3 The Fitted Distribution Tails

In this study, the tails of four commonly used distribution in hydrology, i.e., the tails of the Weibull (W), lognormal (LN), pareto type II (PII) and the gamma (G) distributions are fitted and compared for the performance. The reason behind the selection of the above probability distributions is their simplicity, superiority and popularity in the literature for frequency analysis of extreme events (tail-equivalent with complicated distributions). The pareto and the lognormal distributions belong to the subexponential class and are considered heavy-tailed distributions; Weibull distribution is a versatile distribution. Based on the value of shape parameter (α) it can take on the characteristics of other types of distributions.

The Weibull distribution, also known as extreme value type III distribution (a generalization of exponential distribution), is a commonly used two-parameter distribution in the field of hydrology (Heo et al. 2001a, b). It is defined by scale parameter ($\beta > 0$) and shape parameter ($\alpha > 0$). The PDF and EPF are given, respectively, as

$$f_W(x) = \frac{\alpha}{\beta}\left(\frac{x}{\beta}\right)^{\alpha-1} exp\left(-\left(\frac{x}{\beta}\right)^{\alpha}\right) \tag{10.6}$$

$$\overline{F}_W(x) = exp\left(-\left(\frac{x}{\beta}\right)^{\alpha}\right) \tag{10.7}$$

This distribution is categorized into a subexponential family with a tail heavier than the exponential one for ($\alpha < 1$), while for ($\alpha > 1$) it belongs to hyperexponential family with a tail thinner than the exponential.

Lognormal distribution, as may be summarized by the name, has certain similarities to the normal distribution. A random variable is lognormally distributed if the logarithm of the random variable is normally distributed. The PDF and EPF for lognormal distribution specifically with scale ($\beta > 0$) and shape parameter ($\alpha > 0$) are given by,

$$f_{LN}(x) = \frac{1}{\sqrt{\pi}\alpha x}\exp\left(-\ln^2\left(\frac{x}{\beta}\right)^{\frac{1}{\alpha}}\right) \tag{10.8}$$

$$\overline{F}_{LN}(x) = \frac{1}{2} erfc\left(\ln\left(\frac{x}{\beta}\right)^{\frac{1}{\alpha}}\right) \quad (10.9)$$

where the complementary error function is denoted as $erfc(x) = 2\pi^{-1/2} \int_0^\infty e^{-t^2} dt$.

This distribution can approximate power law distributions for a large portion of distribution body (Mitzenmacher 2004).

Pareto distribution is a probability distribution of power law, which is used in the description of many types of observations. The hierarchy of pareto distribution is as follows: Pareto types I, II, III, IV and Feller-Pareto distributions. Pareto type II distribution is the simplest power-type distribution defined by the scale parameter $(\beta) > 0$ and the shape parameter $(\alpha) \geq 0$ in the range $[0,\infty)$ (Zhanling et al. 2014) with PDF and EPF given by

$$f_{PII}(x) = \frac{1}{\beta}\left(1 + \alpha\frac{x}{\beta}\right)^{-\frac{1}{\alpha}-1} \quad (10.10)$$

$$F_{pii}(x) = \left(1 + \alpha\frac{x}{\beta}\right)^{-\frac{1}{\alpha}} \quad (10.11)$$

For $\alpha = 0$, pareto tail degenerates to the exponential tail, while it becomes heavier as the value for shape parameter increases. Burr type XII (Burr 1942), the two- and three-parameter Kappa (Mielke 1973), the log-logistic (e.g., Ahmad et al. 1988) are some of the other power-type distributions that have asymptotic behavior similar to pareto type II tail.

The gamma distribution is a commonly used two-parameter distribution, distinctly skewed to the right. It suits the distribution of daily rainfall and accommodates the lower limit of zero, which constrains rainfall values (Stacy 1962). Its PDF and EPF are given, respectively, by

$$f_G(x) = \frac{1}{\beta\Gamma(\alpha)}\left(\frac{x}{\beta}\right)^{\alpha-1} \exp\left(-\frac{x}{\beta}\right)$$

$$\overline{F}_G(x) = \Gamma\left(\alpha, \frac{x}{\beta}\right)/\Gamma(\alpha)$$

where $\Gamma(\alpha)$ is a standard mathematical function called the gamma function and is defined by:

$$\Gamma(\alpha) = \int_0^\infty t^{\alpha-1} e^{-t} dt$$

The shape parameter ($\alpha > 0$) essentially determines the level of positive skew. For $0 < \alpha < 1$, the gamma distribution has a "slightly lighter" tail than the exponential tail as it decreases faster. For $\alpha = 1$, the gamma tail behaves similar to the exponential tail, while for $\alpha > 1$ it exhibits a "slightly heavier" tail as it decreases more slowly than the exponential tail (Embrechts and Goldie 1982). The scale parameter (β) determines the spread of values, stretching or squeezing the distribution when large or small, respectively.

The four distributions which we compared in this study have two parameters, namely, one-scale parameter (β) and one-shape parameter (α). Despite of same parameter structure, these distributions differ in terms of flexibility. A quantitative measure for flexibility includes a comparison of various shape measures (e.g., skewness, kurtosis, etc.). With a feasible range of $(-1.14, \infty)$ Weibull distribution seems to be the most "flexible", while pareto with skewness range of $(2, \infty)$ is the least flexible. However, this argument is not valid when we focus on the tail because the general shape of the tail is basically similar and what differs is the rate at which the tail approaches zero.

10.4 Results and Discussion

The basic statistical results obtained by fitting the four distribution tails to the 4949 daily rainfall records after following the methodology described in Sect. 10.3 are given in Table 10.1. The MSE is calculated by forming a fitness function (known as modified mean square error) and then optimizing the parameter associated with distribution using genetic algorithm. The MSEs of these distributions are then compared to reveal the overall performance of the fitted tails. The best fitted tail is found in each record based on their average rank. The tails with smaller MSE is ranked as 1, while the one with the largest is ranked as 4. Figure 10.3 depicts the best-fitted tails based on various distributions for 4949 grids. The figure indicates that there is no regular pattern regarding the best-fitted distribution.

Out of 4949 daily rainfall record, the lognormal fits on a maximum number of stations, followed by pareto, Weibull, and gamma. The percentages of each distribution tail that was best fitted are 44.27% for lognormal, 33.21% for pareto, 18.51% for Weibull and 4.01% for gamma. The lognormal distribution comes out to be the best according to these percentages, and gamma distribution, the most popular model for rainfall, performed the worst.

Figure 10.4 depicts the histogram (empirical distributions) of the shape parameters of the different fitted tails. It is well known that the most probable values are the ones around the mode, which for the pareto shape parameter is 0.18. It implies the nonexistence of moments higher than the fifth order (Papalexious et al. 2013). Regarding the Weibull tail, the estimated mode of its shape parameter is 0.740, implying a much heavier tail compared to the exponential one. The shape parameter of gamma distribution mainly controls the behavior of the left tail; the right tail mainly behaves like the exponential tail. A low mode value of 0.076 results in a

Table 10.1 Summary statistics from the fitting of the four distribution tails into the 4949 tail-samples of daily rainfall (expressed in mm)

Pareto				Lognormal		
	MSE	β	α	MSE	β	α
Min	0.0022	0.8120	0.0150	0.0025	0.0346	0.5774
Mean	0.0879	8.7492	0.2424	0.0490	10.1954	1.2534
Max	0.7299	26.0950	0.6195	0.5769	23.2873	4.0090
Median	0.0355	8.2975	0.2292	0.0267	9.8432	1.2199
SD	0.1134	4.1463	0.1076	0.0619	4.6165	0.2544
Skew	2.0359	0.5325	0.5284	3.1640	0.2774	0.8938
Weibull				Gamma		
	MSE	β	α	MSE	β	α
Min	0.0029	0.5187	0.3100	0.0024	5.3317	0.0766
Mean	0.1047	12.2512	0.8641	0.1527	17.9654	1.0916
Max	0.9911	28.5969	15.7427	0.6253	31.7335	8.9049
Median	0.0794	12.7321	0.8295	0.1413	18.2577	0.9144
SD	0.0991	4.9375	0.5183	0.0875	4.1906	0.7343
Skew	2.8237	0.2171	18.8477	0.7728	0.2060	2.4030

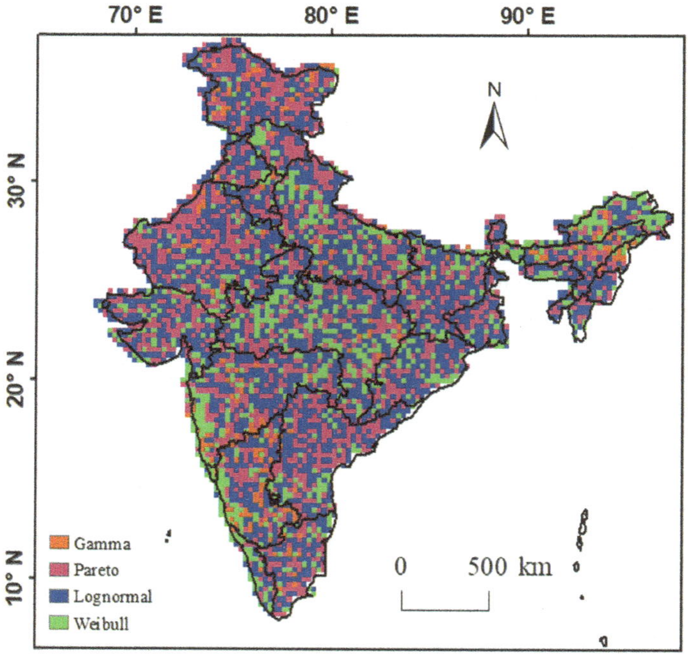

Fig. 10.3 The best-fitted tail showed at every station. Out of 4949 stations the best fit for (**a**) lognormal is 2191, (**b**) pareto is 1644, (**c**) Weibull is 916, and (**d**) gamma is 198

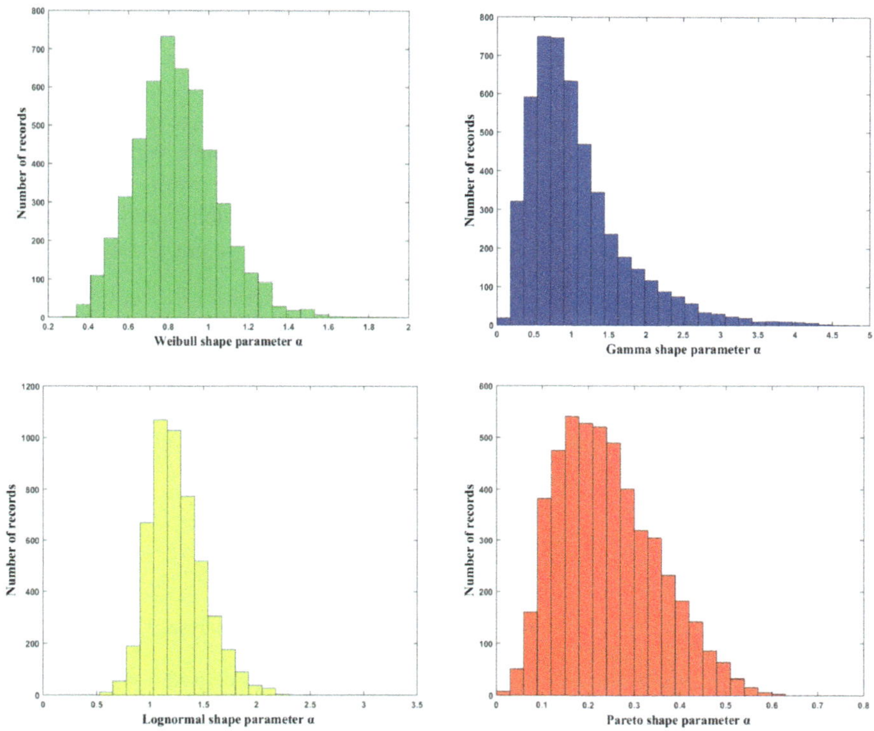

Fig. 10.4 Histograms of the shape parameters of the fitted tails

J- or bell-shaped densities which is inconsistent in describing the whole body of distribution of daily rainfall. Since most of the extreme events are associated with the right tail of the distribution, gamma distribution will probably underestimate the behavior of extremes.

10.5 Conclusions

In this paper, we compared the tails of empirical distributions of daily precipitation data and tails of fitted theoretical distributions (e.g., pareto, lognormal, Weibull and gamma distributions). Gridded precipitation data prepared by the India Meteorological Department (IMD) having a resolution of 0.25° was used for the analysis. In general, the results indicated that heavy-tailed distributions describe the observed precipitation extremes more effectively than the light-tailed distributions. In addition, the following observations are inferred from the analysis.

(i) The ranking of distributions from best to worst in terms of their performance to adequately capture the response of extreme events is (a) the lognormal, (b) pareto, (c) the Weibull and (d) the gamma distributions.
(ii) Heavy-tailed distributions adequately capture the tail behavior of precipitation data than their counterpart light-tailed distributions. Hence, we recommend using heavy-tailed distributions to model extreme rainfall events worldwide.
(iii) Gamma distribution, in general, underestimates the frequency and the magnitude of extreme events. Hence, we should reconsider using it for modeling extreme events.

This study reveals the importance of considering heavy-tailed distributions for reliable estimation of extreme precipitation events required for the design of major civil engineering infrastructure. The results from the study can find use in the design of urban storm drainage network, spillways of major water control structures, etc.

References

Ahmad MI, Sinclair CD, Werritty A (1988) Log-logistic flood frequency analysis. J Hydrol 98:205–224. https://doi.org/10.1016/0022-1694(88)90015-7
Ben-Zvi A (2009) Rainfall intensity–duration–frequency relationships derived from large partial duration series. J Hydrol 367:104–114. https://doi.org/10.1016/j.jhydrol.2009.01.007
Beguería S, Vicente-Serrano SM, López-Moreno JI, García-Ruiz JM (2009) Annual and seasonal mapping of peak intensity, magnitude and duration of extreme precipitation events across a climatic gradient, northeast Spain. Int J Climatol 29:1759–1779
Burr IW (1942) Cumulative frequency functions. Ann Math Stat 13:215–232
Cunnane C (1973) A particular comparison of annual maxima and partial duration series methods of flood frequency prediction. J Hydrol 18:257–271. https://doi.org/10.1016/0022-1694(73)90051-6
El Adlouni S, Bobée B, Ouarda TBMJ (2008) On the tails of extreme event Distributions in hydrology. J Hydrol 355:16–33. https://doi.org/10.1016/j.jhydrol
Embrechts P, Goldie CM (1982) On convolution tails. Stoch Proc Appl 13:263–278. https://doi.org/10.1016/0304-4149(82)90013-8
Gupta SK (2011) Modern hydrology and sustainable water development. John Wiley & Sons
Goldberg DE (1989) Genetic algorithms in search, optimization, and machine learning. Addison-Wesley, Reading, MA
Goldie CM, Klüppelberg C (1998) Subexponential distributions. In: Adler R, Feldman R, Taggu MS (eds) A practical guide to heavy tails: statistical techniques and applications. Birkhäuser Boston, pp 435–459
Heo JH, Boes DC, Salas JD (2001a) Regional flood frequency analysis based on a Weibull model: part 1. Estimation and asymptotic variances. J Hydrol 242:157–170
Heo JH, Salas JD, Boes DC (2001b) Regional flood frequency analysis based on a Weibull model: Part 2. Simulations and applications. J Hydrol 242:171–182
Klüppelberg C (1988) Subexponential Distributions and Integrated Tails. J Appl Probab 25:132–141. https://doi.org/10.2307/3214240
Koutsoyiannis D (2004a) Statistics of extremes and estimation of extreme rainfall, 1, Theoretical investigation. Hydrolog Sci J 49:575–590
Koutsoyiannis D (2004b) Statistics of extremes and estimation of extreme rainfall, 2, Empirical investigation of long rainfall records. Hydrolog Sci J 49:591–610

Kroll CN, Stedinger JR (1996) Estimation of moments and quantiles using censored data. Water Resour Res 32:1005–1012
Mielke PW Jr (1973) Another family of distributions for describing m and analyzing precipitation data. J Appl Meteorol 12:275–280
Michalewicz Z (1992) Genetic algorithms + data structures = evolution programs, 3rd edn. Springer-Verlag
Mitzenmacher M (2004) A brief history of generative models for power law and log-normal distribution. Internet Math 1(2):226–51. https://doi.org/10.1080/15427951.2004.10129088
National disaster Management Guidelines (2008) Managements of floods
Pai D, Sridhar L, Rajeevan M, Sreejith OP, Satbhai NS, Mukhopadhyay B (2014) Development of a new high spatial resolution (0.25° × 0.25°) long period (1901–2010) daily gridded rainfall data set over India and its comparison with existing data sets over the region. Mausam 65:1–18.1
Papalexiou SM, Koutsoyiannis D, Makropoulos C (2013) How extreme is extreme? An assessment of daily rainfall distribution tails. Hydrol Earth Syst Sci 17:851–862
Papalexiou SM, Koutsoyiannis D (2013) Battle of extreme value distributions: a global survey on extreme daily rainfall. Water Resour Res. https://doi.org/10.1029/2012wr012557
Stacy EW (1962) A generalization of the gamma distribution. Ann Math Stat 33:1187–1192
Teugels J (1975) Class of subexponential distributions. Ann Probab 3:1000–1011. https://doi.org/10.1214/aop/1176996225
Wang QJ (1996) Using partial probability weighted moments to fit the extreme value distributions to censored samples. Water Resour. Res 32:1767–1771
Zhanling Li, Zhanjie Li, Zhao W, Wang Y (2014) Probability modeling of precipitation extremes over two river basins in northwest of china. Advances in Meteorology, vol 2015. https://doi.org/10.1155/2015/374127

Chapter 11
Benefit of Time-Varying Models Developed Using Graphical Modeling Approach for Probabilistic Prediction of Monthly Streamflow

Riya Dutta and Rajib Maity

Abstract Hydroclimatic systems consist of various interacting processes/components. The variables influencing a process and the interaction among the variables is dynamic in nature. Streamflow is one such component of the hydrologic cycle influenced by a large pool of several influencing hydroclimatic variables in different complex ways. There is a plethora of models for streamflow prediction; however, temporal variation in cause–effect relationship may lead to a decaying performance of the developed model. Existing prediction models are mostly stationary in nature or at most update the model parameters retaining the same predictors. This study demonstrates a recently developed concept of time-varying models using graphical modeling (GM) approach to capture the temporal variation in streamflow modeling considering the Upper Mahanadi river basin. This approach provides a detailed conditional independence structure and quantifies the association of the predictors with the predictand that can be utilized for the development of prediction models. The time-varying, GM-based approach shows the need to update the set of input variables for streamflow. The performance of the time-varying model is contrasted with a time-invariant model and support vector regression (SVR)-based models, well-established in the field of hydroclimatology. For the considered study area, the proposed model is found to capture the anomaly in streamflow variation and provides satisfactory prediction performance, at a lead time of one month, for normal as well as extreme flow events. Updating the set of potential predictors and the corresponding model parameters help to improve the predictability of monthly streamflow.

Keywords Graphical model · Conditional independence structure · Vine-copula · Probabilistic prediction · Time-varying model · Streamflow prediction · Mahanadi river basin

---

R. Dutta · R. Maity (✉)
Department of Civil Engineering, Indian Institute of Technology Kharagpur, Kharagpur 721302, West Bengal, India
e-mail: rajib@civil.iitkgp.ac.in; rajibmaity@gmail.com

R. Dutta
e-mail: riyadutta.iitkgp@gmail.com

R. Jha et al. (eds.), *Climate Change Impacts on Water Resources*, Water Science and Technology Library 98, https://doi.org/10.1007/978-3-030-64202-0_11

11.1 Introduction

Spatio-temporal variations of hydrologic variables, such as streamflow, soil moisture, precipitation, evapotranspiration etc., are associated with numerous inter-related hydrologic and climatic (collectively known as hydroclimatic) processes as well as several geomorphological characteristics of the watershed. The dependence among all the associated variables needs to be taken into account to develop effective prediction models, which necessarily involves a multivariate approach. Inclusion of all possible hydroclimatic variables may lead to a prohibitively large number of variables in the predictor set resulting in highly complex prediction models and poses serious challenges in parameter estimation. Thereby, selection of competent predictors is a vital part for the development of prediction models and poses a challenge for statistical modeling. There is a plethora of approaches to tackle the problem of multi-dimensionality, including dimensionality reduction of the input variables such as data compression using principal component analysis (PCA) and supervised PCA (Byrne et al. 1980; Bair et al. 2006; Price et al. 2006; Maity et al. 2013; Kim and Park 2015) and importance analysis technique of input variables (Wood et al. 1992; Ouarda et al. 2001; Burn and Hag Elnur 2002; van Griensven et al. 2006; Maity and Kashid 2011). In the process of dimensionality reduction, PCA or importance analysis techniques are not capable of revealing the true dependence structures; rather compress the information (in case of PCA) or ignore/declare the so-called important/unimportant variables (in case of importance analysis). In addition to these, gamma test (GT) and forward selection (FS) are other popular techniques employed to reduce the dimensionality of an input variable set (Moghaddamnia et al. 2009; Noori et al. 2011). These techniques are good in case of predefined set of input variables. However, in many cases knowledge on the set of input variables is vaguely known or incomplete. Moreover, it also remains unknown whether the same information is provided by more than one variable or not, which is known as redundancy in information. Such information can be obtained through the conditional independence between two variables given another (other) variable(s). Thus, complete information on the dependence structure is essential in order to develop a parsimonious multivariate model. Here lies the potential of graphical model (GM) approach, as it provides a means of representing dependence structure among variables (Ihler et al. 2007). The most distinctive feature of GM approach is formulating probabilistic models of complex phenomena in applied fields with detailed dependence structure, while reducing the complexity of the models (Jordan 2004). When multiple predictors are likely to govern the response of the hydrological systems, the probabilistic GMs offer a conditional independence structure for parsimonious predictor selection.

Some recent studies in the field of hydrology and hydroclimatology utilize the concepts of GM (Robertson and Wang 2009; Nagarajan et al. 2010; Dyer et al. 2014; Morrison and Stone 2014; Ramadas and Govindaraju 2015; Avilés et al. 2016; Dutta and Maity 2018, 2020a). GM has been used in combination with the latest linear regression techniques to develop prediction models. However, literature suggests, as explored in other fields of study, using copula for development of the prediction model after identification of the competent set of predictors can drastically improve the performance of the statistical models as it can capture the nonlinear relation among the variables (Dutta and Maity 2018, 2020b).

In the context of climate change, an important aspect is the time-variability in the association of the predictors and the predictand. This could be the reason behind the frequent forecast failure despite advancement in physical understanding and development of advanced statistical models. Some major issues and inherent problems in the statistical models such as variation in the predictor–predictand relationship over time and conditional independence among the predictors show the necessity for constant scrutiny and updating of the models (Rajeevan 2001; Rajeevan et al. 2007, 2012; Wang et al. 2015). In this context, it is vital to consider two important issues: (i) identifying the most influencing predictors and (ii) time-varying nature of predictor–predictand association (Dutta and Maity 2018, 2020a). Addressing these issues considering different regions and time period of analysis forms the motivation of this study.

The objective of this study is to explore the benefit of the time-varying GM approach in the field of hydrology and hydroclimatology. A time-varying prediction model is developed for 1-month ahead prediction of streamflow, generally influencing by numerous hydrologic and climatic variables for the Upper Mahanadi river basin. The input variables (identified using GM) and parameters (identified using vine copula) of the model are updated to capture the dynamic nature of interaction among the variables. The results are compared with the same obtained using a time-invariant GM approach and a time-varying SVR approach.

11.2 Methodology

The methodology consists of several important aspects. First, it starts with the identification of the time-varying dependence/association of the predictors (pool of influencing variables) with the target variable. The GM approach is used for selection/identification of the conditional independence structure among the input variables and target variable using maximum likelihood approach. The appropriate deviance for testing if any edge can be eliminated from the saturated model (Whittaker 2009) is the edge exclusion deviance (EED). The EED can be considered to follow chi-squared distribution with one degree of freedom (Whittaker 2009), as one edge is removed at a time. At 95% confidence level for one degree of freedom the p-value for chi-squared distribution is 3.84, so the edges for which the EED does not reach the value of 3.84 (threshold value) are to be excluded. The obtained graph structure represents the conditional dependence structure among the random variables and can be used for the development of conditional probabilistic model. The joint probability distributions associated with a given graph can be parameterized as product over functions associated with subsets of nodes. For directed graphs, the function turns out to be the conditional probability of a node given its parent nodes (Ihler et al.

2007). The joint probability distribution can be further used to calculate the conditional probability distribution of X_i given rest. There could be multiple predictors directly associated with the target variable, complicating the evaluation of the conditional distribution of the target variable. Multivariate copulas, like nested copula or vine copula, are the best choice to develop a multivariate probabilistic model. Among different alternatives in vine copulas, canonical vine (C-Vine) is used in this study to develop the probabilistic model. C-Vine can be used for prediction of the predictand given a set of predictors by a sequence of trees (Xiao 2011; Bauer et al. 2012; Righi et al. 2015; Liu et al. 2015; Dalla Valle et al. 2016; Dutta and Maity 2018). These trees are referred as C-Vine and the corresponding multivariate distribution is called C-Vine distribution. The conditional distribution of the target variable given the input variables (potential predictors) is obtained using the final C-Vine tree, which can be utilized for the prediction of target variable.

Secondly, a prediction time horizon (n in years) after which the model needs to be updated to impart the time-varying characteristics needs to be identified. First step is to identify the optimum prediction time horizon (denoted by τ in years) after which the model needs to be updated. Thereby, the time-varying association between the predictors and the predictand is captured by updating the potential predictor after each τ year(s). In order to identify the value of τ, the entire analysis is carried out for different values of n depending on the problem in hand. Development of the time-varying model and optimization of n are described as follows, *let* d years be the model development period which is considered as a moving window in the time-varying approach and t be the first year from which the analysis is to be started. Considering these, the first model development period is from t to t+$d-1$ and the model testing period is from t+d to t+d+($n-1$). As the model is updated after n years the next model development period is shifted by n years. Thereby, the second model development period is considered from t+n to t+$d-1$+n and the model testing period is from (t+$d-1$+n)+1 to (t+$d-1$+n)+1+(n$-$1). Similarly, the model is updated after every n year for the entire time period of the study. This value of n should be optimized such that the time-varying model is capable of capturing major temporal variations that affect the target variable and avoiding the small variations that may overcomplicate the model without adding valuable information. An optimum value of n will ensure the best possible prediction results. For detailed description of each step mentioned in this section, refer to Dutta and Maity (2018, 2020a).

11.3 Application of the Time-Varying GM Approach

11.3.1 Study Area and Data Source

The Upper Mahanadi basin, a part of the Mahanadi river basin in India, is considered as the study area (Fig. 11.1). The spatial extent of the study area is approximately 29,645 km^2. The approximate location of the study area is 20 to 23°N latitude and 80.5

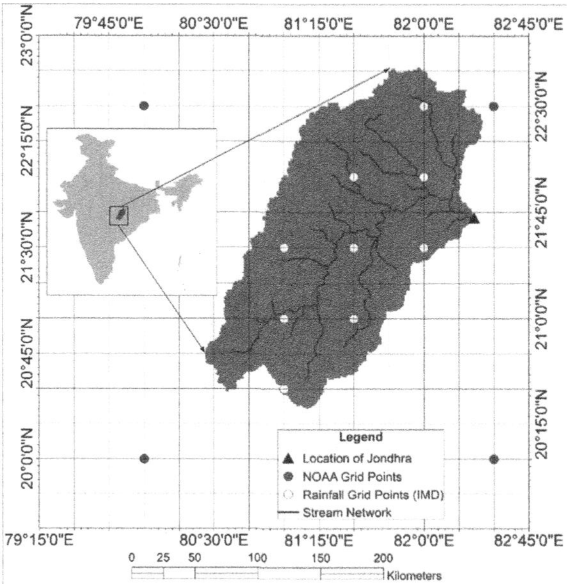

Fig. 11.1 Study area used for the prediction of streamflow at Jondhra gauging site in the Upper Mahanadi basin along with the grid locations of different hydroclimatic data

to 82.5° E longitude. The time-period considered for the study is from January 1980 to December 2005. For this study, streamflow at lead of 1 month (Y) is used as the target variable and the input variables, based on the physical understanding, include rainfall (X_1), precipitable water (X_2), soil moisture (X_3), pressure (X_4), relative humidity (X_5), potential evapotranspiration (X_6), temperature (X_7), U-wind (X_8), V-wind (X_9), geo-potential height (X_{10}) and streamflow (X_{11}). It may be noted that all the input variables are used from the current time-step and 1-month ahead streamflow is the target variable.

Daily streamflow data at the outlet of the basin (Jondhra station) are collected from the Water Resources Information System (India-WRIS version 4 2014) in India. Daily rainfall data for the study area are obtained from India Meteorological Department (IMD) (Rajeevan et al. 2008). Daily rainfall data at each grid point is converted to monthly rainfall depth by accumulating it over the month. The data for rest of the input variables are obtained from the Climate Prediction Centre (CPC) of the National Oceanic and Atmospheric Administration (NOAA) (Fan and van den Dool 2004; CPC 2014). All the data are taken from grid points lying within the study basin as shown in Fig. 11.1.

11.3.2 Model Performance and Discussion

At the outset, optimum prediction time horizon (τ in years) has to be decided for the development of the time-varying model. Toward this, a range of prediction time horizons (n in years), starting from 0.5 to 2.5 years (i.e. $n = 6, 12, 18, 24$ and 30 months) are utilized. For each case, the model development period is considered as a moving 15 years (180 data points) window. The conditional independence structure is developed for each model development period. Next, the probabilistic model is developed using the conditional independence structure to predict rainfall anomaly during the model testing period. Finally, the prediction results are obtained for all the testing periods that constitute 11 years of the study (1995–2005) using the time-varying model. The analysis is repeated for each prediction time horizons and the results are compared for identification of the optimum value of n. Correlation coefficient (R), index of agreement (Dr), Nash–Sutcliffe efficiency (NSE) and coefficient of determination (R^2) are used as performance statistics to identify the value of optimum prediction time horizon (τ). The performance statistics comparing the observed and the predicted streamflow for different prediction time horizons are shown in Fig. 11.2. It is noticed that the model performance becomes poorer with an increase in the value of n beyond 1 year.

On the other hand, the model performance does not show any improvement for a 6-month prediction time horizon (Fig. 11.2). The optimum value of n, i.e. τ, for the time-varying model is obtained as 1 year. Thus, the association between predictor set of hydroclimatic variables and streamflow is recommended to check and update the model parameter after every year.

Considering the value of τ as 1 year, the different model development periods and corresponding testing periods are shown in Table 11.1. Based on the obtained graph structure, the monthly streamflow for the first model development period is

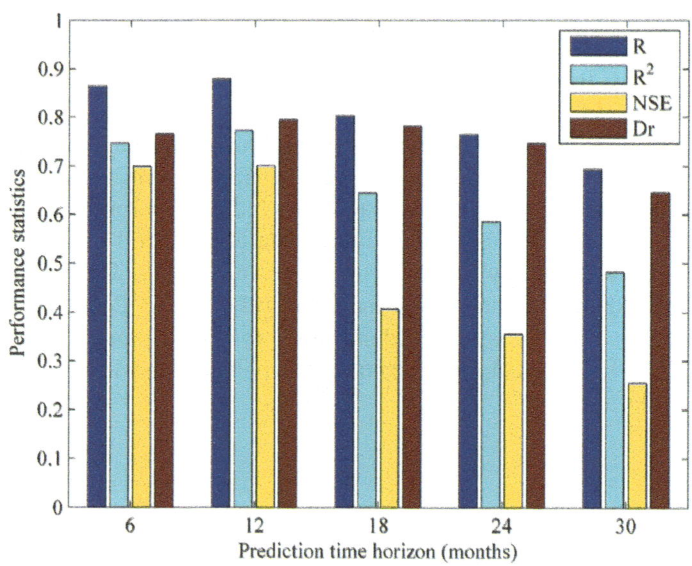

Fig. 11.2 Comparison of the performance statistics for selection of τ

Table 11.1 Details of the probabilistic models developed considering the different model development periods

S. No.	Development period	Testing period	Probabilistic model with inputs
1	1980–1994	1995	$f(Y/X_1, X_3, X_{11})$
2	1981–1995	1996	$f(Y/X_1, X_3, X_7, X_{11})$
2	1982–1996	1997	$f(Y/X_3, X_7, X_{11})$
3	1983–1997	1998	$f(y/X_1, X_3, X_{11})$
4	1984–1998	1999	$f(y/X_1, X_3, X_5, X_{11})$
5	1985–1999	2000	$f(y/X_1, X_5, X_{11})$
6	1986–2000	2001	$f(y/X_1, X_4, X_{11})$
7	1987–2001	2002	$f(y/X_4, X_7, X_{11})$
8	1988–2002	2003	$f(y/X_2, X_4, X_7, , X_{11})$
9	1989–2003	2004	$f(y/X_4, X_5, X_{11})$
10	1990–2004	2005	$f(y/X_1, X_4, X_5, X_{11})$

found to be directly dependent on X_1, X_3 and X_{11} (parents of Y) and conditionally independent of other variables.

Thereby, the conditional independence structure for the first model development period suggests that the information of rainfall, soil moisture and streamflow from the current month is perhaps sufficient for 1-month ahead streamflow prediction. It is further noticed that lagged streamflow and soil moisture have stronger association with streamflow as compared to rainfall as reflected by the edge strength. These results can be mainly associated with the memory characteristics of streamflow and soil moisture. Precipitable water, relative humidity and temperature are conditionally independent to streamflow and can be ignored if the information of rainfall, soil moisture and streamflow are available. Thus, it is not only the parsimonious model; GM ensures a better performance avoiding the presence of potentially "unwanted" or "conditionally independent" variables in the input set. Similar graph structures are obtained for each model development period, and the conditional distributions of the target variable for each model development period are shown in Table 11.1. It is interesting to note that the association among the hydroclimatic variables and streamflow changes with time. For instance, the conditional independence structure for the last model development period shows precipitable water, pressure, temperature and streamflow from the current month as the parents of the target variable. Thereby, it is expected that the time-varying characteristics of the model will improve the predictive performance. To assess the improvement gained through the time-varying GM approach against time-invariant approach, the model developed for the time period of 1980–1994 is used for prediction of streamflow during the entire testing period (1995–2005). Further, the time-varying concept is also used in a linear regression-based model to determine the benefit of adopting C-Vine copula as a probabilistic prediction model. The scatter plot for the observed versus predicted streamflow at lead time of 1 month obtained using the above-mentioned methodologies is shown in Fig. 11.3.

The first row of the plots shows the results obtained using the proposed approach and the second and third rows show the results obtained using the time-varying SVR approach and time-invariant GM approach, respectively. These plots depict good performance of the proposed approaches in predicting monthly streamflow as the majority of the observed/predicted data are distributed around the 1:1 line. The scatter plots between anomaly values are also presented in Fig. 11.3. It can be observed that the proposed approach is able to capture the actual range of anomaly values. The model performance clearly depicts that compared to the time-varying SVR approach and time-invariant GM approach, significantly better performance is obtained using the proposed approach during model testing period. It may further be noted that the (time-varying) model performance using prediction time horizon of 30 months is also better than the time-invariant model. Thereby even a higher prediction time horizon may also be selected that allow less frequent model updating. Figure 11.4 presents the predicted uncertainty intervals obtained using the time-varying GM approach and the observed streamflow. The plots also show the estimated 90% confidence band (5th to 95th percentile) of predicted streamflow. Results reveal that the uncertainty intervals can capture the variations of actual monthly flow during the model testing period.

11.4 Conclusions

In the context of climate change, time-varying modeling approaches are found to be more suitable to capture the relationship that vary with time considering the Upper Mahanadi river basin. Moreover, hydrologists often come across situations where the knowledge of conditional independence becomes very important to develop a parsimonious prediction model. In these contexts, the study focuses on the effectiveness of the GM approach in many applications. It helps to develop a parsimonious probabilistic prediction model using the effective information from the large set of input variables. The time-varying nature of such association that may throttle the consistency of any prediction model in the context of climate change may also be tackled using the time-varying GM approach. In this study, time-variation in the predictand–predictor relationship is considered for the proposed time-varying model. It is found that the model should be re-calibrated after an optimum prediction time horizon of 1 year after which the input set and model parameters need to be updated to account for the time-varying characteristics. This study sheds light on temporal variation in the association of different hydroclimatic variables with streamflow and the role of conditional independence structure for selection of predictors for a time-varying model. Better performance of the time-varying GM approach is realized as compared to its time-invariant counterpart. These finding are in line with the finding made by Dutta and Maity (2018; 2020b) considering Indian Summer Monsoon Rainfall and

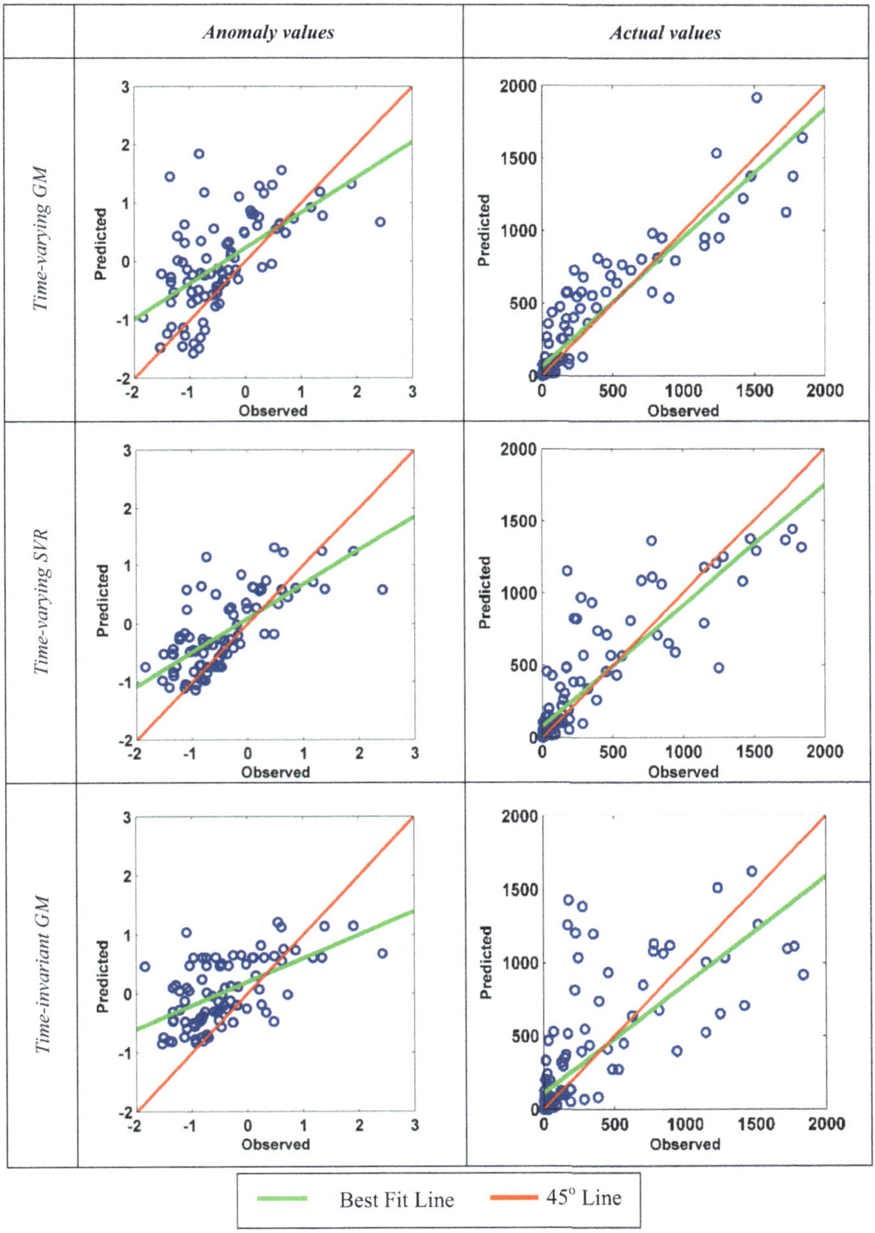

Fig. 11.3 Scatter plot between observed and predicted monthly streamflow during model testing period (1995–2005). The first to third row shows plots obtained using three different approaches

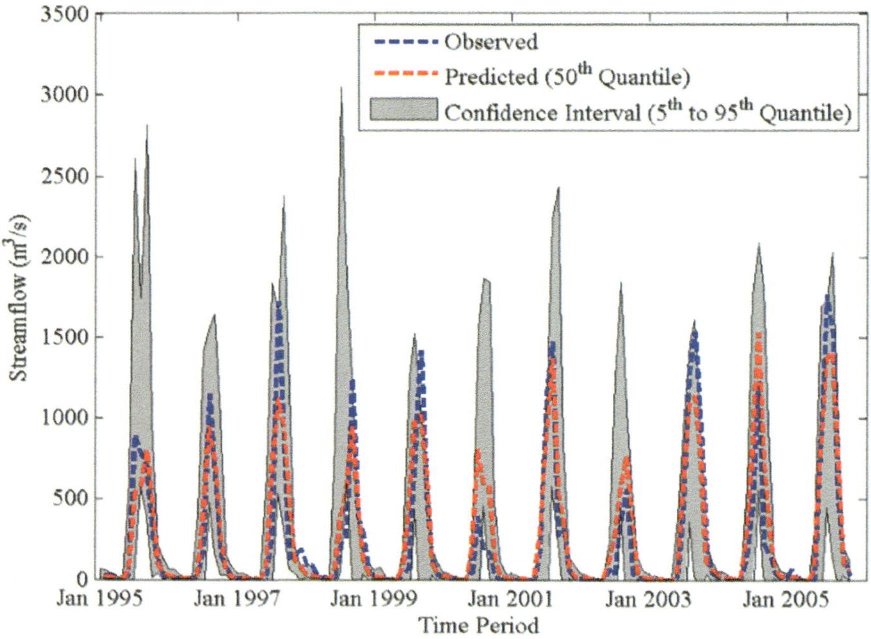

Fig. 11.4 Uncertainty band of the predicted streamflow for model testing period using the time-varying GM approach. The upper limit shows the predicted value at the 95th quantile and the lower limit shows the same at the 5th quantile

Ropero et al. (2017) and Dutta and Maity (2020a) considering reservoir management and monthwise streamflow for the Bhadra river basin, respectively. Thus, the time-varying GM approach ensures a parsimonious model by avoiding the presence of potentially "unwanted" or "conditionally independent" variables in the input set and capturing time-variability in the association for different hydroclimatic variables. Further, the robustness of the time-varying GM-based modeling approach is proven by a better or significantly better performance across different regions of analysis. This information in hydrologic or hydroclimatic problems is very important where the possible pool of influencing variables could be numerous in the absence of clear knowledge of dependence. Thus, the time-varying GM approach seems to be promising in many similar studies to achieve insights and for better model performance by avoiding the unimportant/less important/conditionally independent variables.

Acknowledgements The is study is partially supported by a research project sponsored by the Space Application Centre (SAC), Indian Space Research Organization (ISRO), Govt. of India (Ref No.: IIT/SRIC/CE/VIR/2016-17/88).

References

Avilés A, Célleri R, Solera A, Paredes J (2016) Probabilistic forecasting of drought events using Markov chain-and Bayesian network-based models: a case study of an Andean regulated river basin. Water (Switzerland) 8. https://doi.org/10.3390/w8020037

Bair E, Hastie T, Paul D, Tibshirani R (2006) Prediction by supervised principal components. J Am Stat Assoc 101:119–137. https://doi.org/10.1198/016214505000000628

Bauer A, Czado C, Klein T et al (2012) Pair-copula constructions for non-Gaussian DAG models. Can J Stat 40:86–109. https://doi.org/10.1002/cjs.10131

Burn DH, Hag Elnur MA (2002) Detection of hydrologic trends and variability. J Hydrol 255:107–122. https://doi.org/10.1016/S0022-1694(01)00514-5

Byrne GF, Crapper PF, Mayo KK (1980) Monitoring land-cover change by principal component analysis of multitemporal landsat data. Remote Sens Environ 10:175–184. https://doi.org/10.1016/0034-4257(80)90021-8

CPC (2014) Soil Moisture (V2), NOAA/OAR/ESRL PSD, Boulder, Colorado, USA. http://www.esrl.noaa.gov/psd/data/gridded/data.cpcsoil.html. Accessed 1 Nov 2014

Dalla Valle L, De Giuli ME, Tarantola C, Manelli C (2016) Default probability estimation via pair copula constructions. Eur J Oper Res 249:298–311. https://doi.org/10.1016/j.ejor.2015.08.026

Dutta R, Maity R (2018) Temporal evolution of hydroclimatic teleconnection and a time-varying model for long-lead prediction of Indian summer monsoon rainfall. Sci Rep 8:10778. https://doi.org/10.1038/s41598-018-28972-z

Dutta R, Maity R (2020a) Temporal networks based approach for non-stationary hydroclimatic modelling and its demonstration with streamflow prediction. Water Resour Res 0–1. https://doi.org/10.1029/2020wr027086

Dutta R, Maity R (2020b) Spatial variation in long-lead predictability of summer monsoon rainfall using a time-varying model and global climatic indices. Int J Climatol 1–16. https://doi.org/10.1002/joc.6556

Dyer F, ElSawah S, Croke B et al (2014) The effects of climate change on ecologically-relevant flow regime and water quality attributes. Stoch Environ Res Risk Assess 28:67–82. https://doi.org/10.1007/s00477-013-0744-8

Fan Y, van den Dool H (2004) Climate Prediction Center global monthly soil moisture data set at 0.5° resolution for 1948 to present. J Geophys Res D: Atmos 109:1–8. https://doi.org/10.1029/2003JD004345

Ihler AT, Kirshner S, Ghil M et al (2007) Graphical models for statistical inference and data assimilation. Phys D Nonlinear Phenom 230:72–87. https://doi.org/10.1016/j.physd.2006.08.023

India-WRIS version 4 (2014) Non classified hydro observation data at jondhra Station, India

Jordan MI (2004) Graphical Models. Stat Sci 19:140–155. https://doi.org/10.1214/088342304000000026

Kim J, Park J (2015) Mathematical modeling of coastal marine environments using observational data for coastal management. Ocean Coast Manag 116:396–403. https://doi.org/10.1016/j.ocecoaman.2015.08.007

Liu Z, Zhou P, Chen X, Guan Y (2015) A multivariate conditional model for streamflow prediction and spatial precipitation refinement. J Geophys Res Atmos 120:10-116–10-129. https://doi.org/10.1002/2015jd023787

Maity R, Kashid SS (2011) Importance analysis of local and global climate inputs for basin-scale streamflow prediction. Water Resour Res 47:1–17. https://doi.org/10.1029/2010WR009742

Maity R, Ramadas M, Govindaraju RS (2013) Identification of hydrologic drought triggers from hydroclimatic predictor variables. Water Resour Res 49:4476–4492. https://doi.org/10.1002/wrcr.20346

Moghaddamnia A, Ghafari Gousheh M, Piri J et al (2009) Evaporation estimation using artificial neural networks and adaptive neuro-fuzzy inference system techniques. Adv Water Resour 32:88–97. https://doi.org/10.1016/j.advwatres.2008.10.005

Morrison R, Stone M (2014) Spatially implemented Bayesian network model to assess environmental impacts of water management. Water Resour Res 50:8107–8124. https://doi.org/10.1002/2014WR015600.Received

Nagarajan K, Krekeler C, Slatton KC, Graham WD (2010) A scalable approach to fusing spatiotemporal data to estimate streamflow via a bayesian network. IEEE Trans Geosci Remote Sens 48:3720–3732. https://doi.org/10.1109/TGRS.2010.2049115

Noori R, Karbassi AR, Moghaddamnia A et al (2011) Assessment of input variables determination on the SVM model performance using PCA, Gamma test, and forward selection techniques for monthly stream flow prediction. J Hydrol 401:177–189. https://doi.org/10.1016/j.jhydrol.2011.02.021

Ouarda TBMJ, Girard C, Cavadias GS, Bobée B (2001) Regional flood frequency estimation with canonical correlation analysis. J Hydrol 254:157–173

Price AL, Patterson NJ, Plenge RM et al (2006) Principal components analysis corrects for stratification in genome-wide association studies. Nat Genet 38:904–909. https://doi.org/10.1038/ng1847

Rajeevan M, Bhate J, Jaswal AK (2008) Analysis of variability and trends of extreme rainfall events over India using 104 years of gridded daily rainfall data. Geophys Res Lett 35:1–6. https://doi.org/10.1029/2008GL035143

Rajeevan M, Pai DS, Anil Kumar R, Lal B (2007) New statistical models for long-range forecasting of southwest monsoon rainfall over India. Clim Dyn 28:813–828. https://doi.org/10.1007/s00382-006-0197-6

Rajeevan M, Unnikrishnan CK, Preethi B (2012) Evaluation of the ENSEMBLES multi-model seasonal forecasts of Indian summer monsoon variability. Clim Dyn 38:2257–2274. https://doi.org/10.1007/s00382-011-1061-x

Rajeevan MN (2001) Prediction of Indian summer monsoon: Status, problems and prospects. Curr Sci 81:1451–1457

Ramadas M, Govindaraju RS (2015) Probabilistic assessment of agricultural droughts using graphical models. J Hydrol 526:151–163. https://doi.org/10.1016/j.jhydrol.2014.09.026

Righi MB, Schlender SG, Ceretta PS (2015) Pair copula constructions to determine the dependence structure of Treasury bond yields. IIMB Manag Rev 27:216–227. https://doi.org/10.1016/j.iimb.2015.10.008

Robertson DE, Wang QJ (2009) Selecting predictors for seasonal streamflow predictions using a Bayesian joint probability (BJP) modelling approach. 18th World IMACS/MODSIM Congr Cairns, Aust 13–17 July 2009 376–382

Ropero RF, Flores MJ, Rumí R, Aguilera PA (2017) Applications of hybrid dynamic Bayesian networks to water reservoir management. Environmetrics 28:1–11. https://doi.org/10.1002/env.2432

van Griensven A, Meixner T, Grunwald S et al (2006) A global sensitivity analysis tool for the parameters of multi-variable catchment models. J Hydrol 324:10–23. https://doi.org/10.1016/j.jhydrol.2005.09.008

Wang B, Xiang B, Li J et al (2015) Rethinking Indian monsoon rainfall prediction in the context of recent global warming. Nat Commun 6:1–8. https://doi.org/10.1038/ncomms8154

Whittaker J (2009) Graphical Models in Applied Multivariate Statistics. Wiley Publishing

Wood EF, Lettenmaier DP, Zartarian VG (1992) A land-surface hydrology parameterization with subgrid variability for general circulation models. J Geophys Res 97:2717. https://doi.org/10.1029/91JD01786

Xiao H (2011) Pair-copula Construction for Non-Gaussian Graphical Models. Can J Stat 40:86–109

Chapter 12
Determination of Effective Discharge Responsible for Sediment Transport in Cauvery River Basin

Shobhit Maheshwari and Sagar Rohidas Chavan

Abstract The mechanism of sediment transport is mainly governed by surface water flow within the river basins. Excessive sediment transport plays an important role in reducing the carrying capacity of channel networks and storage capacity of reservoirs/dams. An important task for most of the hydrologists is to determine the reliable stream flow estimate which causes majority of the sediment transport within river basins/stream channels. The transport effectiveness of a stream flow event of particular magnitude in carrying a sediment load is defined as the product of the effect of that event (i.e. sediment transport rate corresponding to the stream flow event) and the frequency with which the event occurs. This approach is famously known as magnitude frequency analysis (MFA). MFA has been widely used to compute "effective discharge" which is considered as the stream flow that is responsible for transportation of majority of the sediments from a river basin or catchment over a long period of time. In MFA, the stream flow at a location is assumed to follow a continuous probability distribution (e.g., normal, lognormal, exponential, gamma, generalized pareto and Poisson) whereas the sediment transport is described by a power law function between stream flow and sediment rate. Subsequently, a transport effectiveness function is constructed by taking product of stream flow distribution with power law function. Finally, the effective discharge can be obtained by maximizing the transport effectiveness function with respect to stream flow. In this paper, effective discharge estimates were determined for 12 stream gauges in Cauvery river basin by fitting appropriate continuous probability distributions (normal, lognormal, exponential, gamma, generalized pareto and Poisson) and assuming power law relationship for sediment transport. Kolmogorov–Smirnov test (KS test) at 1% significance level was tested for fitting probability distributions to daily stream flow data at each of the gauges. Results indicated that all of the above distributions failed to fit stream flow data at all the gauges. However, following the previous literature, the daily stream flow data at every gauge was assumed to follow lognormal distribution and the corresponding effective discharge was determined. Further, recurrence interval was calculated for the effective discharge estimate at the each of the gauge. The results from this study can find use in effective planning and functioning of dams/reservoirs.

S. Maheshwari (✉) · S. R. Chavan
Department of Civil Engineering, Indian Institute of Technology, Ropar, India
e-mail: 2017cez0003@iitrpr.ac.in

R. Jha et al. (eds.), *Climate Change Impacts on Water Resources*, Water Science and Technology Library 98, https://doi.org/10.1007/978-3-030-64202-0_12

Keywords Effective discharge · Sediment transport · Magnitude and frequency analysis · KS test · Cauvery river basin

12.1 Introduction

Reservoirs/dams are usually used as a consistent source of water in arid as well as semiarid areas and their sustainability is of huge importance for the society (Araújo et al. 2006). However, various natural and anthropogenic changes in the river basin lead to affect the sediment transport in the stream flow. Excess sediment transport can affect the carrying capacity of channel networks as well as the storage capacity of reservoirs/dams and reduce the availability of water in the reservoirs. In addition, the sediment content in water affects the quality of river water. An important task for most of the hydrologists is to determine the reliable stream flow estimate which causes majority of the sediment transport within river basins/stream channels. Hudson and Mossa (1997) proposed that stream flow magnitude and its frequency can help in calculating the timing and fluctuation of sediment transports from river basins. Marginal changes in stream flow to sediment transport ratio over the channel networks modify the channel morphology considerably which alters the stream flow event.

Wolman and Miller (1960) presented the concept of effective discharge, which can be described as a product of the load of sediment carried by a given stream flow and the frequency of the stream flow. They introduced the concept of magnitude frequency analysis (MFA) to determine effective discharge. The effective discharge is considered as the stream flow which is responsible for transportation of majority of the sediments from a river basin or catchment over a long period of time. In MFA, the stream flow at a location is assumed to follow a continuous probability distribution (e.g., normal, lognormal, exponential, gamma, generalized pareto and Poisson) whereas the sediment transport is described by a power law function between stream flow and sediment rate. Subsequently, a transport effectiveness function is constructed by taking product of stream flow distribution with power law function. Finally, the effective discharge can be obtained by maximizing the transport effectiveness function with respect to stream flow. The determination of this effective discharge can help in assessment of stream flow for the consistency of channel networks and reservoirs/dams (Wahl et al. 1995; Hester et al. 2006), water structures construction, ecological refurbishment and environmental augmentation of rivers (Shields et al. 2003). Leopold (1994) advocated that small stream flow events have higher frequency, but these events do not have sufficient capacity to transport higher quantity of sediments. Contrary to this, catastrophic or large stream flow events have enough capacity to transport higher quantities of sediments but have less probability of occurrence. Consequently, the effective discharge is the stream flow that is in-between the small and large stream flows, which are capable of transporting most of the sediment loads on long-term basis (Wolman and Miller 1960; Lenzi et al. 2006; Ma et al. 2010). Many researchers have also suggested that stream flow events of a particular recurrence interval cannot be considered to be characteristic of effective discharge for all rivers because stream flow values are influenced by basin geology, catchment area, hydrologic system and sediment transportation (Ashmore and Day 1988; Nash 1994; Whiting et al. 1999; Phillips 2002; Lenzi et al. 2006). The aim of the present study is to check the validity of various probability distributions to calculate the frequency of stream flow event and provide estimates of effective discharge and recurrence interval based on MFA for stream gauges in Cauvery river basin.

12.2 Methodology

12.2.1 Goodness-of-Fit Test for Stream Flow Data

In this study, the stream flow data corresponding to various gauges in Cauvery basin were assessed to follow different frequency distributions using the Kolmogorov–Smirnov test (KS test) at 1% significance level. It is a nonparametric check of the equality of continuous probability distributions which can be used to compare data samples with a reference continuous probability distribution. Various probability distributions considered in the study are normal, lognormal, exponential, gamma, generalized pareto and Poisson.

12.2.2 Determination of Effective Discharge Using MFA

MFA involves assumptions about daily stream flow frequency and sediment transport rate. For stream flow, log transformed daily stream flow is assumed to follow continuous normal distribution as shown in Eq. (12.1).

$$f(Ql) = \frac{1}{\beta\sqrt{2\pi}} e^{-(Ql - \alpha)^2/2\beta^2} \tag{12.1}$$

where Ql is the log transformed daily stream flow, α and β are the mean and standard deviation of the log transformed stream flow, respectively.

On the other hand, the sediment load is assumed to exhibit a power law function of daily stream flow as given by Eq. (12.2).

$$S = aQ^b \tag{12.2}$$

where S and Q are sediment load and daily stream flow, respectively; a and b are empirically calculated coefficient and exponent.

Based on stream flow frequency and empirically derived sediment load, sediment transport effectiveness (E) is determined as,

$$E = \frac{aQ^b}{\beta\sqrt{2\pi}} e^{-(Ql-\alpha)^2/2\beta^2} \tag{12.3}$$

By maximizing the effectiveness function with respect to stream flow Q, "effective discharge", Q_e is obtained (Nash 1994).

$$\begin{aligned}
&For,\ \tfrac{dE}{dQ} = 0\ and\ Ql = \ln(Q)\\
&\tfrac{d}{dQ}\left[\tfrac{a}{\beta\sqrt{2\pi}}\left\{Q^b.e^{-(\ln Q-\alpha)^2/2\beta^2}\right\}\right] = 0\\
&\tfrac{a}{\beta\sqrt{2\pi}}\left[Q^b.e^{-(\ln Q-\alpha)^2/2\beta^2}.\left\{\tfrac{-2(\ln Q-\alpha)}{2\beta^2}.\tfrac{1}{Q}\right\} + e^{-(\ln Q-\alpha)^2/2\beta^2}.bQ^{b-1}\right] = 0\\
&Q^b.e^{-(\ln Q-\alpha)^2/2\beta^2}.\left\{\tfrac{-2(\ln Q-\alpha)}{2\beta^2}.\tfrac{1}{Q}\right\} = -e^{-(\ln Q-\alpha)^2/2\beta^2}.bQ^{b-1}\\
&\tfrac{Q_b}{Q^{b-1}.Q}\left\{\tfrac{(\ln Q-\alpha)}{\beta^2}\right\} = b\\
&(\ln Q - \alpha) = b\beta^2 \Rightarrow \ln Q = b\beta^2 + \alpha\\
&Q_e = \exp(b\beta^2 + \alpha)
\end{aligned} \tag{12.4}$$

12.2.3 Recurrence Interval Prediction

Recurrence interval of effective discharge Q_e is defined as the inverse of probability of stream flow equal or greater than the effective discharge. The predicted recurrence interval T_P of a particular stream flow Q, in years, is given as,

$$T_P = \frac{1/365.25}{1 - \phi\left[\frac{Ql-\alpha}{\beta}\right]} \tag{12.5}$$

where $\phi[\cdot]$ represents the cumulative distribution function (CDF) of an assumed standard normal variable.

12.3 Study Area and Data Considered for MFA

The Cauvery river is one of the major rivers of the peninsula (Fig. 12.1). The river is also the main source for drinking, irrigation and electricity facilities to most part of southern Karnataka specifically Mysore region. Several dams have been built on the river. The Cauvery river originates from Brahmagiri Range of hills in the Western Ghats at an elevation of 1341 m (above MSL) at Talakaveri near Cherangala village of Kodagu District of Karnataka and has catchment area of 81,155 km^2 which is nearly 2.7% of the total geographical area of the country. The basin lies between 10°9′ to 13°30′ N latitudes and 75°27′ to 79°54′ E longitudes. It is circumscribed on the west by the Western Ghats, on the east by the Eastern Ghats and in north by the ranges separating it from Krishna River basin and Pennar River basin. The total length of the river from source to drain is 800 km. It has important tributaries joining from the left are the Arkavati, the Harangi, the Hemavati and the Shimsha. The river drains into the Bay of Bengal. Most of the basin is covered by agricultural land which accounts up to 66.21% of the total area, and waterbodies also have considerable coverage accounting for 4.09% of the total area (http://www.india-wris.nrsc.gov.in/wrpinfo/index.php?title=Cauvery).

Gauge-discharge data and sediment concentration data were obtained from Water Resources Information System (WRIS), Government of India. Sediment load in terms of tons per day was calculated from sediment concentration (gram per liter) and gauge-discharge (cubic meter per second) using a suitable constant factor (86.4). In order to evade the problem of gaps in the data, only years with complete accounts of gauge-discharge and sediment concentrations were included in the analysis. Finally, gauge-discharge and sediment concentrations data of 12 gauges were selected for the study. River basin and stream flow characteristics are presented in Table 12.1.

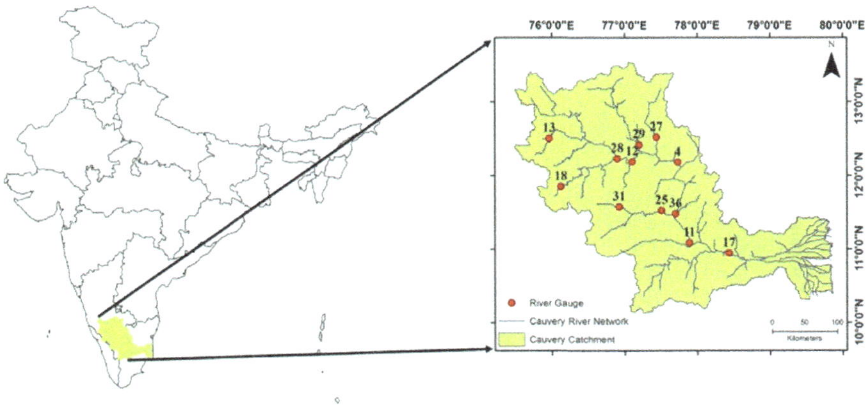

Fig. 12.1 Locations of the 12 gauges of Cauvery river basin considered in this study (numbers correspond to IDs of the stream gauges)

Table 12.1 River basin and stream flow characteristics

ID #	River Gauge	Elevation (m)	Area (km^2)	Length of record (days)	Mean Q (m^3/s)	α	β	Skew	Kurtosis
4	"Biligundulu"	255	36682	13149	238.5	4.83	1.08	0.43	2.80
11	"Kodumudi"	130	53233	13148	275.4	4.89	1.35	−0.33	2.46
12	"Kollegal"	622	21082	11689	199.9	4.55	1.16	0.18	3.82
13	"Kudige"	809	1934	6209	89.6	3.18	1.64	0.17	2.75
17	"Musiri"	89	66243	12418	260.8	4.44	1.80	−0.29	1.82
18	"Muthankera"	705	1260	13879	77.4	2.92	2.12	−1.20	6.06
25	"Savandapur"	189	5776	10958	20.3	2.72	0.77	−1.04	8.36
27	"Tbekuppe"	600	3500	365	6.0	1.72	0.38	0.95	4.50
28	"Tnarasipur"	635	7000	10957	92.1	3.69	1.26	0.02	3.88
29	"Tkhalli"	580	7890	6209	23.3	1.92	1.94	−1.40	5.94
31	"Thengumarahada"	345	1370	2557	12.3	2.12	0.81	0.11	5.79
36	"Urachikottai"	168	44100	3652	193.7	1.79	4.52	−0.37	1.25

ID: River Gauge ID according to WRIS
Elevation: Height of a river gauge above MSL (Mean Sea Level)
Area: Total area in river gauge in km^2, including non-contributing areas
Length of record: Number of days for which the continuous data is available for years
Mean Q: Average of daily stream flow in m^3/sec for the length of record
α: Mean of log-transformed daily stream flow
β: Standard deviation of log-transformed daily stream flow
Skew: Skewness of log-transformed daily stream flow
Kurtosis: Kurtosis of log-transformed daily stream flow

Table 12.2 KS test for fitting different probability distributions

ID #	River Gauge	Normal	Lognormal	Exponential	Gamma	Generalized Pareto	Poisson
4	"Biligundulu"	Fail	Fail	Fail	Fail	Fail	Fail
11	"Kodumudi"	Fail	Fail	Fail	Fail	Fail	Fail
12	"Kollegal"	Fail	Fail	Fail	Fail	Fail	Fail
13	"Kudige"	Fail	Fail	Fail	Fail	Fail	Fail
17	"Musiri"	Fail	Fail	Fail	Fail	Fail	Fail
18	"Muthankera"	Fail	Fail	Fail	Fail	Fail	Fail
25	"Savandapur"	Fail	Fail	Fail	Fail	Fail	Fail
27	"Tbekuppe"	Fail	Fail	Fail	Fail	Fail	Fail
28	"Tnarasipur"	Fail	Fail	Fail	Fail	Fail	Fail
29	"Tkhalli"	Fail	Fail	Fail	Fail	Fail	Fail
31	"Thengumarahada"	Fail	Fail	Fail	Fail	Fail	Fail
36	"Urachikottai"	Fail	Fail	Fail	Fail	Fail	Fail

12.4 Results and Discussions

12.4.1 KS Test for Fitting Probability Distributions to Daily Stream Flow Data

In this study, KS test at 1% significance level was considered for fitting probability distributions to daily stream flow data at each of the gauges. The distributions considered included normal, lognormal, exponential, gamma, generalized pareto, Poisson, etc. The stream flow at each of the gauges failed to pass the KS test for all distributions. Results are presented in Table 12.2. However, following the previous literatures (Ashmore and Day 1988; Nash 1994; Whiting et al. 1999; Phillips 2002; Vogel et al. 2003; Lenzi et al. 2006, Sichingabula 1999), the daily stream flow data was assumed to follow lognormal distribution.

12.4.2 Stream Flow Distribution and Sediment Transport Analysis

As discussed earlier, log-transformed stream flow data was considered in this study. Subsequently, that transformed stream flow data of each gauge was fitted with normal probability distribution. Sediment transport rate and the streamflow data was found to exhibit power law as shown in Fig. 12.2. Log–log plot of discharge and sediment transport rate for each of the 12 gauges was prepared. The plots are found to be fairly

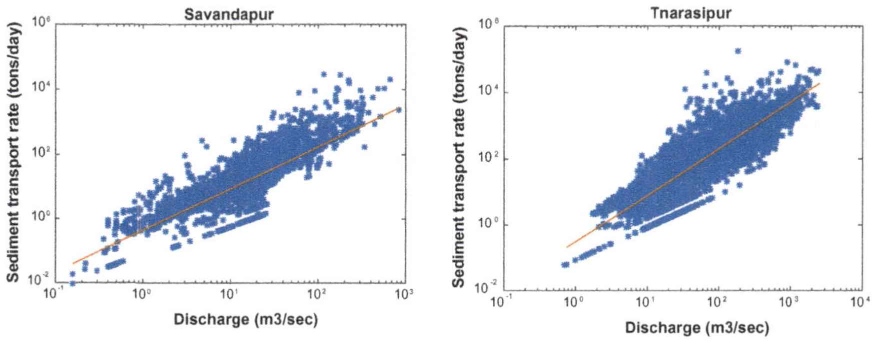

Fig. 12.2 Power law relationship between sediment transport rate and discharge observed at two typical stream gauges

Table 12.3 Effective discharge and recurrence intervals determined for 12 stream gauges

River Gauge	a	b	R^2	Predicted Q_e (m^3/s)	Predicted RI (years)	Observed Q_e (m^3/s)	Observed RI (years)
"Biligundulu"	0.15	1.53	0.891	739.9	0.055	739.9	0.065
"Kodumudi"	0.32	1.32	0.930	1484.6	0.074	1485.0	0.587
"Kollegal"	0.15	1.44	0.923	657.3	0.057	657.4	0.092
"Kudige"	0.60	1.22	0.983	656.2	0.124	656.0	0.783
"Musiri"	0.19	1.46	0.863	9544.8	0.640	6400.0	34.999
"Muthankera"	0.97	1.30	0.964	6378.4	0.922	2050.0	38.999
"Savandapur"	0.69	1.29	0.957	32.9	0.017	32.9	0.087
"Tbekuppe"	4.87	1.00	1.000	6.4	0.008	6.4	0.019
"Tnarasipur"	0.35	1.41	0.938	379.0	0.073	379.1	0.128
"Tkhalli"	0.51	1.38	0.938	1211.0	0.727	971.2	17.999
"Thengumarahada"	0.74	1.48	0.868	22.0	0.024	22.0	0.038
"Urachikottai"	0.13	1.10	0.988	37030426837.5	8957.526	5854.5	10.999

where a and b are empirically calculated constant coefficient and exponent

linear for all the gauges. The correlation coefficients (R^2) for power function of daily stream flow and sediment transport rate are provided in Table 12.3.

12.4.3 *Effective Discharge and Recurrence Interval Determination*

Based on MFA, effective discharge, Q_e, was determined for 12 stream gauges in the Cauvery basin based on effectiveness function at each of the gauges (Fig. 12.3). In

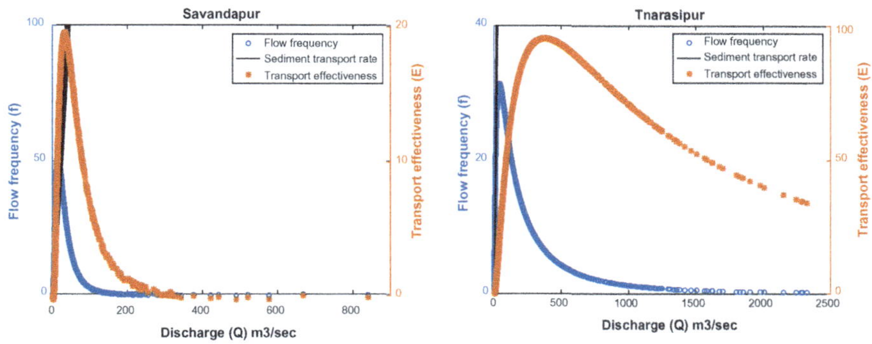

Fig. 12.3 Transport effectiveness curves of two typical stream gauges

addition, the effective discharge estimates were also computed by finding the peak of transport effectiveness function and designated as observed Q_e. Finally, recurrence intervals were calculated for predicted Q_e and observed Q_e (by ranking the daily stream flows, dividing the rank of observed Q_e by the length of record in years). They were named as predicted RI and observed RI, respectively. Results are tabulated in Table 12.3. Predicted effective discharges and observed effective discharges are found to have significant correlation. This implies fit between the observed daily stream flow frequency and the log-transformed normal distribution is good.

12.4.4 Duration of Effective Discharge

Duration curves for discharge and sediment load show that they are quite similar in shape. The curves indicate that most of the discharges are able to transport sediments through the channels at each gauge, but the magnitude of sediment transport varies with the varying discharges consistently. The steep portion of the duration curve is the zone of effective sediment transport. Sediment load duration curves show that the extreme low discharges do not have the capacity for sediment transport (Fig. 12.4).

12.5 Conclusions

In this paper, effective discharge and recurrence interval based on MFA were determined for 12 stream gauges in Cauvery river basin. The major observations from this study are as follows:

1. None of the (normal, lognormal, exponential, gamma, generalized pareto and Poisson) probability distribution was able to pass the KS test at 1% significance level.

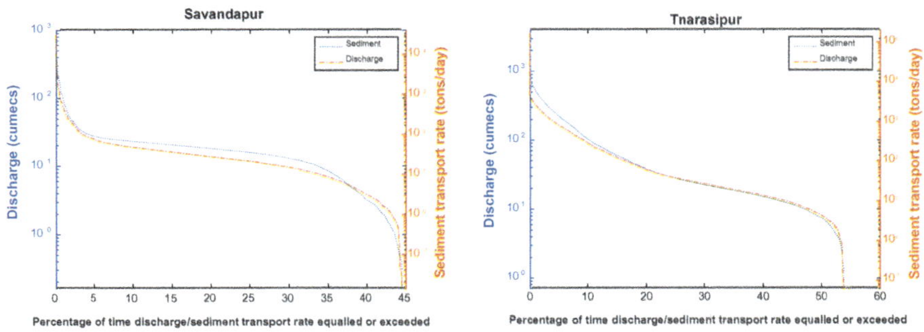

Fig. 12.4 Discharge and sediment load duration curves of two typical stream gauges

2. Predicted effective discharges and observed effective discharges are found to be in good correlation. This implies fit between the observed daily stream flow frequency and the log-transformed normal distribution is good.
3. The correlation coefficients (R^2) for power function of daily stream flow and sediment transport rate were found to be in the range 0.86–0.99. So, the assumption that sediment transport rate is a power function of daily stream flow is also valid.

References

Ashmore PE, Day TJ (1988) Effective discharge for suspended sediment transport in streams of the Saskatchewan River basin. Water Resour Res 24(6):864–870

De Araújo JC, Güntner A, Bronstert A (2006) Loss of reservoir volume by sediment deposition and its impact on water availability in semiarid Brazil. Hydrol Sci J 51(1):157–170

Hester G, Carsell K, Ford D (2006) Benefits of USGS streamgaging program—users and uses of USGS streamflow data. National Hydrologic Warning Council

Hudson PF, Mossa J (1997) Suspended sediment transport effectiveness of three large impounded rivers, US Gulf Coastal Plain. Environ Geol 32(4):263–273

Lenzi MA, Mao L, Comiti F (2006) Effective discharge for sediment transport in a mountain river: computational approaches and geomorphic effectiveness. J Hydrol 326(1–4):257–276

Leopold LB (1994) A view of the river. Harvard University Press

Ma Y, Huang HQ, Xu J, Brierley GJ, Yao Z (2010) Variability of effective discharge for suspended sediment transport in a large semi-arid river basin. J Hydrol 388(3–4):357–369

Nash DB (1994) Effective sediment-transporting discharge from magnitude-frequency analysis. J Geol 102(1):79–95

Phillips JD (2002) Geomorphic impacts of flash flooding in a forested headwater basin. J Hydrol 269(3–4):236–250

Shields FD, Knight SS, Morin N, Blank J (2003) Response of fishes and aquatic habitats to sand-bed stream restoration using large woody debris. Hydrobiologia 494(1–3):251–257

Sichingabula HM (1999) Magnitude-frequency characteristics of effective discharge for suspended sediment transport, Fraser River, British Columbia, Canada. Hydrol Process 13(9):1361–1380

Vogel RM, Stedinger JR, Hooper RP (2003) Discharge indices for water quality loads. Water Res Res 39(10)

Wahl KL, Thomas WO, Hirsch RM (1995) The stream-gaging program of the US Geological Survey, (No. 1123). US Geological Survey
Whiting PJ, Stamm JF, Moog DB, Orndorff RL (1999) Sediment-transporting flows in headwater streams. Geol Soc Am Bull 111(3):450–466
Wolman MG, Miller JP (1960) Magnitude and frequency of forces in geomorphic processes. J Geol 68(1):54–74

Chapter 13
A Comparative Study of Potential Evapotranspiration in an Agroforestry Region of Western Ghats, India

Pandu Narayana and K. Varija

Abstract Evapotranspiration accounts for a major part of the water budget. Estimation of the evapotranspiration is complex because it involves too many parameters. Over many decades various methodologies have been introduced and applied to the wide range of climatic conditions to predict the water loss from vegetation. Among them FAO-56 Penman-Monteith equation is known to be the widely accepted and it has been validated to the wide range of vegetation and climatic conditions across the globe. Agroforestry is an important and one of the high water-demanding agricultural practices. Due to the presence of multiple crops, the rate of evapotranspiration is generally expected to be high. Therefore, it is important to understand the evaporative demand in the agroforestries to reduce the water consumption in irrigation planning. Due to the lack of measured meteorological data with many parameters, it is important to compare the performances of the methods which are less data-intensive in nature. For this purpose, alternatively Priestley-Taylor, Hargreaves and Turc methods have been used and compared with Penman-Monteith equation. A study carried out in Seegodu watershed consists of coffee plantation, which is one of the major practices in agroforestries to compare the estimates of different methods. The results showed that the Hargreaves and Turc method predicts better results on a temporal scale with only temperature data.

Keywords Evapotranspiration · Priestley-Taylor · Hargreaves · Turc · Penman-Monteith

13.1 Introduction

Evapotranspiration (ET) is a major component of the water balance than the discharge in many watersheds (Bevan Beven 2011). In agriculture and irrigation planning it becomes important to know the water loss that occurs due to the atmospheric demand for the loss of water from the agricultural fields. The actual ET occurring from the

P. Narayana · K. Varija (✉)
Department of Applied Mechanics and Hydraulics, National Institute of Technology Karnataka, Surathkal, India
e-mail: pandubikkodu@gmail.com

R. Jha et al. (eds.), *Climate Change Impacts on Water Resources*, Water Science and Technology Library 98, https://doi.org/10.1007/978-3-030-64202-0_13

fields can be estimated from two steps. First step is to determine the atmospheric demand for the water to vaporize from the plant and water surface and the next step is the behavior or response from the crops to the atmospheric demand through their plant physiological mechanisms such as stomatal conductance and the available water content from the soil. To find out the atmospheric demand is an important step in calculating the total water loss from the crops. Observed climatic parameters can be used to estimate the atmospheric demand for ET. Various methods have been proposed to determine the atmospheric demand for different climatic and vegetation in the world. Among them Penman-Monteith equation is taken as the standard method for determining the crop water requirements. The atmospheric demand is determined for any climatic conditions in terms of a reference crop, i.e., a standard grass with sufficient availability of water, and the water demand for any other crop is compared with the suitable crop coefficient to the reference crop. Penman-Monteith method is a physically based and data-intensive method. It can be applied only when there is availability of meteorological data with all the four basic climatic parameters. The calculation procedures are difficult as compared to other methods (Xu and Singh 2002). But meteorological data with all the basic parameters (George et al. 2002) are often difficult to measure under strict instrumentation. So, it becomes important to validate the alternative methods with minimum meteorological parameters which results in similar results to that of the standard Penman-Monteith method. Many researches have been carried out as an alternative; among them Hargreaves, Turc and Priestley-Taylor methods were analyzed for the study. The main agroforestries are coffee, tea, rubber and arecanut in the humid regions of the Western Ghats. Coffee plantation has been chosen for the present study.

13.2 Materials and Methods

13.2.1 Penman-Monteith Method (PM)

Crop evapotranspiration depends on the climatological parameters and each crop behaves based on soil moisture content, nature of vegetation and management practices. To determine the crop evapotranspiration, FAO-56 Penman-Monteith (Allen et al.) equation is widely accepted and validated for wide range of climatic conditions. In the FAO-56 the atmospheric demand of evapotranspiration is referenced to a hypothetical grass of 0.12 m height, surface resistance of 70 sm^{-1} and an albedo of 0.23. Reference evapotranspiration is given by

$$ET_o = \frac{0.408\Delta(R_n - G) + \gamma \frac{900}{T+273} u_2(e_s - e_a)}{\Delta + \gamma(1 + 0.34u_2)}$$

ET_o = reference evapotranspiration [mm day^{-1}]
Rn = net radiation at the crop surface [MJ m^{-2} day^{-1}],

G = soil heat flux density [MJ m^{-2} day^{-1}],
T = mean daily air temperature at 2 m height [°C], u_2 is wind speed at 2 m height [m s^{-1}],
e_s = saturation vapor pressure [kPa],
e_a = actual vapor pressure [kPa],
e_s–e_a = saturation vapor pressure deficit [kPa],
Δ = slope of vapor pressure curve [kPa $°C^{-1}$] and
γ = psychrometric constant [kPa $°C^{-1}$].

13.2.2 Priestley-Taylor Method (PT)

Priestley and Taylor (1972) suggested an equation for wet regions. It is the simplified version of the combination equation which satisfies the condition for potential evaporation. The aerodynamic component was ignored and the energy component was multiplied by a factor, α = 1.26, when the general surrounding areas were wet or under humid conditions.

$$ET = \alpha \frac{\Delta}{\Delta + \gamma} \frac{Rn}{\lambda}$$

where Rn = the net radiation (cal/cm^2 day) is the difference between the net short-wave and long-wave radiations. The equations pertaining to the calculation of net radiation is presented in Chap. 3 of FAO-56 (Allen et al. 1998).

13.2.3 Hargreaves Method (HG)

The Hargreaves method (Hargreaves and Samani 1982) estimates ET_o based on maximum and minimum air temperature, and is written as

$$ET_o = 0.0023\text{Ra}\left[\frac{T\text{max} + T\text{min}}{2} + 17.8\right]\sqrt{(T\text{max} - T\text{min})}$$

where T_{max} = maximum air temperature (°C), T_{min} = minimum air temperature (°C), Ra = extra-terrestrial radiation (MJ·m^{-2}), and 0.408 is a factor to convert MJ m^{-2} to mm of water. Extra-terrestrial radiation, Ra, is estimated based on the location's latitude and the calendar day of the year by

$$\text{Ra} = \frac{24(60)}{\pi}\text{Gscdr}\ [\omega s \sin(\phi)\sin(\delta) + \cos(\phi)\cos(\delta)\sin(\omega s)]$$

where

Ra= extra-terrestrial radiation (MJ m^{-2} day^{-1});
Gsc= solar constant = 0.0820 MJ m^{-2} min^{-1};
dr = inverse relative distance Earth–Sun;
ωs= sunset hour angle;
ϕ= latitude (rad);
δ= solar decimation.

Further equations for calculating the parameters dr, ωs, ϕ and δ are given in Chap. 3 of FAO-56 (Allen et al. 1998).

13.2.4 Turc's Method (TC)

The Turc's method estimates monthly ET_o based on measurements of maximum and minimum air temperature and solar radiation using the equation

$$ET_o = 0.40\left[\frac{T_{mean}}{T_{mean} + 15}\right](\mathrm{Rs} + 50)$$

where ET_o = reference evapotranspiration (mm·mon^{-1}), Rs = solar radiation (MJ·m^{-2}), and T_{mean} = average air temperature (°C) calculated as $(T_{max} + T_{min})/2$. To estimate ET_o on a daily basis, the factor 0.40 is divided by 30 (average days per month), and the equation can be written as

$$ET_o = 0.0133\left[\frac{T_{mean}}{T_{mean} + 15}\right](\mathrm{Rs} + 50)$$

where ET_o = reference evapotranspiration (mm·day^{-1}).

The relationship to determine solar radiation from maximum and minimum temperature is given by

$$Rs = 0.16(T_{max} - T_{min})^{0.5} Ra$$

Rs and *Ra* have the same notations as in Hargreaves equation.

13.3 Study Area

The Seegodu climatological station in the state of Karnataka in India was used in this study. It is in the Western Ghats region of the west coast of the Deccan Plateau, which separates the coastal plain along the Arabian Sea. The region falls under tropical monsoon climate in Köppen climate classification. Weather data from 2012 to

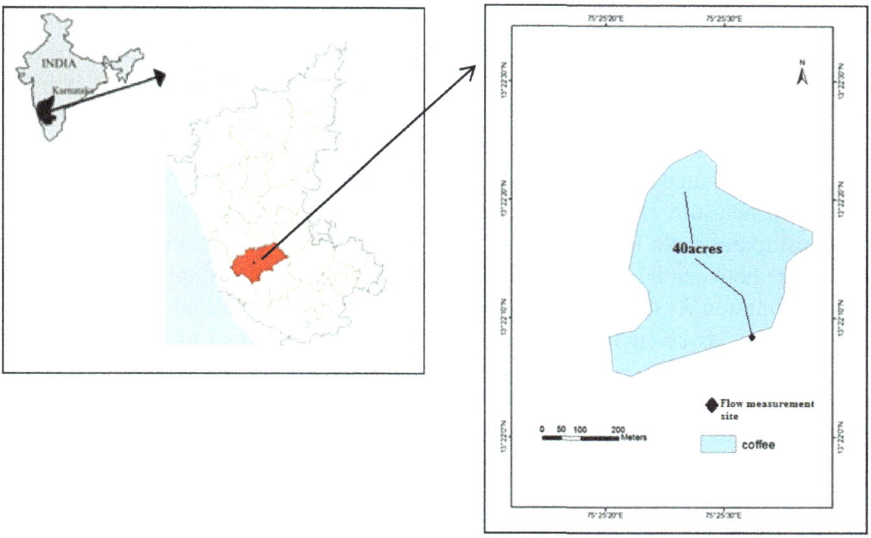

Fig. 13.1 Study Area

2015 was obtained from Central Coffee Research Institute Seegodu, a coffee plantation area. Coffee plantation under the tree shade is major practice in agroforestry and patches of arecanut plantations are common. Various climatological parameters including air temperature, sunshine hours and wind speed are used for the analysis. The location map of study area is presented in Fig. 13.1.

13.4 Results and Discussion

13.4.1 Comparisons of Daily ET_0 Methods

The meteorological data from 2012 to 2015 for the study were used to estimate the evapotranspiration by four ET_0 methods on daily basis. In order to compare the results, the mean daily ET_0 values were obtained by averaging daily results across the period of record and are tabulated in Table 13.1. Calculations on a daily basis showed that Hargreaves method makes better prediction of the reference ET_0 than the other

Table 13.1 Mean daily ET_0 estimates

Methods	Mean daily ET_0 Estimates (mm/day)
Penman-monteith method (PM)	4.19
Hargreaves method (HG)	4.04

two methods. In order to have the comparison between the methods, regression and error statistics were carried out to the whole set of data from all the three methods with Penman-Monteith method. Three error matrices, namely standard error of estimate (SEE),

standard deviation of the estimates (STDEV) and coefficient of determination (R^2), were computed for both daily and monthly estimates. Standard error of estimate (SEE) is a measure of precision, and standard deviation represents the dispersion of the estimates from FAO-56 method. Coefficient of determination represents the covariance between the two variables (Shreedhar et al. 2016). Maximum coefficient of determination R^2 of 0.63 for both Hargreaves and Turc method and a minimum of 0.36 were observed for Priestley-Taylor method. Standard errors of 1.51 and 1.58 mm/day were observed for Hargreaves and Turc method, respectively. Standard deviation of estimates (STDEV) of 1.69 and 0.91 mm/day were observed for HG and TC methods. In both the cases PT method predictions were not near to the standard PM method. Ranking of the equations for different methods are done based on least standard error of estimates (SEE), standard deviation of the estimates and highest coefficient of determination (Nandagiri and Kovoor 2006). Least average value of all the rankings is considered to be the best predictor of the ET_0 values and the methods are ranked according to their average rank. Scatter plots for different methods are

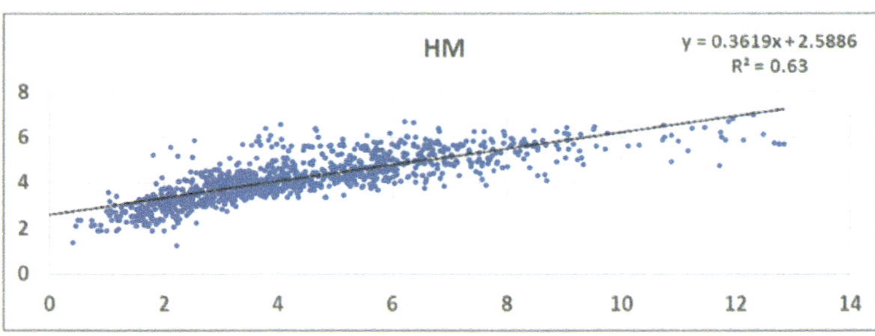

Fig. 13.2 Scatter plot of ET_0 values of PM method with Hargreaves method

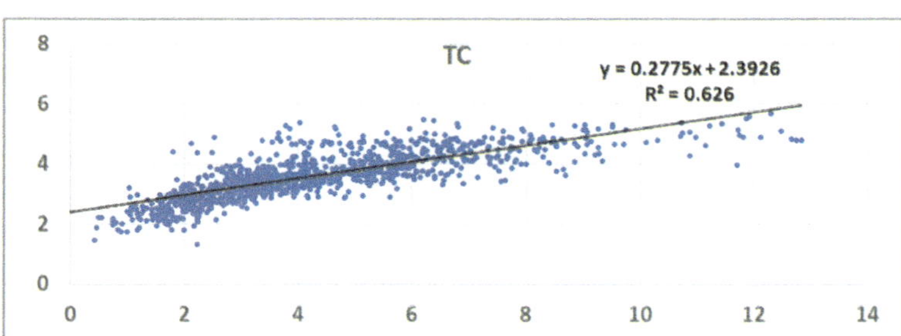

Fig. 13.3 Scatter plot of ET_0 values of PM method with Turc method

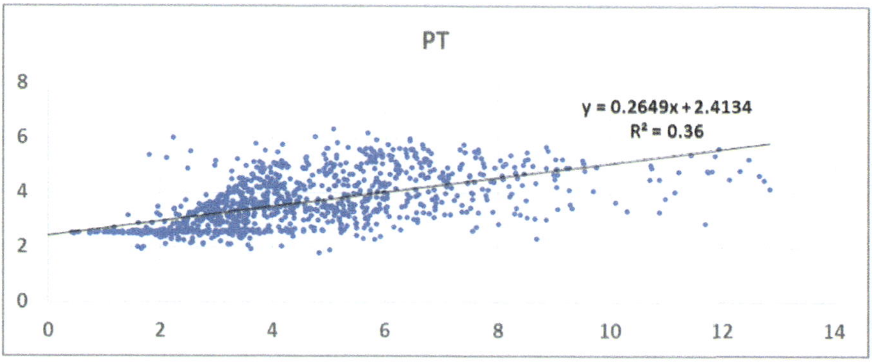

Fig. 13.4 Scatter plot of ET_0 values of PM method with Priestley-Taylor method

Table 13.2 Regression statistics for daily ET_0 comparisons

Station	Method	R^2	STDEV	SEE	Rank
Seegodu	HG	0.63(1)	1.69(3)	1.51(1)	1
N = 1461 (days)	PT	0.36(3)	1.09(2)	1.70(2)	3
	TC	0.63(2)	0.91(1)	1.58(3)	2

shown in Figs. 13.2, 13.3 and 13.4 and regression statistics for the daily values are presented in Table 13.2.

13.4.2 Comparisons of Monthly ET_0 Methods

Monthly predictions made by the other methods are very near to the standard method during the rainy seasons and vary to a large extent during the non-rainy seasons. This can be related to the effects of the other parameters which are not considered in the alternative methods, namely wind speed, sunshine hours and the dew point temperature. Overall monthly variations are shown in Fig. 13.5 and average monthly variations are presented in Fig. 13.6. The correlation between Penman-Monteith equation and other three methods is presented in Figs. 13.7, 13.8 and 13.9. In most of the monthly values, all the three methods underestimated the ET_0 values compared to Penman-Monteith method. Regression statistics for the monthly values are presented in Table 13.3.

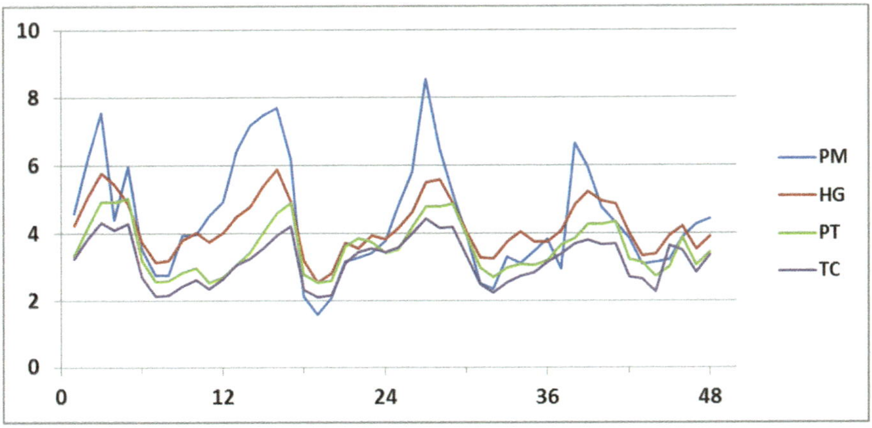

Fig. 13.5 Comparison of monthly ET_0 (2012–2015) computed by four different methods

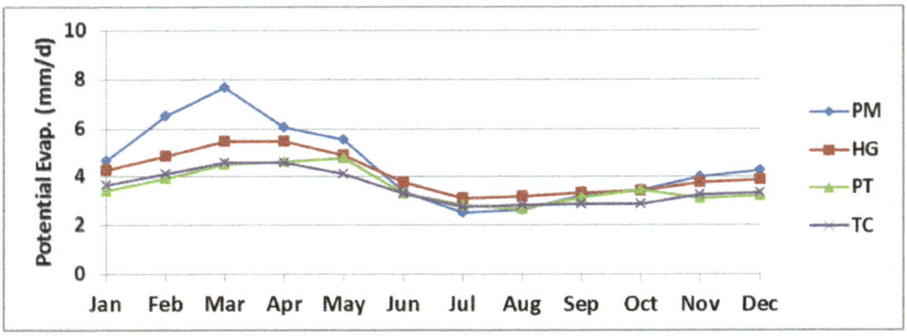

Fig. 13.6 Comparison of mean monthly ET_0 computed by four different methods

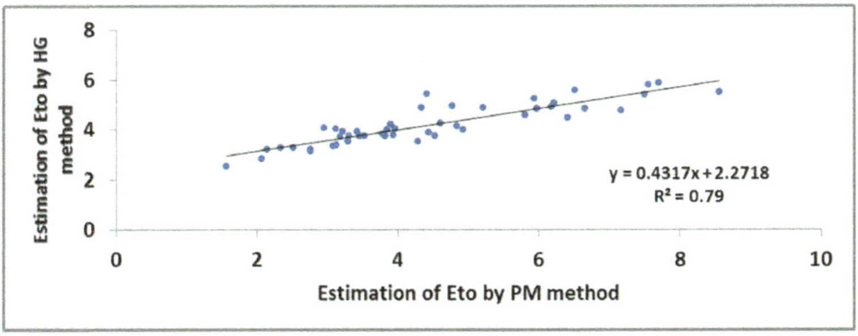

Fig. 13.7 Scatter plot of ET_0 values of PM method with Hargreaves method

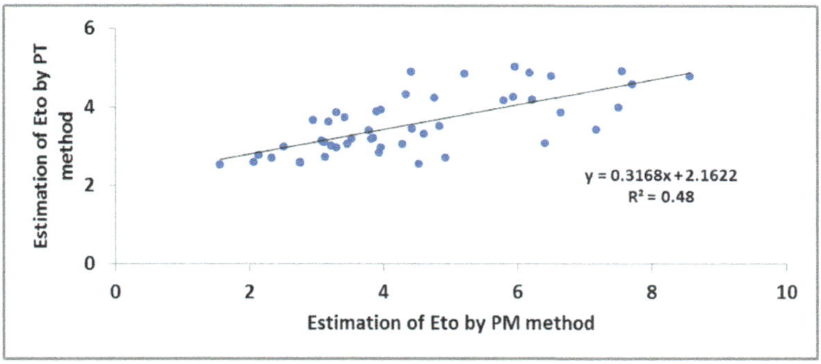

Fig. 13.8 Scatter plot of ET_0 values of PM method with Priestley-Taylor method

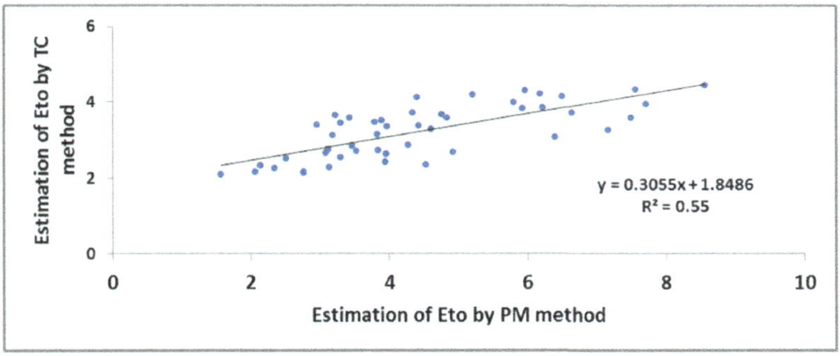

Fig. 13.9 Scatter plot of ET_0 values of PM method with Turc's method

Table 13.3 Regression statistics for monthly ET_0 comparisons

Station	Method	R^2	STDEV	SEE	Rank
Seegodu	HG	0.80(1)	1.32(3)	0.77(3)	2
N = 48	PT	0.48(3)	0.84(2)	0.42(2)	3
	TC	0.55(2)	0.75(1)	0.26(1)	1

13.5 Conclusion

The present study was carried out in Seegodu coffee plantation area which is a major agroforestry in the region to identify the best alternative method to FAO-56 Penman-Monteith method with least climatological parameters. The observed data from 2012 to 2015 were used to estimate ET_0 values from three different alternative methods. The following observations were made from the study:

1. Hargreaves method makes better prediction of ET_0 near to Penman-Monteith method than other two methods on a daily scale and Turc method makes better prediction for the monthly values than other two methods.
2. In the summer and winter season, the estimates of the other methods are less than that of the Penman-Monteith method. Therefore, the other climatological parameters such as wind speed and sunshine hours play a major role in ET_0 of Penman-Monteith equation.

In conclusion, one can choose either Hargreaves or Turc's method based on the temporal scale in the study area considered.

Acknowledgements The meteorological data was procured from Central Coffee Research Institute (CCRI), Seegodu, India. The authors gratefully acknowledge D'Souza George Felix, Patil Somashekhargouda and Deepika Kumar Umesh for their kind support in carrying out this study.

References

Allen RG, Pereira LS, Raes D, Smith M (1998) Crop evapotranspiration: Guidelines for computing crop water requirements. FAO Irrigation and Drainage Paper No. 56, FAO, Rome
Beven KJ (2011) Rainfall-runoff modelling: the primer. John Wiley & Sons
George BA, Reddy BRS, Raghuwanshi NS, Wallender WW (2002) Decision support system for estimating reference evapotranspiration. J Irrigat Drainage Eng 128(1):1–10
Hargreaves GH, Samani ZA (1982) Estimating potential evapotranspiration. J Irrigat Drainage Eng 108:223–230
Nandagiri L, Kovoor GM (2006) Performance evaluation of reference evapotranspiration equations across a range of Indian climates. J Irrigat Drainage Eng 132(3):238–249
Priestley CHB, Taylor RJ (1972) On assessment of surface heat flux and evaporation using large-scale parameters. Mon Weather Rev 100:81–92
Shreedhar S, Venkatesh B, Purandara BK (2016) A comparison of five potential evapotranspiration methods for biligihole watershed in western ghats of india. Int J Scient Eng Res 7(4)
Xu CY, Singh VP (2002) Cross comparison of empirical equations for calculating potential evapotranspiration with data from Switzerland. Water Resour Manage 16:197–219

Chapter 14
Influence of Air Temperature on Local Precipitation Extremes Across India

Sachidanand Kumar, Kironmala Chanda, and Srinivas Pasupuleti

Abstract In the present study, the influence of maximum temperature on rainfall events at the daily scale is explored at four locations—Chennai, Kolkata, Mumbai and New Delhi, which represent different climatic regions across India. The widely accepted binning technique is used to pair the daily temperature and rainfall (95th and 50th percentile). Two different data products—global gridded data (1979–2017) and station-based global dataset (1991–2017) are used and the scaling relationships are compared alongside the Clausius-Clapeyron scaling of 6.8% K^{-1}. Results indicate negative scaling, ranging from −0.4 to 22% per degree Celsius (for 95th percentile rainfall) and −3 to −41% per degree Celsius (for 50th percentile rainfall) at all studied locations. It is also noticed that rainy days (> 0.3 mm rainfall) are mostly observed when temperature is above 30 °C in Kolkata and New Delhi, but in Chennai and Mumbai, rainy days are fairly common even when the temperature is below 30 °C. The evolution of the scaling relationship is studied considering four sequential time periods: 1979–1988, 1989–1998, 1999–2008 and 2009–2017. All the locations indicate negative scaling for all decades except New Delhi which indicates a very mild positive scaling during 1979–1988. Moreover, the relationship of daily rainfall with preceding days' maximum temperature is also studied for Kolkata and Chennai, with the former showing positive scaling at three-day lag and beyond.

Keywords Extreme precipitation · Atmospheric temperature · Clausius-clapeyron scaling · India

---

S. Kumar · K. Chanda (✉) · S. Pasupuleti
Department of Civil Engineering, Indian Institute of Technology (Indian School of Mines), Dhanbad 826004, Jharkhand, India
e-mail: kironmala.iitkgp@gmail.com; kironmala@iitism.ac.in

S. Kumar
e-mail: sachincit70@gmail.com; sachidanand.17dr000533@cve.ism.ac.in

S. Pasupuleti
e-mail: vasu77.p@gmail.com; srinivas@iitism.ac.in

R. Jha et al. (eds.), *Climate Change Impacts on Water Resources*, Water Science and Technology Library 98, https://doi.org/10.1007/978-3-030-64202-0_14

14.1 Introduction

Globally, extreme precipitation and hydrological events are on the rise over the past 50 years. Such increase in extreme events is often linked to global warming which causes an increase in the atmospheric moisture-holding capacity (Trenberth et al. 2003; Shaw et al. 2011; Utsumi et al. 2011; Shiu et al. 2012; Mishra et al. 2012; Panthou et al. 2014; Westra et al. 2012, 2014; Wasko and Sharma 2014; Wasko et al. 2015; Ali and Mishra 2017; and Mukherjee et al. 2018). As per the Clausius-Clapeyron (C-C) relationship, atmospheric moisture content increases approximately by 6.8% K^{-1} increase in temperature at a constant relative humidity (Hardwick Jones et al. 2010). Moreover, the entire atmospheric moisture content is generally considered to be converted to rain in case of extremes in rainfall. Hence, the relationship of rainfall extremes with temperature may be expected to be similar to Clausius-Clapeyron (C-C) scaling. However, the C-C scaling is found to decrease with temperature: 7.3% K^{-1} at 0 °C, 6.4% K^{-1} at 15 °C and 6.0% K^{-1} at 25 °C (Panthou et al. 2014). In fact, C-C scaling values could range between 7% K^{-1} (Lenderink and van Meijgaard 2008, 2010; Lendrink et al. 2011; Shaw et al. 2011; Trenberth 2011; Shiu et al. 2012; and Panthou et al. 2014), 6–7% K^{-1} (Schneider et al. 2010) and 6.8% K^{-1} (Hardwick Jones et al. 2010). Further, the relationship between rainfall and temperature may vary substantially depending on the range of temperature, latitude and convection characteristics among other factors (Hardwick Jones et al. 2010; Lenderink and van Meijgaard 2010).

For example, a decreasing scaling relationship was observed between the median rainfall intensity and temperature in Australia and Europe but for heavy rainfall intensity, an increasing scaling was observed (Hardwick Jones et al. 2010; Lenderink and van Meijgaard 2008, 2010 and Wasko et al. 2015). The scaling may vary seasonally also as observed in Australia (Wasko and Sharma 2014) and Europe (Berg et al. 2009; Shaw et al. 2011; and Mishra et al. 2012). The C-C scaling was found to decrease with increase in the duration of rainfall events in Canada (Panthou et al. 2014). Similar findings were also reported in Japan (Utsumi et al. 2011), Hong Kong (Lenderink et al. 2011), India (Vittal et al. 2016; Ali and Mishra 2017) and the USA (Shaw et al. 2011; Mishra et al. 2012). In Australia, the rainfall-temperature relationship is found very strong for shorter duration rainfall events but weakens for longer duration events (Herath and Sarukkalige 2018). At higher temperature in the tropics, a negative C-C scaling relationship between rainfall extreme and surface air temperature is often attributed to local cooling effect (Vittal et al. 2016; Ali and Mishra 2017; Zhang et al. 2017; Wang et al. 2017; Mukherjee et al. 2018). Such decrease in C-C scaling could also be accounted for inadequate availability of moisture in the atmosphere which is evident from the decreasing relative humidity at such temperatures (Hardwick Jones et al. 2010). In addition to surface air temperature, the dew point temperature is also sometimes used for temperature-extreme rainfall scaling in the tropics (Lenderink and van Meijgaard 2010; Ali and Mishra 2017).

In the present study, the influence of maximum temperature on extreme precipitation events at the daily scale is explored for a number of locations over different climatic zones in India. Apart from the same-day relationship, the influence of preceding days' temperature on rainfall extremes is also investigated. The evolution of the scaling relationship over time is studied by splitting the available dataset decade-wise. Two different data products—global gridded data and station-based global dataset are used to perform the analysis and compare the scaling relationships.

14.2 Study Area and Data

The four locations—Chennai, Kolkata, Mumbai and New Delhi—are chosen as the study locations to represent different climatic regions across India. Figure 14.1 shows a map of India with approximate positions of the study locations.

Gridded daily maximum temperature data (°C/day) and daily rainfall data (mm/day) are acquired from the Climate Prediction Center (CPC), National Oceanic and Atmospheric Administration (NOAA), Earth System Research Laboratory (ESRL), Physical Sciences Division (PSD), Boulder, Colorado, USA, for the period 1979–2017 at a spatial resolution of 0.5° latitude × 0.5° longitude. Apart from the gridded data, station-based daily rainfall along with daily maximum temperature are also acquired from the Global Summary of the Day (GSOD) dataset which is derived from the Integrated Surface Hourly (ISH) obtained from the United States Air Force (USAF) climatology data. This station-based data is used for the period 1991–2017 as the data was found to be inconsistent before 1991 with several days data missing in each month. For Chennai and Kolkata, the analysis is performed with both gridded dataset and the GSOD station-based data for a comparison. However, the same was not performed for Mumbai and New Delhi as the GSOD data was unavailable for several months for these two locations. The details of the data used for the analysis are summarized in Table 14.1.

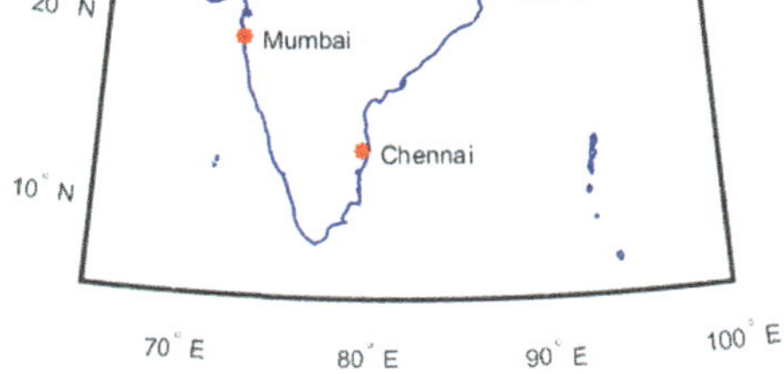

Fig. 14.1 Study locations

14.3 Methodology

14.3.1 Pairing the Maximum Daily Temperature and Daily Rainfall of Wet Days

For each location, the days recording a daily rainfall value of greater than 0.3 mm is considered as wet days and these are identified for the analysis. The corresponding daily maximum temperature is also noted and these are paired together and then the pairs are sorted based on the ascending order of the daily maximum temperatures. This sorted dataset is then segregated into several bins considering a certain range of temperature interval for each bin. This technique is known as the *binning technique* and has been used in several studies (Panthou et al. 2014; Lenderink and Van Meijgaard 2008, 2010; Hardwick Jones et al. 2010; Herath and Sarukkalige 2018). The aforementioned studies used equal-width bins which tend to produce very less number of data points in the bins at the extreme ends. This limitation was addressed by Wasko and Sharma (2014) who proposed unequal bin sizes, keeping the number of datapoints equal in each bin. This modified approach has been adopted in the present study. The temperature-rainfall pairs are sorted and segregated in a suitable number of bins and the median temperature of the datapoints from each bin is considered as the representative temperature from that bin. Within each bin, the rainfall intensities are fitted to an empirical cumulative distribution function and the extreme rainfall (95th percentile) and median rainfall (50th percentile) are obtained for each temperature bin.

In order to analyze the influence of the preceding day's maximum daily temperature on daily rainfall, the aforementioned technique is also applied to pairs consisting of daily rainfall and preceding day's (1-day, 3-day, 7-day and 14-day) maximum temperature. This is explained further in Sect. 14.4.3.

14.3.2 Clausius-Clapeyron (C-C) Scaling

As per the Clausius-Clapeyron (C-C) relationship, the atmospheric moisture content increases approximately by 6.8% K^{-1} at a constant relative humidity. Though some studies report the C-C scaling to range from 6 to 7%, we use the generally accepted value of 6.8% K^{-1} in the present study. Since the intensity of rainfall is expected to increase with available moisture content, their relationship with temperature may be expected to be similar to C-C scaling. Literature (Hardwick Jones et al. 2010; Utsumi et al. 2011) suggests that the precipitation-temperature relationship may be given by:

$$P_2 = P_1(1 + \alpha)^{\Delta T} \tag{14.1}$$

Table 14.1 Details of study locations and data records used

S. no.	Station name	Coordinates	Gridded NOAA data period	Station-based GSOD data period
1	Chennai	13.08 °N, 80.27 °E	1979–2017	1991–2017
2	Kolkata	22.57 °N, 88.36 °E	1979–2017	1991–2017
3	Mumbai	19.07 °N, 72.87 °E	1979–2017	–
4	New Delhi	28.70 °N, 77.10 °E	1979–2017	–

where P_1 and P_2 represent the precipitation percentiles and T_1 and T_2 denote corresponding temperatures in degree Celsius. α is the precipitation-temperature scaling coefficient which is considered 6.8% °C^{-1} at 25 °C as per C-C scaling. ΔT is the temperature difference between T_1 and T_2.

In the present study, the extreme precipitation is represented by 95th percentile and the median precipitation is represented by 50th percentile. The relationships of these with maximum daily temperature are determined (Table 14.1).

14.4 Results and Discussions

14.4.1 Relationship Between Rainfall and Daily Maximum Temperature

14.4.1.1 Using CPC Gridded Dataset

The CPC gridded daily temperature and rainfall data for the period 1979–2017 is used for this analysis. Figure 14.2 shows the variation in the logarithms of 95th and 50th percentiles of daily precipitation with paired daily maximum temperature for each of the four study locations. The linear regression line (in solid red and blue for 95th and 50th percentile, respectively) is indicated alongside the C-C relationship (in dashed red and blue for 95th and 50th percentile, respectively). Negative scaling is evident for all the studied locations. The slope of the regressed line is slightly negative for New Delhi, whereas the negative slope is quite strong for the remaining locations, varying between 11 and 22% in case of 95th percentile rainfall and between 19 and 41% in case of 50th percentile rainfall. These are summarized in Table 14.2.

Another feature to be noted from the plots in Fig. 14.2 is that altogether 14 bins (not necessarily equal temperature intervals) are considered for each location to allow equal number of datapoints in each bin. The distribution of scatter points indicates that the rainy days are mostly observed when temperature is above 30 °C in Kolkata and New Delhi, but in Chennai and Mumbai, rainy days are fairly common even when temperature is below 30 °C.

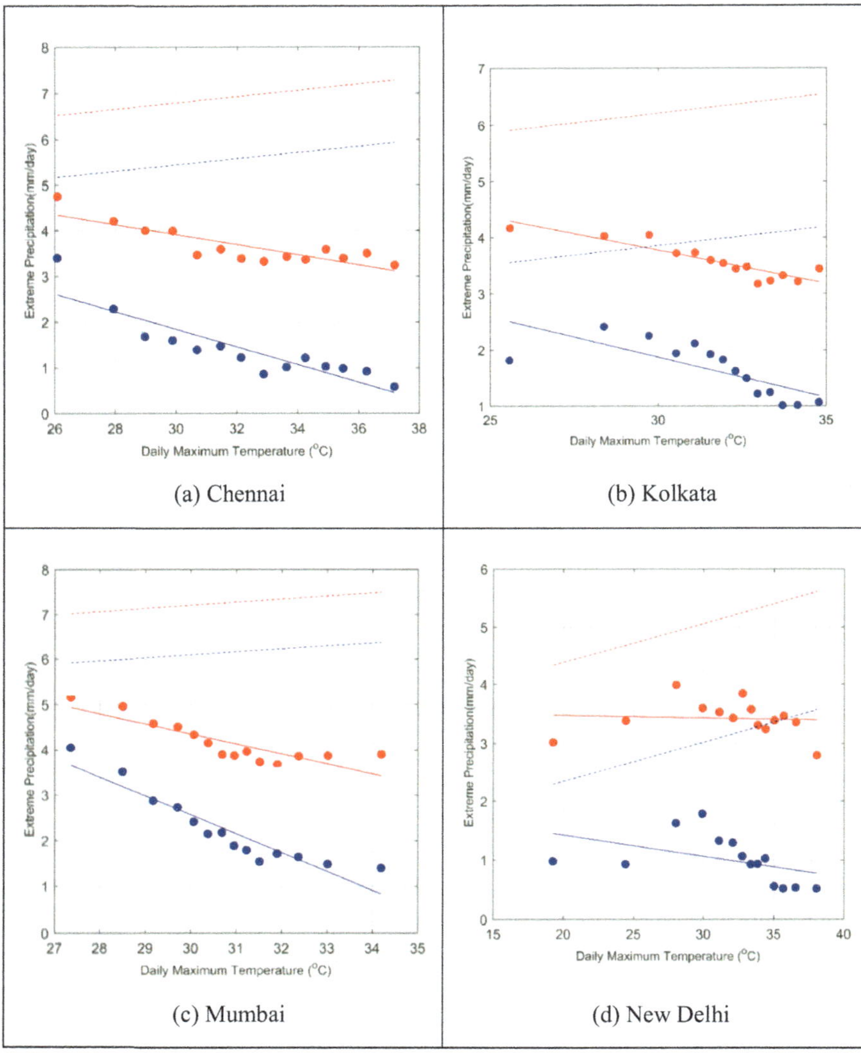

Fig. 14.2 Relationship between logarithm of extreme daily rainfall (95th percentile) and maximum daily temperature are shown in red while the relationship between logarithm of median rainfall (50th percentile) and maximum daily temperature are shown in blue for the locations **a** Chennai, **b** Kolkata, **c** Mumbai and **d** New Delhi. The C-C scaling is represented by the dashed red line for 95th percentile and dashed blue line for 50th percentile. The CPC gridded dataset is used for all the plots

Table 14.2 Comparison of scaling relationship

Location	Scaling relationship			
	CPC gridded data		Station-based GSOD data	
	95th Percentile	50th Percentile	95th Percentile	50th Percentile
Chennai	−0.111	−0.194	−0.135	−0.179
Kolkata	−0.119	−0.144	−0.126	−0.143
Mumbai	−0.222	−0.414	–	–
New Delhi	−0.004	−0.036	–	–

14.4.1.2 Using GSOD Data

Using the station-based GSOD dataset, the variation of the logarithms of 95th and 50th percentiles of daily precipitation with paired daily maximum temperature for Kolkata and Chennai is presented in Fig. 14.3. In general, the relationships are in agreement with the corresponding ones in Fig. 14.2. Negative scaling with similar scaling values as in the case of CPC gridded data is also observed for GSOD data. As observed earlier, rainy days are mostly detected when temperature is above 30 °C in Kolkata. As mentioned before, the GSOD data was not found to be consistent for Mumbai and New Delhi and hence these are not included in the analysis.

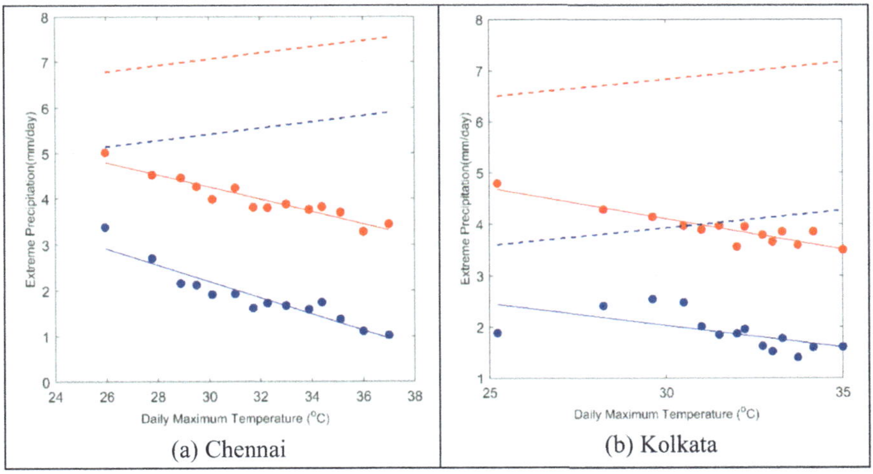

Fig. 14.3 Relationship between logarithm of extreme daily rainfall (95th percentile) and maximum daily temperature is shown in red, while the relationship between logarithm of median rainfall (50th percentile) and maximum daily temperature is shown in blue for the locations **a** Chennai and **b** Kolkata. The C-C scaling is represented by the dashed red line for 95th percentile and dashed blue line for 50th percentile. The station-based GSOD dataset is used for all the plots

14.4.2 Evolution of the Scaling Relationship Over Time

The evolution of the scaling relationship between rainfall and maximum temperature is studied considering four sequential time periods: 1979–1988, 1989–1998, 1999–2008 and 2009–2017. The CPC dataset is used for this analysis as it has sufficiently long and consistent records. The decade-wise scaling relationships for the study locations are presented in Fig. 14.4 and summarized in Table 14.3. All the locations indicate negative scaling for all decades except New Delhi which has a very mild positive scaling during 1979–1988. In general, the scaling coefficients are small (between 0.03% and −5.9%) for New Delhi and the strongest for Mumbai (in the range of −19% to −48%). It is observed that in Chennai scaling relationship is most strong (about −19% for 95th percentile and −24% for 50th percentile) during 1989–1998, while in Kolkata it is least strong (about −7.6% for 95th percentile and −9.6% for 50th percentile) during the same decade.

14.4.3 Relationship Between Rainfall and Maximum Temperature of Preceding Days

The scaling relationships between daily rainfall with daily maximum temperature of preceding days (1, 3, 7 and 14-day) are investigated. The station-based GSOD data is used for this analysis. The plots are presented in Fig. 14.5 (Chennai) and Fig. 14.6 (Kolkata) and the scaling coefficients are summarized in Table 14.4. For Chennai, negative scaling relationship is observed with all preceding day temperatures and the coefficients generally decrease with increase in lag. However, in Kolkata, the scaling relationship turns positive at 3-day lag and continues to be so till 14 days lag.

14.5 Concluding Remarks

In this study, the relationship between rainfall and daily maximum temperature is examined at four study locations representing different climatic regions across India using two global datasets: gridded CPC data and station-based GSOD data. The scaling relationship is found to deviate substantially from the Clausius-Clapeyron scaling of 6.8% K^{-1}. In fact, negative scaling is observed at all locations, with scaling coefficients as high as −22% for 95th percentile daily rainfall and −41% for 50th percentile daily rainfall for Mumbai. It is also noticed that rainy days (> 0.3 mm rainfall) are mostly observed when temperature is above 30 °C in Kolkata and New Delhi, but in Chennai and Mumbai, rainy days are fairly common even when temperature is below 30 °C. The observations using CPC data and GSOD data are found to be in agreement with each other. Using the CPC dataset, the evolution of the scaling relationship is studied considering four sequential time periods: 1979–1988,

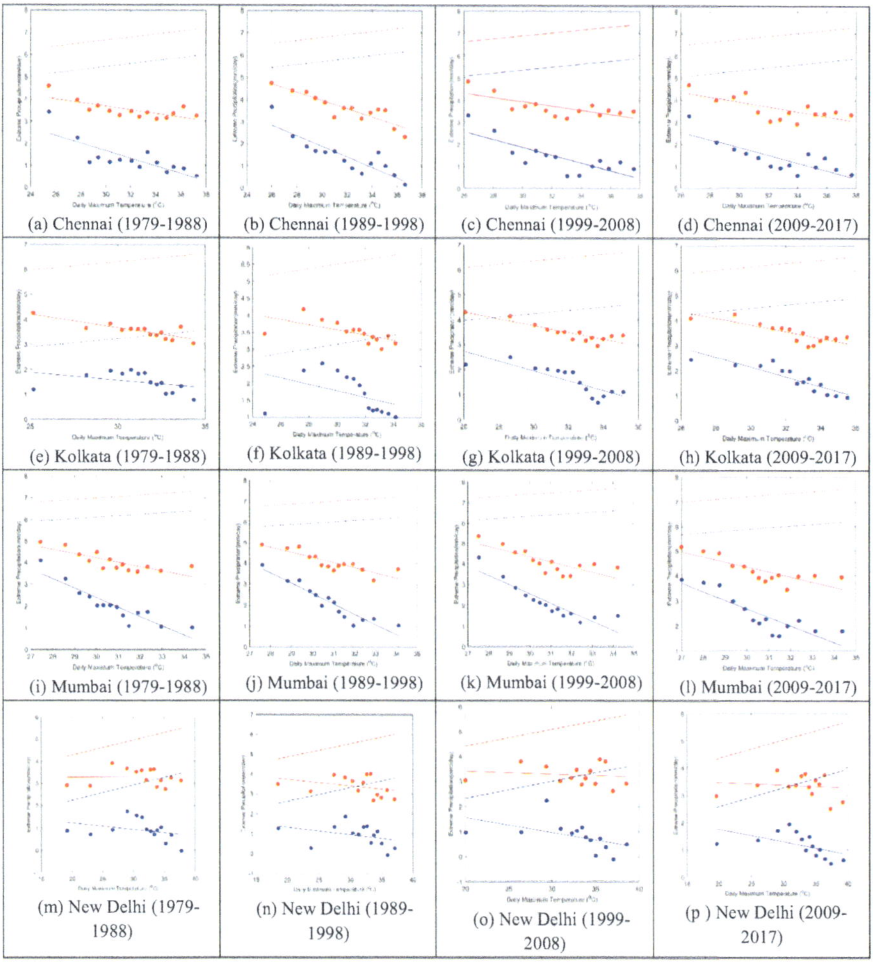

Fig. 14.4 Relationship between extreme daily rainfall (95th percentile) and maximum daily temperature is shown in red, while the relationship between median rainfall (50th percentile) and maximum daily temperature is shown in blue for the locations Chennai (topmost panel), Kolkata (2nd panel from top), Mumbai (3rd panel from top) and New Delhi (bottom-most panel). The C-C scaling is represented by the dashed red line for 95th percentile and dashed blue line for 50th percentile. The CPC gridded dataset is used for all decadal time periods: 1979–1988 (left most), 1989–1998 (2nd from left), 1999–2008 (3rd from left) and 2009–2017 (right most)

1989–1998, 1999–2008 and 2009–2017. All locations (except New Delhi, during 1979–1988) show negative scaling for both 95th and 50th percentile daily rainfall. However, the scaling coefficients fluctuate during the different temporal windows. While studying the scaling relationship between daily rainfall and maximum daily temperature of preceding days, it is observed that for Kolkata, the relationship is

Table 14.3 Comparison of scaling relationship for different temporal windows

	1979–1988		1989–1998		1999–2008		2009–2017	
	95th	50th	95th	50th	95th	50th	95th	50th
Chennai	−0.082	−0.172	−0.186	−0.240	−0.095	−0.177	−0.112	−0.178
Kolkata	−0.106	−0.061	−0.076	−0.097	−0.135	−0.195	−0.135	−0.196
Mumbai	−0.120	−0.427	−0.249	−0.488	−0.250	−0.437	−0.203	−0.342
New Delhi	0.0003	−0.029	−0.032	−0.038	−0.011	−0.060	−0.010	−0.046

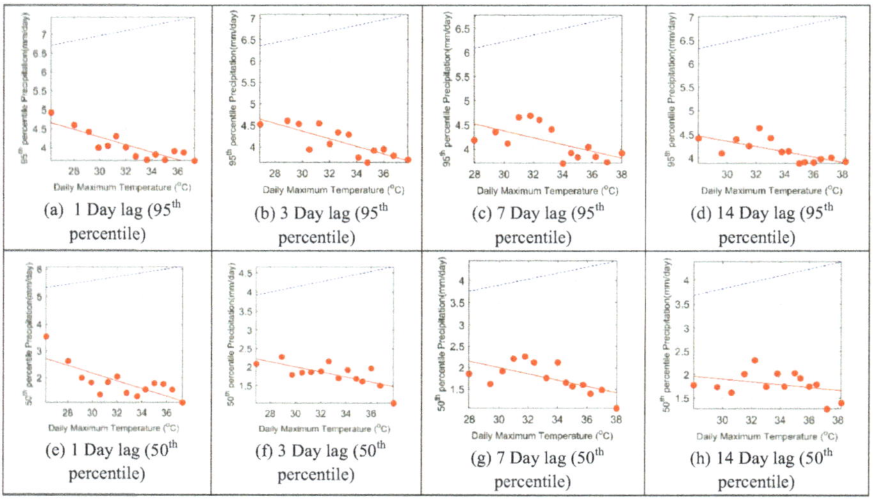

(a) 1 Day lag (95th percentile) (b) 3 Day lag (95th percentile) (c) 7 Day lag (95th percentile) (d) 14 Day lag (95th percentile)

(e) 1 Day lag (50th percentile) (f) 3 Day lag (50th percentile) (g) 7 Day lag (50th percentile) (h) 14 Day lag (50th percentile)

Fig. 14.5 Relationship between extreme daily rainfall (95th percentile) and maximum daily temperature as well as the relationship between median rainfall (50th percentile) and maximum daily temperature are shown in red for the location Chennai. The C-C scaling is represented by the dashed blue line. The station-based GSOD dataset is used. The lag considered is mentioned alongside each subplot

negative up to 1-day lag but turns positive at 3-day lag and beyond. However, in Chennai, the scaling continues to be negative at all lags considered.

While some of the aspects of the temperature-precipitation scaling are assessed in this study, further analysis may be interesting, particularly the investigation of sub-daily storm events. These studies are expected to help in estimating the uncertainty in the frequency and intensity of extreme precipitation events due to anthropogenic climate change.

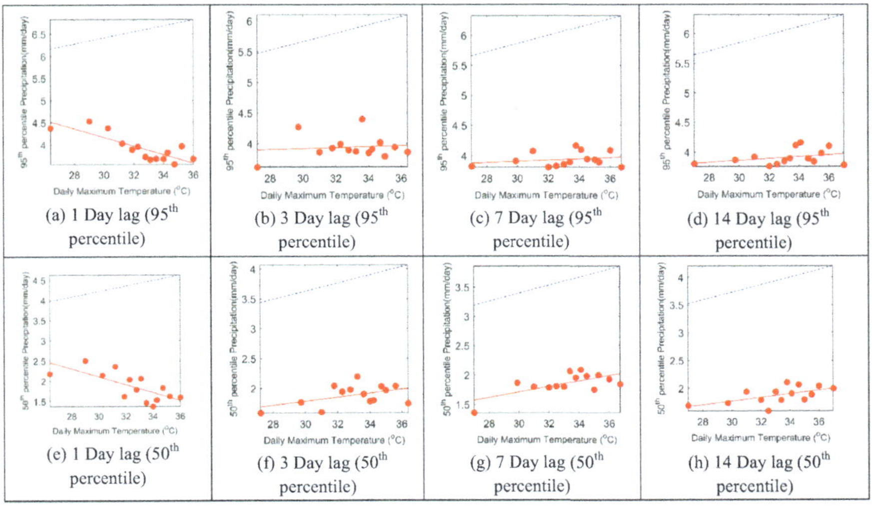

(a) 1 Day lag (95th percentile) (b) 3 Day lag (95th percentile) (c) 7 Day lag (95th percentile) (d) 14 Day lag (95th percentile)

(e) 1 Day lag (50th percentile) (f) 3 Day lag (50th percentile) (g) 7 Day lag (50th percentile) (h) 14 Day lag (50th percentile)

Fig. 14.6 Relationship between extreme daily rainfall (95th percentile) and maximum daily temperature as well as the relationship between median rainfall (50th percentile) and maximum daily temperature are shown in red for the location Kolkata. The C-C scaling is represented by the dashed blue line. The station-based GSOD dataset is used. The lag considered is mentioned alongside each subplot

Table 14.4 Comparison of the scaling relationship between rainfall and maximum temperature of preceding days

Preceding days	Chennai		Kolkata	
	95th	50th	95th	50th
1-Day lag	−0.101	−0.141	−0.096	−0.098
3-Day lag	−0.089	−0.068	0.008	0.034
7-Day lag	−0.071	−0.073	0.0010	0.046
14-Day lag	−0.057	−0.030	0.016	0.033

Acknowledgements The authors acknowledge the data from NOAA/OAR/ESRL PSD Boulder, Colorado, USA (https://www.esrl.noaa.gov/psd/) for gridded data and station-based GSOD data from the synoptic/hourly observations contained in USAF DATSAV3 surface data and Federal Climate Complex Integrated Surface Data (ISD) (ftp://ncdc.noaa.gov/pub/data/gsod/).

References

Ali H, Mishra V (2017) Contrasting response of rainfall extremes to increase in surface air and dewpoint temperatures at urban locations in India. Scient Rep 7(1):1228

Berg P, Haerter JO, Thejll P, Piani C, Hagemann S, Christensen JH (2009) Seasonal characteristics of the relationship between daily precipitation intensity and surface temperature. J Geophys Res Atmosp 114(D18)

Hardwick Jones R, Westra S, Sharma A (2010) Observed relationships between extreme sub-daily precipitation, surface temperature, and relative humidity. Geophys Res Lett 37(22)

Herath SM, Sarukkalige R (2018) Evaluation of empirical relationships between extreme rainfall and daily maximum temperature in Australia. J Hydrol 556:1171–1181

Lenderink G, Van Meijgaard E (2008) Increase in hourly precipitation extremes beyond expectations from temperature changes. Nat Geosci 1(8):511

Lenderink G, Van Meijgaard E (2010) Linking increases in hourly precipitation extremes to atmospheric temperature and moisture changes. Environ Res Lett 5(2):025208

Lenderink G, Mok HY, Lee TC, Van Oldenborgh GJ (2011) Scaling and trends of hourly precipitation extremes in two different climate zones–Hong Kong and the Netherlands. Hydrol Earth Syst Sci 15(9):3033–3041. https://doi.org/10.5194/hess-15-3033-2011

Mishra V, Wallace JM, Lettenmaier DP (2012) Relationship between hourly extreme precipitation and local air temperature in the United States. Geophys Res Lett 39(16)

Mukherjee S, Aadhar S, Stone D, Mishra V (2018) Increase in extreme precipitation events under anthropogenic warming in India. Weather and Climate Extremes

Panthou G, Mailhot A, Laurence E, Talbot G (2014) Relationship between surface temperature and extreme rainfalls: A multi-time-scale and event-based analysis. J Hydrometeorol 15(5):1999–2011

Schneider T, O'Gorman PA, Levine XJ (2010) Water vapor and the dynamics of climate changes. Rev Geophys 48(3)

Shaw SB, Royem AA, Riha SJ (2011) The relationship between extreme hourly precipitation and surface temperature in different hydroclimatic regions of the United States. J Hydrometeorol 12(2):319–325

Shiu CJ, Liu SC, Fu C, Dai A, Sun Y (2012) How much do precipitation extremes change in a warming climate?. Geophys Res Lett 39(17)

Trenberth KE (2011) Changes in precipitation with climate change. Climate Res 47(1–2):123–138

Trenberth KE, Dai A, Rasmussen RM, Parsons DB (2003) The changing character of precipitation. Bull Am Meteor Soc 84(9):1205–1218

Utsumi N, Seto S, Kanae S, Maeda EE, Oki T (2011) Does higher surface temperature intensify extreme precipitation?. Geophys Res Lett 38(16)

Vittal H, Ghosh S, Karmakar S, Pathak A, Murtugudde R (2016) Lack of dependence of Indian summer monsoon rainfall extremes on temperature: an observational evidence. Scient Rep 6:31039

Wang G, Wang D, Trenberth KE, Erfanian A, Yu M, Bosilovich MG, Parr DT (2017) The peak structure and future changes of the relationships between extreme precipitation and temperature. Nature Climate Change 7(4):268

Wasko C, Sharma A (2014) Quantile regression for investigating scaling of extreme precipitation with temperature. Water Resour Res 50(4):3608–3614

Wasko C, Sharma A, Johnson F (2015) Does storm duration modulate the extreme precipitation-temperature scaling relationship? Geophys Res Lett 42(20):8783–8790

Westra S, Fowler HJ, Evans JP, Alexander LV, Berg P, Johnson F, Roberts NM (2014) Future changes to the intensity and frequency of short-duration extreme rainfall. Rev Geophys 52(3):522–555

Westra S, Mehrotra R, Sharma A, Srikanthan R (2012) Continuous rainfall simulation: 1. A regionalized subdaily disaggregation approach. Water Resour Res 48(1)

Zhang X, Zwiers FW, Li G, Wan H, Cannon AJ (2017) Complexity in estimating past and future extreme short-duration rainfall. Nat Geosci 10(4):255–259

Chapter 15
Effect of Spatial and Temporal Land Use-Land Cover Change on the Rainfall Trend: A Case Study in Kerala

Lini R. Chandran and P. G. Jairaj

Abstract Climate change is a global phenomenon which is mainly caused due to anthropogenic activities. Land use-land cover (LULC) changes occur in river basins due to human interventions and change in climate. The change in LULC causes change in evapotranspiration and contributes to further climatic change in the area. Hydrologic extremes occur in a river basin due to the combined effect of climatic extremes and LULC changes. Both climate and LULC change scenarios are usually used for estimating the variation in the hydrological response of a watershed. Hence LULC change detection and estimation is very essential for any watershed prior to the estimation of hydrologic response variation from it. The influence of LULC change on the rainfall variability of a region was studied taking the Bharathapuzha river basin which is spread over Tamil Nadu and Kerala. Bharathapuzha river is the second longest river in Kerala. Drastic changes have taken place in the river basin which caused the degradation of the river. Hence the study of the effect of LULC change in the study region on the rainfall trend is of great significance. The LULC change in the area was estimated both spatially and temporally using remote sensing-based techniques. The LANDSAT1 and LANDSAT8 satellite images for the years 1973 and 2018 were used for determining the land cover change in the study region. Necessary corrections were applied on the mosaicked images and these satellite images were classified using supervised classification. The land use change analysis was done on the classified images and the land use classes used were waterbodies, mixed vegetation, agricultural field, cultivable land, dense trees, barren land/rocky outcrop, built-up area and sandy area. It is seen that there has been severe reduction in dense trees and large increase in barren land and built-up areas over time. The rainfall trend in the study region was estimated using the data obtained from India Meteorological Department. The study indicates that LULC change is associated to the decrease in the rainfall over the area. It denotes the necessity of controlling the

L. R. Chandran (✉) · P. G. Jairaj
Department of Civil Engineering, College of Engineering, Trivandrum, Kerala, India
e-mail: linisam@cet.ac.in

P. G. Jairaj
e-mail: jairaj@cet.ac.in

R. Jha et al. (eds.), *Climate Change Impacts on Water Resources*, Water Science and Technology Library 98, https://doi.org/10.1007/978-3-030-64202-0_15

LULC changes in the basin for preventing further deterioration of the basin. The rainfall trend analysis also helps to devise proper management plans for the basin.

Keywords Climate change · Land use-land cover change · Rainfall trend · Mann–kendall test · Sen's slope estimator

15.1 Introduction

Climate change is of great concern during present time. Even though there are many factors that affect the climate change in a region, many anthropogenic activities leading to land use-land cover change are seen to have an impact in the climate change of an area. Understanding the rainfall pattern in a local scale is important for water resources planning in an area. It is of great importance in a region where agriculture and many hydroelectric projects are dependent on the rainfall. The land use-land cover (LULC) in an area is seen to influence the local climate. Even though the LULC change is largely attributed to urbanization, the change in the climatic conditions is also seen to influence the land cover in a region. Urbanization transforms land use from agriculture and forests into industry, residential and commercial buildings and associated infrastructure. Land use change also takes place when forest areas are cleared and converted to other types of vegetation. These changes bring about change in the local rainfall pattern, thereby providing a cyclic effect by further altering the land cover of the region.

Severe water shortage is being experienced in regions where land cover change has resulted in reduction in forest areas. The forests are referred to as the lungs of the earth: taking in carbon dioxide and providing oxygen, whereas wetlands are referred to as kidneys of the landscapes absorbing harmful pollutants like carbon dioxide and nitrogen. They help in controlling the local climate and hydrological processes. Drastic reduction of them affects the water-retention capacity of soils of the river basins (Nikhil Raj and Azeez 2010). India's agricultural output, economy and industrial activities depend considerably on the variability and extremes of summer monsoon rainfall. 80% of the rainfall received by the southern parts of India is contributed by the Indian summer monsoon rainfall (ISMR) during the months of June, July, August and September (Webster et al. 1998). Kerala is the southernmost state of India, bounded on the east by a geological escarpment running roughly northwest to southeast parallel to the coast known as the Western Ghats and on the west by the Arabian Sea. Kerala is characterized by an asymmetrical topography with three distinct elevation zones known as the lowland, midland and highland regions. Major land use changes have taken place in Kerala, and even though it is claimed that the forest area have not changed much, the remote sensing images shows that thick vegetation has reduced across the state (George and Chattopadhyay 2001). Studies in the river basins of south Kerala have revealed that the area under forests and agriculture drastically decreased since 1970 (Krishnakumar and Dhanya 2017; Chattopadhyay 1985). The land use-land cover change is seen to impact the rainfall characteristics of watersheds (Xue et al. 1996). Urbanization-induced rainfall change is also investigated by many researchers (Shepherd 2005). Remote sensing-based techniques are usually used for preparing the land use-land cover map of the areas and assessing their changes (Dewan and Yamaguchi 2009; Mamun et al. 2013; Islam et al. 2018).

Indian summer monsoon rainfall trends and their spatial variability over India have been studied and found that significant decrease is found in monsoon rainfall in many water surplus basins over India (Ghosh et al. 2009). The land use-land cover change in the Western Ghats has also altered the rainfall pattern along the region (Kale et al. 2016).

In this context this study is aimed at understanding the land use-land cover changes that has taken place over the Bharathapuzha river basin of Kerala state since 1973. An attempt is made to understand the rainfall trend over the basin and how it is impacted by the change in land cover of the area.

15.2 Study Area and Description

The Bharathapuzha river also known as the River Nila is the second longest west-flowing river in Kerala state that drains into the Arabian Sea. The basin lies approximately between 10° 26' and 11° 13' North latitudes and 75° 53' to 77° 13' East longitudes. The study area is shown in Fig. 15.1. About 5818 km^2 area drains into the river of which about 70% is in Kerala and remaining 30% is in Tamil Nadu. It rises in the eastern slopes of Anamalai hills of the Western Ghats at an elevation

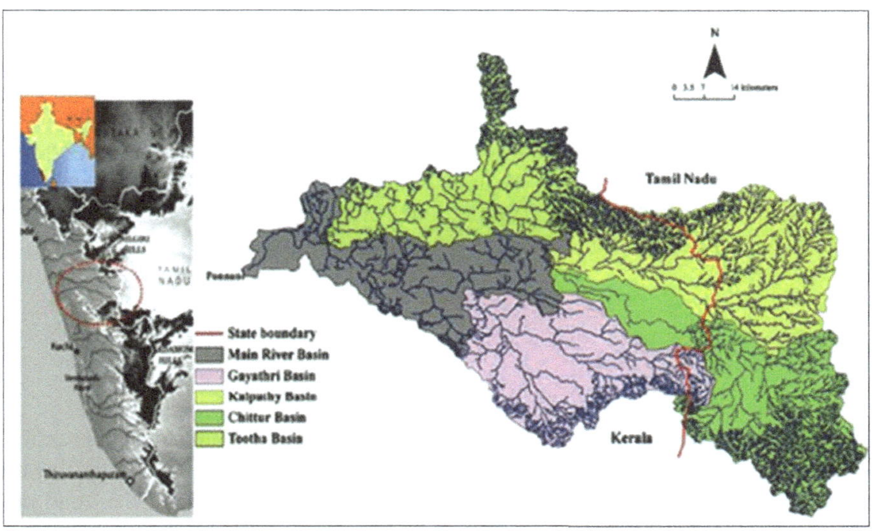

Fig. 15.1 Map of Bharathapuzha basin

of 2,250 m above MSL. The total length of the river from its origin to outfall is about 209 km. It enters the Palghat district of Kerala state through the Palghat gap. The Gayathripuzha, the Kalpathipuzha and the Pulanthode are the three important tributaries. All the three tributaries rise in the Western slopes of the different ranges of the Western Ghats and drains major parts of the Palghat, Trichur and Malappuram districts.

The Bharathapuzha river is extensively dammed. The river has undergone extensive degradation due to sand mining and large-scale environmental changes brought about by human intervention.

15.3 Data and Methodology

15.3.1 Rainfall Data

The rainfall data used for the study is the daily rainfall data prepared by the India Meteorological Department for the entire Indian region. It is the gridded data provided at a resolution of $0.25^\circ \times 0.25^\circ$. The daily gridded data for the study region was extracted and the monthly rainfall for the study region was computed. The rainfall data for the period from 1973 to 2013 was used for the study.

15.3.2 Land Use-Land Cover Maps

The land use-land cover maps were prepared from the satellite images of LANDSAT series for the years 1973 and 2018. LANDSAT 1 series for the year 1973 and LANDSAT 8 series for the year 2018 were used. The images covering the study area were downloaded, layer stacked and mosaicking of images were done. The shape file for the region was prepared and used for generating area of interest. Necessary corrections were performed on the image. Classification of the images was done using supervised classification. Training sites were created using signature editor and maximum likelihood rule was used to prepare the output file.

15.3.3 Rainfall Trend

The temporal trend in the rainfall was calculated using simple linear regression, which is a parametric test. The regression was performed taking year (time) as the independent variable and rainfall as the dependent variable. Slope of the trend line shows the decrease or increase in the rainfall over time. The trend in the annual precipitation as well as the precipitation for various seasons like pre-monsoon, South-west monsoon, post-monsoon and winter rainfall was estimated. The pre-monsoon period is from March to May, South-west monsoon contributes maximum rainfall to the state from June to September, post-monsoon period is from October to November and winter rainfall is from December to February.

15.4 Results and Discussion

15.4.1 Land Use-Land Cover Change

The land use-land cover (LULC) maps were prepared for the periods 1973 and 2018 from the LANDSAT images. The classification of LULC was done under the following categories: waterbody, mixed vegetation, agricultural field, cultivable land, dense trees, barren land/rocky outcrop and built-up areas. The classified image for the year 1973 is given in Fig. 15.2. Similarly, the LULC map for the year 2018 was prepared from the LANDSAT 8 image and the classified image is shown in Fig. 15.3.

The LULC change statistics for the period computed from the maps is detailed in Table 15.1. A comparison of these maps shows significant changes in the land use-land cover pattern in this region. The most predominant change seen during this period is that the built-up area has increased by 13.1% and that the dense trees has decreased by 10.5%. This may be attributed to the increased population and related urbanization. The waterbodies have reduced by 2.2%. Even though the agricultural field areas have almost remained constant, the cultivable land area has been reduced

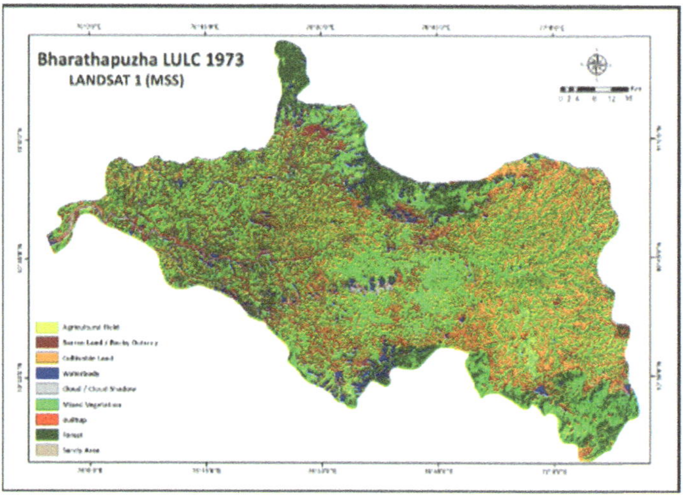

Fig. 15.2 LULC during 1973

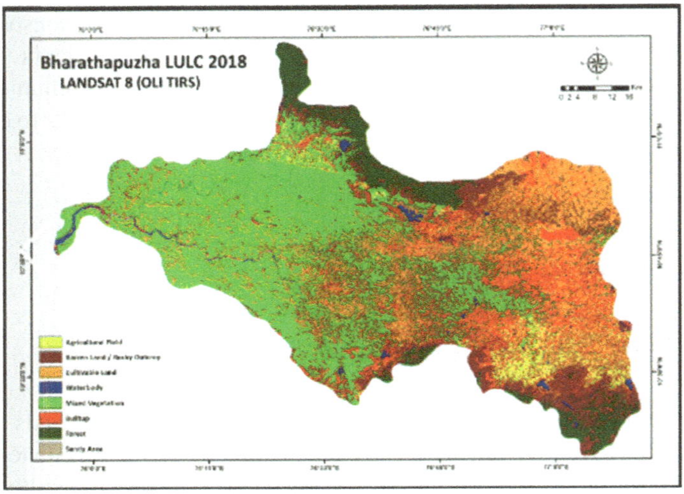

Fig. 15.3 LULC during 2018

Table 15.1 LULC change statistics during study period

Description	1973 Area (in sq km)	2018 Area (in sq km)	% change in area
Waterbody	211.94	85.38	−2.2
Mixed vegetation	1733.68	1732.66	0.0
Agricultural field	258.79	255.76	−0.1
Cultivable land	1541.88	1341.34	−3.4
Dense trees	967.39	357.59	−10.5
Barren land/rocky outcrop	900.35	1127.85	3.9
Built-up	148.20	910.70	13.1
Sandy area	42.79	6.48	−0.6
Clouds	12.93	0.00	
Total area	5818	5818	

by 3.4%. The Bharathapuzha river basin is alleged to have undergone degradation due to uncontrollable sand mining. The LULC change statistics shows that the sandy areas have reduced from 42.786 to 6.48 km^2. The increase in barren land is the result of human interference with the natural vegetation and ecology.

15.4.2 Rainfall Trend Estimation

The rainfall data obtained from the India Meteorological Department for the period 1973–2012 was used for calculating the trend. Even though the land use change was calculated for the year 2018, the rainfall data was available only up to 2012. Hence rainfall up to 2012 was used for calculating the trend. The trend in the annual rainfall over the study region was estimated using simple linear regression. The trend line fitted over the annual rainfall data and the regression equation is shown in Fig. 15.4. It is evident that there are wide variations in the annual rainfall over time. The fitted trend line shows that there is a decreasing trend in the annual rainfall over the region.

Kerala experiences moderate climate with the South-west summer monsoon rainfall as the predominant rainfall over the region. The rainfall for the entire year was categorized as pre-monsoon rainfall, South-west monsoon rainfall, post-monsoon rainfall and winter rainfall. It is seen that the South-west (SW) monsoon rainfall contributed to about 64% of the total rainfall in the region followed by post-monsoon or North-east monsoon contributing to about 21%. The winter rainfall is only about 3% of the total rainfall. The pre-monsoon rainfall from March to May, post-monsoon rainfall from October to November and the winter rainfall from December to February are having slightly increasing trend. The trend in the rainfall during the pre-monsoon, SW monsoon, post-monsoon and winter season are shown in Figs. 15.5, 15.6, 15.7 and 15.8, respectively. Rainfall trends for all the four seasons plotted in Fig. 15.9 show that the increase in rainfall trends for these seasons are negligible when compared to SW monsoon trend.

The nonparametric trend detection method was also used for identifying the rainfall trends for all the seasons. Mann–Kendall test was done to test the trend of seasonal rainfall. It indicated that there is a significant decreasing trend for the South-west monsoon in the river basin during 1973–2012. No significant trend in rainfall is shown in all the other seasons. The Theil–Sen estimator, also known as Sen's slope estimator or the Kendall robust line-fit method is a method for linear regression that

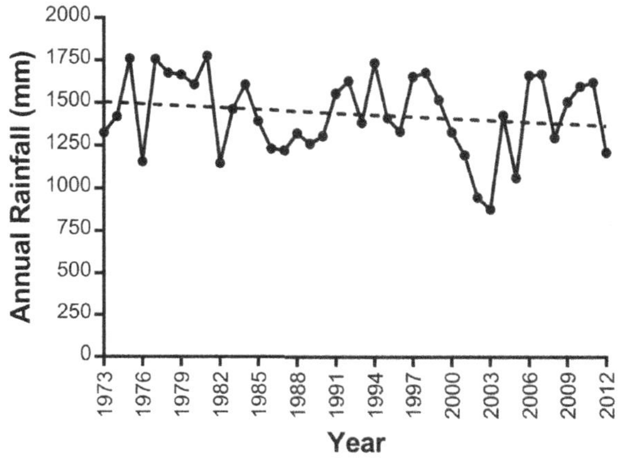

Fig. 15.4 Variation of annual rainfall with time

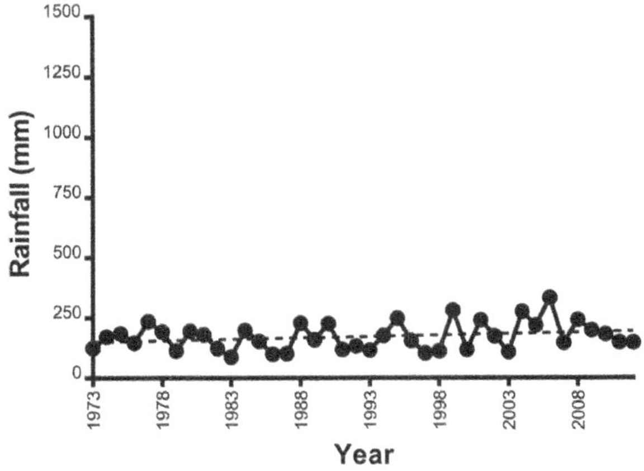

Fig. 15.5 Pre-Monsoon rainfall trend

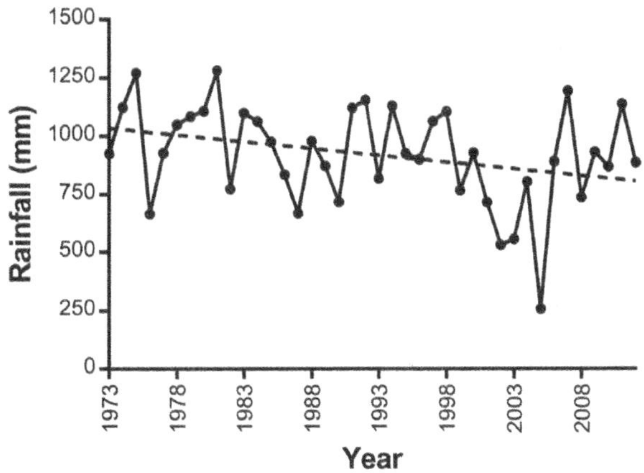

Fig. 15.6 South-West monsoon rainfall trend

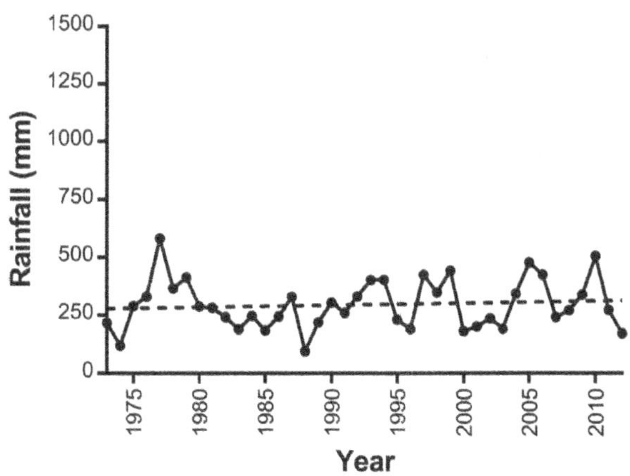

Fig. 15.7 Post-Monsoon rainfall trend

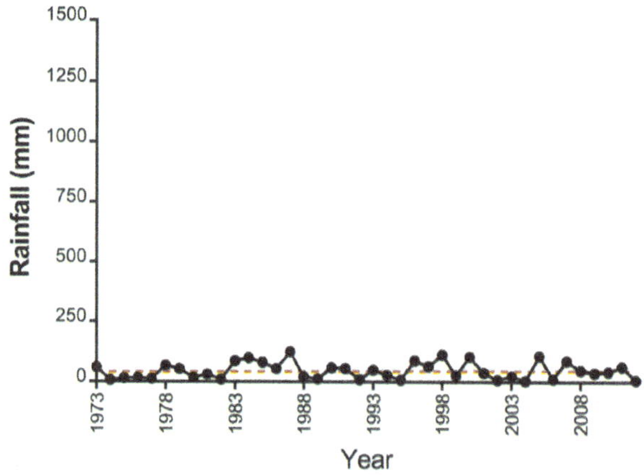

Fig. 15.8 Winter rainfall trend

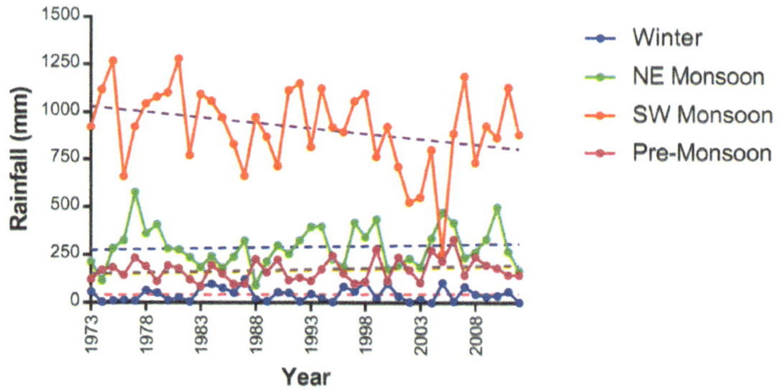

Fig. 15.9 Rainfall trend over various seasons

chooses the median slope among all lines through pairs of two-dimensional sample points. The Theil–Sen line is a nonparametric alternative to the parametric ordinary least square regression line. It is considered as the most popular nonparametric technique for estimating a linear trend. Sen's slope was estimated for both annual rainfall and SW monsoon rainfall over these years. The slope values computed by ordinary linear regression and Kendall robust line-fit method are given in Table 15.2.

Table 15.2 Comparison of the slope of the trend lines

Method	Annual rainfall	SW monsoon rainfall
Simple linear regression	−2.9	−5.8
Kendall robust line-fit method	−3.2	−5.5

15.4.3 Relating LULC Modification to Rainfall Change

The exchange of greenhouse gases and sensible heat between the land surface and the atmosphere and the roughness of the land surface associates the land cover in a region with the weather and climate of the region. The change in the rainfall pattern over the Kerala state is found to be due to the change in the biophysical resources of the state which was the result of unplanned anthropogenic activities in recent years. These activities indirectly affect the physical processes between the earth-atmosphere continuum and influences the distribution of local rainfall during winter and pre-monsoon season (Krishnakumar et al. 2009).

As discussed in Sect. 15.4.1, there has been reduction in the areal extent of dense trees and an increase in the built-up area during the years from 1973 to 2018. The reduction of the thick forests is found to be more on the upstream reaches of the study region which is mainly part of the Western Ghats. Many researches have shown that there has been considerable reduction in thick forests in Kerala especially along the Western Ghats. Feedback of dense vegetation along the Western Ghats on the South-west monsoon rainfall had been studied by performing regional climate simulations and found that evapotranspiration from the vegetation of Western Ghats contributes moisture for about 25–40% of the South-west monsoon rainfall (Paul et al. 2018). Hence, it is inferred that the reduction in the dense trees coupled with the increase in the built-up area is attributed to the decreasing trend in the rainfall over the region.

15.5 Conclusion

The land use-land cover change over the Bharathapuzha river basin which had been undergoing many environmental changes due to urbanization and unplanned activities over a period of 40 years was studied. The study revealed that the areal extent of dense trees has diminished over the region and built-up area has increased. The findings match similar studies performed over the Western Ghats region that states that considerable reduction in dense forest has taken place over the area and that it has altered the hydrological cycle in the region. The rainfall pattern and trend over the area during the study period was also analyzed both annually and seasonally. It is seen that the annual rainfall shows a decreasing trend. Seasonal split-up of the rainfall indicates that the predominant SW monsoon rainfall has a decreasing trend, whereas the rainfall over the other seasons has a marginal increasing trend.

It is inferred that the rainfall decrease over the region is attributed to the land use changes that have happened in the region over the period. It also suggests the significance of devising proper developmental plans for the watershed by protecting the biodiversity of the region.

Acknowledgements The authors are thankful to the India Meteorological Department for providing the gridded rainfall data at various spatial resolutions over India.

References

Chattopadhyay S (1985) Deforestation in parts of Western Ghats region (Kerala), India. Int J Environ Manag 20:219–230

Dewan AM, Yamaguchi Y (2009) Using remote sensing and GIS to detect and monitor land use and land cover change in Dhaka Metropolitan of Bangladesh during 1960–2005. Environ Monit Assess 150(2009):237–249

George PS, Chattopadhyay S (2001) *Population and Land Use in Kerala*. Growing Populations, Changing Landscapes: Studies from India, China and the United States. National Academy of Sciences, Washington DC, pp 79–106

Ghosh S, Luniya V, Gupta A (2009) Trend analysis of Indian summer monsoon rainfall at different spatial scales. Atmos Sci Let 10:285–290

Islam K et al (2018) Land use classification and change detection by using multi-temporal remotely sensed imagery. Egypt J Rem Sens Space Sci 21:37–47

Kale MP et al (2016) Land-use and land-cover change in Western Ghats of India. Environ Monit Assess 188:387

Krishnakumar KN, Prasada Rao GSLHV, Gopakumar CS (2009) Rainfall trends in twentieth century over Kerala, India. Atmos Env 43:1940–1944

Krishnakumar A, Revathy D, Dhanya T (2017) Land cover change analysis with special reference to forests and paddy wetlands of neyyar and karamana river basins, Kerala, SW India using GIS and remote sensing. Int J Scient Res Publicat 7(11):190–198

Mamun A Al, Mahmood A, Rahman M (2013) Identification and Monitoring the Change of Land Use Pattern Using Remote Sensing and GIS: A Case Study of Dhaka City. IOSR J Mechan Civil Eng 6(2):20–28

Nikhil Raj P, Azeez PA (2010) Land use and land cover changes in a tropical river basin: a case from Bharathapuzha River Basin, Southern India. J Geogr Inform Syst 185–193

Paul S, Ghosh S, Rajendran K, Murtugudde R (2018) Moisture Supply From the Western Ghats Forests to Water Deficit East Coast of India. Geophys Res Lett 45(9):4337–4344

Shepherd JM (2005) A review of current investigations of urban-induced rainfall and recommendations for the future. Earth Interactions 9 paper 12, p 27

Webster PJ, Magana VO et al (1998) Monsoons: Processes, predictability, and the prospects for prediction. J Geophys Res 103(C7):14451–14510

Xue Y, Fennessy MJ, Sellers PJ (1996) Impact of vegetation properties on U.S. summer weather prediction. J Geophys Res 101(D3):7419–7430

Chapter 16
Innovations and Application of Operational Ocean Data Products for Security of Marine Environment

Madhulika Sinha and Shrikant Charhate

Abstract The rapid development of operational oceanography in recent years has led to improved access to real-time data and products generated from in situ and satellite observations as well as ocean modelling. Operational oceanography is like weather monitoring and forecasting for the ocean. It can provide estimates of essential ocean variables, for example, sea level, temperature and currents. Ocean observations are required in real time and near real time (within a few days or minutes of collection) and sourced from various national and international programs. A number of large open data sets and metadata from observations (in situ and remote sensing) and from model outputs exist which have application in real-time problem solutions in coastal environments. However, these resources have not been used optimally due to limited capacities, and lack of information on their availability and applicability. In this study attempt has been made to discuss available information to improve safety of life at sea, help create wealth and assist in the security and protection of the marine environment through good measuring networks, and systems for making data available swiftly. Timely prediction of storms and other unfavourable weather can be done by having knowledge of meteorological conditions, and how they are developing above the oceans. Outputs can be used to generate data products, applications and services through national authorities, as well as in some cases through other organizations such as metocean service providers and environmental consultants and for these measurements to be of use, good data management, quality control and fast data availability are essential.

Keywords Oceanography · Ocean data · Meteorological conditions

---

M. Sinha (✉)
Pillai HOC College of Engineering and Technology, Rasayani 410207, India
e-mail: madhusinha13@gmail.com

S. Charhate
Amity University Maharashtra, Panvel, Mumbai, Panvel, Mumbai 410206, India
e-mail: sbcharate@yahoo.co.in

R. Jha et al. (eds.), *Climate Change Impacts on Water Resources*, Water Science and Technology Library 98, https://doi.org/10.1007/978-3-030-64202-0_16

16.1 Introduction

Operational oceanography is an activity of conducting regular oceanographic studies towards providing information services to various sectors, viz., fisherman, shipping, ports, disaster management, environment, coastal states, navy, coast guard, offshore industries for conducting their day-to-day operations. It depends on availability of ocean observations systems and super-computer facilities. It uses computer models and mathematical techniques and is designed to deliver stakeholder-defined outputs and products on a routine basis, using robust and fully supported production and dissemination techniques. As per Schiller and Brassington (2011), ocean observations are required in real time and near real time (within a few days or minutes of collection) and sourced from various national and international programmes; therefore, a systematic approach to secure and disseminate in situ observations is required. Ocean observing system, data management, ocean modelling and deriving different products and services are the essential elements of operational oceanography (Fig. 16.1). In this paper an effort is made to provide the status of the various in situ as well as satellite observing programmes that can supply information about the ocean surface and interior in near real time and that can be used in various operational systems and services. Nowadays, operational oceanography has many applications; it is continuously used to improve safety of life at sea, help create wealth, and assist in the security and protection of the marine environment. Some of the services are providing warnings about coastal floods, storm impacts, harmful algal blooms and contaminants, electronic charts, sea state conditions, optimum routes for ships, prediction of primary productivity, ocean currents, ocean climate variability, and modelling of and response to oil spills and dredging. Operational oceanographic programs are well established in countries such as USA, Japan and France, and

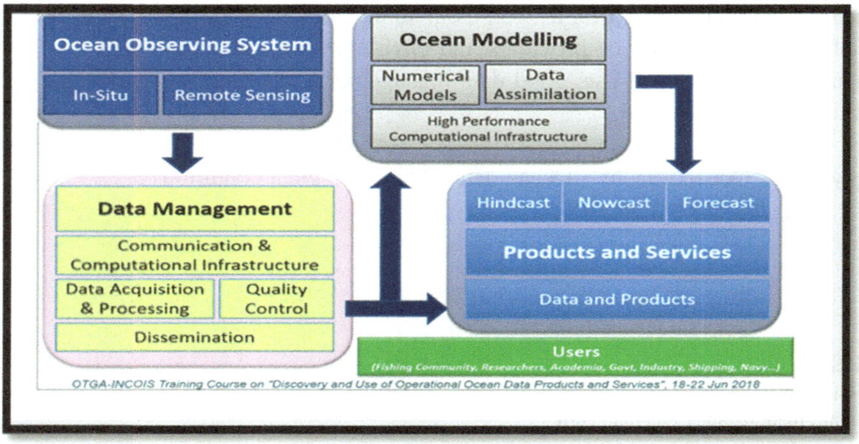

Fig. 16.1 Essential components of operational oceanography (www.incois.gov.in)

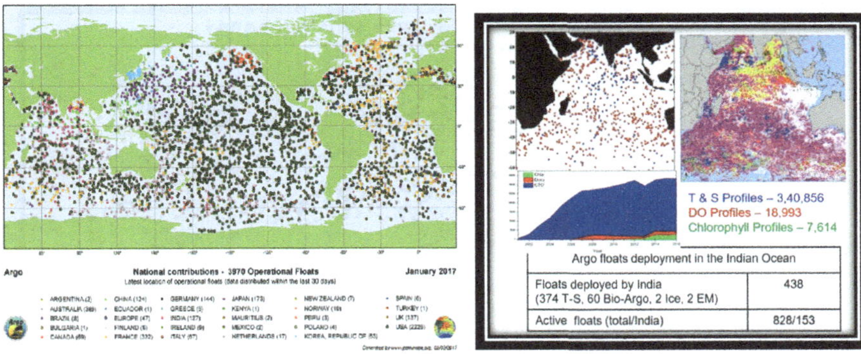

Fig. 16.2 Argo national contributions, as of January 2017; colour coded according to country (*Source* JCOMMOPS)

now in India also this programme is continued at INCOIS. The European component of the Global Ocean Observing System (EuroGOOS) has defined operational oceanography 'as the activity of systematic and long-term routine measurements of the seas and oceans and atmosphere, and their rapid interpretation and dissemination'. The real-time in situ observing system has been reviewed many times before. The two most recent reviews by Legler et al. (2015) and Ravichandran (2011) show that this is a rapidly changing field and an update is required continuously. The need for careful monitoring of climate and ocean research and changing conditions over time have motivated for the development of the Global Ocean Observing System. In situ ocean observations available at present address these needs as well as are vital to ocean, marine, weather and climate forecasts to initialize today's forecasts, verify previous forecasts, improve forecasting systems to be used in the next decade, calibrate remote-sensing data and contribute towards tailored analyses, products and services. Thus, today's ocean observations are critical to many aspects of the operational oceanography enterprise (Figs. 16.2 and 16.3).

Our Earth consists of 70% of ocean, and they are the vital sources for resources. Monsoon driving force acts as medium of transport and land of oceanic disasters. The occurrence of 1982/1983 El Niño was the alarm that provided the motivation to monitor the oceans globally, including weather above the oceans as given by McPhaden et al. (2010) although the development of the ocean observing system had started in the early 1980s. Observations of sea level have been available for centuries, but it was only in 1985 that the Global Sea-Level Observing System (GLOSS) was established. GLOSS was established by the Intergovernmental Oceanographic Commission (IOC) to provide oversight and coordination for sea-level networks and now supplies data from 70-member states. The design of GOOS addresses two broad areas, i.e., oceans and climate, and coastal ocean. The objectives include monitoring and detection of climate change, seasonal to inter-annual climate prediction, marine and weather forecasts, short-range ocean forecasts, understanding decadal variations, routine ocean state determination and scientific research. The climate module of GOOS is the ocean module of the Global Climate Observing System (GCOS) of

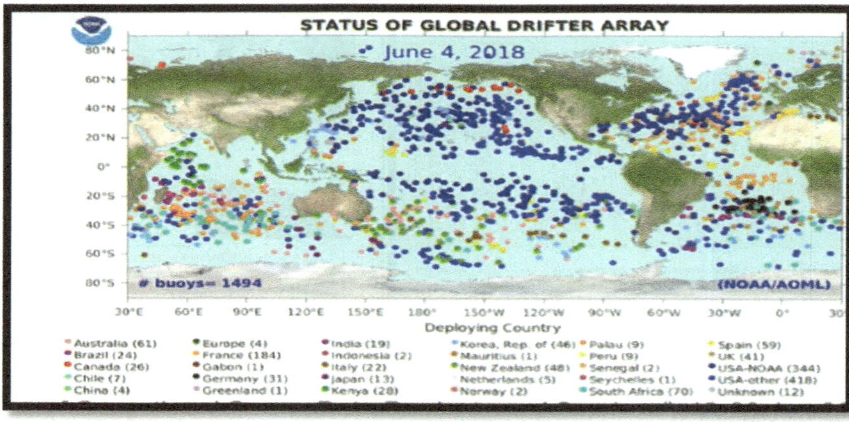

Fig. 16.3 Global drifter as on June 4, 2018 (*Source* JCOMMOPS)

WMO. The Joint WMO/IOC Technical Commission for Oceanography and Marine Meteorology (JCOMM) is a panel for implementation of an operational, integrated ocean observing and data management system for climate. The Ocean Information Bank provides information on physical, chemical, biological and geological parameters of ocean and coasts on spatial and temporal domains that is vital for both research and operational oceanography. In India, INCOIS receives voluminous oceanographic data in real time, from a variety of in situ and remote-sensing observing systems and serves as the central repository for marine data in the country. Further, INCOIS has been designated as the National Oceanographic Data Centre by the International Oceanographic Data Exchange Programme (IODE) of International Oceanographic Commission (IOC). INCOIS also serves as the National Argo Data Centre, Regional Argo Data Centre, and also the regional data centre and clearing house for the Indian Ocean region for the IOGOOS programme.

16.1.1 Argo Floats

As per (Argo Science Team 1998), the primary goal of Argo was to create a systematic global network of instruments, integrated with other elements of the Global Ocean Observing System, for a broad range of applications including climate-relevant variability on seasonal to decadal time-scales, multi-decadal climate change, and improved initialization of ocean and coupled ocean/atmosphere climate models (Freland et al. 2016). The first Argo float was deployed in 1999. The International Argo programme is a collaborative partnership of more than 30 nations to provide a unique opportunity to map ocean temperature, salinity and biogeochemical structure on $3° \times 3°$ coverage. To monitor the thermohaline field of the upper ocean for seasonal, annual and decadal time scale, it is important to sustain the Argo float array, and they are important constituents of the Global Ocean Observing System (GOOS).

In Indian Ocean there is plan to deploy 450 floats by various participating countries out of which India committed to deploy 150 floats. The AIC website gives the general information on the Argo project (participating countries, contacts, real-time status of the network, status of DACs developments, maps, news, etc.). The application is also used for deployment strategy. The Argo Data Management System, as finalized by the International Argo Science Team (IAST) and its Data Management Group, constitutes three levels of data centres, i.e., National Data centres, Regional Data Centres and Global Data Centres located in France and USA.

The Climate Variability and Predictability (CLIVAR) is the most recent and broadest component of the World Climate Research Programme (WCRP) aimed at extending the range and accuracy of seasonal and inter-annual climate prediction through the collection and analysis of observations as well as development and application of models of the coupled climate system.

16.1.2 Drifter Program

The displacement of ships by ocean currents had been used even earlier. In 1852, Maury work on 'the Wind and Current Chart of the North Atlantic' helped sailors to decrease the length of their journey by observing the ocean's currents and winds. In 1982, the World Climate Research Program (WCRP) recognized that a global array of drifting buoys ('drifters') would be invaluable for oceanographic and climate research, but there were large differences in the costs and water-following properties of various designs (World Climate Research Program 1988). The Tropical Ocean Global Atmosphere (TOGA) and World Ocean Circulation Experiment (WOCE) projects required a standardized, low-cost, lightweight, easily deployed drifter as per WCRP recommendations. According to Niiler (2001) the standardized design uses a holey-sock drogue at 15 m depth to follow mixed-layer currents with little influence from the winds. The surface drifters are largely deployed from ships of opportunity (SOOP). Different parts of the ocean show differing characteristics for dispersal. It is well known (because of Ekman layer divergence) that it is extremely hard to sustain an array of surface drifters close to the equator. Lumpkin and Elipot (2010) studied and concluded that the dispersion of surface drifters has yielded a major insight into the eddying and dispersion characteristics of the ocean surface layers. The main purpose of drifting buoys is to collect surface meteorological (atmospheric pressure and winds) and oceanographic (SST, surface velocities, and subsurface temperature) data using the satellite tracked drifting buoys and provide near real time data (SST, sea level pressure and surface winds on GTS) for operational weather analysis and prediction. It also develops monthly mean mixed-layer velocities in the Indian Ocean on $1^\circ \times 1^\circ$ resolution as well as provide data sets as 'sea truths' for validation of remotely sensed ocean surface parameters. INCOIS aims to build an Indian Ocean drifter data archival.

16.1.3 Soop/XBT

The XBT network constituted more than 50% of the global ocean thermal observations before the introduction of Argo profiling floats. With the Argo array now in place, currently XBT observations represent approximately 15% of temperature-profile observations. Levitus et al. (2012) in his paper has informed about the upper-ocean heat content, and global heat content variability before the Argo array was implemented. Goni et al. (2014) in his work mentioned that after the full implementation of the Argo array, XBT transects are mostly used to monitor and investigate variability of surface and subsurface currents and the Meridional Overturning Circulation. XBT/XCTD observations are long-term monitoring of upper-ocean thermal fields in the seas around India and is supported by ESSO-INCOIS. The major XBT transects are Mumbai–Mauritius, Chennai–Port Blair, Port Blair–Kolkata, Chennai–Singapore, and Kochi–Lakshadweep. The data from this system transfer to ESSO-INCOIS and made available to the public in user-friendly data format. In the last one year, a total of 257 XBT profiles and 92 XCTD profiles pertaining to Indian Ocean have been archived at ESSO-INCOIS.

16.1.4 Moored Buoy

This is critical element of the ocean observing system, may be floating or submerged platforms equipped with measurement sensors, moored to anchors on the seafloor through cables. Global Tropical Moored Buoy Array (GTMBA) is a coordinated and sustained multi-national programme to develop and implement moored buoy observing systems for climate research and forecasting throughout the tropics (McPhaden et al. 2009a). It comprises the TAO/TRITON array in Pacific, the PIRATA in the Atlantic and RAMA in the Indian Ocean. The TRITON array has also declined with sites being decommissioned owing to lack of research funding and changes in JAMSTEC priorities. The status of PIRATA is much better, with data return maintaining historical levels. The National Data Buoy Programme (NDBP) of India started in 1996 for in situ meterological-ocean measurements in Arabian Sea and Bay of Bengal (DOD/NIOT). It provides data on hourly basis for pressure, temperature, wind, water temp, salinity, currents and wave from August 1997. At present there are 47 buoys maintained by NIOT. The Moored buoy observations are stable and have proven their reliability: ($N * 24 * 365 * \times \sim 10^6$) of data messages per year. It gives high-quality, real-time data in scheduled time, having unlimited lifespan, and can be recovered and refurbished. The drawbacks of buoys are high operating costs, expensive to build a network with no contingency and rapid deployment. Ocean moored buoy network for northern Indian Ocean (OMNI) buoys is managed by ESSO-INCOIS. There are 12 active deep sea met-ocean buoys deployed in northern Indian Ocean by the INCOIS and NIOT (5 in Arabian Sea and 7 in Bay of Bengal) giving data from October 2010 to present. The variables measured by OMNI buoys are surface meteorological, surface ocean parameters, subsurface parameters. 98Research Moored Array for African-Asian-Australian Monsoon Analysis and Prediction (RAMA) in the Indian Ocean is an observational network designed to address outstanding scientific questions related to the Indian Ocean variability and the monsoons. As per the MoU between MoES/ESSO-INCOIS and NOAA/PMEL, 38 operations at 18 sites were carried out, which include deployments, recovery and repair of the ATLAS, CONE type and ADCP moorings, during the past one year. At present, 31 out of 46 (67%) RAMA sites were covered. Approximately 13 cruises (320 sea days) for RAMA mooring operations for deployment, recovery and maintenance are available. All data from RAMA are available in GTS and INCOIS website in near real time. RAMA data is utilized for data assimilation, validation of ocean model output and satellite parameter (SST, wind etc.), bias corrected ocean model forcing field (Tropflux), to understand air–sea interaction processes in the Tropical Indian Ocean.

16.1.5 Satellite Oceanography

Satellite oceanography is a key component of operational oceanography. Satellites provide long-term, continuous, global, high space and time resolution data for key ocean parameters such as sea level and ocean circulation, sea surface temperature (SST), ocean colour, sea ice, waves and winds. These are the core variables observations required to constrain global, regional and coastal ocean monitoring and forecasting systems. They are also needed to validate them. Only satellite measurements can, in particular, provide observations at high space and time resolution to partly resolve the mesoscale variability and coastal variability. Schiller and Brassington (2011) has specified that satellite data can be directly used for applications such as SAR for sea ice and oil pollution monitoring, ocean colour for water quality monitoring. Remote sensing data are received and processed at ground station by Oceansat-2, NOAA 18&19, METOP A&B, Terra and Aqua (MODIS), NPP 3.

16.2 Ocean Data and Information System (ODIS)

ODIS is a fully automated system for data reception, processing, quality control, database loading and dissemination. It is having a unique integrated in situ database for heterogeneous data received from variety of platforms through various communication channels and assists in data discovery, on-the-fly visualization and data downloads in different formats. It has been developed using open source software. Figure 16.4 shows the flowchart of ocean data and information system. The ODIS is supported by the data received from both the in situ platforms and satellites, global telecommunication system (GTS), projects/experiments, data from other sources and the data exclusively retrieved for the Indian Ocean from historical data sets. INCOIS

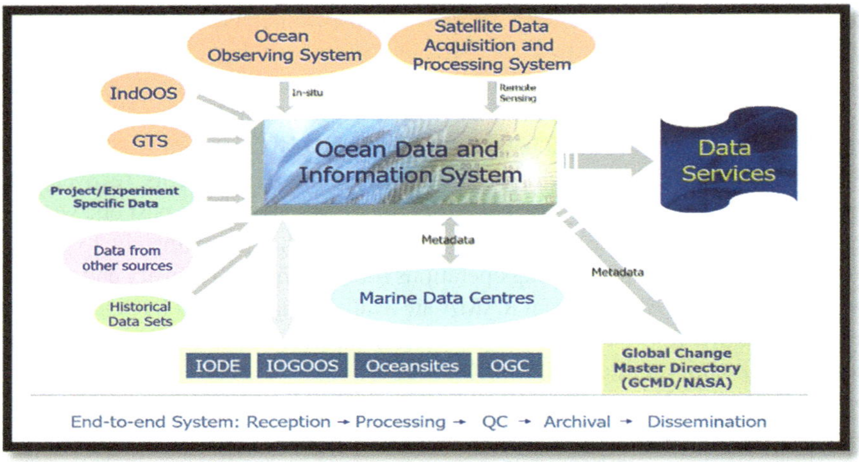

Fig. 16.4 Flowchart showing ocean data and information system (www.incois.gov.in)

is having active collaboration with IODE, Indian Ocean Global Ocean Observation System (IOGOOS) and Open Geospatial Consortium (OGC) programmes on data and information management-related activities. It is estimated that about 5 TB per year of both in situ and remote sensing data is managed by ODIS. It is important to blend the data in standard formats, apply quality control procedures, generate metadata and database, while adopting international standards for continuous exchange of data to make it available for different types of users.

16.3 Ocean Information and Advisory Services

The data received from various observing systems in real time at different communication systems are assembled and standardized. The data go through the quality control procedures for each of the observing system separately as per the internationally adopted quality control procedures and standards. The quality-controlled data are then loaded to the database for providing web-based data services. A number of web-based services are provided after analysis of the ocean information. The Marine Fishery Advisory Service provides daily advisories on operational basis except during fishing ban period, inclusion of altimeter data for detection of eddies, fronts and surface currents, potential fishing zone advisory services. Ocean State Forecast and Information Services/System: It gives quantitative, location-specific, user-customised, multi-parameter, multi-lingual, impact-based, and end-to-end ocean state forecast and information operational daily and package-based services, ocean forecasts (general, daily) (wave, swell, SST, currents, MLD, D20, tides [5–7 days]) and specific products (HWA, Joint IMD-INCOIS bulletins, oils pill trajectory, search and rescue aid, etc.). High-wave alerts specifically notify the areas along the Indian coast that will experience the high waves (i.e. >3.0 m or rapid increase in wave heights) due to bad weather, approaching cyclone or freak waves originated elsewhere and traveling towards the Indian coast. The Indian Tsunami Early Warning Centre (ITEWC) services for an event commence whenever an earthquake is recorded with $M \geq 6.5$ within the Indian Ocean and $M \geq 8.0$ outside of the Indian Ocean.

16.4 Conclusions

From the above discussion, we conclude that operational oceanography is very young relative to weather forecasting. It has many products and applications and is growing. Recent developments and pattern change in oceanography have been stimulated by observations of various processes; in particular, theories and models have typically been developed to explain, quantify, incorporate or parameterize these processes using balance equations. The individual networks of the present sustained ocean observing system for climate, including tropical moorings, XBTs, surface drifters, ship-based meteorology, tide gages, Argo floats, repeat hydrography, and satellite observations, have been developed largely independent of one another. Satellite data also play a fundamental role for operational oceanography. Operational oceanography is both a national and international endeavour, benefits from partnerships and collaboration and provides an array of societal and economic benefits. By integrating developments in data quality, coverage, and delivery and from addition of multi-disciplinary measurements, the present global observing system will be more benefitted.

References

Argo Science Team (1998) On the design and implementation of Argo—An initial plan for a global array of profiling floats. GODAE international project office, c/o Bureau of meteorology, Melbourne, Australia. International CLIVAR project office ICPO report No 21. GODAE Report No 5. 32pp

Freland H et al (2016) Fifteen years of ocean observations with the global argo array, nature climate change, Feb 2016, vol 6 Issue 2, Macmillan Publishers Limited, pp 145–153. https://doi.org/10.1038/NCLIMATE2872, https://archimer.ifremer.fr/doc/00310/42106/©2016

Goni G, Sprintall J, Roemmich D, Gronell-Thresher A, Couley R, Baringer M (2014) In: Samuel D (ed) The global network of XBT temperature sections in support of oceanographic and climate studies. Cambridge Scholars Publishing, Cambridge. Chapter 5, Oceans and society: Blue planet. ISBN: 978-1-4438-5639-3

Legler DM, Freeland HJ, Lumpkin R, Ball G, McPhaden MJ, North S, Crowley R, Goni GJ, Send U, Merrifield MA (2015) The current status of the real-time in situ Global Ocean Observing System for operational oceanography. J Oper Oceanogr 8(sup2):s189–s200. https://doi.org/10.1080/1755876X.2015

Levitus S, Antonov JI, Boyer TP, Baranova OK, Garcia HE, Locarnini RA, Michnov AV, Reagan JR, Seidov D, Yarosh ES, Zhang MM (2012) World ocean heat content and thermosteric sea-level change (0–2000m), 1955–2010. Geophys Res Lett 39:L10603. https://doi.org/10.1029/2012GL051106

Lumpkin R, Elipot S (2010) Surface drifter pair spreading in the North Atlantic. J Geophys Res-Oceans 115. https://doi.org/10.1029/2010JC006338

McPhaden MJ, Busalacchi AJ, Anderson DLT (2010) A TOGA retrospective. Oceanography. 23:86–103

McPhaden MJ, Meyers G, Ando K, Masumoto Y, Murty VSN, Ravichandran M, Syamsudin F, Vialard J, Yu L, Yu W (2009) RAMA: the research moored array for African Asian-Australian monsoon analysis and prediction. Bull Am Meteorol Soc 90:459–480

McPhaden MJ et al (2009a) The global tropical moored buoy array. Community White Paper, Oceanobs'09

Ravichandran M (2011) In situ ocean observing system. In: Schiller A, Brassington GB (eds) Operational oceanography in the 21st century. Springer, Dordtrecht, pp 55–90

Schiller A, Brassington GB (eds) (2011) Operational oceanography in the 21st century. https://doi.org/10.1007/978-94-007-0332-2_2, © Springer Science+Business Media BV

World Climate Research Program (1988) WOCE surface velocity program planning committee report of first meeting: SVP-1 and TOGA pan-Pacific surface current study. Wormley: World Meteorological Organization. 33 pp WMO/TDNo 323, WCRP-26. www.incois.gov.in

Chapter 17
Statistical Downscaling of Sea Level by Support Vector Machine and Regression Tree Approaches

S. Sithara, S. K. Pramada, and Santosh G. Thampi

Abstract Projections of future climate from various climate models indicate that global temperatures are continuously rising. This, in turn, may result in a significant rise in the water levels in the oceans, adversely impacting coastal aquifers. Apart from temperature, some other climatic variables also influence sea level. For better management of coastal aquifers, it is necessary to predict the sea level with a reasonable degree of accuracy. The repercussions of projected climate change on sea level rise can be investigated by projecting future sea level values for different representative concentration pathways (RCPs) using global climate models (GCMs). GCMs are run at a coarser scale; hence for regional-scale studies these projections have to be downscaled before being input into hydrologic models. This paper presents the details and results of a study in which support vector regression (SVR) and regression tree (RT) techniques were applied for statistical downscaling of sea level using climatic variables. The results of both these techniques were compared. It was observed that the performance of the SVR model was better than that of the RT technique.

Keywords Climate change · Sea level rise · Global climate models · Statistical downscaling · Support vector regression · Regression tree

17.1 Introduction

The importance of climate has increased with the realization that climate change has serious repercussions on the environment. Human activities can also change climate, which is referred to as anthropogenic climate change. Some of the examples

S. Sithara (✉) · S. K. Pramada · S. G. Thampi
Department of Civil Engineering, National Institute of Technology Calicut, Kozhikode 673601, Kerala, India
e-mail: sitharasasidharan5@gmail.com

S. K. Pramada
e-mail: pramada@nitc.ac.in

S. G. Thampi
e-mail: santosh@nitc.ac.in

R. Jha et al. (eds.), *Climate Change Impacts on Water Resources*, Water Science and Technology Library 98, https://doi.org/10.1007/978-3-030-64202-0_17

of anthropogenic changes are ozone hole, acid rain, and global warming (David Neelin 2010). Climate change has serious repercussions on various components of the environment including the oceans. The Intergovernmental Panel on Climate Change (IPCC) has reported rise in sea level due to global warming. Climate change is the consequence of increased greenhouse gas emissions, which in turn causes global warming and consequent thermal expansion of water, and melting of ice caps and glaciers (Mimura 2013). Relative to 1850–1900, global surface temperature change by the end of the twenty-first century (2081–2100) is projected to likely exceed 1.5 °C (IPCC 2014). Among the various greenhouse gases, CO_2 is the major contributor to this rise in temperature (Maddah 2016). Emissions of CH_4 and other short-lived anthropogenic greenhouse gases such as CFCs, HCFCs, and HFCs also contribute to sea level rise (Zickfeld et al. 2017). Apart from temperature, there are many other climatic variables which influence the sea level. Li et al. (2016) and Sturges and Douglas (2011) observed that wind also contributes to sea level changes.

GCMs are the main tools used to assess climate change under different scenarios of greenhouse gas emissions (Haylock et al. 2006). Since GCM projections run at coarser resolution, it is necessary to downscale the GCM output to obtain regional-scale information. Approaches adopted for downscaling are broadly classified into two, viz., dynamical and statistical downscaling. Dynamical downscaling refers to the nesting of high-resolution regional climate models (RCMs) in the GCMs. Statistical downscaling involves establishing statistical relationships between large-scale (predictors) variables and small-scale variables (predictands) of the current climate and then applying it to derive regional-scale information (Boé et al. 2007; Haylock et al. 2006). A wide variety of statistical downscaling techniques are available. SDSM is a widely used downscaling technique, which is a hybrid of the conditional weather generator and regression-based methods (Elhakeem et al. 2015).

Classical regression techniques are extensively used for statistical downscaling of local variables from GCM projections. Machine learning techniques have far-reaching applications in the field of water resources. It is possible to extend its applications to the field of climate change for statistical downscaling of climate variables. In this background, the following goals were set for this study: (i) to identify climatic variables which influences sea level the most, (ii) to statistically downscale sea level projections from GCMs using SVR and RT techniques and (iii) to compare the performance of both these methods.

17.2 Methods

The overall methodology consists of three phases. In the first phase, the predictors are identified. The second phase involves establishing statistical relationships between the predictors and the predictand. In the last phase, the predictand is downscaled from bias corrected GCM outputs. The predictors were identified using correlation analysis between the dependent variable (sea level) and the independent variables (climatic variables). Machine learning algorithms such as SVR and RT were employed for statistical downscaling of sea level from the GCM output. Support vector machine (SVM) is a widely used machine learning algorithm for classification analysis; later it was realized that the same principle can be adopted for regression also, resulting in the so-called support vector regression (SVR) technique. SVM has been employed for downscaling projections of precipitation (Tripathi et al. 2006) and is reported to be a powerful tool which can be employed in prediction problems (Voyant et al. 2017). Regression tree is a data mining technique which can also be employed in prediction problems; this technique is a better replacement for statistical regression (Mangai and Gulyani 2016). In this study, the GCM output was corrected for inherent biases by linear scaling technique.

17.2.1 Support Vector Regression

The objective function is to minimize

$$J = C\sum_{n=1}^{N} E[y(x_n) - y_{dn}] + \frac{1}{2}\|w^2\| \tag{17.1}$$

where C is the inverse weight penalty parameter, E is an error function and $\frac{1}{2}\|w^2\|$ is a weight penalty term.$y(x_n)$ is the estimator output and y_{dn} is the target data. In order to retain the sparseness property of the SVM classifier, E is defined as:

$$E_\epsilon(z) = \begin{cases} |z| - \epsilon, & \text{if}|z| > \epsilon \\ 0, & \text{Otherwise} \end{cases} \tag{17.2}$$

This is called ϵ-insensitive error function; it ignores errors of size smaller than ϵ. Two slack variables ξ_n and $\xi_n^{'}$ are introduced such that $\xi_n \geq 0$ and $\xi_n^{'} \geq 0$. $\xi_n > 0$ denotes data points above the ϵ-tube and $\xi_n^{'} \geq 0$ denotes those below the ϵ-tube.

$$y_{dn} > y(x_n) + \epsilon, \quad \text{if}\, \xi_n > 0$$

$$y_{dn} < y(x_n) + \epsilon, \ \text{if}\ \xi_n > 0$$

The objective function can be written as

$$\text{Minimize} J = C\sum_{n=1}^{N} (\xi_n + \xi_n^{'}) + \frac{1}{2}\|w^2\| \tag{17.3}$$

subject to $\xi_n \geq 0$, $\xi_n^{'} \geq 0$, $y_{dn} \leq y(x_n) + \epsilon + \xi_n$, $y_{dn} \geq y(x_n) - \epsilon - \xi_n^{'}$.

In order to handle the constraints, Lagrange multipliers are introduced.

SVR is performed in a feature space,

$$y(x) = w^T \Phi(x) + w_0 \tag{17.4}$$

where ϕ is the feature map and w_0 is the bias

$$w = \sum_{n=1}^{N} (\lambda_n - \lambda_n^{'}) \Phi(x_n) \tag{17.5}$$

where λ_n and $\lambda_n^{'}$ are the Lagrangian multipliers.

Substituting (17.5) into (17.4), we get

$$y(x) = \sum_{n=1}^{N} \left(\lambda_n - \lambda_n^{'}\right) K(x, x_n) + w_0 \tag{17.6}$$

$K(x, x_n) = \Phi^T(x)\Phi(x_n)$, which is the kernel function. The data points which contribute in the above summation (Eq. 17.6) are called support vectors (Hsieh 2009). The commonly used kernel functions are linear, polynomial, sigmoid, splines, and radial basis function (RBF) (Tripathi et al. 2006). Polynomial kernel was used in this study.

17.2.2 Regression Tree

These are piecewise-constant regression models in the form of a decision tree with leaf nodes. These leaf nodes give numerical values, whereas classification trees give categorical values. Regression trees are constructed based on the recursive partition algorithm. A set of training data was recursively partitioned into smaller sets till certain termination criteria are satisfied. A suitable local criterion was chosen to obtain the optimum split at each of the nodes. The usual approach is to build the tree in such a way that the parameters minimize the least squares error (Cichosz 2014; Mangai and Gulyani 2016).

$$\text{Least Square error} = \frac{1}{n} \sum_{i=1}^{n} \left(y_i - \hat{y}_i\right)^2 \tag{17.7}$$

where n is the number of training data points, y_i is the actual value, and $\hat{y}_i$ is the predicted value. The value of leaf node 1 is the mean of the target values of all data points that meet this node as given by Eq. (17.8)

$$\text{Value}_l = \frac{1}{n_l}\sum_{i=1}^{n_l} y_l \tag{17.8}$$

where n_l is the number of data points in the leaf node l. The error at leaf node l can be computed as

$$\text{Error}(l) = \frac{1}{n_l}\sum_{i=1}^{n_l}(y_i - \text{Value}_l)^2 \tag{17.9}$$

Let the probability of a leaf node be P(l) and the number of leaf nodes be nl, then the error of a tree T can be computed as the weighted mean of the error in all its leaves as given in the following equation

$$\text{Error}(T) = \sum_{i=1}^{nl} P(l) \times \text{Error}(l) \tag{17.10}$$

The error of split 's' at a node 't' is given by Eq. (17.11).

$$\text{Error}(s,t) = \frac{n_{t_{\text{left}}}}{n_t} \times \text{Error}(t_{\text{left}}) + \frac{n_{t_{right}}}{n_t} \times \text{Error}\left(t_{\text{right}}\right) \tag{17.11}$$

where t_{left} and t_{right} are the left and right subtrees of node 't' after the split. $n_{t_{\text{left}}}$ and $n_{t_{\text{right}}}$ are the cardinality of t_{left} and t_{right}, respectively. The best split is the one which maximizes

$$\Delta\text{Error}(s,t) = \text{Error}(t) - \text{Error}(s,t) \tag{17.12}$$

This criterion is employed to choose the best split for all internal nodes. The recursive partition algorithm selects the one with the best ΔError.

After building the tree, in order to check the goodness of fit of the tree, the RT algorithm employs its pruning algorithm by either cross-validation or an independent test sample. Mean squared error (MSE) is the commonly adopted pruning criterion. Once a minimal-cost tree (tree with least MSE) is identified, an optimal tree is selected by applying the one-standard-error rule to the minimal-cost tree (Yohannes and Webb 1999).

17.3 Data Collection

Data were retrieved for the grid point (9.75° N, 76.25° E), which is close to the Willingdon island (Cochin) tidal gauge station, Ernakulam, Kerala. Both reanalysis and satellite data were used for the analysis. Reanalysis data were obtained from the European Centre for Medium-Range Weather Forecasts (ECMWF) and satellite data were acquired from the Copernicus Marine Environment Monitoring Service (CMEMS) for the period 2000 to 2014. Sea level (ssh) is the predictand and the independent variables are sea surface temperature (sst), mean sea level pressure (psl), surface pressure (sp), evaporation (e), ocean current (oc), and sea surface salinity (ss). Data pertaining to sea surface temperature, mean sea level pressure, surface pressure, and evaporation were retrieved from ECMWF, whereas data regarding sea level, ocean current, and salinity were from satellite altimetry observations obtained from CMEMS. The GCM data for the historical period (2000–2005) were acquired from Coupled Model Intercomparison Project (CMIP) phase 5.

17.4 Identification of Predictors

In order to identify the predictors, correlation analysis was performed between the dependent (sea level) and independent variables (sea surface temperature, mean sea level pressure, surface pressure, evaporation, ocean current, and sea surface salinity). From this analysis, the variables which are more correlated to sea level were identified. These variables which influence sea level the most were subsequently employed as predictors for sea level. Results of the correlation analysis are presented in Table 17.1.

From the correlation matrix, it is evident that the variables most correlated to sea level are mean sea level pressure, surface pressure, sea surface temperature, ocean current, and sea surface salinity. Apart from this, surface pressure reveals a strong correlation with the mean sea level pressure. Similarly, ocean current exhibits good correlation with sea surface temperature and mean sea level pressure. These intercorrelated variables were discarded from the list of predictors. Thus, the variables which influence sea level the most are sea surface temperature, sea level pressure, and sea surface salinity.

Table 17.1 Correlation matrix

Variables	ssh	sst	psl	sp	e	ss	oc
ssh	1	**0.552**	**0.622**	0.618	−0.378	**−0.743**	−0.644
sst	0.552	1	0.036	0.049	−0.540	−0.384	−0.565
psl	0.622	0.036	1	0.999	−0.088	−0.484	−0.497
sp	0.618	0.049	0.999	1	−0.088	−0.495	−0.510
e	−0.378	−0.540	−0.088	−0.088	1	0.179	0.152
ss	−0.743	−0.384	−0.484	−0.495	0.179	1	0.369
oc	−0.644	−0.565	−0.497	−0.510	0.152	0.369	1

17.5 Development of Statistical Downscaling Model

The first step in the development of the downscaling model is to establish a statistical relationship between the predictand and predictors. The present study employed two machine learning techniques, viz., SVR and RT. The SVR downscaling model was trained using 70% of the data (January 2000 to June 2010) and tested using the remaining 30% data (July 2010 to December 2014). First-degree polynomial was used as the kernel function. Model parameters obtained after training are $C = 10$ and $\epsilon = 0.001$. Root mean square error was taken as the validation criteria. In order to test the model, the remaining data were input to the trained model to predict sea level values. The sea level values obtained from satellite altimetry and SVR downscaling model during the testing period were compared (Fig. 17.1).

In order to develop the RT downscaling model, 70% of the data (January 2000 to June 2010) was used to train the model. MSE was adopted as the split criterion. Additional constraints imposed on the model were a minimum parent size of 10, minimum leaf size of 1, and maximum splits of 125. During training, the dataset was recursively divided according to the split criterion until the optimal condition was attained. In order to avoid overtraining, pruning was employed, and the criterion used being MSE. During testing, the remaining 30% of the data was input to the trained RT model. Sea level values obtained from satellite altimetry and the RT downscaling model during the testing period were compared (Fig. 17.2).

The performance of these downscaling models was evaluated by statistical measures such as coefficient of determination (R^2) and percentage bias (PBIAS). $R^2 > 0.5$ are considered acceptable. Percentage bias measures the average tendency of the computed data to be larger or smaller than their observed counterparts. Positive values represent underestimation, while negative values represent overestimation. For a good model, the PBIAS value should be between ±10 and ±15% (Moriasi et al. 2007). Results of performance evaluation are presented in Table 17.2.

It is evident that the predictions of the SVR downscaling model are closer to the altimeter observations when compared to predictions of the RT downscaling model. Also, the R^2 value is larger for the SVR model. Therefore, the SVR model was

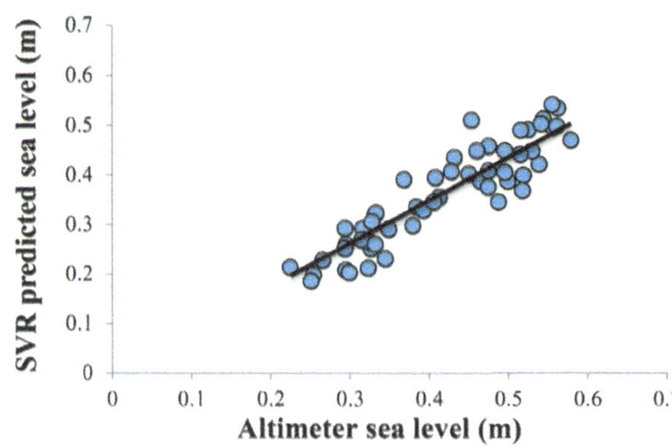

Fig. 17.1 Altimeter versus SVR sea levels during testing period

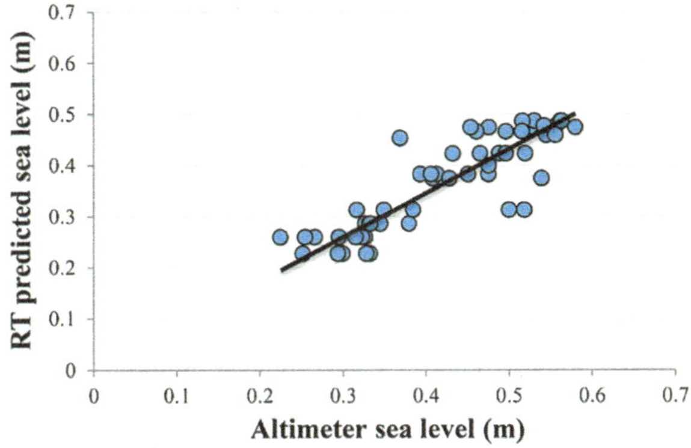

Fig. 17.2 Altimeter versus RT sea levels during the testing period

Table 17.2 Model performance

Parameter	R^2	PBIAS (%)
Model		
SVR	0.839	13.4
RT	0.763	13.27

identified as a better performing downscaling model. This model can be used to predict the sea level for future years.

The GCM data of predictors for the historical period (2000–2005) were acquired from CMIP5. These data were corrected for inherent biases by linear scaling technique. Then, the bias-corrected data pertaining to predictors such as mean sea level pressure, sea surface temperature, and sea surface salinity were input to the SVR model to predict sea level values during the historical period (Fig. 17.3).

From Fig. 17.3, it is clear that a reasonably good match is obtained between the SVR-downscaled sea level and the altimeter observations. Hence, this model can be

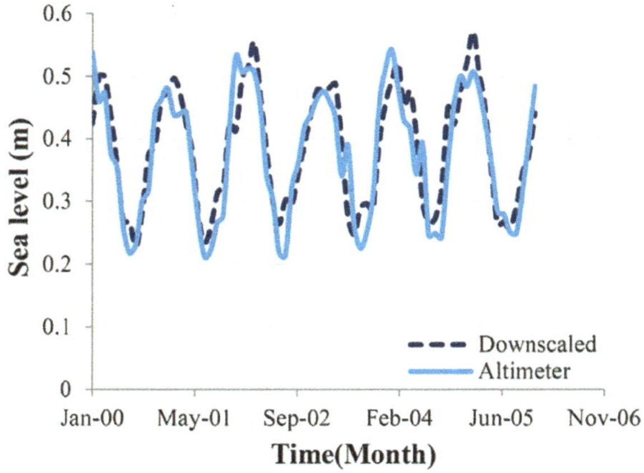

Fig. 17.3 Downscaled versus altimeter sea levels

used for future projection of sea level under different representative concentration pathways (RCPs).

17.6 Conclusions

Climate change can have serious repercussions on sea level changes. In this study, climatic variables which influence sea level the most were identified; these are mean sea level pressure, sea surface temperature, and sea surface salinity. The performances of two statistical downscaling models, viz., SVR and RT models, were compared based on standard statistical measures. It was observed that support vector regression performed better than the regression tree technique. Results indicated a reasonably good match between the altimeter and downscaled sea level values. Based on the results, it is observed that the SVR technique can be used for statistical downscaling of sea level projections from GCMs.

References

Boé J, Terray L, Habets F, Martin E (2007) Statistical and dynamical downscaling of the Seine basin climate for hydro-meteorological studies. Int J Climatol 27:1643–1655

Cichosz P (2014) Data mining algorithms: explained using R, 1st edn. John Wiley & Sons Ltd., UK, pp 261–272

David Neelin J (2010) Climate change and climate modeling, 1st edn. Cambridge University Press, USA, pp 1–7

Elhakeem A, Elshorbagy WE, AlNaser H, Dominguez F (2015) Downscaling global circulation model projections of climate change for the United Arab Emirates. J Water Resour Plan Manag 141(9):1943–5452

Haylock MR, Cawley GC, Harpham C, Wilby RL, Goodess CM (2006) Downscaling heavy precipitation over the United Kingdom: a comparison of dynamical and statistical methods and their future scenarios. Int J Climatol 26:1397–1415

Hsieh WW (2009) Machine learning methods in the environmental sciences: neural networks and kernels, 1st edn. Cambridge University Press, UK, pp 196–200

IPCC (2014) Climate change 2014: synthesis report. Contribution of working groups I, II and III to the fifth assessment report of the intergovernmental panel on climate change [Core Writing Team, RK Pachauri and LA Meyer (eds)]. IPCC, Geneva, Switzerland, p 151

Li Y, Zuo J, Lu Q, Zhang H, Chen M (2016) Impacts of wind forcing on sea level variations in the East China Sea: local and remote effects. J Mar Syst 154:172–180

Maddah HA (2016) Modeling the relation between carbon dioxide emissions and sea level rise for the determination of future (2100) sea level. Am J Environ Eng 6(2):52–61

Mangai JA, Gulyani BB (2016) Induction of model trees for predicting BOD in river water: A data mining perspective. Industrial conference on data mining. Springer, Cham, pp 1–13

Mimura N (2013) Sea-level rise caused by climate change and its implications for society. Proc Jpn. Acad Ser B-Phys Biol Sci 89(7):281–301

Moriasi DN, Arnold JG, Van Liew MW, Bingner RL, Harmel RD, Veith TL (2007) Model evaluation guidelines for systematic quantification of accuracy in watershed simulations. Trans ASABE 50(3):885–900

Sturges W, Douglas BC (2011) Wind effects on estimates of sea level rise. J Geophys Res 116(C6):C06008
Tripathi S, Srinivas VV, Nanjundiah RS (2006) Downscaling of precipitation for climate change scenarios: A support vector machine approach. J Hydrol 330:621–640
Voyant C, Notton G, Kalogirou S, Nivet ML, Paoli C, Motte F, Fouilloy A (2017) Machine learning methods for solar radiation forecasting: a review. Renew Energy 105:569–582
Yohannes Y, Webb P (1999) Classification and regression trees, CART: a user manual for identifying indicators of vulnerability to famine and chronic food insecurity. International Food Policy Research Institute, USA, pp 21–28
Zickfeld K, Solomon S, Gilford DM (2017) Centuries of thermal sea-level rise due to anthropogenic emissions of short-lived greenhouse gases. Proc Natl Acad Sci 114(4):657–662

Chapter 18
Assessing the Impacts of Climate Change on Crop Yield in Upper Godavari River Sub-basin Using H08 Hydrological Model

Pushpendra Raghav and T. I. Eldho

Abstract World population is growing continuously that lead to increase in demand for food. Despite of increase in land area under agriculture production since last decades, one-seventh of people today still do not get sufficient nutrients in their diet. Climate change is well recognized by the scientific community and it further threats the food security for the society. Different parts of the world undergo different level of warming, so it becomes more important to know the regional impact of climate change on the food production. This study aims to find out the impact of climate change on the crop yield of wheat, sorghum, and millet in the upper Godavari river sub-basin using H08 hydrological model. For future simulations, the bias-corrected climate data were taken from five CMIP5 GCMs for RCP4.5 climate scenario from Inter-Sectoral Impact Model Intercomparison Fast Track (ISIMIP-FT) data archive. Our results show that under RCP4.5, wheat yield is projected to decrease by ~37.0% by the end of mid-future (2040–2059) period with respect to present (1980–2001) period. The wheat yield is projected to decrease by ~49.23% by the end of twenty-first century (2080–2099) under RCP4.5. No significant change was observed for the yield of millet and sorghum under RCP4.5. So, sorghum and millet were found to be the crops having no or less impact due to global warming, and wheat production was found to be affected badly due to change in climate. This study demonstrates the impact of climate change on the yield of major crops in the upper Godavari river sub-basin due to temperature stress and water stress in the basin, which is very useful to implement the adaptive means to deal with the possible climate scenarios.

Keywords Crop yield · Climate change · Regional impact · H08 hydrological model · RCP · GCM · CMIP5

P. Raghav
Department of Civil Engineering, Indian Institute of Technology Bombay, Powai, Mumbai 400076, India
e-mail: praghav444@gmail.com

T. I. Eldho (✉)
Interdisciplinary Program in Climate Studies, Indian Institute of Technology Bombay, Powai, Mumbai 400076, India
e-mail: eldho@civil.iitb.ac.in

R. Jha et al. (eds.), *Climate Change Impacts on Water Resources*, Water Science and Technology Library 98, https://doi.org/10.1007/978-3-030-64202-0_18

18.1 Introduction

The change in the global average temperature and precipitation rates was observed during the twentieth century, and this change is accelerating in the twenty-first century also globally (Alexander et al. 2006; Oki and Kanae 2006; Pachauri et al. 2014). The greenhouse gases (e.g. CO_2, CH_4, N_2O, O_3, H_2O, CFCs, etc.) drive the global climate, and due to increase in GHGs concentration in the atmosphere, it becomes important to yearn for change in future climate and its possible impacts on water availability for humanity and agricultural production (Rosenzweig and Parry 1994; Fischer et al. 2002). Different parts of the world are experiencing different rates of climate change, hence it becomes essential to find out the regional impacts of climate change on agriculture production so that the required adaptive measures could be implemented for the food security of the people living in that region (Watson et al. 1998; Pachauri et al. 2014).

The crop growth is driven by various factors (e.g. CO_2 concentration in atmosphere that is related to stomatal conductance and photosynthesis efficiency of plant, decisions on fertilizer applications, irrigation management, climate, technology, etc.) that makes plant growth a very complex process (Lawlor et al. 2004). Among these factors, weather and climate are prominent factors affecting the agricultural production and despite advancement in technology and other fronts, it has been observed that recent trends in change in climate variables may be responsible for significantly affecting crop yield trends (Kukal and Irmak 2018). Various studies have been carried out to get insight about how these factors correlate with the plant growth (Rosenzweig and Parry 1994; Siebert and Ewert 2014; Donatelli et al. 2015; Heino et al. 2018).

In this paper, we analyze the impact of water stress and temperature stress on the yield of three major crops in the upper Godavari river basin, namely sorghum, millet, and wheat. Although nitrogen and phosphorous play a very important role in plant growth but due to non-availability of data on fertilizer application, these factors were not considered in the present study. To reduce uncertainties in the future projections of hydro-meteorological variables, the hydrological model known as H08 was forced with climate data obtained from ISIMIP-FT project where bias-corrected climate data from five CMIP5 GCMs are available. Representative concentration pathway 4.5 (RCP4.5) is taken as a possible climate scenario for the present study. The present study has been carried out in three time slices, namely historical period (1980–2001), mid-future period (2040–2059), and far–future period (2080–2099).

18.2 Study Area and Database Development

The river Godavari ranks 32nd and 34th in terms of water flow and catchment area, respectively, among the 60 major rivers worldwide and flows through six states, viz., Madhya Pradesh, Maharashtra, Karnataka, Andhra Pradesh, Chhattisgarh, and Orissa. The river Godavari originates at the Sahyadri hills near Triambakeshwar in

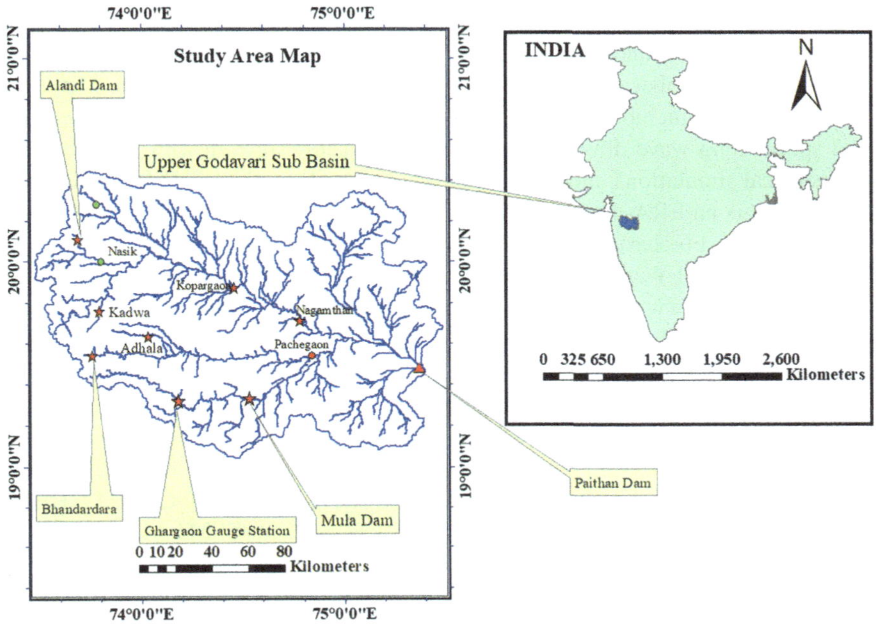

Fig. 18.1 Geographical settings of upper Godavari river basin

Nasik district of Maharashtra state in India at an altitude of 1,065 m in the Western Ghats, about 80 km east of Arabian Sea. After descending from Western Ghats, it traverses with a south-easterly course across the southern part of Indian peninsula, and after flowing about 1,230 km, it drains into Bay of Bengal. About 49% (~152,199 km^2) of total drainage area of Godavari basin falls into Maharashtra and in Maharashtra Godavari basin is divided into 27 sub-basins. Figure 18.1 shows the location map of the present study region: upper Godavari river sub-basin (UGRSB). UGRSB is a sub-basin of Godavari river basin having its whole catchment area in Maharashtra state. Generally, the catchment area upstream of Paithan dam where rivers Mula, Pravara and all other tributaries which joins the Godavari river is been designated as UGRSB. UGRSB is one of the developed basins in terms of industry, urban, and agriculture. Its total catchment area is ~21,774 km^2 and about 8.6 million population lives in 45 towns and 1883 villages. The Godavari basin has a tropical climate characterized by cold weather, hot weather, Southwest monsoon, and post monsoon. The UGRSB receives major part of its rainfall (around 85%) during July to September months during Southwest monsoon period. The sub-basin can be shared into two regions, the upstream reaches in higher elevation areas along the Western Ghats, and the downstream reaches with lower elevation where Jayakwadi reservoir and its command zones lie. When spatial variations are considered, the upstream reaches receive heavy rainfall while lower reaches lie in rain shadow zone with much lower rainfall. The average annual rainfall in the UGRSB is 760 mm. The mean annual surface temperature in the Western Ghats side is about 24 °C, and it increases gradually toward the east. The land use in UGRSB is dominated by barren land and agricultural land.

Table 18.1 describes different types of input data used in this study. Meteorological input data include rainfall, air temperature, specific humidity, air pressure, wind speed, long-wave downward radiation, and short-wave downward radiation. For historical simulations, these variables were taken from the WFD (Weedon et al. 2011) which is an ERA-40 reanalysis product of the European Centre for Medium Range Weather Forecasting (ECMWF). The albedo values were taken from the Second Global Soil Wetness Project (GSWP2). Data on harvested area was downloaded from http://www.ramankuttylab.com/data.html where inventory on contemporary and historical global land use practices is available. For detailed information on this data, see Ramankutty et al. (2008) and Monfreda et al. (2008). The model parameters related to soil and vegetation were calibrated catchment-wise. The gauged streamflow records were collected from Central Water Commission of India (http://www.cwc.nic.in/, Accessed on: 25/08/2017). For impact analysis, bias-corrected climate data from five CMIP5 GCMs, namely MIROC-ESM-CHEM, HadGEM2-ES, GDFL-ESM2M, IPSL-CM5A-LR, and NorESM1-M were downloaded from ISIMIP-ft archives (http://dataservices.gfz-potsdam.de/pik/showshort.php?id=escidoc:1478169). ISIMIP is a community-driven modeling effort that brings together impact models across sectors and scales to generate consistent and comprehensive projections and impacts on the agriculture, water, health, biome, and infrastructure sectors at different level of global warming. For detailed information on bias correction methodology applied to different variables, see Hempel et al. (2013).

18.3 Methodology and Hydrological Model: H08

H08 is a grid-based, distributed hydrological model developed by Hanasaki et al. (2008). It consists of six sub-models of major human activities: land surface hydrology, river routing, crop growth, anthropogenic water withdrawal, crop growth, reservoir operation, and environment flow requirement. For this study, only three modules, the land surface module, river routing module, and crop growth module were used. Land surface module simulates water balance components and energy balance components above and beneath the land surface when forced by the high-temporal-resolution climate data. As a water balance component, runoff (surface and sub-surface runoff), evapotranspiration, and potential evapotranspiration are the outputs and soil moisture acts as state variable. River routing module routs the runoff from upstream to downstream. Crop growth module calculates crop calendar, crop yield, and irrigation water requirement.

The land surface hydrology module is based on a bucket model (Manabe 1969 and Robock et al. 1995) but differs from original formulation in the following aspects. First, soil temperature is calculated using the force restore method (Bhumralkar 1975; Deardorff 1978) to simulate the diurnal cycle of surface temperature reasonably using three-hourly meteorological forcing inputs. Second, a simple subsurface runoff

Table 18.1 Basic input data used in the present study

Data type	Description	Source/reference(s)	Original spatial resolution	Period	Comments
Physical data	DEM[1]	SRTM[2]	1 Arc-second (~30 m)	–	Global
	Crop areas	Monfreda et al. (2008)	5′ (~10 km)	2000	Global
	Soil	–	–	–	Soil parameters were calibrated catchment-wise
	Vegetation	–	–	–	Vegetation parameter (bulk transfer coefficient) was calibrated catchment-wise
Meteorological data	Rainfall, near surface pressure, air temperature, specific humidity, wind speed, long-wave downward radiation, short-wave downward radiation	WFD[3] (Weedon et al. 2011)	0.5°	1980–2001	5′ (~10 km mesh) data was prepared by linear interpolation for this study
	Albedo	GSWP2[4] (Dirmeyer et al. 2006)	1°	1980–1995	Mean monthly 5′ data was prepared for this study

(continued)

Table 18.1 (continued)

Data type	Description	Source/reference(s)	Original spatial resolution	Period	Comments
Hydrologic data	Discharge	CWC[5]	Gauged	1980–2014 with missing data	Period may differ with different gauge stations
GCM data	Rainfall, near surface pressure, air temperature, specific humidity, wind speed, long-wave downward radiation, short-wave downward	MIROC-ESM-CHEM	2.81° × 2.77°	1980–2001, 2040–2059, 2080–2099	ISIMIP-ft[6] project provide bias-corrected climate data for all five CMIP5 GCMs (Hempel et al. 2013)
		HadGEM2-ES	1.875° × 1.25°		
		GFDL-ESM2M	2.5° × 2.0225°		
		IPSL-CM5A-LR	3.75° × 1.8947°		
		NorESM1-M	2.5° × 1.8947°		

[1]DEM is digital elevation map
[2]SRTM is shuttle radar topography mission
[3]WFD is WATCH forcing data
[4]GSWP2 is second global soil wetness project
[5]CWC is Central Water Commission of Gov. of India
[6]ISIMIP-ft is Inter-Sectoral Impact Model Intercomparison Fast Track project

parameterization scheme is added to the model. Third, two independent land surface conditions can be simulated within a single grid that is intended to separate irrigated cropland from other land types.

Potential evaporation E_P is calculated as

$$E_P(T_S) = \rho C_D U(q_{SAT}(T_S) - q_a) \quad (37.1)$$

where ρ is the air density, C_D is the bulk transfer coefficient, U is the wind speed, $q_{SAT}(T_S)$ is the saturated specific humidity at surface temperature, and q_a is the specific humidity.

Evaporation from the surface is calculated as

$$E = \beta E_P(T_S) \quad (37.2)$$

where

$$\beta = \begin{cases} 1\, if\, 0.75W_f \le W \\ \frac{W}{W_f}\, if\, W < 0.75W_f \end{cases}$$

where W is the soil water content and W_f is the soil water content at field capacity.

Surface runoff is calculated as

$$Q_S = \begin{cases} W - W_f\, if\, W_f < W \\ 0\, if\, W \le W_f \end{cases} \quad (37.3)$$

Subsurface runoff is calculated as

$$Q_{sb} = \frac{W_f}{\tau}\left(\frac{W}{W_f}\right)^{\gamma} \quad (37.4)$$

where τ and γ are time constant and shape parameter, respectively, and these two parameters were calibrated in the present study.

River routing module is based on Oki et al. (1999) and crop growth module is based on soil and water integrated model (Krysanova et al. 2000).

The crop growth module estimates the cropping period necessary to obtain mature total plant biomass and crop yield and estimates crop growth using heat unit theory.

During the cropping period, plant biomass is calculated using a simple photosynthesis model as

$$\Delta B = BE \times PAR \times REGF \quad (37.5)$$

$$PAR = 0.02092RAD[1 - \exp(-0.65LAI)] \quad (37.6)$$

$$REGF = \min(TS, WS, NS, PS) \quad (37.7)$$

where ΔB is the daily increase in total biomass, BE is a crop-specific parameter, PAR is photosynthetically active radiation, $REGF$ is the crop-regulating factor, RAD is short-wave radiation, and LAI is leaf area index. The four stress factors which affect crop growth are temperature stress (TS), water stress (WS), nitrogen stress (NS), and phosphorous stress (PS). The NS and PS were not considered in the present study because of the lack of information on fertilizer application.

The crop yield is estimated as

$$YLD = HVSTI \times \frac{WSF}{WSF + \exp(6.117 - 0.086 \times WSF)} \times BAG \quad (37.8)$$

where

$$BAG = 1 - (0.4 - 0.2 \times IHUN) \sum \Delta B \quad (37.9)$$

$$IHUN = \frac{\sum_t (T - T_B)}{PHUN} \quad (3.10)$$

$$WSF = \frac{SWU}{SWP} \times 100 \quad (3.11)$$

$HVSTI$ is a crop-specific parameter, $PHUN$ is potential heat units required for the maturity of the crop, T_B is plant's specific base temperature, SWU (SWP) is the accumulated actual (potential) plant transpiration in the second half of the growing season.

Figure 18.2 shows the flowchart of methodology used in the present study from model setup to future simulations. Four model parameters, namely root-zone soil depth (d), bulk transfer coefficient (C_D), time constant (τ), and shape parameter (γ) were calibrated in the present study. WFD forcing data is given as the input to the land surface module that outputs gridded runoff that is further routed to get a hydrograph at desired locations by using river routing module of H08 model. Crop growth model takes potential evapotranspiration (PET), actual evapotranspiration (AET), downward short-wave radiation, and temperature as an input to simulate crop biomass. PET and AET are simulated by land surface module of H08 model. Once the model is calibrated and validated, then forcing data of the GCMs is given as the input to the model to investigate the impact of climate change on crop yield. The model is run for five GCMs individually for both historical period and future periods. The results were averaged over the five GCMs to deal with the uncertainty due to each GCM's unique structure.

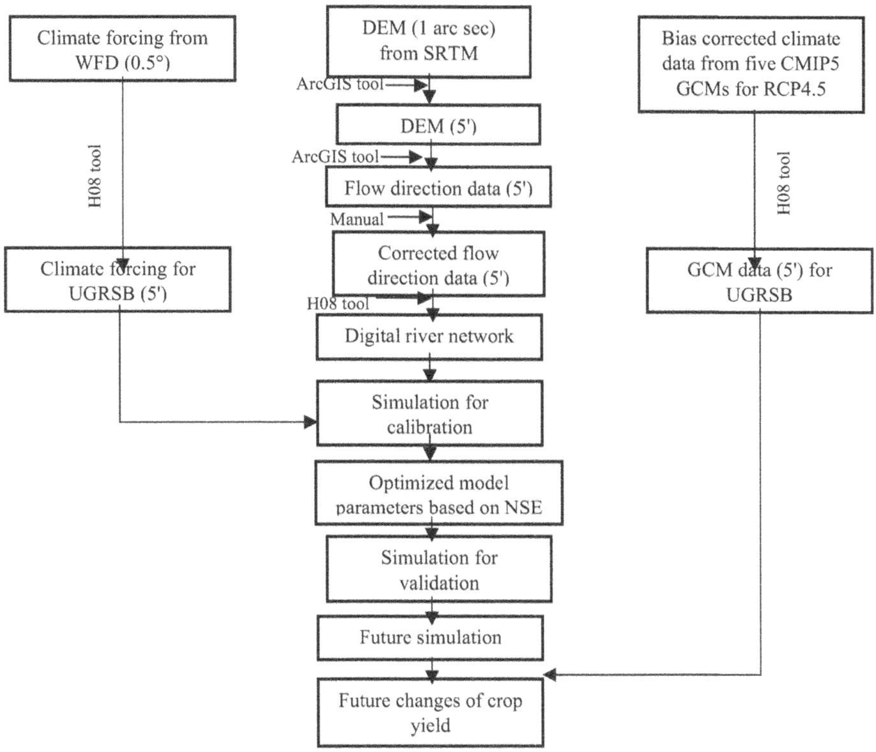

Fig. 18.2 Flowchart of methodology used for the study

18.4 Results and Discussion

Table 18.2 presents the model performance indices for both calibration and validation period which indicate that model simulated streamflow is in good agreement with the historical observed streamflow for all two stations.

Figures 18.3 and 18.4 plot the comparison of simulated and observed streamflow at Mula and Bhandardara stations, respectively. The Nash–Sutcliffe efficiency varies between 0.80 and 0.84 for calibration period and 0.71 and 0.75 for validation period. Other model performance indices like refined index of agreement (d_r), percentage

Table 18.2 Model calibration and validation

Routing station	Time period	NSE	d_r	PBIAS (%)	RMSE (m^3/s)	R^2
Mula	1984–1994	0.84	0.87	−5.2	16.7	0.84
	1995–2001	0.75	0.82	−14.1	17	0.79
Bhandardara	1985–1993	0.80	0.87	17.4	10.1	0.85
	1994–2001	0.71	0.82	32.1	14.1	0.83

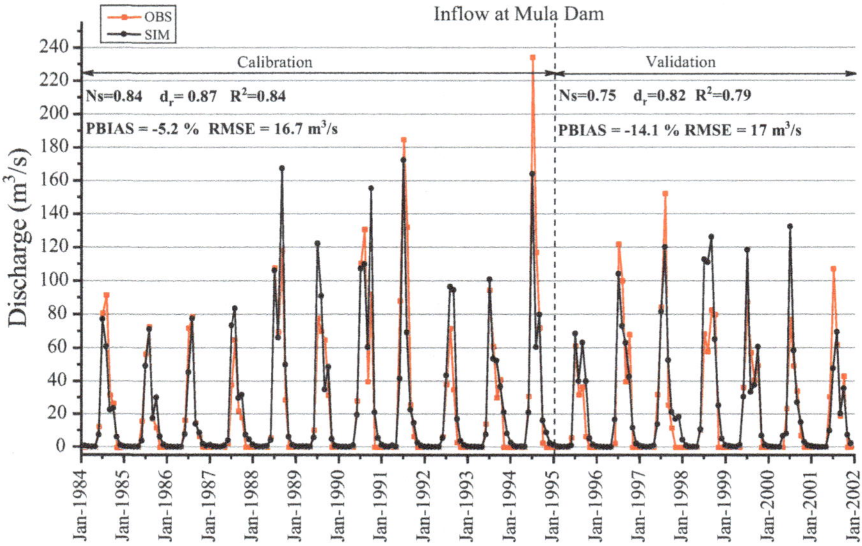

Fig. 18.3 Comparison of simulated streamflow and observed streamflow during calibration period (1984–1994) and validation period (1995–2001) at Mula station

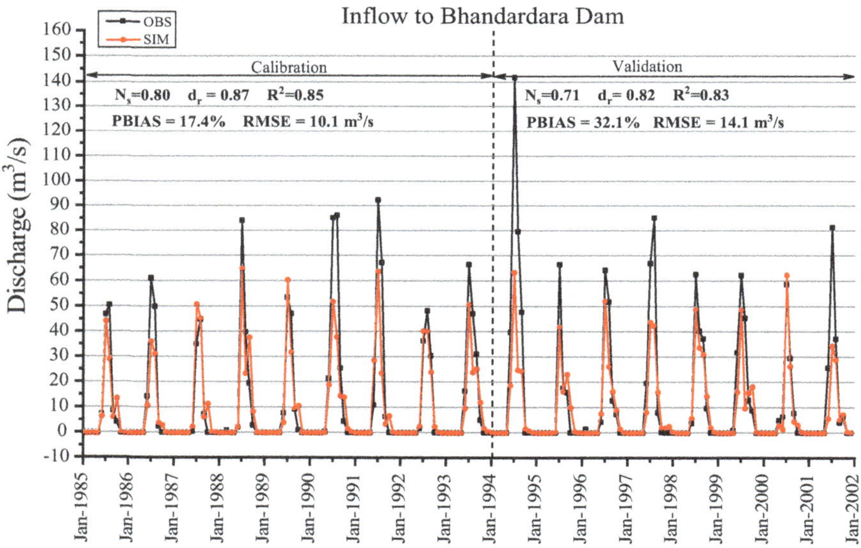

Fig. 18.4 Comparison of simulated streamflow and observed streamflow during calibration period (1985–1993) and validation period (1994–2001) at Bhandardara station

Table 18.3 Annual mean and percentage change of yield of major crops with respect to historical period in UGRSB under RCP4.5

Crops	Time period	Annual mean (kg/ha)	% change
Millet	Historical (1980–2001)	8704	–
	Mid-future (2040–2059)	8678	−0.30
	Far-future (2080–2099)	8856	1.75
Sorghum	Historical (1980–2001)	7582	–
	Mid-future (2040–2059)	7586	0.05
	Far-future (2080–2099)	7764	0.02
Wheat	Historical (1980–2001)	2584	–
	Mid-future (2040–2059)	1628	−37.0
	Far-future (2080–2099)	1312	−49.23

bias (PBIAS), and coefficient of determination (R^2) were also within satisfactory limits.

Table 18.3 shows the annual mean and percentage change in yield of various crops with respect to historical period. It was observed that yield of wheat was adversely affected due to change in USRSB climate that is attributed to the increase in future temperature. Millet and sorghum production was also not affected much. Generally, millet and sorghum are considered photoperiodic and belong to the Hatch-Slack (or C4) class of photosynthesis which allows photosynthesis to continue at higher temperatures. Millet also has high level of resistance to drought; because of that millet has high potential productivity and millet is capable of adjusting to changing climate.

Figure 18.5 shows the inter-annual variations in the yield of major crops in UGRSB. It is clear from Figure 18.5 that yield of wheat is decreasing in the near-future to end of twenty-first century.

18.5 Conclusions

This study attempted to investigate on the impact of climate change on the yield of major crops in the UGRSB. Water and temperature are considered the only crop growth deciding factors. Following are the main findings of this study:

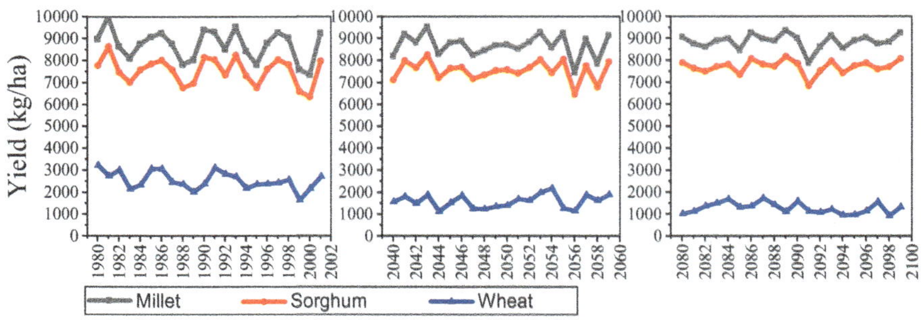

Fig. 18.5 Inter-annual variations in yield of major crops in UGRSB under RCP4.5. Yield of each crop in each panel represents the mean of yields from five GCMs used in this study

1. The H08 model performed very well during calibration and validation periods at all stations with Nash–Sutcliffe efficiency ranging from 0.80 to 0.84 during calibration period and 0.71–0.75 during validation period.
2. Yield of wheat crop was projected to change by −37.0 and −49.23% by the mid-future and end of twenty-first century, respectively under RCP4.5 climate scenario.
3. No significant trend was observed in case of millet and sorghum's yield because of their ability to adjust their biomass in changing climate.

Acknowledgments The authors sincerely thank Dr. Naota Hanasaki for providing us the source code of H08 model. The climate data was downloaded from the data distribution center (http://h08.nies.go.jp) which is developed and maintained by H08 development team and S-8 project team.

References

Alexander LV, Zhang X, Peterson TC, Caesar J, Gleason B, Tank AK, Haylock M, Collins D, Trewin B, Rahimzadeh F, Tagipour A (2006) Global observed changes in daily climate extremes of temperature and precipitation. J Geophys Res: Atmos 111(D5)

Bhumralkar CM (1975) Numerical experiments on the computation of ground surface temperature in an atmospheric general circulation model. J Appl Meteorol 14(7):1246–1258

Deardorff JW (1978) Efficient prediction of ground surface temperature and moisture, with inclusion of a layer of vegetation. J Geophys Res: Ocean 83(C4):1889–1903

Dirmeyer PA, Gao X, Zhao M, Guo Z, Oki T, Hanasaki N (2006) GSWP-2: multimodel analysis and implications for our perception of the land surface. Bull Am Meteorol Soc 87(10):1381–1398

Donatelli M, Srivastava AK, Duveiller G, Niemeyer S, Fumagalli D (2015) Climate change impact and potential adaptation strategies under alternate realizations of climate scenarios for three major crops in Europe. Environ Res Lett 10(7):075005

Fischer G, Shah MM, Van Velthuizen HT (2002) Climate change and agricultural vulnerability

Hanasaki N, Kanae S, Oki T, Masuda K, Motoya K, Shirakawa N, Shen Y, Tanaka K (2008) An integrated model for the assessment of global water resources–Part 1: model description and input meteorological forcing. Hydrol Earth Syst Sci 12(4):1007–1025

Heino M, Puma MJ, Ward PJ, Gerten D, Heck V, Siebert S, Kummu M (2018) Two-thirds of global cropland area impacted by climate oscillations. Nat Commun 9(1):1257

Hempel S, Frieler K, Warszawski L, Schewe J, Piontek F (2013) A trend-preserving bias correction–the ISI-MIP approach. Earth Syst Dyn 4(2):219–236

Krysanova FW, Arnold J, Srinivasan R, Williams J (2000) PIK report Nr 69 SWIM (Soil and Water Integrated Model), User Manual, 239p

Kukal MS, Irmak S (2018) Climate-Driven crop yield and yield variability and climate change impacts on the us great plains agricultural production. Sci Rep 8(1):3450

Lawlor DW, Mengel K, Kirkby EA (2004) Principles of plant nutrition

Manabe S (1969) Climate and the ocean circulation. Mon Wea Rev 97(11):739–774

Monfreda C, Ramankutty N, Foley JA (2008) Farming the planet: 2. Geographic distribution of crop areas, yields, physiological types, and net primary production in the year 2000. Glob Biogeochem Cycles 22(1)

Oki T, Kanae S (2006) Global hydrological cycles and world water resources. Science 313(5790):1068–1072

Oki T, Nishimura T, Dirmeyer P (1999) Assessment of annual runoff from land surface models using Total Runoff Integrating Pathways (TRIP). J Meteorol Soc Jpn Ser II, 77(1B):235–255

Pachauri RK, Allen MR, Barros VR, Broome J, Cramer W, Christ R, Church JA, Clarke L, Dahe Q, Dasgupta P, Dubash NK (2014) Climate change 2014: synthesis report. Contribution of working groups I, II and III to the fifth assessment report of the intergovernmental panel on climate change. IPCC, p 151

Ramankutty N, Evan AT, Monfreda C, Foley JA (2008) Farming the planet: 1. Geographic distribution of global agricultural lands in the year 2000. Glob Biogeochem Cycles 22(1)

Robock A, Vinnikov KY, Schlosser CA, Speranskaya NA, Xue Y (1995) Use of midlatitude soil moisture and meteorological observations to validate soil moisture simulations with biosphere and bucket models. J Clim 8(1):15–35

Rosenzweig C, Parry ML (1994) Potential impact of climate change on world food supply. Nature 367(6459):133–138

Siebert S, Ewert F (2014) Future crop production threatened by extreme heat. Environ Res Lett 9(4):041001

Watson RT, Zinyowera MC, Moss RH, Dokken DJ (1998) The regional impacts of climate change. An assessment of vulnerability: a special report of IPCC working group iI, p 517

Weedon GP, Gomes S, Viterbo P, Shuttleworth WJ, Blyth E, Österle H, Adam JC, Bellouin N, Boucher O, Best M (2011) Creation of the WATCH forcing data and its use to assess global and regional reference crop evaporation over land during the twentieth century. J Hydrometeorol 12(5):823–848

Chapter 19
Evaluation of Time Discretization of Daily Rainfall From the Literature for a Specific Site

R. Harshanth, Saha Dauji, and P. K. Srivastava

Abstract Design of hydraulic structures involves usage of the hydrographs, which are derived from the information of rainfall history at the site. It is well recognized that the shape of the design hydrograph depends on the hyetograph used to generate it. However, for derivation of hyetograph, information regarding temporal distribution of rainfall should be available. It remains a fact that in many places of India and other developing countries, the information on temporal discretization of rainfall is not available. This has led to the adoption of curves from the literature, using which the time distribution of rainfall is derived. Recently, IMD-CWC had published temporal distribution curves for various river basins in India. The curves considered for purposes of comparison in this study include those provided by the NRCS-TR55, US Army Corps and IMD-CWC. The study indicates that the time distribution curve derived for the site from the recorded continuous hourly rainfall data for many years is significantly different from those obtained for the existing curves in the literature. It is also shown that this difference leads to either over-design or under-design of hydraulic elements. The study underlines the importance of obtaining the site-specific curves for important projects so as to carry out economical and safe designs.

Keywords Time discretization · Hydrograph · Hyetograph · Site-specific · Design storm

R. Harshanth
Nuclear Power Corporation India Limited, Kudankulam, India
e-mail: harshanthr@npcil.co.in; harshanthr@gmail.com

S. Dauji (✉)
Nuclear Recycle Board, Bhabha Atomic Research Centre, Mumbai, India
e-mail: acad.dauji@gmail.com; dauji_saha@yahoo.com

Homi Bhabha National Institute, Mumbai, India

P. K. Srivastava
Nuclear Recycle Board, Bhabha Atomic Research Centre, Tarapur, India
e-mail: pansriv@gmail.com

R. Jha et al. (eds.), *Climate Change Impacts on Water Resources*, Water Science and Technology Library 98, https://doi.org/10.1007/978-3-030-64202-0_19

19.1 Introduction

For design of hydraulic structures such as weirs, check dams, culverts and storm water drainage systems, it is required to arrive at the design discharge at the point of interest. The importance of the structure defines the mean recurrence interval, also known as return period, which needs to be considered for the evaluation of the extreme rainfall for that purpose. In India, most of the rainfall records are daily records, from 08:30 h on a day to 08:30 h the next day, and this is recorded as the rainfall for that day. Extreme value analysis of this data yields the desired daily rainfall extreme. However, for hydraulic structures having much less time of concentration, information about finer discretization in time is required to arrive at the deign rainfall extreme, and subsequently runoff calculations.

Time distribution curves illustrate the variation of the hourly rainfall as a function of time. This distribution is valuable for hydrologic design of structures as it is possible to obtain the maximum rainfall value corresponding to the time of concentration specified for the structure. It may also be recognized here that the design hyetograph directly depends on the hyetograph used to generate it. Time distribution curves also form a basis of generating the time distribution of an n-hour rainfall hyetograph when the maximum daily rainfall value is known. Knowledge of the time distribution is useful for analyzing the rainfall characteristics and forms a rational basis for runoff prediction. In the absence of well-established time distribution curves for India, designers mostly resort to such curves from foreign locations.

Recently, IMD-CWC had published such time distribution information for few river basins in India. In this paper, the time distribution curves available in the literature are applied to a western coastal site in India and the results are compared to those obtained from site-specific analysis. Details on the site, rainfall data and the methodology are presented in the sections to follow.

19.2 Site Description

The site under consideration is located about 200 km northwards of Mumbai. The rainfall season is dominated by the South-west monsoon. The average rainfall of about 2000 mm occurs between the months of June to September. The rainfall data was obtained in a chronological manner in the form of hourly rainfall for a period of 17 years from 1997 to 2013. The unit of measurement of rainfall depth was in millimeters. This data was used for further analysis.

19.3 Time Distribution Curves from the Literature

In USA, research was carried out by several agencies like NRCS—National Resources Conservation Services TR 55 (NRCS TR-55 1986) and the USAC (EM1110-2–1411 1965) to provide design basis spatial rainfall distribution. The results were provided in the form of dimensionless curves for various regions of the country. The zonation of the country was based on the hydro-meteorological characteristics and rainfall patterns of USA. The curves in TR-55 also include hourly rainfall intensities for selected design frequency arranged in a sequence that would produce the maximum runoff. The TR-55 curves are shown in Fig. 19.1.

But it is well recognized that preparation of such curves requires continuous rainfall data over many years. However, most countries, due to the lack of continuous rainfall information, resort to these models for arriving at the design rainfall hyetograph. As TR-55 curves were originally obtained by denaturing the rainfall so that its maximum value of rainfall occurs toward the mid time of the total duration of rainfall, it is imperative that its suitability in Indian monsoon rainfall conditions, particularly for this site is yet to be established. In order to assess the applicability of methods for a western coastal site in India, the dimensionless rainfall time distribution curves were evaluated from several established curves obtained from the literature, namely TR55, SPS and IMD-CWC (PMP Atlas 2015). Herein, the maximum n-hour fraction of 24-h rainfall obtained from these curves are compared with the maximum n-hour fraction of the 24-h devised from continuous rainfall data of several years recorded at the site. In case of SPS curves, it is required to select the index rainfall in inches for

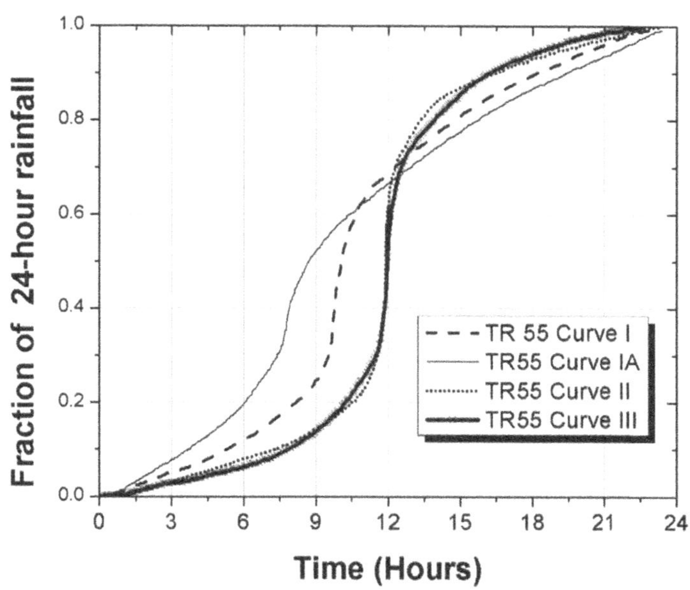

Fig. 19.1 NRCS TR-55 dimensionless urves (NRCS TR-55 1986)

the site under consideration. This is normally done as per the guidelines suggested by EM1110-2–1411 (1965). It is presented Fig. 19.2. However, the guideline is applicable only for USA. In order to assess its applicability in India, the extreme values (minimum and maximum) of index rainfall are chosen to arrive at the possible range of values for comparison.

Studies were also conducted by the India Meteorological Department and the Central Water Commission (PMP Atlas 2015) for different catchments in India subjected to monsoon rainfall through the probable maximum precipitation or the PMP method using the physical or meteorological approach. This involves the analysis of the maximum rainfall events in the given catchment area along with the neighboring catchments. Consequently, the depth-area-duration (DAD) curves are developed to understand the temporal and areal distribution of extreme storms in the catchment. This is followed by application of statistical techniques like frequency analysis to determine the maximum storm. There are separate sets of documents for Brahmaputra basin, Cauvery basin, Ganga basin, Godavari basin, Mahanadi basin, Narmada basin and for the west-flowing basin. For west-flowing rivers, PMP Atlas (2015) has been developed by IMD-CWC for two different catchments, the northern and the southern, and time distribution of rainfall for different durations of storm events is suggested, such as 12 h, 24 h, 48 h, etc. with the temporal axis in hours. For intermediate duration of storms, choice of the time distribution is left to the judicious decision of the designer. For example, the 24 h time distribution applicable for the site from PMP Atlas (2015) is reproduced in Fig. 19.3.

The methodology adopted for arriving the maximum n-hour fraction of 24-h rainfall from NRCS and SPS curves is described in the following sections. Also,

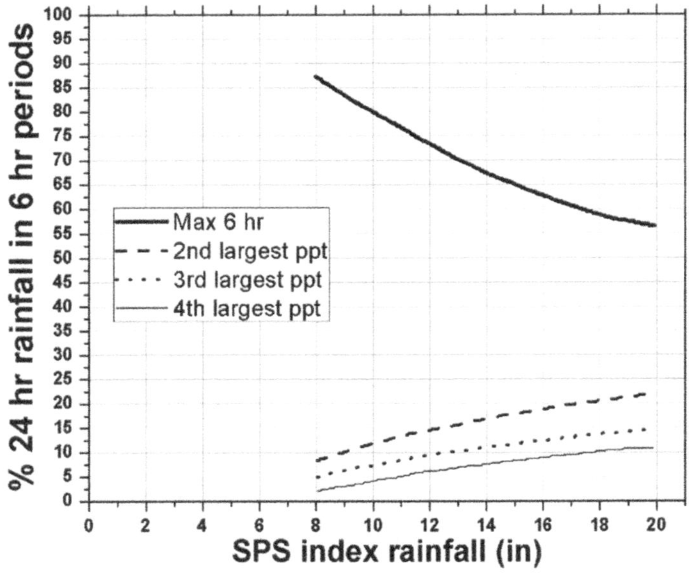

Fig. 19.2 US Army Corps curves (EM1110-2–1411 1965)

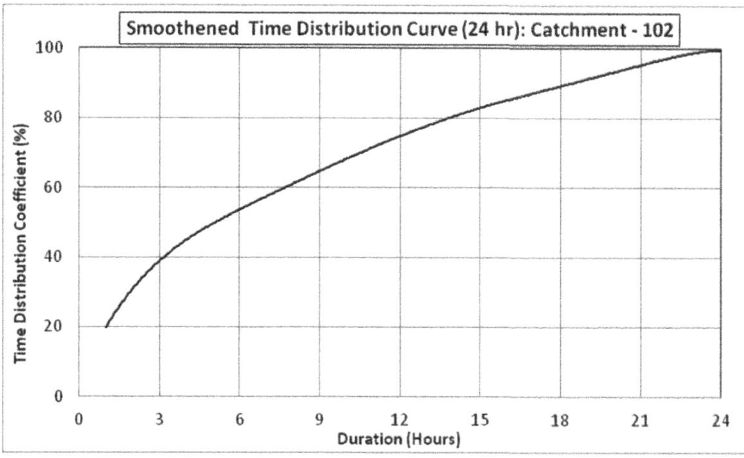

Fig. 19.3 IMD-CWC curve for 24-h rainfall for the Catchment 102: Northern basin for west-flowing rivers (PMP Atlas 2015)

the results of its comparison with the results obtained with Tarapur site data and IMD-CWC curves for western coastal site are presented.

19.4 Data and Methodology

The data for the study was continuous hourly rainfall records for 17 years, from 1997 to 2013. The maximum fraction of daily rainfall occurring in 1 h, continuous 2 h, continuous 3 h, and so on till continuous 24 h was calculated from these data. In India, for most of the stations where daily rainfall data is recorded, hydrologic data begins at 08:30 h on a particular and ends at 08:30 h on the next day. Calculations are carried out to find out the maximum 1 h, maximum continuous 2 h, and so on up to maximum continuous 24-h rainfall value of a hydrologic day. The calculated values are normalized with respect to the total rainfall value of that particular day.

In case of NRCS TR-55 curves, the graphs are available in the form of dimensionless continuous mass curves for a period of 24 h. In order to obtain the time distribution curves, hyetographs are sought. So initially, the mass curves were converted to corresponding hyetographs by taking the derivative of the mass curve. Once the digitized data of hyetograph were available, it was observed that the data was available in the form of distinct peaks for discrete time instants. These time instants range from 0 to 24 h with spacing between each time instant being maintained between 0.05 and 0.1 h. In order to make the data comparable with other curves in literature, the values of rainfall were lumped so as to obtain hourly rainfall values. This means that the total amount of rainfall in the interval 0 to 1 h is lumped at 1st hour. Similarly, the total rainfall occurring in the interval 1 to 2 h is lumped at the 2nd h and so on. The hyetographs so obtained for each of the TR-55 curves are shown in Figs. 19.4, 19.5, 19.6 and 19.7.

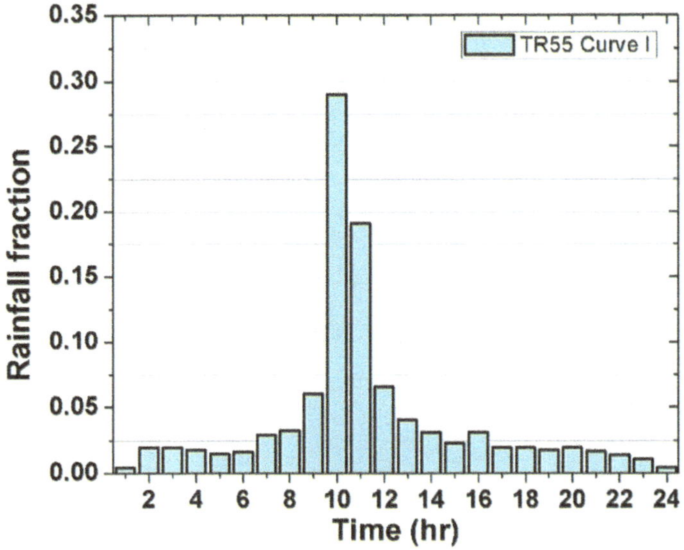

Fig. 19.4 Hyetograph of TR-55 Curve I (NRCS TR-55 1986)

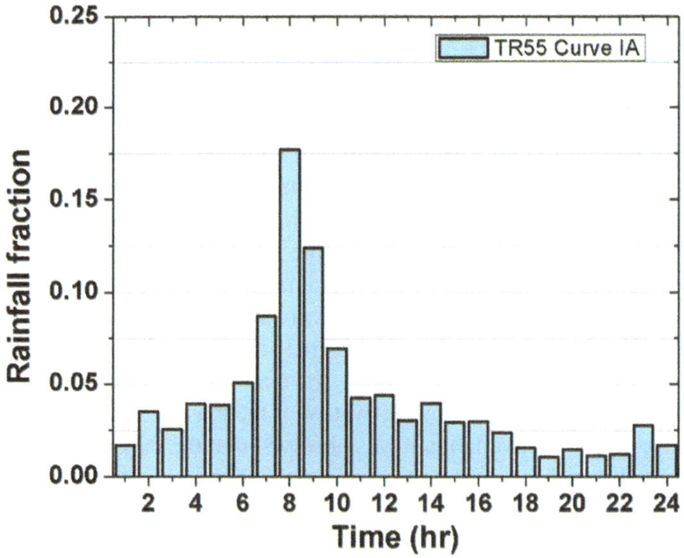

Fig. 19.5 Hyetograph of TR-55 Curve IA (NRCS TR-55 1986)

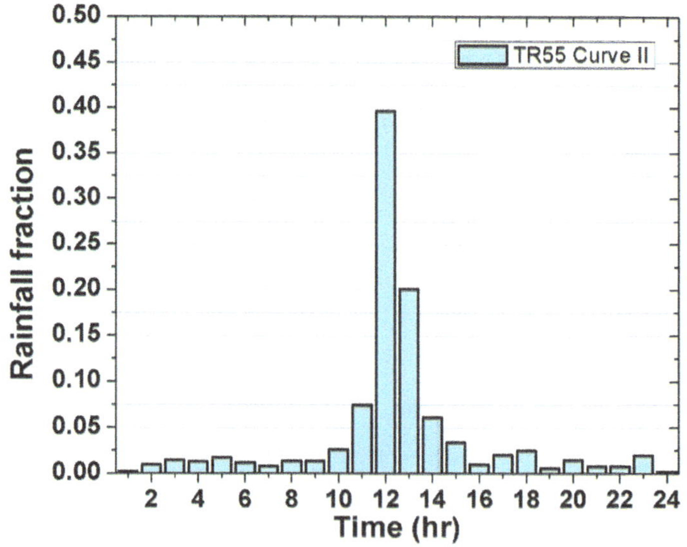

Fig. 19.6 Hyetograph of TR-55 Curve II (NRCS TR-55 1986)

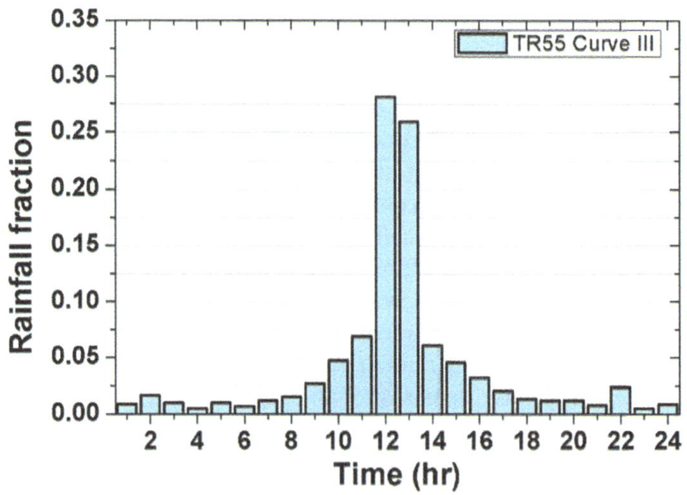

Fig. 19.7 Hyetograph of TR-55 Curve III (NRCS TR-55 1986)

It can be observed from these hyetographs that most of the rainfall is concentrated toward the mid duration of the entire rainfall. The typical hyetograph for tropical monsoon, applicable for India, is from Curve III as depicted in Fig. 19.7. Later, this will be compared with the typical hyetographs obtained from the site. Once the hyetograph is obtained, the values of continuous n-hour maximum rainfall as a fraction of 24-h rainfall were calculated for each of the NRCS TR-55 hyetograph.

Table 19.1 Hourly rainfall distribution for 6-h segment (EM1110-2–1411 1965)

Rainfall period (sub-division of 6-h period)	Time distribution of 6-h SPS rainfall, expressed in percent of total 6-h rainfall			
	Selected unit rainfall duration			
	6-h	3-h	2-h	1-h
1st	100	33	26	10
2nd		67	53	12
3rd			21	15
4th				38
5th				14
6th				11
Total	100	100	100	100

In the case of SPS (Standard Project Storm) curves, the original curve is presented in the form of various curves corresponding to different 6-h periods of a 24-h storm as shown in Fig. 19.2. Having analyzed various storms in the USA, the USAC (EM1110-2–1411 1965) concluded that for maximum runoff the maximum proportion of rainfall should occur in the third 6-h period (13–18 h). The USAC also provides the 1-h distribution of rainfall in each 6-h rainfall period as a percentage of 24-h rainfall (see Table 19.1 for reference).

However, as already mentioned earlier, in order to develop the hyetograph, it is required to give the index rainfall in inches as input. This index rainfall is site specific. The USAC presents the value of index rainfall for various sites in USA. Thus, index rainfall values are not available for Indian sites. However, only the hyetographs corresponding to extreme values of index rainfall (8 inches: SPSmin and 20 inches: SPSmax) are generated. This is done to capture the range of curves that can be generated through the USAC method. Hyetographs corresponding to the aforementioned index values are shown in Figs. 19.8 and 19.9.

In case of Tarapur site, data was obtained in the form of hourly data for 17 years. The continuous maximum n-hour rainfall (where $n = 1, 2, 3, \ldots, 24$) normalized to the maximum 24-h rainfall was calculated for all hydrologic days in a year. The plot of the variation of the maximum fraction of 24-h rainfall in continuous n-hour rainfall is shown in Fig. 19.10. Along with this, such fractions for the area under consideration developed by IMD-CWC (PMP Atlas 2015) were also compared. From Fig. 19.10, it can be inferred that for the maximum 2-h continuous rainfall contributes to about 40% of the 24-h rainfall. Similarly, we have about 65% in continuous 6 h and about 80% in continuous 9 h.

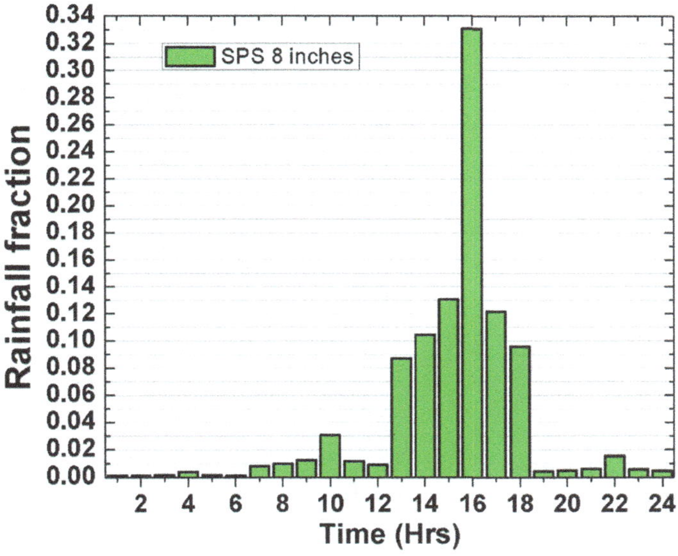

Fig. 19.8 Hyetograph of SPS index rainfall of 8 inches (EM1110-2–1411 1965)

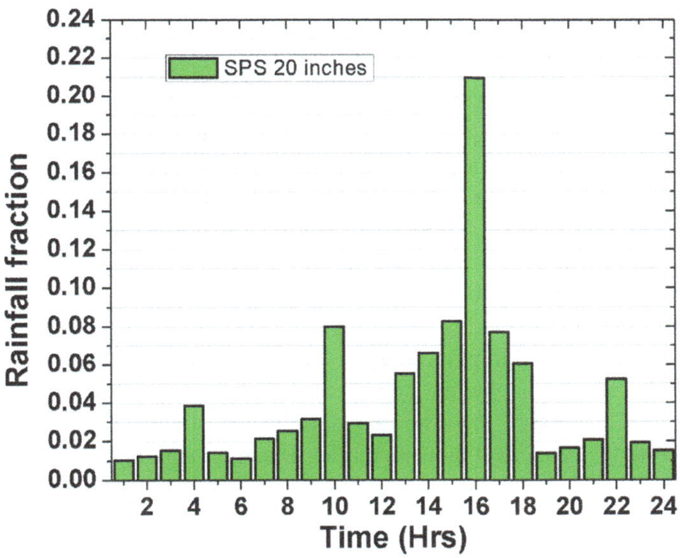

Fig. 19.9 Hyetograph of SPS index rainfall of 20 inches (EM1110-2–1411 1965)

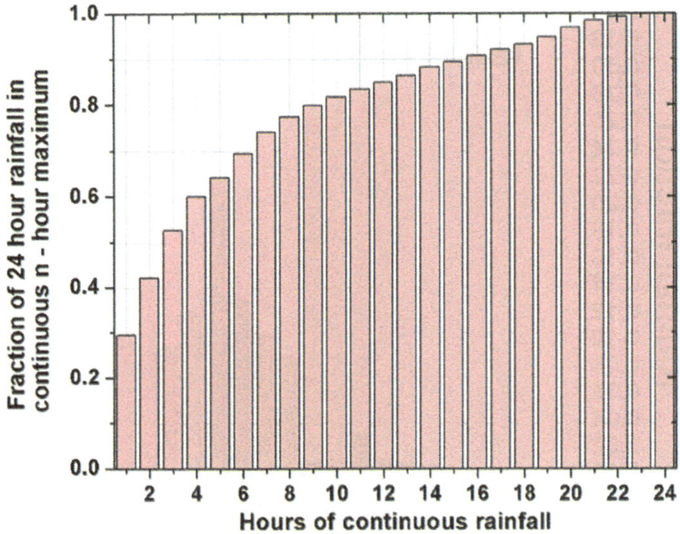

Fig. 19.10 Maximum n-hour fraction of 24-h rainfall for the site

19.5 Results and Discussion

The results of comparative analysis of maximum n-fraction obtained from the above-mentioned sources with Tarapur site data are presented in the form of column plots of percentage deviation of Tarapur calculated values. The percentage deviation of Tarapur site values from the values predicted by different curve types of TR-55 (NRCS TR-55 1986) are shown in Figs. 19.11, 19.12, 19.13 and 19.14. Similarly, the percentage deviation of Tarapur site values from those predicted by SPS 8 inches and SPS 20 inches (EM1110-2–1411 1965) are shown in Figs. 19.15 and 19.16, respectively. Finally, considering the available Indian temporal distribution curve, the percentage deviation of Tarapur values from IMD-CWC (PMP Atlas 2015) values for catchment in an applicable region is presented in Fig. 19.17.

In case of TR-55 Curve I, the values are slightly over-predicted up to 5 h continuous rainfall and slightly under-predicts thereafter as can be seen in Fig. 19.11. It can be observed in Fig. 19.12 that the values obtained from site data is much higher than those predicted TR-55 Curve IA especially in the initial hours. A maximum deviation of as much as 40% is observed. However, as can be seen in Fig. 19.13, it is observed that TR 55 Curve II grossly over-predicts the values up to about 45%. It can also be seen from Fig. 19.14 that TR55 Curve III, the applicable curve for India, mostly over-predicts the n-hour maximum fraction with the maximum deviation of about 30%. As the typical time distribution patterns were developed for USA, they fail to capture the rainfall characteristics of the Indian site adequately, resulting in these deviations.

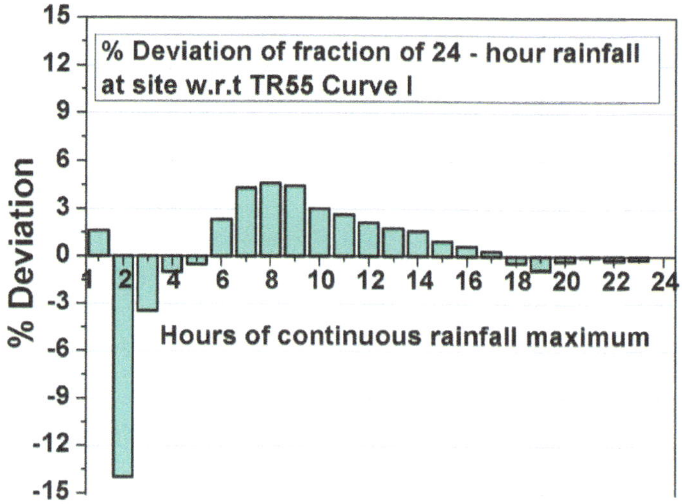

Fig. 19.11 Percentage deviation of Tarapur site values from TR-55 Curve I

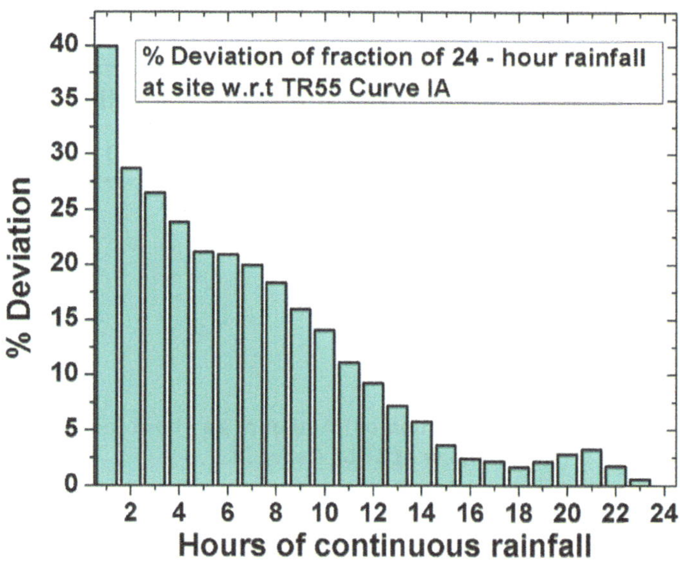

Fig. 19.12 Percentage deviation of Tarapur site values from TR-55 Curve IA

Similarly, it can be inferred from Figs. 19.15 and 19.16 that while SPS (8 inches) over-predicts of the n-hour continuous maximum of about 25%, SPS (20 inches) under-predicts by an amount of 30%. Since index rainfall values have not been defined in (EM1110-2–1411 1965) for Indian conditions, this variation is expected. Interestingly, the IMD-CWC predicted values for a similar western coastal site are

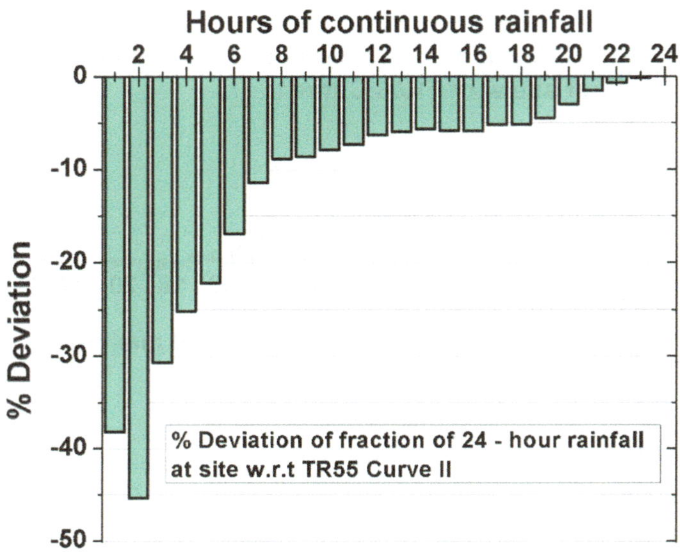

Fig. 19.13 Percentage deviation of Tarapur site values from TR-55 Curve II

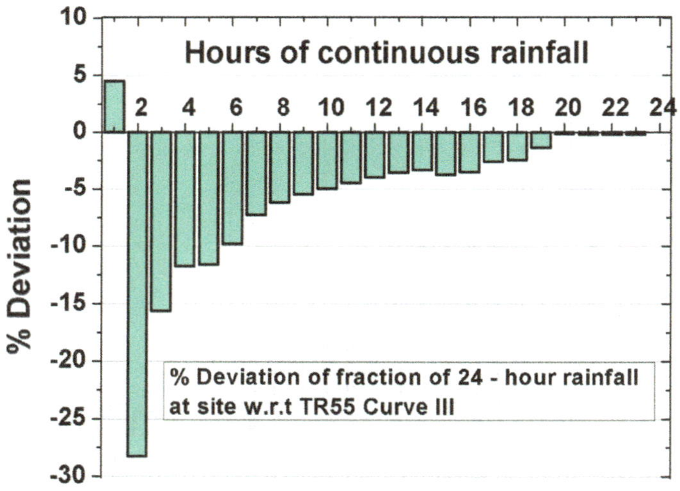

Fig. 19.14 Percentage deviation of Tarapur site values from TR-55 Curve III

also lower than Tarapur site calculated values by a maximum of 35%, as shown in Fig. 19.17. This could be due to the areal averaging performed for the entire catchment-2 to arrive at the time distribution given in IMD-CWC curve (PMP Atlas 2015). This deviation, in similar fashion to other curves in literature, is especially pronounced for smaller time of concentration.

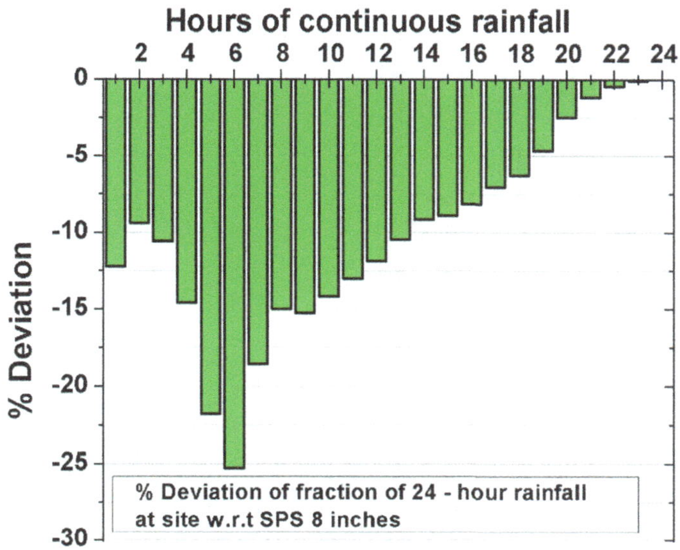

Fig. 19.15 Percentage deviation of Tarapur site values from SPS-8 inches

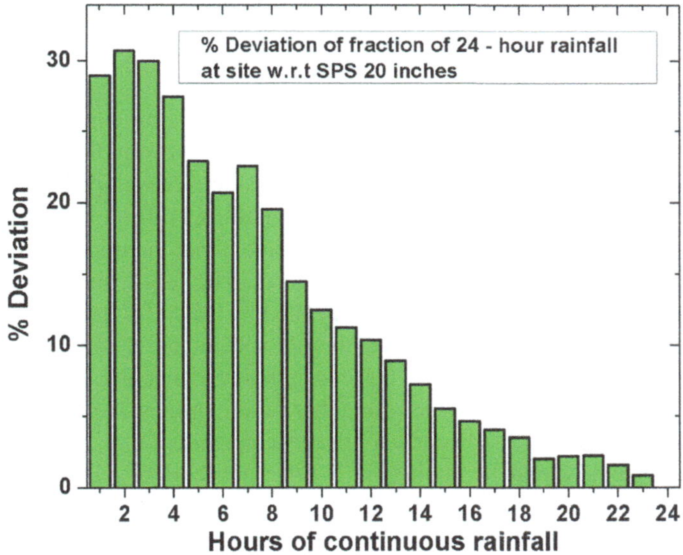

Fig. 19.16 Percentage deviation of Tarapur site values from SPS 20-inches

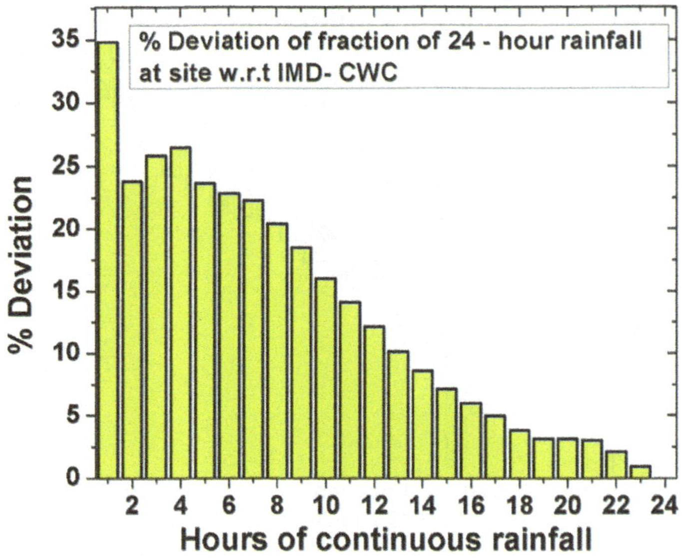

Fig. 19.17 Percentage deviation of Tarapur site values from PMP Atlas (2015)

19.6 Conclusions

Thus, to summarize, results indicate that the site-specific n-hour fraction of 24-h rainfall is quite different in nature when compared to those from the literature, even when the applicable generic curve for the particular region developed by IMD-CWC is considered. The American curves depict a rainfall over the storm duration such that the maximum proportion of the rainfall occurs about the mid time of the total rainfall duration. For Indian monsoon rainfall, it could be very different. The analysis carried out also indicated that the values tabulated by IMD-CWC for western coasts provide a less conservative estimate than the site-specific curve warranting the need for detailed location-focused study. Thus, it can be stated that usage of literature that provided generalized curves may lead to either unsafe or uneconomical designs.

Future work would be directed toward the development of time distribution curve for the site, which would be applicable for the surrounding region as well. The site-specific time distribution curve would be better suited to represent the temporal distribution of rainfall at the site under consideration, and would be very useful for planning and design of the hydraulic structures such as storm water drainage systems, culverts, bridges, etc. and would facilitate safe hydraulic designs when compared to the time distribution curves from literature.

Acknowledgements The authors express their sincere gratitude to the ESL, Tarapur for sharing the data used in this study.

References

1 PMP Atlas for West Flowing Rivers of Western Ghats (2015) Volume 1, Indian Meteorological Department–Central Water Commission
2 EM 1110-2-1411 (1965) Standard project flood determination, Department of Army, US Army Corps of Engineers, Washington DC, USA, Mar 1965
3 Technical Release TR-55 (1986) Urban hydrology for small watersheds. US Department of Agriculture, National Resources Conservation Service, Conservation Engineering Division, June 1986

Chapter 20
Quality Checks on Continuous Rainfall Records: A Case Study

R. Harshanth, Saha Dauji, and P. K. Srivastava

Abstract The records obtained from the rain gauge stations may contain missing information or other forms of human or machine errors, which might hamper with analysis or lead to incorrect inferences drawn from the data. It is pertinent that the data should be screened and suitable quality checks be applied on the records before using them for hydrological analysis. In this paper, the statistical quality checks suitable for checking the quality of hourly time history of rainfall from a site are discussed along with example applications as a case study.

Keywords Quality checks · Hourly rainfall · Screening · Statistical tests

20.1 Introduction

For many hydrological studies, the data is obtained in the form of time series information. For example, we generally obtain the rainfall information in the form of daily data (in mm of rainfall) for several years. In certain cases, hourly data is also available for analysis. However, this data cannot be used for the study before certain statistical quality checks are conducted on the entire dataset in order to ensure the stationarity, homogeneity and consistency of the data (Tung and Yen 2005). In order to study the characteristics of rainfall at a particular site, the rainfall data obtained is in the form of large time series information, consisting of hourly storm value for a period of 17 years (1997–2013).

R. Harshanth
Nuclear Power Corporation India Limited, Kudankulam, India
e-mail: harshanthr@npcil.co.in; harshanthr@gmail.com

S. Dauji (✉)
Nuclear Recycle Board, Bhabha Atomic Research Centre, Mumbai, India
e-mail: dauji_saha@yahoo.com; acad.dauji@gmail.com

Homi Bhabha National Institute, Mumbai, India

P. K. Srivastava
Nuclear Recycle Board, Bhabha Atomic Research Centre, Tarapur, India
e-mail: pansriv@gmail.com

R. Jha et al. (eds.), *Climate Change Impacts on Water Resources*, Water Science and Technology Library 98, https://doi.org/10.1007/978-3-030-64202-0_20

This storm data was obtained from the observed recordings of a tipping bucket-type rain gauge. It works on the principle that the rainfall collection bucket tips over when it collects the rainfall and returns back to its position. Each tip corresponds to a pre-calibrated amount of rainfall. The total rainfall falling in 1 h is calculated based on total number of tips in 1 h. Thereby we obtain continuous rainfall information. Since the data is obtained digitally, the data need to be screened before being put into further use. This is because of the possibility of missing information and human or machine errors during the process of recording of data.

Since we obtain the data in the form of large time series, specific statistical quality checks are employed to study them. Passing such statistical screening tests provides the necessary confidence that the data is free of any instrumental or human errors. Only after screening and correcting the data would the data be used for any further analysis. The procedures used for screening the data are described in the following sections.

20.2 Site Description

The site under consideration is located about 200 km northwards of Mumbai. The rainfall season is dominated by the South-west monsoon. The average rainfall of about 2000 mm occurs between the months of June and September. The rainfall data was obtained in a chronological manner in the form of hourly rainfall for a period of 17 years from 1997 to 2013. The unit of measurement of rainfall depth was in millimeters. This data was used for further analysis.

20.3 Screening of Hydrological Data

In order to check the consistency, homogeneity and stationery properties of the time series data, statistical tests were conducted according to Dahmen and Hall (1990). The test procedures along with their results obtained from the data under study are elucidated in paragraphs that follow.

20.3.1 Check for Trend: Spearman's Rank Correlation Method

This test is carried out to check the absence of trend in time series data (Dahmen and Hall 1990). This test is also conducted to ensure that the data collected is uncorrelated to the order in which they have been collected. The test is conducted on annual total rainfall values. Trend characteristics are noted by testing the null hypotheses $R_{sp} =$

Table 20.1 Results of Spearman's test for trend

Parameter	Value	Remark
Sigma	334	Standard deviation
N	17	No of years
d.o.f.	15	Degrees of freedom
R_{sp}	0.591	Rank correlation coefficient
t_i	2.835	Test statistic
t_{ul}	2.02	Limits for t_i to accept the null hypothesis, i.e., there is no trend
t_{ll}	−2.02	

Since $t_i > t_{ul}$ the test shows an increasing trend of the time series data

0 (which indicates absence of trend), where R_{sp} is the Spearman's rank correlation coefficient, against the test statistic t_t. Here, t_t has a double-tailed Student's t distribution with n − 2 degrees of freedom. These tests were conducted with a significance value of 5%. The test conducted on the time series data indicated a slightly increasing trend as shown in Table 20.1. This nature, however, was also confirmed by several other studies conducted across the country indicating an increasing rainfall trend over the years.

20.3.2 Check for Trend: Kendall's Tau Test

This test also performs the check on presence of trend in the total annual rainfall data. In this test, the Mann Kendall test parameter S, defined as the sum of number of positive and negative deviations of the subsequent value from the current value is determined (Spiegel 1961). From this the test statistic Z_s is calculated. This statistic is tested against the standard test statistic $Z_{0.025}$, considering a significance level of 5% and two-tailed distribution. The null hypothesis is that there is no trend in the dataset, i.e., the data is random. The result of the calculation is shown in Table 20.2. It can be thus inferred from the results that the total annual rainfall values are random, i.e., there is no significant trend in the data.

Table 20.2 Results of Mann Kendall's test for trend

Parameter	Value	Remark
No of data points	17	17 years data
Mann Kendall Test parameter, S	0	Calculated values
Z_s	0	
$Z_{0.025}$	1.96	5% Significance

Since $|Z_s|$ is less than $Z_{0.025}$, the null hypothesis is accepted, i.e., there is no trend in the data

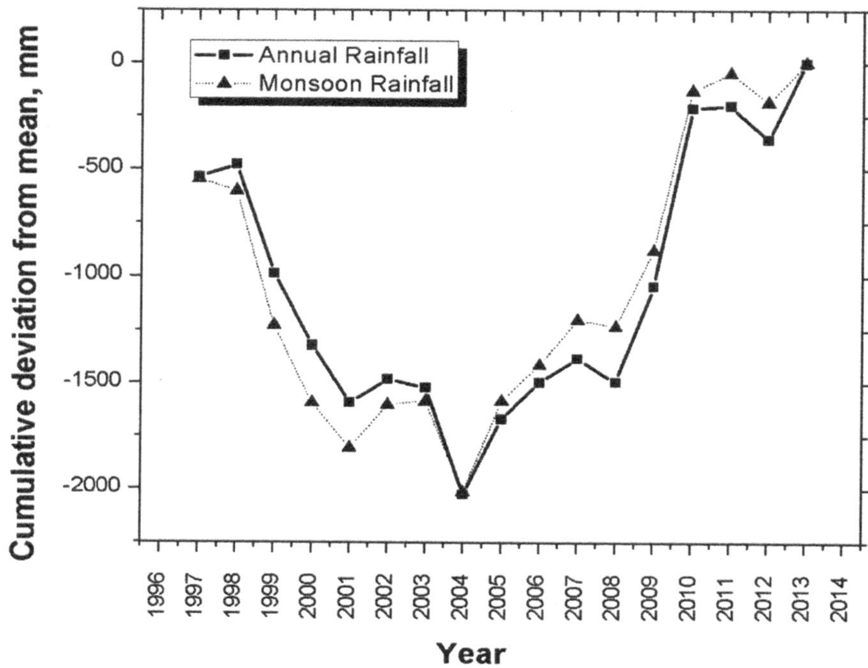

Fig. 20.1 Variation of cumulative deviation from mean curve of yearly and monsoon rainfall through the years

20.3.3 Test for Change Point: Cumulative Deviation from the Mean

This test also evaluates the stability of mean of the time series data (Dahmen and Hall 1990). Here, the deviation from mean of each data is calculated and their cumulative values are calculated. Then these values are plotted against time. Calculations are carried out for total annual rainfall and monsoon rainfall and the curve is shown in Fig. 20.1. It can be seen that there is a break point between the years 2002 and 2004.

20.3.4 Check for Stability of Variance: Fisher's Test

This test is carried out to assess the stability of variances between smaller subsets of a larger time series data. The time series data can be used for further analysis only if the variances are stable. Similar to earlier tests, the time series data used in this study corresponds to the annual rainfall values and subsequently the subsets refer to chronological groupings of annual rainfall values. The choice of number of such

Table 20.3 Results of Fisher's test for two subsets of annual rainfall

Subsets	2
Lower limit (LL)	0.221
Upper limit (UL)	4.9
Statistic, F_t	0.945
Result	Thus, the variance is stable

Table 20.4 Results of Fisher's test for two subsets of annual rainfall

Between subsets 1 and 2		Between subsets 3 and 2		Between subsets 3 and 1	
F_t	0.873	F	1.737	F	1.989
UL	7.15	UL	7.39	UL	7.39
LL	0.14	LL	0.107	LL	0.107
Since LL < F < UL, the variance is stable		Since LL < F < UL, the variance is stable		Since LL < F < UL, the variance is stable	

subsets is arbitrary. However, Dahmen and Hall (1990) advise against very small subsets, i.e., subsets with very small number of values.

In view of the same, the present study adopts a procedure where the annual rainfall values were divided into maximum of three subsets with the condition that the number of values in all subsets be maintained almost to be equal. Thus, we have 17 annual rainfall values corresponding to 17 years.

These 17 annual rainfall values (corresponding to years 1997–2013) were split further into smaller subsets of rainfall, say 9 values in one set (corresponding to years 1997–2006) and 8 values in another (corresponding to years 2007–2013). In case of three subsets as well, the whole time series were split so as to maintain almost equal number of annual rainfall values, i.e., subset 1 (1997–2002), subset 2 (2003–2008), and subset 3 (2009–2013).

The test statistic calculates the ratio of variances of two split datasets, the null hypotheses being the equality of variances. The test is carried out with a confidence level of 95%. Tests were conducted on two subsets and three subsets of the larger dataset, i.e., the values of annual rainfall. Results are presented in Tables 20.3 and 20.4. They indicate that the variances between split datasets are stable.

20.3.5 Check for Stability of Means: T-Test

This test is carried out on the same split datasets used in Fisher's test (see Sect. 20.3.4) and is used for checking the stability of means (Dahmen and Hall 1990). This test is conducted to assess the stability of means between two datasets. The groupings of the data into subsets were carried out in similar fashion to Fisher's test for stability of variance. The null hypothesis tests the equality of means between two datasets against the alternative that they are unequal. The test statistic has a t-distribution. The results are shown in Tables 20.5 and 20.6. The calculations were carried out with a statistical significance level of 5%. Results of the test indicate that the means are unstable when the number of subdivisions of data is 2. Whereas the results also indicate that means are stable for three subdivisions of data. This instability in mean in the larger dataset can be attributed to increasing rainfall pattern over the years.

Fisher's test and t-test were also carried out on annual monsoon rainfall data, wherein the total rainfall occurring through the monsoon months of June to September each year was used for statistical calculations and they yielded similar results. The variation of monsoon rainfall when compared to annual rainfall through the years is shown in Fig. 20.2.

It can be observed from Fig. 20.2 that the difference in the values of total and monsoon rainfall is very less, suggesting that the bulk of the annual rainfall occurs during the months of June to September. Understandably, the statistical tests performed on monsoon rainfall alone did not alter the results of analysis.

Table 20.5 Results of t-test for two subsets of annual rainfall

Parameter	Value	Remarks
t_t	−2.495	Test statistic
V	15	Degrees of freedom
UL	2.12	Limits for acceptance of null hypothesis, i.e., mean is stable
LL	−2.12	

Since t_t < LL, the mean is marginally unstable
Note LL: Lower limit; UL: Upper limit

Table 20.6 Results of t-test for three subsets of annual rainfall

Between subsets 1 and 2		Between subsets 2 and 3		Between subsets 1 and 3	
V	8	v	7	v	7
UL	2.31	UL	2.36	UL	2.36
LL	−2.31	LL	−2.36	LL	−2.36
t_t	−1.327	t_t	−1.284	t_t	−2.396
Since LL < t_t < UL, mean is stable		Since LL < t_t < UL, mean is stable		Since t_t < LL, mean is marginally unstable	

Note LL: Lower limit; UL: Upper limit

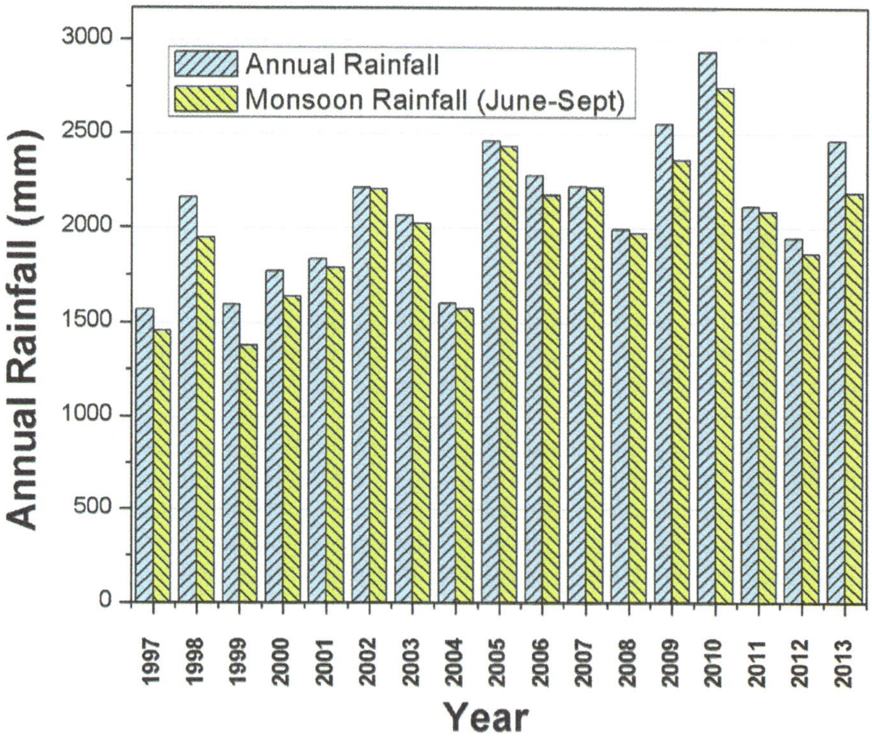

Fig. 20.2 Comparison of monsoon rainfall and annual rainfall over the years

20.3.6 Check for Persistence of Data: Lag 1 Serial Correlation Coefficient Test

This test is carried out as a check for absence of persistence in the dataset (Dahmen and Hall 1990). Essentially, this test also checks the randomness of data. Since it is a possibility that field measurements can lead to patterns in the dataset due to malfunction of the instrument or due to persistent human error, it is obvious we test for the same. Here the null hypothesis is that consequent values of annual rainstorm values are uncorrelated with each other. The test statistic is lag 1 serial correlation coefficient, r_1 as defined by Box et al. (2013). From Table 20.7, it can be seen that the data is non-persistent, i.e., the data is random.

20.3.7 Check for Randomness: Wald–Wolfowitz Test

Wald–Wolfowitz test for randomness (also known as Runs test) was also carried out for individual rainstorms (Spiegel 1961). Around 15 salient rainstorms were picked

Table 20.7 Results of Lag Serial 1 coefficient test

Parameter	Value	Remarks
r_1	0.164	Serial correlation coefficient
LL	−0.537	Limits for acceptance of the null hypothesis, i.e., the dataset is random
UL	0.412	

Since LL < r1 < UL, the data is non-persistent, i.e., the dataset is random
Note LL: Lower limit; UL: Upper limit

at random and were subjected to the Runs test. The typical characteristics of the storms, such as duration, average rainfall intensity, maximum rainfall intensity, and total rainfall for each of the selected rainstorms are also listed in Table 20.8 (salient values in boldface).

The maximum average intensity was 42.13 mm/h, whereas the maximum hourly intensity was 167.7 mm/h, which coincided with the maximum total rainfall of 764 mm. The minimum average rainfall intensity was about 5 mm/h and the minimum total rainfall was 21 mm in any rainstorm. The results of the Run test indicate that the rainstorms pass the test, which means the null hypothesis that the values are random cannot be rejected with a confidence level of 95%. A sample hyetograph passing the Runs test is shown in Fig. 20.3 for completeness.

Table 20.8 Charateristics of the selected storms for Wald–Wolfowitz test (Runs test) for randomness

S. No	Year	Date	Time	Storm characteristics			
				Duration (hours)	Average (mm/h)	Max (mm)	Total (mm)
1	2009	14-Jul	1900 h	**34**	9.75	29	331.5
2	2002	26-Jun	0100 h	**28**	27.28	**167.7**	**764**
3	2009	22-Jul	0200 h	**24**	16.08	36	**386**
4	2012	28-Jun	0600 h	2	**58.9**	**74.8**	117.8
5	2000	12-Jul	0700 h	8	6.11	**104**	368.9
6	2006	04-Sep	0600 h	4	**42.13**	66.5	168.5
7	2012	03-Sep	1100 h	21	21.32	64.3	**447.8**
8	2004	03-Aug	2200 h	13	9.43	38.9	122.6
9	2008	14-Jun	1300 h	4	8.25	16	33
10	2012	28-Jun	0600 h	3	7	10	21
11	2013	15-Jun	1800 h	8	**28.31**	56	226.5
12	1999	19-Jun	0600 h	6	5.7	9.5	34.2
13	1998	28-Jun	0500 h	11	18.2	73	200.2
14	2001	13-Jun	2300 h	3	11.13	19.4	33.4
15	2010	01-Aug	0000 h	6	5.16	9	31

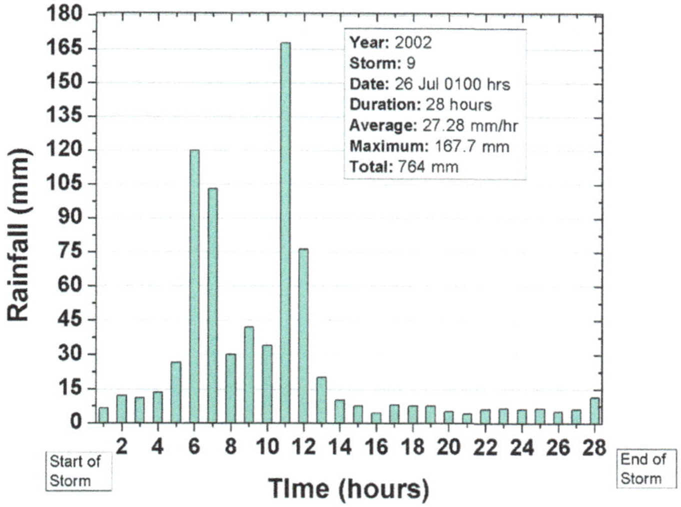

Fig. 20.3 A sample storm hyetograph

20.4 Conclusions

The screening of the rainfall records and statistical checks for quality are important prerequisite for useful inferences from the data. In this paper, few such tests applicable for hourly rainfall records were discussed and their application was presented as a case study. From the results of statistical tests presented in the previous section, it can be concluded that the data follows an increasing trend in rainfall which is consistent with other similar studies carried out across the country. It can also be concluded that the data is mostly stable and homogeneous for most cases. Further, the randomness of the rainfall data was established for the annual rainfall, as well as the individual rainstorms from different years. Hence, the same data may be employed for further analysis without implementing any corrections.

Acknowledgements The authors express their sincere gratitude to the ESL, Tarapur for sharing the data used in this study.

References

Box GEP, Jenkins GM, Reinsel GC (2013) Time series analysis. Wiley, New Jersey, USA. http://dx.doi.org/10.1002/9781118619193

Dahmen ER, Hall MJ (1990) Screening of hydrological data: tests for stationarity and relative consistency. Publication 49, International Institute for Land Reclamation and Improvement, The Netherlands

Spiegel MR (1961) Theory and problems of statistics. Schaum Publishing Co., New York
Tung Y-K, Yen B-C (2005) Hydrosystems engineering uncertainty analysis. McGraw Hill, USA. http://dx.doi.org/10.1036/0071467084

Chapter 21
Assuring Water Intake Sustainability Under Changing Climate

Gaurav Talukdar and Arup Kumar Sarma

Abstract River carrying sediments is a crucial component of the geochemical cycle and depending on the local factors, the sediments may be beneficial or detrimental for the society. Thus, management of river sediment is a challenging task from economic, social and environmental perspective. In today's world due to changing climate the policy makers must come out with a more efficient and better management technique of these problems. The present study involves construction of a channel through huge sandbar developed in the Brahmaputra River of Assam in the Guwahati region, over the time leading to serious water extraction problem to an intake supplying water to its locality. The river has shifted north over time and there aroused a water scarcity problem on the south bank. In the present work, MIKE 21C model is used to simulate the water flowing through a channel constructed between the sandbars linking the main river. Initially, based on the field survey, the grid is generated and the bathymetric data are fed into the model. Surface elevation, discharge and water level are introduced as the initial conditions, and boundary conditions with other parameters like the roughness coefficients and eddy viscosity. The simulated velocity was calibrated at the downstream section with the observed one obtained through survey during that period. After the calibration, the vector plots of the velocity and water depth are produced. Combining the knowledge derived from the hydrodynamic model study and geoinformatics study, it has been found that a 10 m wide channel dredged through the sandbars connecting the river with the intake will be able to draw sufficient water from the main river to the intake. Thus, it can be found that MIKE 21C has performed well in handling the complex situation of water scarcity to sustain its availability for the benefit of the society.

G. Talukdar (✉) · A. K. Sarma
Department of Civil Engineering, Indian Institute of Technology, Guwahati 781039, Assam, India
e-mail: gauravt@iitg.ac.in

A. K. Sarma
e-mail: aks@iitg.ac.in

R. Jha et al. (eds.), *Climate Change Impacts on Water Resources*, Water Science and Technology Library 98, https://doi.org/10.1007/978-3-030-64202-0_21

21.1 Introduction

Water being the most essential substance on Earth is a vital component and the basic prerequisite for all living beings in general. Out of the total available water, only 2.5% are freshwater and of that, roughly one-third is available to human use and ecosystem. Rivers are lifeline of a nation and its exploitation can lead to serious impact on human life. The Brahmaputra river is one among the large braided rivers and seventh largest in the world (Tandon and Sinha 2007) flowing through China, India and Bangladesh. The river is governed by high and low flow resulting in tremendous modifications in the bed form and is characterized by its high discharge, varying morphology, high bed aggradations and severe erosion. The dwellers residing in the floodplains rely on the normal flood to bring fresh sediments to the floodplains favoring agriculture and farming. The people living in the floodplains and chars utilize these sandbars for cultivating crops during various seasons and make their living. However, the river can be widening and changing its course with slow migration significantly with time (Gilfellon et al. 2003) because of climate change. The morphodynamic behavior of the rivers in terms of plan form and land use change is crucial and affects various processes like the ecological activities, sedimentation and erosion processes.

Although flood and river bank erosion are the major problems in the river, high seasonal variation creates drought-like situation during the non-monsoon period. These leads to the formation of large masses of sandbar in certain reaches in the entire river. During the lean period in the months of January to May, the water level in the river recedes and exposed areas can be observed. This situation leads to a problem of extracting the surface water. Though utilizing the surface water of the river is essential for meeting the water need (Sarma 2005), the problem of sandbar formation has led to several inconveniences to the daily cycle of human activities. During this period, the interruption of the regular water supply, water requirement for agriculture, diminished industrial goods and so on are some of the major problems faced by the community. A study needs to be done to understand these problems and come up with a suitable and practical alternative. This study presents the planform analysis of the study area to identify spatiotemporal changes of the river in the study area that has affected the water intake for supply to various organizations and individuals. Based on the understanding we have applied MIKE 21C model to identify the portion where a guide channel can be dredged to overcome the problem of flow into the intake location.

21.2 Motivation and Objective

The global climate change has seen some adverse effects on the environment in several parts of the globe. The glaciers have shrunk, ice started melting earlier, accelerating sea level rise, untimed drought situations and intense heat waves. According to the United States, National Climate Assessment, the intensity of extreme events like the sea level rise, ice and snow melt has already disrupting and damaging the live and sectors in the U.S. economy (Karl et al. 2009). Climate change encircles the shifts in precipitation, increasing and decreasing trend of temperature leading to changes in several features of the climatic system. As a society, we are habituated to the normal day-to-day situations but are sensitive to extremes. These extremes lead to tremendous loss of lives, property and economy. The rivers, lakes and reservoirs have depleted due to its overuse. The crisis of water has prevailed globally, where people do not have sufficient access to water for their basic needs. WHO and UNICEF has reported that, worldwide, around 2 billion people still do not have access to clean drinking water in their homes and nearly 1 billion defecate outside. In India, the undersupply of water will be increasing with the growing population of 1.4 billion at a rate of 1%. According to Niti Aayog, the demand for water is projected to be twice the available supply. In account of those effects, the problem of water supply to various organizations is also of utmost importance, since the industries are expanding and the need for water is speeding up at a higher rate. Keeping all the prospects in view, it necessitates the importance of a mathematical model study that will help in understanding the processes taking place in the rivers and will contribute in knowing few aspects associated with it.

21.3 Methodology

To study the spatiotemporal variation of the bed forms of the river through model study, and to have a proper understanding of the river course, sediment deposition and movement, satellite data are required. Topographic maps and satellites imageries are very much helpful in understanding the area and relative changes in the plan form. Hence, for the study, U.S. Army topographic map (1955), SOI topographic map (1972), Landsat MSS and TM, Google imagery and LISS IV imageries (2018) have been used. From the imageries (Fig. 21.1), it can be seen that there is formation of a sandbar in the study area and also shifting of sandbars toward the point of interest, i.e., the intake point. From the cycle of changes, it is observed that for the Brahmaputra river, due to high sediment yield and deposition, there are sandbar formations in different locations during different time period. Therefore, for analyzing the problem aroused due to the formation of huge sandbars converging near the intake, the need and scope of creating an artificial cross channel needs to be explored through model study. The model will help in understanding different situations that may arise during the practical application in the field.

In view of the model development, the bathymetry of the river is essential, and hence a field survey within the river reach is carried out. A detailed hydrographic survey has been conducted in the Brahmaputra river for about 11 km upstream and downstream of the intake location to determine the bathymetry of the river. Benchmarks were installed near the bank and water level is monitored daily. The instruments used during survey were a DGPS, total station and echo sounder. With the

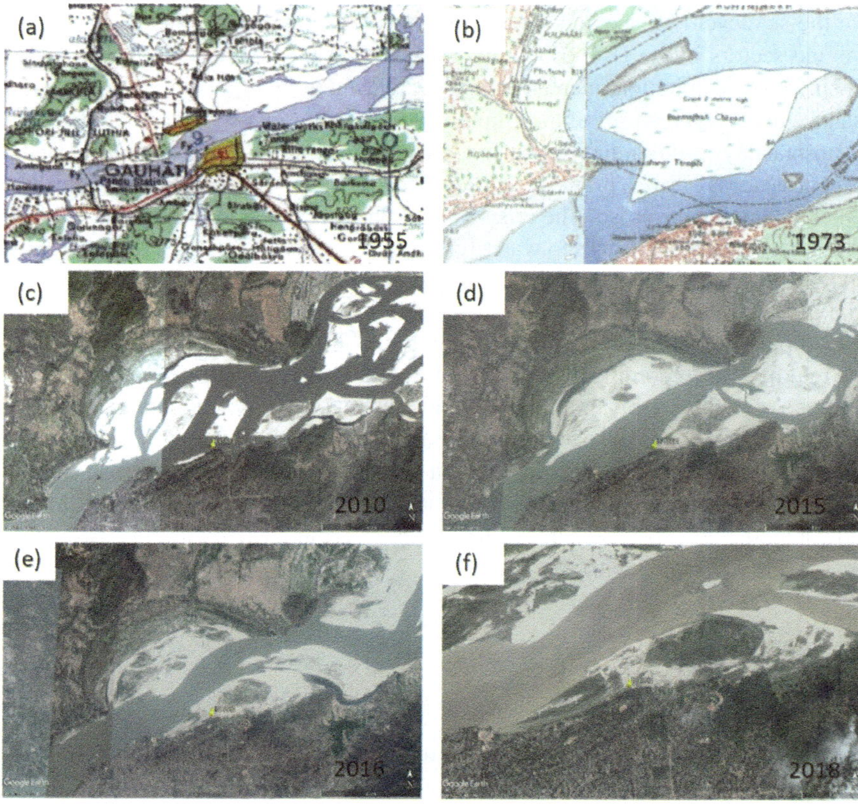

Fig. 21.1 Satellite imageries from **a** 1955, U.S. Army topographic map to **b** 2018, Google Earth Image, showing the formation of sandbars near the intake point resulting in restricting the normal flow toward the point

acquired data, a model study was carried out to examine the situation. The hydrodynamic model, MIKE 21C (Guide 2005) was applied and simulated for the conditions representing different seasons. Based on the points surveyed through the reach, the computational grid is generated in the curvilinear grid generator and the boundary conditions are specified at the upstream and downstream portion.

21.4 Basic Governing Equations

With the advancement in mathematical modeling technique and computational facility, governing equations of two-dimensional unsteady flow can now be solved in its fully dynamic form considering all significant forces that governs river hydraulic. 2D hydrodynamic model available in MIKE 21C has been found to give acceptable result, and the design carried out on the basis of these results was found to give successful solution to a real problem of Brahmaputra river (Sarma and Rao 2005). Therefore, MIKE 21C has been used for flow simulation in this study.

The hydrodynamic model solves the fully dynamic and vertically integrated equations of conservation of mass and momentum in two-dimensional form (MIKE 21C, 2011). The curvature of the grid lines gives rise to additional terms in the partial differential equations for the flow. The equations solved in MIKE 21C are:

$$\frac{\partial p}{\partial t}+\frac{\partial}{\partial s}\left(\frac{p^2}{h}\right)+\frac{\partial}{\partial n}\left(\frac{pq}{h}\right)-2\frac{pq}{hR_n}+\frac{p^2-q^2}{hR_n}+gh\frac{\partial H}{\partial s}+\frac{g}{C^2}\frac{p\sqrt{p^2+q^2}}{h^2}=\text{RHS} \tag{21.1}$$

$$\frac{\partial q}{\partial t}+\frac{\partial}{\partial n}\left(\frac{q^2}{h}\right)+\frac{\partial}{\partial s}\left(\frac{pq}{h}\right)+2\frac{pq}{hR_s}-\frac{q^2-p^2}{hR_n}+gh\frac{\partial H}{\partial n}+\frac{g}{C^2}\frac{q\sqrt{p^2+q^2}}{h^2}=\text{RHS} \tag{21.2}$$

$$\frac{\partial H}{\partial t}+\frac{\partial p}{\partial s}+\frac{\partial q}{\partial n}-\frac{q}{R_s}+\frac{p}{R_n}=0 \tag{21.3}$$

where s, n = coordinates in the curvilinear coordinate system; p, q = mass fluxes in the s and n direction, respectively; H = water level; h = water depth; g = gravitational acceleration; C = Chezy roughness coefficient; R_s, R_n = radii of curvature of s- and n-lines, respectively, and RHS describes the Reynold stresses.

The curvilinear MIKE 21C model requires less computational points and hence smaller capacity than the rectilinear grid, thus providing better flow resolutions. In application to river study, an accurate boundary resolution is required and hence the curvilinear grid is favorable in such situations.

21.5 Results and Discussion

The HD model study has revealed that in the absence of an artificial channel, the water will not reach the intake portion during the lean period of the year. This is because through the satellite images we have found that the sandbar is developing in the specified location and growing and moving southwest near the intake, giving no sign of river flowing near it. Hence, a channel of 10 m wide and 1.2 km long is dredged between the sandbars and flow simulated through the channel. The model output comprised the 2D water depth, flux, current speed and flow velocity in both the directions. The validation of the model has been done with the observed velocities at the downstream portion. The observed and simulated velocities were found to be in good agreement (Fig. 21.2). The simulated depth along with the velocity vectors computed through MIKE 21C model are shown in Fig. 21.3.

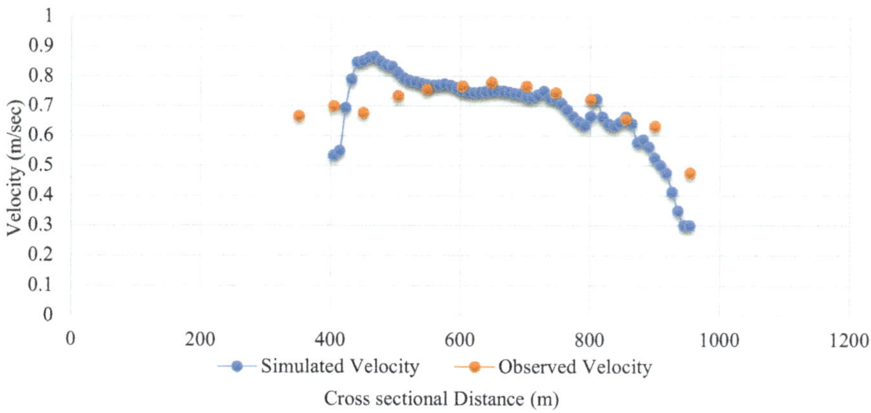

Fig. 21.2 Comparison of computed velocity and observed velocity

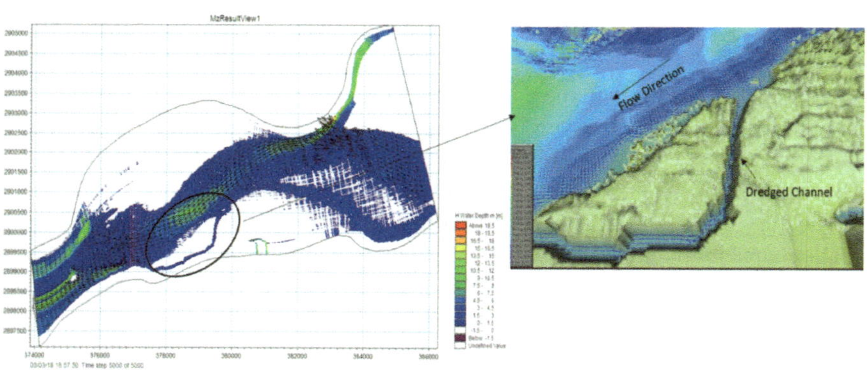

Fig. 21.3 Flow depth simulation using MIKE 21C (left) and velocity vectors in the dredged channel shown as animated view (right)

21.6 Conclusions

From the simulation study it has been found that with the dredged channel between two sandbars, the flow passes near the intake point. The simulation was done for a 10 m wide channel with a length approximately 1.2 km from the main channel to the intake. The mouth of the entry has been made wider to draw sufficient water to the proposed excavated channel. Based on the simulation study, the channel is dredged and through the satellite imagery (Fig. 21.4), it has been found that with the dredged channel, the water has taken diversion and started flowing through it. Thus, it can be concluded that the model study performed using MIKE 21C has been successful in simulating the flow.

Fig. 21.4 Satellite imagery of the dredged channel (2019) constructed after conducting the model study which shows the flow of water in the channel passing through the intake point

References

Gilfellon GB, Sarma JN, Gohain K (2003) Channel and bed morphology of a part of the Brahmaputra River in Assam. J Geol Soc India 62(2):227–236

Guide DU (2005) Reference manual of Mike21

Karl TR, Melillo JM, Peterson TC, Hassol SJ (eds) (2009) Global climate change impacts in the United States. Cambridge University Press

Sarma AK (2005) Need of utilizing surface water for water supply in greater Guwahati. In: Proceedings of national convention on urban water supply and chemical contamination of ground water, Institute of Public Health Engineers (INDIA), Guwahati, 20th–21st Feb, 2005. IPHE, pp 87–89

Sarma AK, Rao NP (2005) Simulation study for suggesting bank protection measures in the river Brahmaputra. In: Proceedings of national conference on advances in water engineering for sustainable development, IIT Madras, Chennai, 16th to 17th May 2005, pp 3–9

Tandon SK, Sinha R (2007) Geology of large river systems. Large rivers: geomorphology and management, pp 7–28

Chapter 22
Characteristics of Gldas Evapotranspiration and Its Response to Climate Variability Across Ganga Basin, India

Lalit Pal, C. S. P. Ojha, and Amit Kumar

Abstract Evapotranspiration is a key component of hydrological cycle which together with precipitation determines the water availability within a region. The present study aims to investigate spatiotemporal patterns in evapotranspiration (ET) over the period 1981–2015. The monthly time series of ET obtained from Noah Land Surface Model of GLDAS was aggregated to seasonal and annual time series to carry out the analyses. The response of ET to climate variability is examined by analysing the relationship between trend in ET and key climate parameters (temperature and precipitation) at annual and seasonal timescale. The significance of trend is tested using nonparametric Mann–Kendall test at 5% significance level. Magnitude of trends is defined by slope parameter of linear regression line. The results show an increase in ET over major portion of the basin except for the upper reaches for all the seasons. In the upper reaches (Himalayan region), negative trends prevail in ET at both seasonal and annual temporal scale. However, the trends in precipitation and mean temperature are found to be increasing over the upper Ganga basin. The results reveal the existence of factors other than climate parameters controlling the variations in ET, especially in the upper reaches of Ganga basin located in the Himalayan region.

Keywords Evapotranspiration · GLDAS · Trend · Climate variability · Ganga basin

L. Pal (✉) · C. S. P. Ojha
Department of Civil Engineering, Indian Institute of Technology Roorkee, Roorkee 247667, Uttarakhand, India
e-mail: lalitpl4@gmail.com

C. S. P. Ojha
e-mail: cspojha@gmail.com

A. Kumar
School of Geography, University of Nottingham, Nottingham NG7 2RD, UK
e-mail: amit1792kumar@gmail.com

R. Jha et al. (eds.), *Climate Change Impacts on Water Resources*, Water Science and Technology Library 98, https://doi.org/10.1007/978-3-030-64202-0_22

22.1 Introduction

Evapotranspiration (ET) is an important component of hydrological cycle as it constitutes major fraction of total water loss to the atmosphere. ET, together with precipitation, is a primary determinant of water availability within a region (Long et al. 2014). Accurate knowledge of ET plays a crucial role in understanding water budget and developing efficient water resource management plans. In recent times, the focus of hydrologic modelers has shifted from mere rainfall-runoff modelling to accurate estimation of all major water balance components. Accurate estimation of water balance components, especially ET, requires validation with ground observed datasets. However, limited network of ET monitoring stations limits the quantification of actual ET worldwide. In such scenario, alternate sources such as global ET dataset simulated using land surface models (LSMs) (e.g., Noah, VIC, Mosaic and CLM) available from Global Land Data Assimilation System (GLDAS) and ET datasets generated using satellite remote sensing (e.g., MODIS ET dataset) provides an opportunity to monitor spatiotemporal variability in ET for various basin of the world (Xu et al. 2006; Mueller et al. 2011). ET varies both regionally and seasonally controlled by various hydro-meteorological factors such as precipitation, vegetation index, temperature, soil moisture, sunshine hours, etc. Knowledge of spatial and temporal variations in evapotranspiration provides an insight to the changes in outgoing water flux, which may serve as an indicator of climate change (Peters-Lidard et al. 2011; Park and Choi 2015).

Ganga river basin is the largest river basin in India which is nurturing a population of about 165 million people. A major fraction of this population depends on agriculture for its livelihood. Subsequently, fate of these people is dependent on the state of water availability in the region. Variability in ET in combination with precipitation provides an estimate of variability in water availability. The present study aims to analyse spatial and temporal patterns in ET obtained from Noah LSM simulations of GLDAS over Ganga basin in India. An attempt has been made to identify relationship between observed variations in ET estimates with the variations in rainfall and temperature. The study outcomes help in determining the existing patterns in ET over the study region and identifying dominant hydro-climatic parameter controlling the observed changes.

22.2 Study Area

Ganga river basin is the largest river basin in India extending over an area of about 0.86 Mkm^2 (in India), originates in the western Himalayas and finally meets the Bay of Bengal in the east. The basin covers the latitudes of 22°30′ and 31°30′ North and the longitudes of 73°30′ and 89°00′ East. The total basin area in India is shared among 11 major states, viz., Himachal Pradesh, Uttarakhand, Uttar Pradesh, Madhya Pradesh, Chhattisgarh, Bihar, Jharkhand, Punjab, Haryana, Rajasthan, West Bengal

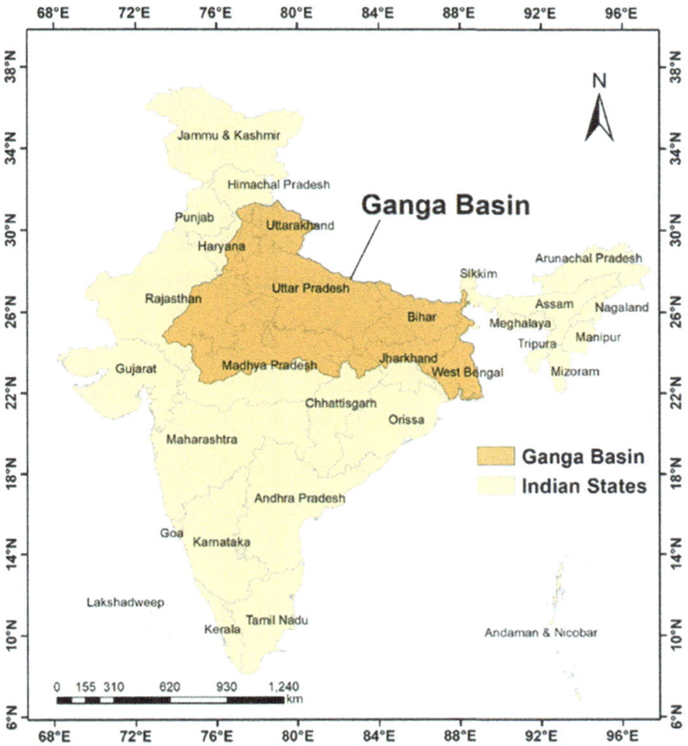

Fig. 22.1 Location map of study area

and the Union Territory of Delhi. Precipitation and snow melt from Himalayas are the major sources of water in the river. The basin experiences a gradient in precipitation along east to west with eastern parts receiving higher precipitation than the western parts of the basin. Major fraction of total rainfall is received during monsoon months (June–September) with annual rainfall values ranging between 300 and 2000 mm. The temperature during winter season ranges from 2 to 15 °C, while that during summer season varies from 25 to 45 °C. The annual ET values lies in the range from 236 to 1271 mm. The geographical location and other details of the study area of the Uttarakhand state are given in Fig. 22.1.

22.3 Datasets

22.3.1 Evapotranspiration

ET output from Noah LSM in GLDAS was used in the study to carry out various analyses over the period 1981–2015. In GLDAS, datasets from four LSMs are available for various parameters such as atmospheric pressure, atmospheric radiation, atmospheric temperature, precipitation, evapotranspiration, etc. In the study, ET dataset for the period 1981–2000 was obtained from Noah LSM V2.0 (GLDAS_NOAH025_M.2.0) and for the period 2001–2015 ET values were obtained from Noah LSM V2.1 (GLDAS_NOAH025_M.2.1) outputs at monthly time steps. In the dataset, monthly ET values are available at a resolution of $0.25° \times 0.25°$ resolution. Originally, ET values are provided as flux with unit $kg/m^2/s$ which is converted to monthly ET values in mm for the study period. The dataset can be downloaded from the link: https://disc.sci.gsfc.nasa.gov/datasets?keywords=GLDAS.

22.3.2 Rainfall

Daily rainfall gridded dataset at the resolution of $0.25° \times 0.25°$ was collected from India Meteorological Department (IMD). Daily rainfall values were aggregated to monthly rainfall for the grids falling within the study region for the period 1981–2015.

22.3.3 Temperature

Daily temperature dataset acquired from IMD at a resolution of $1° \times 1°$ was converted to monthly mean temperature data to carry out the analyses. All monthly time series of ET, precipitation and temperature were aggregated to seasonal (pre-monsoon, monsoon, post-monsoon and winter) and annual time series to carry out various analyses.

22.4 Methods

The present work aims to analyse spatial and temporal trends in ET over Ganga river basin. Since precipitation and temperature are the two most dominant climatic parameters which controls ET in a region, spatiotemporal patterns are also analysed in precipitation and temperature. An attempt has been made to identify the parameter controlling behaviour observed in ET. Temporal trends in various time series were computed using nonparametric Mann–Kendall (MK) test (Mann 1945; Kendall 1975). Series effected with the presence of serial correlation was tested using modified version of MK test proposed by Hamed and Rao (1998).

22.4.1 Mann–Kendall Test

$$S = \sum_{i=1}^{n-1} \sum_{j=i+1}^{n} \operatorname{sgn}(x_j - x_i) \tag{22.1}$$

$$\operatorname{sgn}(x_j - x_i) = \begin{cases} 1, & \text{if } (x_j - x_i) > 0 \\ 0, & \text{if } (x_j - x_i) = 0 \\ -1, & \text{if } (x_j - x_i) < 1 \end{cases} \tag{22.2}$$

$$\operatorname{Var}(S) = \frac{n(n-1)(2n+5) - \sum_{i=1}^{m} t_i(t_i - 1)(2t_i + 5)}{18} \tag{22.3}$$

$$Z_S = \begin{cases} \frac{S-1}{\sqrt{\operatorname{Var}(S)}}, & \text{if } S > 0 \\ 0, & \text{if } S = 0 \\ \frac{S+1}{\sqrt{\operatorname{Var}(S)}}, & \text{if } S < 0 \end{cases} \tag{22.4}$$

At 5% significance level, the null hypothesis of no trend is rejected, if $|Z_S| > 1.96$.

22.4.2 Modified Mann–Kendall Test

The empirical variance of test statistic S is computed as:

$$V^*(S) = \operatorname{Var}(S).\frac{n}{n_S^*} = \frac{n(n-1)(2n+5)}{18}.\frac{n}{n_S^*} \tag{22.5}$$

The $\frac{n}{n_S^*}$ is calculated as:

$$\frac{n}{n_S^*} = 1 + \frac{2}{n(n-1)(n-2)} \sum_{i=1}^{n-1} (n-i)(n-i-1)(n-i-2)\rho_s(i) \tag{22.6}$$

22.5 Results and Discussion

The present study focuses on understating the patterns in spatial and temporal distribution evapotranspiration over Ganga basin using its global estimates from GLDAS Noah land surface model. Figure 22.2 shows the spatial variation of annual and seasonal (monsoon, pre-monsoon, post-monsoon and winter) ET over Ganga basin for the period 1981–2015. As can be seen in the figure, portion of basin along Himalayan ranges experiences higher ET losses as compared to parts of basin with

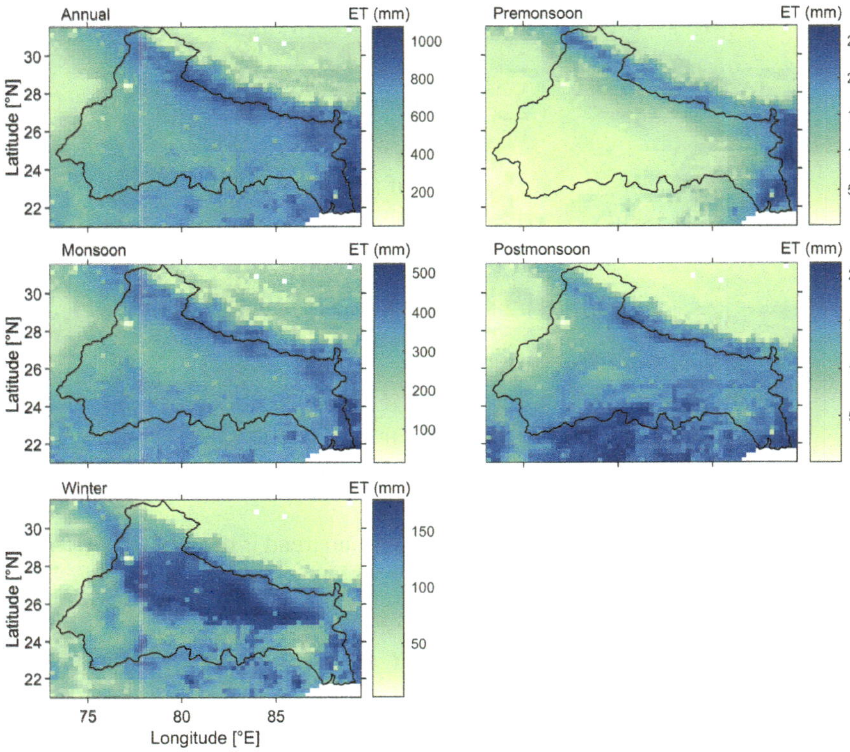

Fig. 22.2 Spatial distribution of annual and seasonal accumulated evapotranspiration across Ganga basin

flatter terrain. A high variability in annual ET can be observed in Ganga basin with northern and eastern parts having higher ET (700–1000 mm) than western and southern parts of the basin (200–500 mm). However, the observed spatial variability in ET is not consistent across all the seasons as in winter season, and higher ET losses are experienced over middle parts compared to rest of the basin. In Fig. 22.2, it can also be observed that highest amount of ET occurs during monsoon season, followed by post-monsoon, pre-monsoon and winter season, respectively. This high seasonal variability in ET is possibly due to seasonal variability in rainfall in terms of Indian summer monsoon (ISM) rainfall over the Indian sub-continent. Due to ISM rainfall, major portion of annual rainfall over the basin is received during monsoon season, i.e., from June to September. A variation in space of annual rainfall over the Ganga basin is shown in Fig. 22.3. As can be seen in the figure, northern and eastern parts of the basin receive higher rainfall than the other, which is evidently reflected in the spatial variation of ET across the basin. However, the spatial patterns in distribution of rainfall received over the basin are more or less the same for different seasons. On the other hand, significant variations in distribution of ET across different season

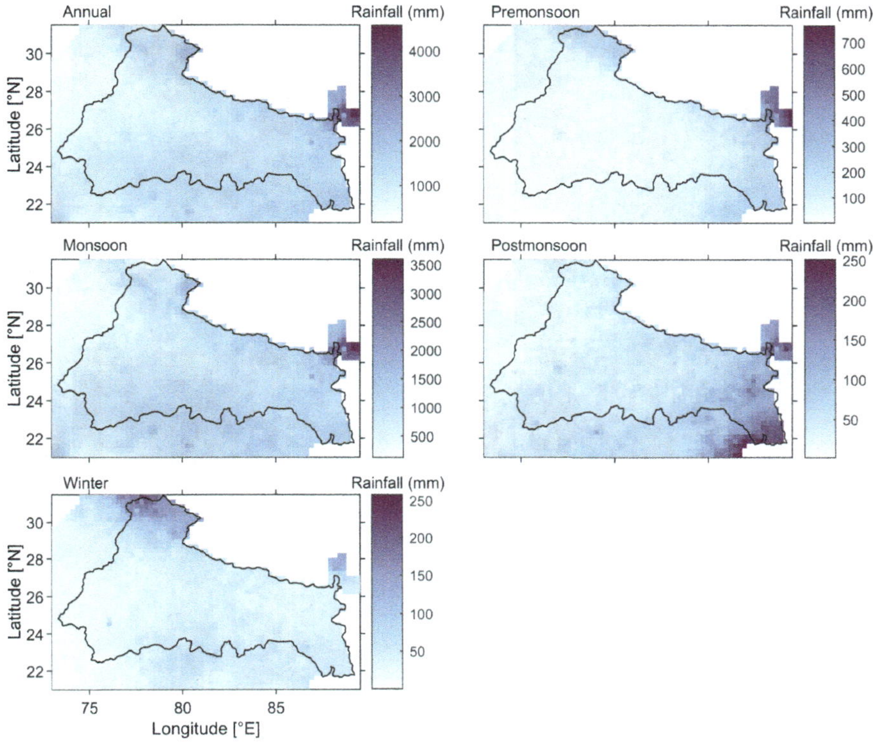

Fig. 22.3 Spatial distribution of annual and seasonal accumulated rainfall across Ganga basin

can be observed, which shows the existence of factors than precipitation and thus resulting in altered ET response of the basin.

Seasonal and annual time series of ET obtained from GLDAS over Ganga river basin were analysed for trend over the period 1981–2015. Figure 22.4 shows the spatial distribution of trend in annual time series of ET from MK test at different grid points over Ganga basin. As can be seen from the figure, ET over majority of Ganga basin is following an increasing trend (significant at most grids) expect for the upper region. In the upper region, trends in ET are found to be decreasing. The analysis of trends in ET for different season has shown that spatial distribution of these trend changes across different season. As evident form Fig. 22.4, spatial variations of trend are fairly similar for annual, monsoon and pre-monsoon ET, whereas space distribution of trend in post-monsoon and winter ET is significantly inconsistent.

The trends in annual and seasonal time series of rainfall at various grid points falling in Ganga basin are shown in Fig. 22.5. Results show no uniform spatial pattern in temporal trend of rainfall for annual and seasonal time series except for post-monsoon season. For post-monsoon season, a consistent negative trend has been noticed in rainfall for western half portion of the basin. On the other hand, a few grid points falling along the eastern coastal belt are found to be experiencing significant

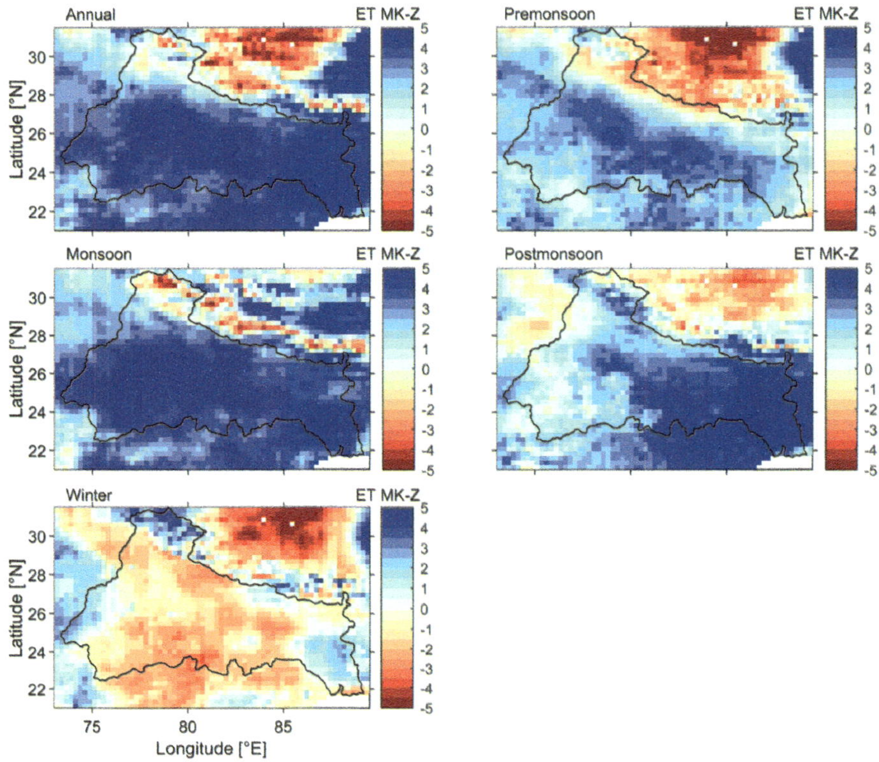

Fig. 22.4 Spatial distribution of MK-Z values; annual and seasonal accumulated evapotranspiration across Ganga basin

increasing trend. Trends in annual rainfall over central region are decreasing, whereas upper and lower reaches of basin are experiencing an increasing trend. Figure 22.6 presents the trend in annual mean temperature over the basin. As evident from the figure, the upper and lower middle regions of basin are experiencing an increasing trend in mean temperature. On the other hand, the middle and lower reaches of Ganga basin is found to be experiencing a falling trend in annual mean temperature. The observations for upper Ganga basin are contradictory as both rainfall and temperature in the upper region are found to be increasing with time, yet negative trends are observed in annual ET time series. Also, no definite relationship can be observed in spatial distribution of trends in ET with rainfall and temperature. However, the spatial patterns in temperature trends are correlating with the patterns in rainfall trends (Fig. 22.6). The results show the time lag in response of ET to the variations in temperature and rainfall. It is possible due to the dependence of ET on soil moisture. The seasonal variations observed in spatial distribution of ET and trends in ET may be attributable to the different land cover classes that exist in Ganga basin. Various land cover classes existing in Ganga basin are shown in Fig. 22.7.

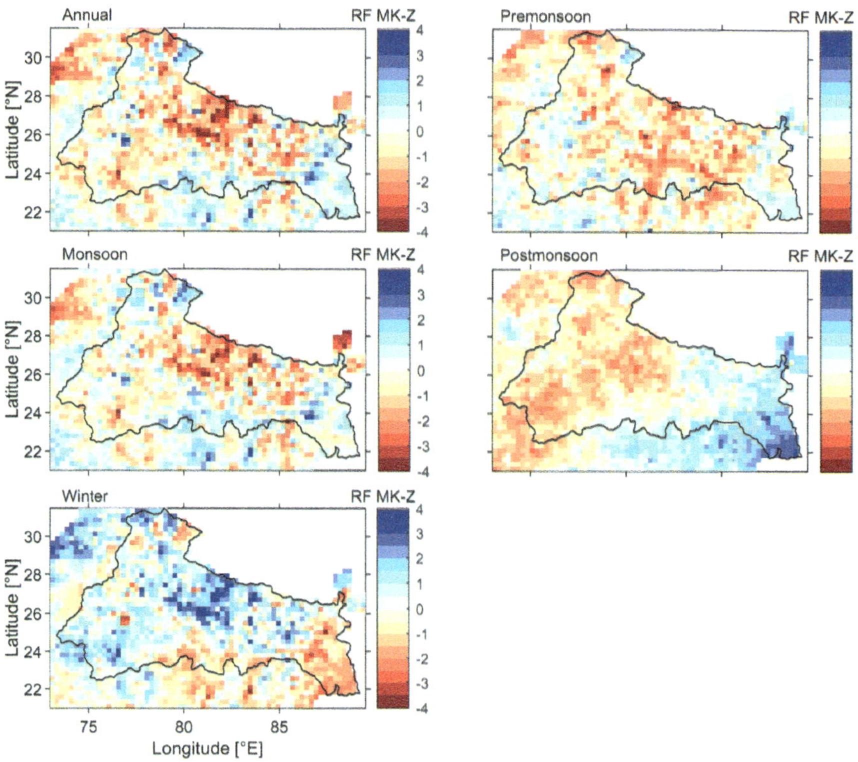

Fig. 22.5 Spatial distribution of MK-Z values; annual and seasonal accumulated rainfall across Ganga basin

22.6 Conclusions

The study aims to analyse spatiotemporal patterns in seasonal and annual time series of ET dataset obtained from GLDAS Noah model simulations. The observed variations in ET are also analysed for relationship with changes in precipitation and temperature over Ganga river basin. Study results show increasing trend in ET over majority of region except for upper reaches of Ganga basin. Moreover, a definite relationship can be seen between patterns in ET and patterns in rainfall and temperature. Further, decreasing trend in annual ET in the upper reaches of Ganga basin may not be attributable to the either precipitation or temperature. This could be due to fact that the upper reaches of basin are located in Himalayan region where the interaction between climatic and hydrologic parameters are much more complex and their relationships are yet not well established. Moreover, land use-land cover patterns may majorly affect the ET response of the basin. The seasonal variations observed in spatial distribution of ET and trends in ET may be attributable to the different land cover classes exist in Ganga basin. Overall, the study provides the trends observed

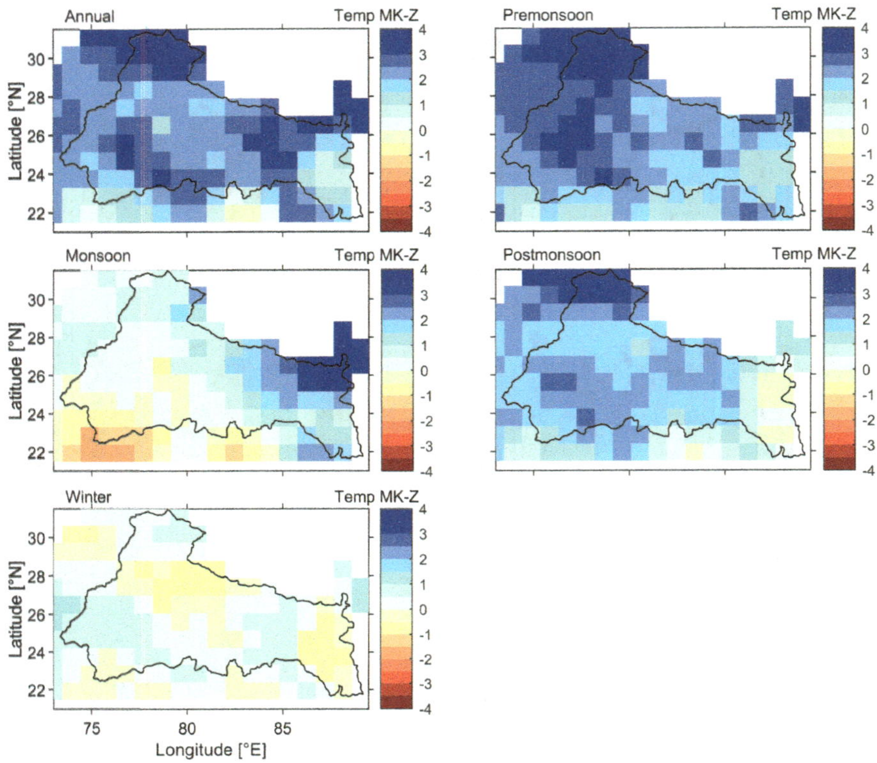

Fig. 22.6 Spatial distribution of MK-Z values; annual and seasonal mean temperature across Ganga basin

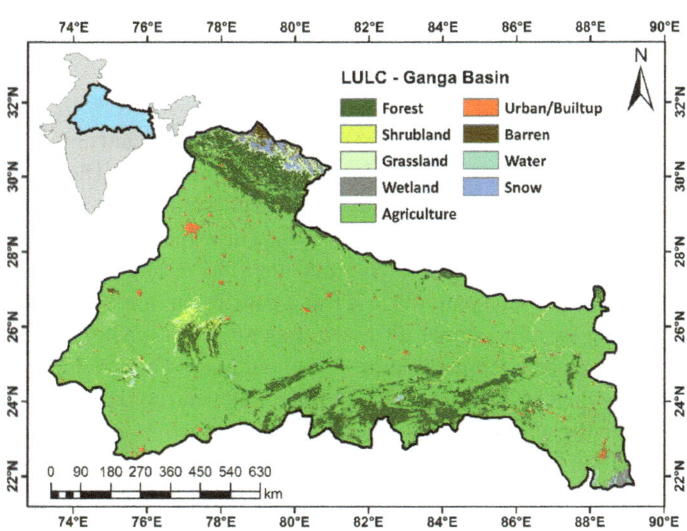

Fig. 22.7 Land use-land cover map of Ganga basin for the year 2015

in ET over Ganga basin and its spatial distribution which may prove to be a useful input for efficient planning and execution sustainable water resources management in the basin.

Acknowledgements The authors gratefully acknowledge the infrastructural and financial support provided by Ministry of Human Resources Development (MHRD), Government of India. The authors are also thankful to the anonymous reviewers for their constructive suggestions to improve the quality of present work.

References

Hamed KH, Rao AR (1998) A modified Mann-Kendall trend test for autocorrelated data. J Hydrol 204(1–4):182–196
Kendall MG (1975) Rank correlation methods. Griffin, London, UK
Long D, Longuevergne L, Scanlon BR (2014) Uncertainty in evapotranspiration from land surface modeling, remote sensing, and GRACE satellites. Water Resour Res 50(2):1131–1151
Mann HB (1945) Nonparametric tests against trend. Econometrica, 245–259
Mueller B, Seneviratne SI, Jimenez C, Corti T, Hirschi M, Balsamo G, Jung M (2011) Evaluation of global observations-based evapotranspiration datasets and IPCC AR4 simulations. Geophys Res Lett 38(6):L06402
Park J, Choi M (2015) Estimation of evapotranspiration from ground-based meteorological data and global land data assimilation system (GLDAS). Stoch Environ Res Risk A 29(8):1963–1992
Peters-Lidard CD, Kumar SV, Mocko DM, Tian Y (2011) Estimating evapotranspiration with land data assimilation systems. Hydrol Process 25(26):3979–3992
Xu CY, Gong L, Jiang T, Chen D, Singh VP (2006) Analysis of spatial distribution and temporal trend of reference evapotranspiration and pan evaporation in Changjiang (Yangtze River) catchment. J Hydrol 327(1):81–93

Chapter 23
Seasonal and Inter-Annual Variability of Sea Surface Temperature and Its Correlation with Maximum Sustained Wind Speed in Bay of Bengal

Jiya Albert and Prasad K. Bhaskaran

Abstract Sea surface temperature (SST) and oceanic heat content (OHC) play a significant role in tropical cyclogenesis formation as well as in the intensification of tropical cyclones. Prior studies over the recent past indicate the intensification of tropical cyclones as well as increased cyclone size, which has a strong correlation with increased SST over the global ocean basins. In the context of tropical cyclogenesis, the Bay of Bengal basin, a semi-marginal sea in the North Indian Ocean, is quite active. Many tropical cyclones form over this basin and make landfall either as severe or very severe cyclones in the countries surrounding the Bay of Bengal rim. The SST and sub-surface temperature in the Bay of Bengal basin also showed an increasing trend over the recent past. Interestingly, unlike the other ocean basins, the dynamics in the Bay of Bengal region is primarily governed by the reversing monsoon wind system and enormous freshwater influx through primary riverine sources that regulates the transport and distribution of water mass characteristics and the overall circulation characteristics. Differential circulation characteristics can, in turn, lead to spatial and temporal variability of SST and oceanic heat content responsible for tropical cyclone formation and intensification. The present study performed a comprehensive evaluation on the seasonal and inter-annual variability of SST and determines its correlation to maximum sustained wind speed (V_{max}) for the past two decades. Strategically, keeping in view the water mass distribution characteristics, the geographical domain in the Bay of Bengal region was sub-divided into four sectors (box of $5° \times 5°$) and the dependence of SST versus V_{max} across these sub-domains was investigated. The study considered 124 cyclonic events that include depressions, cyclonic storms, and severe cyclonic storms from 1996 through 2016. Overall an increasing SST trend was observed in climatological SST obtained from the CMIP5 models ACCESS 1.3 and HadGEM2-ES. The above statement substantiated by the correlation factor and increasing trend in all sub-domains was noticed between V_{max}

J. Albert · P. K. Bhaskaran (✉)
Department of Ocean Engineering and Naval Architecture, Indian Institute of Technology Kharagpur, West Bengal, Kharagpur 721 302, India
e-mail: pkbhaskaran@naval.iitkgp.ac.in

J. Albert
e-mail: jiyaalbert2012@gmail.com

R. Jha et al. (eds.), *Climate Change Impacts on Water Resources*, Water Science and Technology Library 98, https://doi.org/10.1007/978-3-030-64202-0_23

obtained from the Joint Typhoon Warning Centre (JTWC) and SST in ERA-Interim data. Interestingly, the central sector in the Bay of Bengal region showed a different correlation, unlike the other sub-domains. Basin-scale SST anomaly varied between 2 °C (maximum) and 1.7 °C (minimum) during the post-monsoon and between 2.4 °C (maximum) and 1.6 °C (minimum) during the pre-monsoon seasons. The findings from the study indicate a definite increase in V_{max} correlated with SST over the recent years. SST variability showed an in-phase correlation with V_{max}, and that is consistent over all the four sub-domains in the study area. A closer examination and analysis of both SST and oceanic heat content is warranted, and their mutual trends need to be evaluated with V_{max} to better understand the process of tropical cyclone intensification. The study is believed to have importance in the research activity of tropical cyclone modelling.

Keywords Sea surface temperature · Tropical cyclogenesis · Ocean heat content

23.1 Introduction

Coupled dynamics of cyclonic formation and its intensification in a changing climate is mostly influenced by the essential factor of sea surface temperature. The pre- and post-monsoon climatology of the recent past have shown that cyclones have grown in both size as well as intensity. Intensification of cyclones is connected to the large thermal inertia of ocean and on the background of its direct reaction response towards the greenhouse forcing which is enormously capturing distant peaks in last decadal periods, and it is going to grow in coming decades too. There are furthermore pieces of evidence of the mutual increase in SST along with landfall rates of cyclonic events and statistically proven to have 45% increment of more TC formation during a study in the Atlantic basin. In this warming world, the indicators and evidence of the strengthening of cyclonic wind speeds along with the rising SST in Northern Hemisphere reported by Arora and Dash (2016) uses an enthalpy exchange process. This projected the importance of a detailed study for the Northern Indian Ocean in which the large land mass cover and high vicinity to tropical and Pacific warm waters can build up the risk in noticeable increment of cyclonic activity having direct socio-economic impacts. In the current study, we perform composite analysis of maximum sustained cyclonic wind velocity (V_{max}) statistics using the SST variability over Bay of Bengal (BoB) domain during the past two decades (1996–2016) as an analogue for climate change. The geographical division of BoB into four square sub-domains (Fig. 23.1) gives out the indirect way to determine the physical response of water mass characteristics under the surface of the ocean towards the formation and intensification of tropical cyclones. However, it also provides an opportunity to explain the influence of various category of water masses especially that differ in their physical properties of temperature and salinity to the cyclonic wind velocity and frequency of cyclonic event generated within the sub-domains.

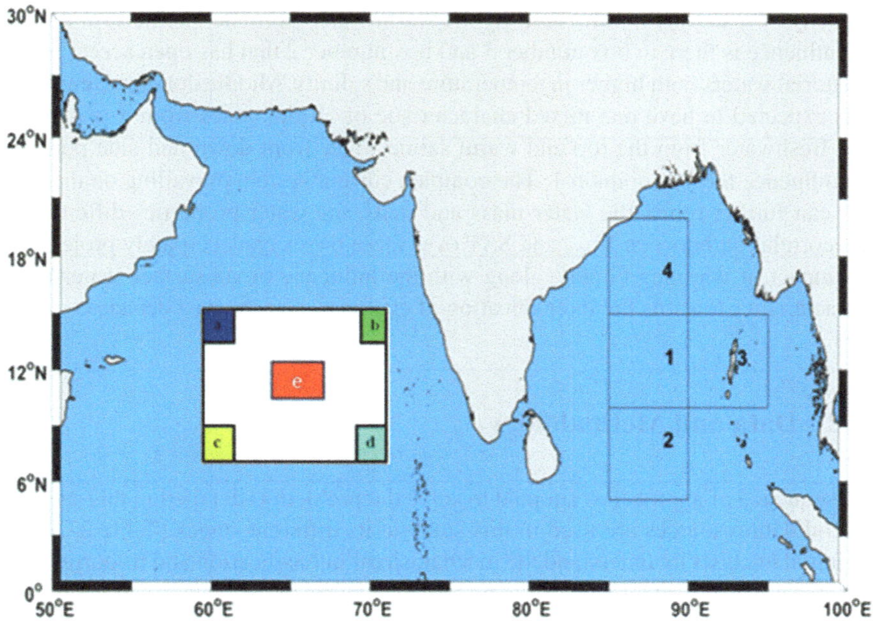

Fig. 23.1 Domain for SST study (Left)

In the context of tropical cyclogenesis, the BoB basin, a semi-marginal sea in the North Indian Ocean, is quite active. Many tropical cyclones form over this basin and make landfall either as severe or very severe cyclones in the countries surrounding the Bay of Bengal rim (numbers shown in Table 23.1). The SST and sub-surface temperature in the Bay of Bengal basin also showed an increasing trend over the recent past. The uppermost sub-domain at head bay is a representation for water masses in which the accumulation and mixing of a considerable quantity of freshwater from the peninsular rivers and rivers from head BoB exists. It represents less saline and low-temperature water characteristics in general. Bottom 2 sub-domain portion represents

Table 23.1 Number of tropical cyclone activities in the domain (Right)

****D- Depression*
CS-Cyclonic Storm
SCS- Severe Cyclonic Storm
****Corner and center points in each box (1-4) marked with a, b, c, d and e*

Section	D	CS	SCS
1	21	13	6
2	9	5	0
3	8	4	5
4	16	7	4

the body of water mass influenced by the warm water from the Pacific. But mostly the influence is there to box number 3 and box number 2 that has open access to the equatorial waters both higher in temperature and salinity. Middle domain (numbered 1) is expected to have one mixed characteristic of all the boxes from 2 to 4. Both cold freshwater from the top and warm saline water from down and side portions can influence the sub-domain 1. The complex current system prevailing on the area also can further remix the water mass and make the water properties differential. The correlation between V_{max} and SST over these boxes can accurately project out the impact of warming climate along with the influence of subsurface water mass characteristics towards the intensification of cyclones over the past decades.

23.2 Data and Methodology

The sequence of approaches adopted to solve the problem is detailed in this section. Several data resources are used in this study at its different stages (Table 23.2). In the initial analysis to understand the mechanism of intensification and its correlation to the ocean heat content and SST, a historical analysis of 20 years was done for the period 1996–2016 in BoB using ERA-Interim daily data. For comparison purpose climatological SST data obtained from three independent datasets of SST, namely Era-Interim real-time data and two CMIP5 (coupled model inter-comparison project phase 5) models ACCESS 1.3 and HadGEM2-ES, are also used for the study.

a. ***CMIP5***: This is multi-climate model dataset based on the CMIP5 archived at the Intergovernmental Panel on Climate Change (IPCC) Data Distribution Centre (DDC). Parts of the CMIP5 multi-model dataset are used in the Fifth Assessment Report (AR5) of the IPCC. This dataset is analyzed by the scientific community worldwide and continues to be available for studying climate, its variability and change, and the societal and environmental implications of climate change regarding impacts, adaptation, and vulnerability (Emori et al. 2016). These models are produced considering the recently presented Representative Concentration Pathways (RCPs) for four distinct scenarios such as RCP 2.6, RCP 4.5, RCP 6.0, and RCP 8.5. In light of these four scenarios, the climate projections for the future accessed from various models are accessible under the CMIP5 venture.
b. ***HadGEM2-ES***: HadGEM2-ES is a coupled earth system model that was used by the Met Office Hadley Centre, U.K. for the CMIP5 centennial simulations. HadGEM2 is a configuration of the Met Office Unified Model (UM) developed

Table 23.2 Description of model

Model	ERA-Interim	HadGEM2-ES	ACCESS-1.0
Data type	Daily	Monthly	Monthly
Resolution	0.25 × 0.25	1.25 × 1.875	1.25 × 1.875

from UM version 6.6. HadGEM2-ES was the first Met Office Hadley Centre model to include earth system components as standard. The unified model is used by some institutions around the world for both operational weather forecasting and climate research. The HadGEM2 family of model configurations includes atmosphere, ocean, and sea-ice components, with and without a vertical extension in the atmosphere model to include a well-resolved stratosphere and earth system components including the terrestrial and oceanic carbon cycle and atmospheric chemistry. The various processes include:

- Troposphere, land surface and hydrology,
- Aerosols, ocean and sea-ice, terrestrial carbon cycle,
- Ocean biogeochemistry, chemistry.

c. ***ACCESS-1.0***: The ACCESS-CM uses the Met Office Unified Model (UM) as its atmospheric component. The atmospheric configuration in ACCESS-1.0 is designed to be the same as that of HadGEM2 version r1.1, which is necessarily the same as that in the HadGEM2 model versions used for CMIP5.

On a monthly basis, data are taken during the period, including historical (1996–2005) and projection (2006–2016) of the model. For the study, the datasets are grouped into post-monsoon (March–April–May) and pre-monsoon (October–November–December) periods. Strategically, keeping in view the water mass distribution characteristics, the geographical domain in the Bay of Bengal region was further divided into four sectors (box of 5° × 5°) and the dependence of SST versus V_{max} across these sub-domains was investigated. The study considered 124 cyclonic events that include depressions, cyclonic storms and severe cyclonic storms that occurred between 1996 and 2016. As per the cyclonic events including the depressions, cyclonic storms and super cyclonic storms that occurred in Bay of Bengal domain during the past 20 years, a total of 125 events have occurred (IMD). Out of that considering the formation location of these, central sections of BoB as shown in Fig. 23.1 hold about 98 events. Section 1 encompassing 85° E–90° E, 10° N–15° N reported 21 depressions along with 13 cyclonic storms and 6 severe cyclonic storms during the past two decades. In Section 2 (85° E–90° E, 5° N–10° N), there has been an occurrence of nine depressions and five cyclonic storms. Section 3 (90° E–95° E, 10° N–15° N) reported eight depressions, four cyclonic storms and five severe cyclonic storms. Section 4 (85° E–90° E, 15° N–20° N) which is near to the head bay region has reported 16 depressions, 7 cyclonic storms, and 4 severe cyclonic storms. The findings obtained are discussed in detail in the next section.

23.3 Results and Discussion

Using the various datasets available for SST, the evaluation was carried out to determine the overall variability exhibited across different sub-domains. The values are examined on a seasonal basis and averaged for pre-monsoon and post-monsoon seasons. Era-Interim data gives out the real-time cyclone-induced SST over the four domains (Table 23.3).

Basin-scale SST anomaly varied between 2 °C (maximum) and 1.7 °C (minimum) during the post- and between 2.4 °C (maximum) and 1.6 °C (minimum) during the pre-monsoon seasons. However, in relation to the freshwater influence, Section 4 showed the maximum variability in the rate of difference between the maximum and minimum set for all the three data but comparatively the average increase is about 30% in pre-monsoon and 35% in case of post-monsoon based on the CMIP5 models, ACCESS-1.0 (Table 23.4) and HadGEN2-ES (Table 23.5). The increment of marginal difference in lower and upper values of SST is bound to be more during post-monsoon because of the 80% of rainfall that occurs in association with the temporal asymmetry of annual mean monsoonal winds over North Indian Ocean and the Indian subcontinent (Parekh and Konwar 2017).

In the proceeding phase, a correlation study of cyclone-induced SST and V_{max} (Fig. 23.2) is performed across the sub-domain and along the corner points for each of the sub-domains. The overall increment in the V_{max} is evident during both the pre- and post-monsoon seasons correlation results even though comparing to pre-monsoon the post-monsoon exhibits a minor hike in the value of trend line slope. Further, a detail evaluation of cyclone-induced SST pattern and its trend across the domain over the past two decades is also studied. Interestingly, the central sector in the Bay of Bengal region showed a different correlation, unlike the other sub-domains. The findings from the study indicate an apparent increase in V_{max} that correlated with SST over the recent years and the cooling of sea surface temperature by the post-cyclonic events. SST variability showed an in-phase correlation with V_{max}, and that is consistent over all the four sub-domains in the study area. Sector-wise correlation values are determined, and all of them are within 0.5 (Table 23.6). It provides a clear indication that along with SST there are more disregarded supporting parameters which can induce the V_{max} to change. Sector 4 negative correlation during pre-monsoon season along with the negative mutuality of post-monsoon season at Sector 1 decipher a clear indication of the freshwater influx from the rivers of the head bay region which in turn cools the domain, and the other sufficient parameters other than SST will dominate to generate the cyclones (Fig. 23.3). A closer examination and analysis of both SST and oceanic heat content is warranted, and their mutual trends need to be evaluated with V_{max} to understand the process of tropical cyclone intensification better. The study is believed to have importance in the research activity of tropical cyclone modelling.

Finally, the overall trend of SST values of the domain, sub-domains and specific points inside the sub-domains for the last two decades are determined based on Era-Interim real-time daily data and findings explained below. All the points in Section 1 (Fig. 23.4, set 1) showed an increasing trend of SST during the 20 years in which the point 'd' showed the maximum increasing rate followed by the middle point 'e.' All the points in Section 2 (Fig. 23.4, set 2) showed an increasing trend of SST during the 20 years in which middle point 'e' showed the maximum increasing rate followed by the first corner point 'a.' All the points in Section 3 (Fig. 23.4, set 3) have shown an increasing trend of SST during the 20 years in which both corner

Table 23.3 Maximum and minimum values of pre–post-monsoon SST using ERA-Interim

Data	Pre-monsoon SST 1996–2016					Post-monsoon SST 1996–2016				
ERA-Interim	Section	Min _SST	Max_SST	ΔSST	Mean SST	Section	Min _SST	Max_SST	ΔSST	Mean SST
	1	30.68	31.162	0.48	30.921	1	29.7764	30.132	0.3556	29.954
	2	31.0769	31.224	0.1471	31.150	2	30.0655	30.232	0.1665	30.649
	3	30.5008	31.394	0.8932	30.947	3	30.0516	30.344	0.2924	31.198
	4	29.0128	30.67	1.6572	29.841	4	28.3393	30.05	1.7107	29.197

Table 23.4 Maximum and minimum values of pre–post-monsoon SST using HadGEN2-ES

Data	Pre-monsoon SST 1996–2016					Post-monsoon SST 1996–2016				
HadGEN2-ES	Section	Min _SST	Max_SST	ΔSST	Mean SST	Section	Min _SST	Max_SST	ΔSST	Mean SST
	1	28.4140	29.1281	0.7141	28.7465	1	26.6572	27.9035	1.2463	27.261
	2	28.2358	31.7250	3.4892	29.2869	2	27.7257	28.5432	0.8175	27.685
	3	27.6453	32.5501	4.9048	30.7560	3	26.0324	27.6764	1.644	26.085
	4	27.8807	33.6272	5.7465	32.8028	4	26.6824	31.5291	4.8467	29.041

Table 23.5 Maximum and minimum values of and pre–post-monsoon SST using ACCESS-1.0

Data	Pre-monsoon SST 1996–2016					Post-monsoon SST 1996–2016				
ACCESS-1.0	Section	Min _SST	Max_SST	ΔSST	Mean SST	Section	Min _SST	Max_SST	ΔSST	Mean SST
	1	28.5160	29.2281	0.7121	28.8720	1	26.4483	27.8185	1.3702	27.133
	2	28.4358	31.6240	3.1882	30.0299	2	27.6146	28.4432	0.8286	28.029
	3	27.7358	31.9501	4.2143	29.8429	3	26.0012	27.5224	1.5212	26.762
	4	27.6814	32.9272	5.2458	30.3043	4	26.6613	31.5111	4.8498	29.086

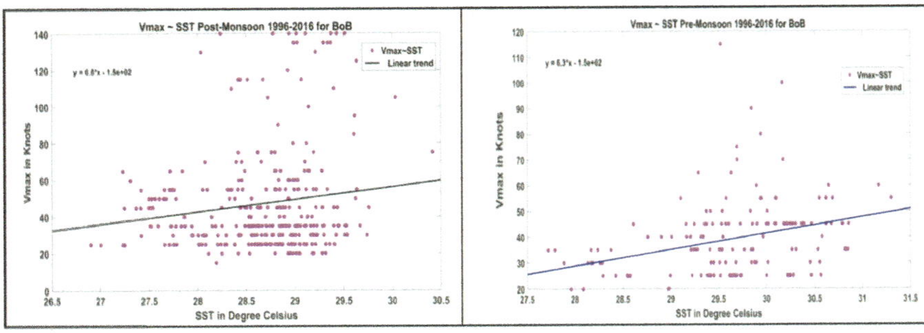

Fig. 23.2 SST~V_{max} for a whole domain on pre- and post-monsoon

Table 23.6 The correlation coefficient of sections in pre- and post-monsoon

Domain	Correlation coefficient R	
	Pre-monsoon	Post-monsoon
Combined	0.30961	0.1302
Sub-domain 1	0.32616	−0.1098
Sub-domain 2	0.33127	0.52370
Sub-domain 3	0.26640	0.11999
Sub-domain 4	−0.05896	0.29204

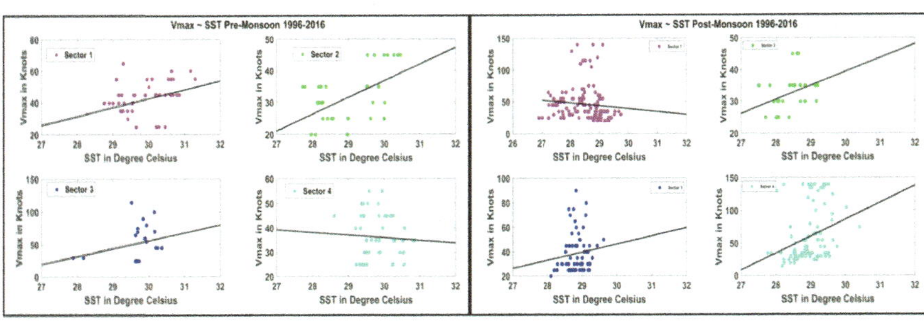

Fig. 23.3 SST~V_{max} for each section from 1 to 4 on pre and post monsoons

points, point ‘c’ and point ‘b’ showed the maximum increasing rate followed by the bottom corner point. All the points in Section 4 (Fig. 23.4, set 4) have shown a mild growing trend for SST during the 20 years and the slope is almost horizontal. During the analysis of post-monsoon seasons, all the points in Section 1 (Fig. 23.5, set 1) showed an increasing trend of SST during the 20 years except point ‘a’ shows least variation and trend line is almost horizontal. Every point in Section 2 (Fig. 23.5, set 2) had shown an increasing trend of SST in which bottom corner points, point ‘d’ showed the maximum increasing rate followed by the middle point ‘e.’ All other points show a similar increase in trend. The recent study for SST over BoB and its mutual dependency towards cyclone-induced cooling along with forwarding motion

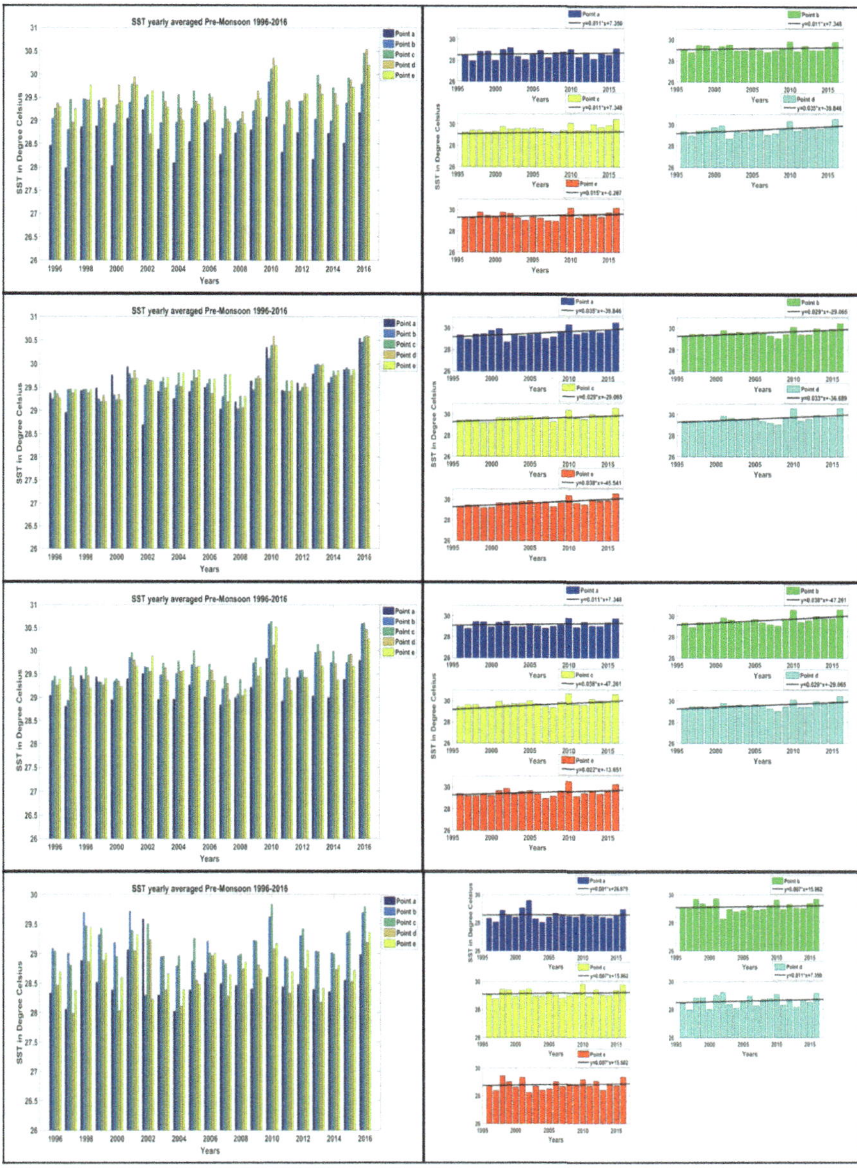

Fig. 23.4 SST yearly averaged pre-monsoon over Sections 1–4 (top to bottom) for points a(indigo), b(green), c(yellow), d(blue) and e(red)

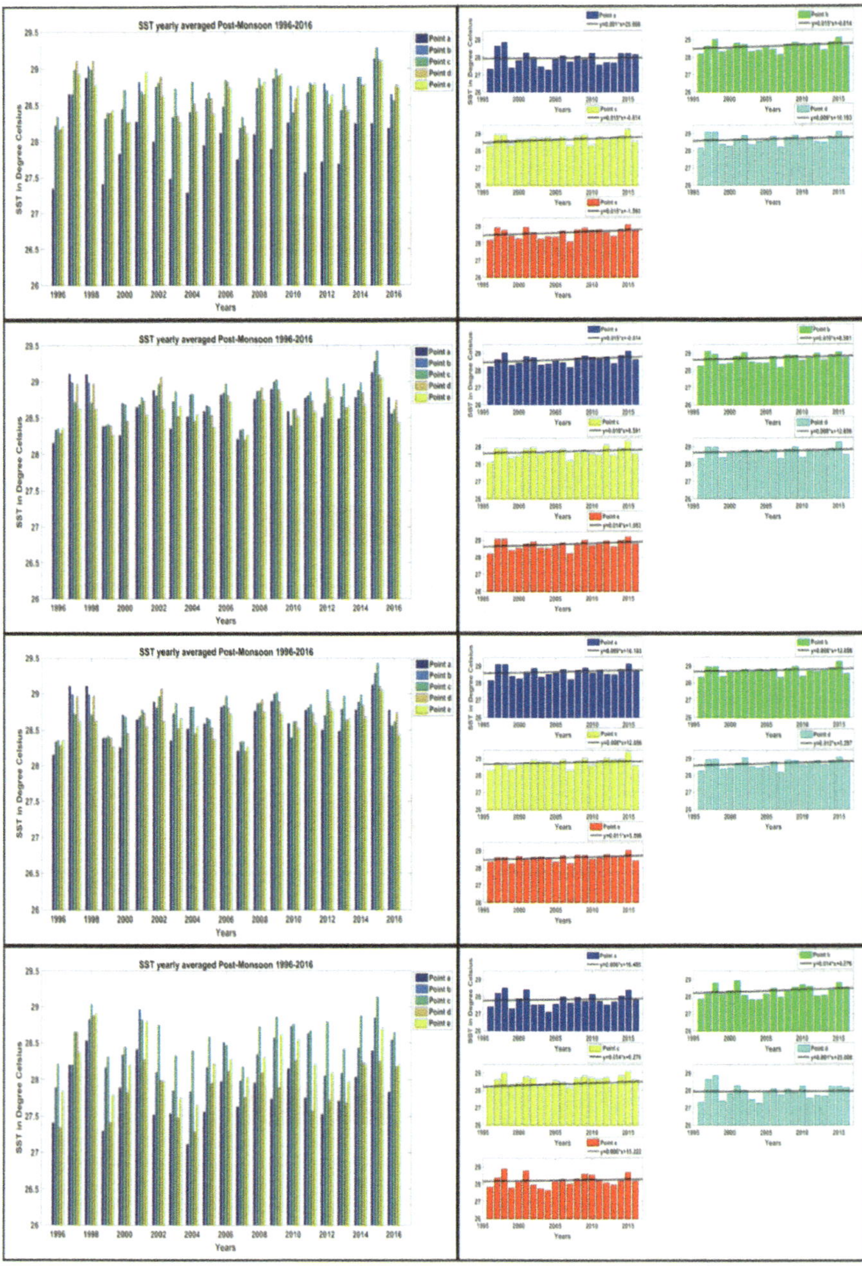

Fig. 23.5 SST yearly averaged post-monsoon over Sections 1–4 (top to bottom) for points a(indigo), b(green), c(yellow), d(blue) and e(red)

speed has quoted a unique behavior of northern part of BoB which corresponds to the sub-domain number four with highly stratified saline characteristics. The reason stated is due to the larger freshwater influx after the monsoon rains in this region, and comparatively, it exhibits a different pattern other than the remaining three regions. Substantiate reason for the anomalous behavior of Section 4 in SST trends in pre- and post-monsoons over the past two decades is explained that has a strong established control from ocean saline stratification (Li et al. 2017).

23.4 Conclusions

Overall, an increasing SST trend was observed in climatological SST obtained from the CMIP5 models ACCESS-1.3 and HadGEM2-ES. The correlation factor substantiates this, a rising trend in all sub-domains noticed between V_{max} obtained from Joint Typhoon Warning Centre (JTWC) and SST in ERA-Interim data. Interestingly, the central sector in the Bay of Bengal region showed a different correlation, unlike the other sub-domains. Basin-scale SST anomaly varied between 2 °C (maximum) and 1.7 °C (minimum) during the post-monsoon and between 2.4 °C (maximum) and 1.6 °C (minimum) during the pre-monsoon seasons. In relation with Section 4 which is under the direct freshwater influence from head bay, exhibits the maximum variability in the value of SST evaluated based on all the three data, but comparatively it has an average increase of 30% in pre-monsoon and 35% in case of post-monsoon based on the CMIP5 models, ACCESS-1.0 and HadGEN2-ES. It gives a clear indication on the enormous cooling during the lifetime of cyclonic events during both pre- and post-monsoon seasons. The findings from the study indicate a definite increase in V_{max} correlated with SST over the recent years. SST variability showed an in-phase correlation with V_{max}, and that is consistent over all the four sub-domains in the study area. A closer examination and analysis of both SST and oceanic heat content is warranted, and their mutual trends need to be evaluated with V_{max} to understand the process of tropical cyclone intensification better. The study is believed to have potential importance in research activity related to tropical cyclone modeling.

References

Arora K, Dash P (2016) Towards dependence of tropical cyclone intensity on sea surface temperature and its response in a warming world. Climate 4(2):30

Emori S, Taylor K, Hewitson B, Zermoglio F, Juckes M, Lautenschlager M, Zermoglio F (2016) CMIP5 data provided at the IPCC Data Distribution Centre

Li Y, Han W, Wang W, Ravichandran M, Lee T, Shinoda T (2017) Bay of Bengal salinity stratification and Indian summer monsoon intraseasonal oscillation: 2. Impact on SST and convection. J Geophys Res Ocean 122(5):4312–4328

Parekh A, Konwar M (2017) The North Indian Ocean. Bull Phys Proj 2:1–3

Chapter 24
Comparison of CMIP5 Wind Speed from Global Climate Models with In-Situ Observations for the Bay of Bengal

Athira Krishnan and Prasad K. Bhaskaran

Abstract Realistic estimates of surface wind speed form an essential prerequisite for process-based air-sea studies and also numerical modeling needs. In this context, historical and projected wind speed estimates obtained from global climate and regional models warrant proper assessment and necessary bias corrections before it can be optimally used for rigorous analysis and research needs. Adequate evaluation and bias correction of wind speed estimates are therefore crucial to understand the extremes. For example, they have direct implications on extreme wind waves that can influence the coastal zone. The present study performed a detailed evaluation of wind speed obtained from the Coupled Model Inter-comparison Project Fifth Phase (CMIP5) products to assess their projections for the Bay of Bengal region. A suite of global climate models is employed to generate the CMIP5 products under four Representative Concentration Pathways (RCPs) of 2.6, 4.5, 6.0, and 8.5 and based on differential CO_2 emission scenarios. The present study also used the Research Moored Array for African–Asian–Australian Monsoon Analysis and Prediction (RAMA) buoys located in the central Bay of Bengal in order to validate and skill assess the CMIP5 wind products under varying RCPs. In addition, an intercomparison exercise that was performed between RAMA buoys data and merged satellite altimeter data from the French Research Institute for Exploitation of the Sea/Laboratory of Oceanography from Space (IFREMER/CERSAT) provided necessary confidence to ascertain the quality of CMIP5 wind speed products. The study signifies that a reasonably good correlation was noticed in the wind speed comparison between CMIP5 GCM products and RAMA buoys (maximum correlation of 0.64), and the correlation factor varied between the suite of models used in CMIP5 experiments. This exercise would provide a detailed know-how on the performance of various GCMs and also provide a basis to select the best-performing GCMs for the Bay of Bengal region. Analysis of the upper 10% (90th percentile) showed a maximum under-estimation/over-estimation of 2.5 m s^{-1} and 1.5 m s^{-1}, respectively

A. Krishnan · P. K. Bhaskaran (✉)
Department of Ocean Engineering and Naval Architecture, Indian Institute of Technology Kharagpur, West Bengal, Kharagpur 721 302, India
e-mail: pkbhaskaran@naval.iitkgp.ac.in

A. Krishnan
e-mail: athirakrishnanambattu@iitkgp.ac.in

R. Jha et al. (eds.), *Climate Change Impacts on Water Resources*, Water Science and Technology Library 98, https://doi.org/10.1007/978-3-030-64202-0_24

for wind speed comparison between CMIP5 and RAMA buoys data. Although the CMIP5 GCMs are not able to represent the contemporary wind speed climatology satisfactorily, the models such as HadGEM2-ES, HadGEM2-CC, CanESM2, and ACCESS-1.3 showed the best performance concerning near-surface wind speed for the Bay of Bengal region.

Keywords CMIP5 · Wind speed · Merged satellite altimeter · RAMA buoys · Bay of Bengal

24.1 Introduction

Climate change can affect the wind patterns and consequently its magnitude and frequency of occurrence. Knowledge of extreme wind and wave climatology are required for practical applications in the marine industries. Therefore, a proper assessment on the potential impact of climate change on the variation of wind regime is very essential (Alinejhad-Tabrizi et al. 2017). The latest findings and knowledge about the scientific, technical and socioeconomical aspects of climate change are well documented in the Fifth Assessment Report by the Intergovernmental Panel on Climate Change (IPCC AR5). Natural as well as anthropogenic factors are contributing to the climate change and subsequent global warming. Human-induced climate change is mainly dominated by the greenhouse gas emissions (mainly CO_2) to the atmosphere. IPCC AR5 relies on the World Climate Research Programme (WCRP) Fifth Coupled Model Inter-comparison Project (CMIP5), a coordinate set of global coupled atmosphere-ocean general circulation model (AOGCM) simulations (Taylor et al. 2012). RCPs represent the socioeconomic pathways that reach a specific radiative forcing by the year 2100, and four RCPs are now available: RCP2.6, RCP4.5, RCP6.0, and RCP8.5 (van Vuuren et al. 2011). RCP-based climate projections are now freely available from many climate models under the CMIP5 family. CMIP5 data allow the evaluation of how the models realistically simulate the recent past and present and provide the projections of future climate changes from the present date up to 2100 (and beyond for some models and experiments).

Climate change mitigation and adaptation strategies are formulated based on the climate model projections of the global mean temperature responses to greenhouse gas emissions in the future (Cowtan et al. 2015). He also pointed out on the ability of models to simulate the transient evolution of climate identified by comparing model projections against the observations. A study by Marena Lin and Peter Huybers (2016) reported on the agreement of recent surface trends with the CMIP5 ensemble and found a model observation consistency at the global scale accompanied by regional discrepancies. From the mean and 90 percentile wind speeds over the global oceans, Young et al. (2011) found an increase by 0.25–0.5% per year in wind speeds over the majority of world's oceans. They also noticed a stronger trend in the Southern hemisphere than the Northern hemisphere. Debernard et al. (2002) assessed the impact of climate change on the regional design winds over the northeast Atlantic based on control climate of 1980–2000 and projected climate of 2030–2050. The climate change caused by global warming has changed the wind conditions at global as well as regional scales noticed by Kulkarni et al. (2013). Their study concluded a rising trend of wind speeds about 6–8% along the Arabian Sea and 2–4% along the Bay of Bengal. The Research Moored Array for African–Asian–Australian Monsoon Analysis and Prediction (RAMA) buoy network over the Indian Ocean, part of the Tropical Atmosphere Observation (TAO) program helps to validate the model data to a good extent over the Bay of Bengal domain. Indira Rani and Das Gupta (2013) used RAMA buoy data to validate sea surface winds from Oceansat-2 scatterometer over the Indian Ocean region. The comparison for 1 year showed that wind speed and direction derived from OSCAT agreed with RAMA buoy winds. Another comparison of RAMA buoy winds with OSCAT winds was reported by Sudha and Prasada Rao (2013) for the Indian and the Pacific Ocean basins. This study aims to compare the wind speeds from CMIP5 GCMs with in-situ data from RAMA buoys in order to obtain the statistics of the two datasets and thereby choose the best-performing RCP for research use.

24.2 Data

The study was conducted for the Bay of Bengal domain. The climate variable considered in this study was the near-surface wind speed. We used observed data as well as the data from global climate models. Observed data obtained from RAMA (Research Moored Array for African–Asian–Australian Monsoon Analysis and Prediction) buoys for Bay of Bengal was available at https://www.pmel.noaa.gov/tao/drupal/disdel/. RAMA surface floats measure the wind speed at a height of 4 m above the mean sea level. Daily averages of the data were used for this study. The Fifth Phase of Coupled Model Inter-comparison Project (CMIP5) and archived at the Intergovernmental Panel on Climate Change (IPCC) Data Distribution Centre at the Program for Climate model diagnosis and Inter-comparison (PCDMI) provides the state-of-art multimodel dataset designed to advance our knowledge of climate variability and climate change. Projected wind speed data from Global Climate Models representing four RCPs, 2.6, 4.5, 6.0, and 8.5 with a temporal resolution of daily means were extracted for the analysis available from https://esgf-node.llnl.gov/projects/esgf-llnl/. The spatial resolution of each model as well the number of models under each RCP were different in number. Table 24.1 provides more details on the family of models under each RCP.

More details regarding the experiments and corresponding models including the participating institutions, model resolutions are available from the URL link https://cmip.llnl.gov/cmip5/guide_to_cmip5.html. The study area and location of RAMA buoys are represented in Fig. 24.1 and Table 24.2 provides information on the period of data available from RAMA buoys.

Table 24.1 Details of the models available under each RCP

RCP 2.6	RCP 4.5	RCP 6.0	RCP 6.5
• CanESMZ • CNRM-CMS • CSIRO-Mk3.6.0 • GFDL-CM3 • GFDL-ESM2G • GFDL-ESM2M • HadGEM2-A0 • HadGEM2-ES • IPSL-CM5A-LR • IPSL-CMSA-MR • MIR0C5 • MIROC-ESM • MIROC-ESM-CHEM • MPI ESM LR • MPI-ESM-MR • MRI-CGCM3	• CanESM2 • ACCESS 1.0 • ACCESS1.3 • CANCM4 • CMCC-CM • CMCC-CMS • CNRM-CM5 • CSIRO Mk3.6.0 • GFDL-CM3 • GFDL-ESM2G • GFDL-ESM2M • HadGEM2-AO • HadGEM2-CC • HadGEM2-ES • INMCM4 • IPSL-CM5A-LR • IPSL-CMSA-MR • IPSL-CM5B-LR • MIR0C4h • MIROC5 • MIROC-ESM • MIROC-ESM-CHEM • MPI ESM-LR • MPI-ESM-MR • MRI-CGCM3	• C5IRO-Mk3.6,0 • GF0L-CM3 • GFDL-ESM2G • GFDL-ESM2M • HadGEM2-AO • HadGEM2-ES • IP5L-CM5A-LR • IPSL-CM5A-MR • MIROC5 • MIROC-ESM • MIROC-ESM-CHEM • MRKGCM3	• CANESM2 • ACCESS1.0 • ACCESS1.3 • CMCC-CESM • CMCC CM • CMCC-CMS • CNRM-CM5 • CSIRO Mk3.6.0 • GFDL-CM3 • GFDL-ESM2G • GFDL-ESM2M • HadGEM2-AO • HadGEM2-CC • HadGEM2 ES . INMCM4 • IPSL-CM5A-LR • IPSL-CM5A MR • IPSL-CM56-LR • Ml ROCS • MIROC ESM • MIROC-ESM-CHEM • MPI-ESM-LR • MPI-ESM-MR • MRI-CGCM3 • MRI-ESM1 • Nor ESM1M

24.3 Methodology

Projected 10 m wind speed available as respective *u* and *v* components were converted into resultant wind speed. Since the family of models under CMIP5 has different spatial resolutions, all the datasets were interpolated to a uniform grid of 0.5° × 0.5°. This study considered the climate data from only one ensemble "r1i1p1." Apart from the model data, RAMA buoy data measure the winds over the ocean surface at the height of 4 m, which is converted to 10 m to conform to the universally accepted standard level. The logarithmic wind profile relation (Manwell et al. 2009) was used to extrapolate the speed using the following relationship:

$$\frac{U(z)}{U(Zr)} = \frac{\ln\left(\frac{Z}{Zo}\right)}{\ln\left(\frac{Zr}{Zo}\right)}$$

where

$U(z)$ = velocity to be estimated at height z
Z = height above the mean level for velocity v

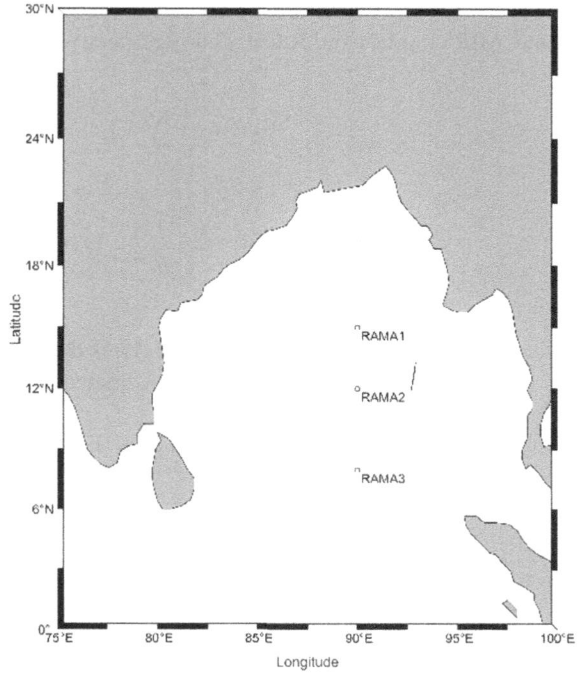

Fig. 24.1 Map of the study area

Table 24.2 Details of the buoy data used

Description	Location	Period of data used
RAMA 1	15° N, 90° E	01 January 2013–31 December 2015
RAMA 2	12° N, 90° E	01 January 2011–31 December 2012
RAMA 3	8° N, 90° E	01 January 2009–31 December 2009

$U(Zr)$ = known velocity at height level Zr
Zr = reference height where v_{ref} is known
Z_0 = roughness length in the current wind direction (0.0002 for water surface).

Statistical parameters like model mean wind speed, buoy mean wind speed, correlation coefficient, wind speed bias between the model and buoy data, root mean square difference for wind speed were estimated. The statistical parameters are calculated as follows:

$$\text{Model wind speed} = \sqrt{Mu^2 + Mv^2}$$

$$\text{Buoy wind speed} = \sqrt{Bu^2 + Bv^2}$$

where *Mu*, *Mv* and *Bu*, *Bv* are the northeast and southwest components of wind speed from CMIP5 models and RAMA buoy respectively.

$$\text{Bias} = \frac{1}{n}\sum_{i=1}^{n}(Ym - Yb)^2$$

$$\text{RMSE} = \sqrt{\frac{1}{n}\sum_{i=1}^{n}(Ym - Yb)^2}$$

where *Ym* represents the modeled value, *Yb* is the observed value (from buoy). The CC (Pearson correlation coefficient) is estimated using the relation:

$$\text{CC} = \frac{\sum_{i=1}^{n}(Yi - Y')(Zi - Z')}{\sqrt{\sum_{i=1}^{n}(Yi - Y')}\sqrt{\sum_{i=1}^{n}(Zi - Z')}}$$

where *Y* and *Z* correspond to the two-time series having equal number of values.

24.4 Results and Discussions

Figure 24.2 illustrates the time series of wind speed from CMIP5 models compared with the buoy observations at representative locations shown in Fig. 24.1. A total of 16 models represent RCP 2.6, 25 for RCP 4.5, 12 for RCP 6.0, and 26 models for RCP 8.5 scenarios. Data from the three buoys for a total duration of 6 years were compared with the time-series data obtained from CMIP5 models corresponding to each RCPs. Statistical parameters were calculated in order to obtain their dependence with observations and also to understand the existing emission scenario in the Bay of Bengal region.

The time-series data obtained from GCMs and RAMA buoy observations for the period of 2009–2015 clearly showed a close match for mean wind speed between GCMs and RAMA data. The maxima and minima in wind speed simulated by the models do not match well with the observations, however, four of the RCPs showed a higher minimum and lower maximum values as compared with RAMA buoys. This variation can be attributed due to bulk parameterization of the drag by climate models. Fitch et al. (2013) analyzed the drag parameterization for wind farms in climate models and observed that for an increase in surface roughness the flow drag was maximum during the day when the near-surface wind is most active.

The variations in wind speed were also examined for the study period. RAMA buoy observations demarcate an increase in mean wind speed from 2009 to 2013 and a sudden decrease toward 2015. Furthermore, all RCPs follow the same trend as seen in the observations. The year 2013 indicates a maximum value for mean wind

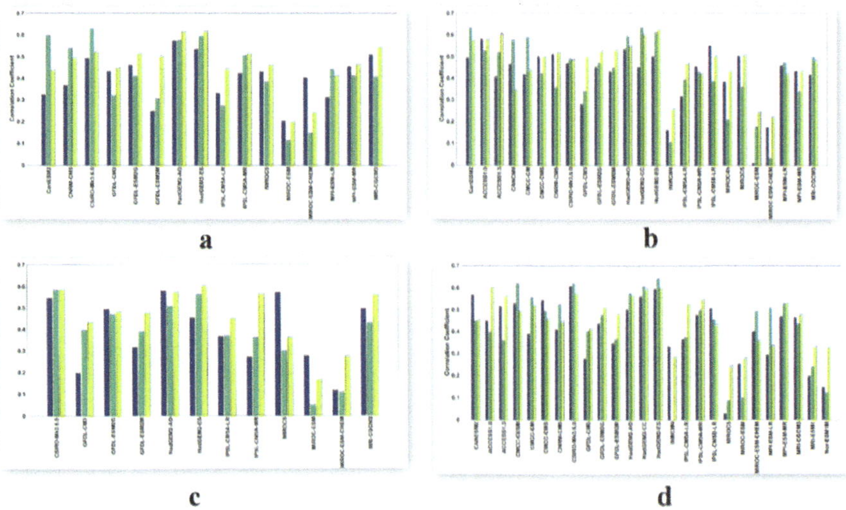

Fig. 24.2 Histograms representing the values of correlation coefficient of GCMs with RAMA buoy data for three locations (blue, green, and yellow) RAMA1, RAMA2, and RAMA3, respectively, for the four RCPs **a** RCP 2.6, **b** RCP 4.5, **c** RCP 6.0, and **d** RCP 8.5

speed among all the years resulting from the formation of eight cyclonic events (sum of depression, cyclonic storm, and severe cyclonic storm) that occurred in the Bay of Bengal during 2013, which is the highest in number considered during the period of study (Cyclone E-Atlas, IMD).

Figures show the comparison of wind speed for CMIP5 models and RAMA buoy observations in the Bay of Bengal. A maximum correlation value of ~0.65 was seen from this study. Considering different locations in the study region that represents the period from 2009 to 2015, the correlation values varied considering different models and various RCPs. The year 2009 showed a maximum correlation for the model CSIRO-Mk3.6.0 (CC~0.60) corresponding to the scenario RCP 8.5. For 2011 and 2012, the correlation value ~0.64 is typical for both RCPs 4.5 and 8.5. For the period 2013–2015, RCP 4.5 has a lead correlation of 0.63 against the RAMA buoy observation. Models from HadGEM2 family showed relatively high correlation values in all RCPs, with no much difference between each RCP. From data analysis, it is evident that MIROC-ESM and MIROC-ESM-CHEM are the least correlated models common in four RCPs. Therefore, these models are not suitable for further analysis in the Bay of Bengal. The study also deciphers the fact that by considering multiple years of data, the HadGEM2-ES, HadGEM2-AO, and ACCESS-1.0 are the models that exhibited maximum correlation irrespective of the individual RCP. Figures 24.3 and 24.4 show the error statistics of wind speed data represented as bias error, and root mean square error. Bias calculation helps to identify the under-estimation or over-estimation in model data compared with the measurements. The

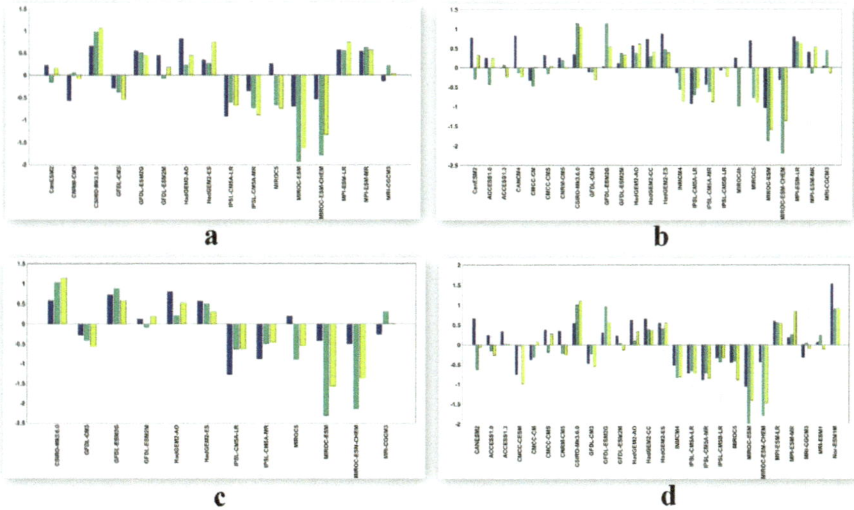

Fig. 24.3 Representation of bias error for each GCMs with RAMA buoy for three locations and four RCPs **a** RCP 2.6, **b** RCP 4.5, **c** RCP 6.0, and **d** RCP 8.5

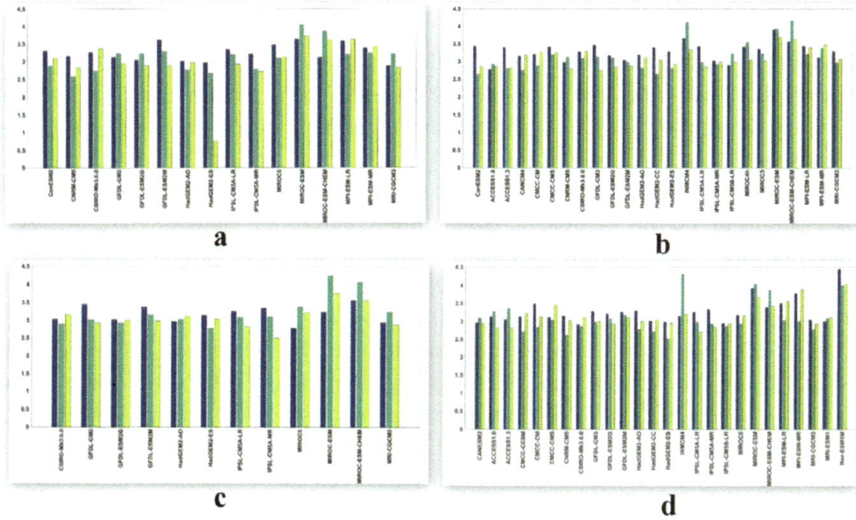

Fig. 24.4 Representation of root mean square error (RMSE) for each GCMs with RAMA buoy for three locations and four RCPs **a** RCP 2.6, **b** RCP 4.5, **c** RCP 6.0, and **d** RCP 8.5

RMSE measures the residuals between the modeled values and observations, where larger numbers indicate a higher variance in the data.

Calculated bias error values from the CMIP5 model predicted wind speed and the one obtained from RAMA buoy explains about the different performance of models in the study domain. Among the total 35 models representing four RCPs, we could find models that overestimate as well as underestimate the buoy observations. Positive values in the diagram indicate an over-estimation by models against RAMA buoy and negative values indicate an under-estimation of wind speed simulated by the climate models. In either case, the bias error highlights the deviation of model predicted data from actual observations.

The overall bias estimate for the period of study in the comparison of wind speed ranges between −2.5 and 2 m/s. Maximum over-estimation of wind speed is observed in models under RCP 8.5, which is believed to be the extreme emission scenario with 8.5 W/m^2 by the end of the year 2100 (van Vuuren et al. 2011). The other three RCPs (2.6, 4.5, and 6.0) performed almost similar with no much difference in bias value for climate models under different RCPs. The models that exhibited higher correlation with observations are those with the least bias. CSIRO-Mk3.6.0, HadGEM2-AO, and HadGEM2-ES are some of the models with least bias error and also overestimate the observations in the order of 0.3–1 m/s. IPSL-CM5A-LR and IPSL-CM5A-MR represent least bias and under-estimates the buoy observations up to −1 m/s. From the overall analysis of wind speed and the bias error between GCMs and RAMA buoy observations, it is evident that RCP 4.5 showed a lead with less bias among the models ACCESS 1.3, HadGEM2-AO, and CNRM-CM5. Besides, the models MIROC-ESM and MIROC-ESM-CHEM exhibited higher bias error for wind speeds in the Bay of Bengal. The root mean square error distribution of various GCMs implies there is no much variation among different models. Similar to the values of bias error, the estimated RMSE shows RCP 8.5 deviated the maximum as compared with observation. Wind speed simulated by the global climate model NorESM1M has maximum RMSE of 4.5 m/s in the study domain. HadGEM2-ES, HadGEM2-AO, and CNRM-CM5 showed less error with values less than 3 m/s compared with observations and irrespective of the emission scenarios.

The study also examined to understand the temporal variability of near-surface wind speeds from climate models and in situ observations using the 90th percentile of wind speed. Young et al. (2011) utilized 23 years of satellite altimeter data to calculate the wind speed and surface gravity significant wave height. There was an increase in mean wind speed, but more pronounced increases were seen in the 90th and 99th percentiles. In the present analysis, the upper 10% (90th percentile) of wind speed from the time series data obtained from RAMA buoy were 10.27, 9.5, 9.94, 10.16, 9.61, and 10.16 m/s for the years 2009, 2011, 2012, 2013, 2014, and 2015 respectively. Figure 24.5 illustrates the 90th percentile wind speed for the CMIP5 models from the four RCPs. The RAMA buoy observation showed an alternate increase and decrease of the 90th percentile wind speed during the period from 2009 to 2015. The study also investigated whether the GCMs followed the observed trend of 90th percentile in wind speed as seen in the RAMA buoy observation. For this period of study, few models were able to replicate the values of 90th percentiles of

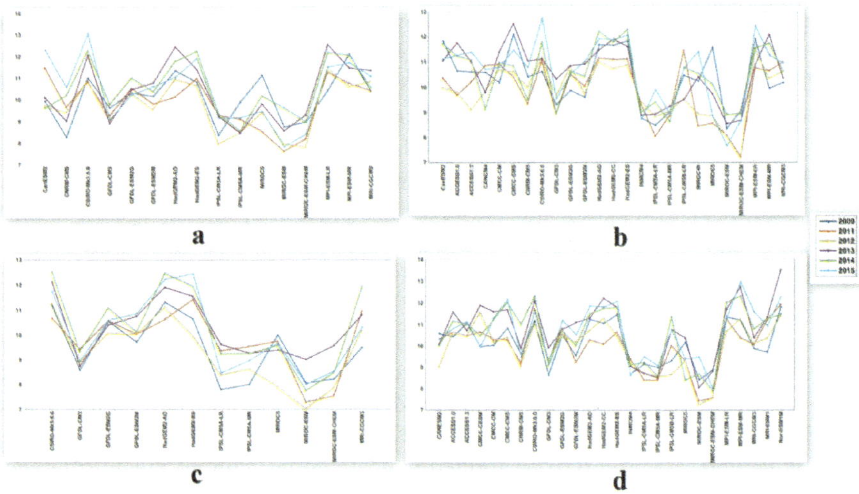

Fig. 24.5 The 90th percentile wind speed distribution by various GCMs under four RCPs **a** RCP 2.6, **b** RCP 4.5, **c** RCP 6.0, and **d** RCP 8.5. (The multiple line segments in figure represent the respective year of analysis)

wind speed similar to the RAMA buoy. It is noted that few models (CNRM-CM5, MRI-CGCM3, CMCC-CM, and MIROC4h) showed a close match compared with observations. It is seen for the RCP 4.5 scenario during the year of 2009. For the same period, RCP 6.0 and RCP 8.5 showed comparatively high values. The increase in 90th percentile wind speed observed in RAMA buoy data from 2011 to 2012 is well represented by ACCESS-1.0 and INMCM4 models that belong to RCP 4.5. For the period of 2013–2015, the RCP 4.5 showed much closer wind speed percentiles against buoy data as indicated by the models GFDL-CM3, MIROC4h, MRI-CGCM3, CNRM-CM5, and IPSL-CM5A-LR. Mean wind speed trend is consistent with the annual mean wind speed percentiles. The study brings to light that the global climate models could represent well the 90th percentile wind speed with a better accuracy. Annual wind speed trends could be attributed to processes that are internal to the climate system.

24.5 Conclusions

The performances of GCMs that participated in the Coupled Model Inter-comparison Phase 5 (CMIP5) were evaluated by comparing the model simulated near-surface wind speed with in-situ observations from RAMA buoy over the Bay of Bengal domain. A fair agreement for wind speed comparisons with a correlation of 0.65, bias error of 0.2 m/s and RMSE of 2.5 m/s was observed in the study domain. Based on statistical analysis and inferences, the best-performing GCMs identified were HadGEM2-ES, HadGEM2-AO, CNRM-CM5, ACCESS 1.0, ACCESS-1.3, and CNRM-CM5. However, the identification of a suitable RCP in the study domain was difficult as all the RCPs almost performed in similar way, with an exception in the extreme value for bias and RMSE found in RCP 8.5. On the other hand, the 90th percentile wind speed from few of the GCMs is found in close proximity with observations. Models that represented well the 90th percentile in wind speed are CNRM-CM5, MRI-CGCM3, and MIROC4h. The discrepancy in surface wind speed between models and observations may arise due to differences in the parameterization schemes used by different models. This is evident by the disparity in surface wind trend as represented by various models used in this study. Thus, there remains considerable uncertainty in the projection of surface wind speeds from CMIP5 models. Although the CMIP5 GCMs are not able to represent the contemporary wind speed climatology satisfactorily, this study can be used as a first-order assessment of near-surface wind speed for the Bay of Bengal region.

Acknowledgements The authors sincerely thank the Department of Science and Technology (DST), Government of India for financial support. This study was conducted under the Centre of Excellence (CoE) in Climate Change studies established at IIT Kharagpur funded by DST, Government of India. The study forms a section of the project "Wind-Waves and Extreme Water Level Climate Projections for East Coast of India" under the CoE in Climate Change at IIT Kharagpur.

References

Alinejhad-Tabrizi T, Hadjizadeh-Zaker N, Kamranzad B (2017) Assessment of CORDEX wind field in the Persian Gulf. In: First international conference on oceanography for West Asia 30–31 October 2017

Cowtan K, Hausfather Z, Hawkins E, Jacobs P, Mann ME, Miller SK, Steinman BA, Stolpe MB, Way RG (2015) Robust comparison of climate models with observations using blended land air and ocean sea surface temperatures. Geophys Res Lett. https://doi.org/10.1002/2015GL064888

Debernard J, Sætra O, Røed LP (2002) Future wind, wave, and storm surge the climate in the northern North Atlantic. Climate Res 23:39–49

Fitch AC, Olson JB, Lundquist JK (2013) Parameterization of wind farms in climate models. Am Meteorol Soc. https://doi.org/10.1175/JCLI-D-12-00376.1

Indira Rani S, Das Gupta M (2013) Oceansat-2 and RAMA buoy winds: a comparison. J Earth Syst Sci 122(6):1571–1582

Kulkarni S, Deo MC, Ghosh S (2013) Impact of climate change on local wind conditions. In: Proceedings Of Hydro 2013 international, 4–6 Dec 2013. IIT Madras, India. https://www.researchgate.net/publication/259865902

Lin M, Huybers P (2016) Revisiting whether recent surface temperature trends agree with the CMIP5 ensemble. Am Meteorol Soc. https://doi.org/10.1175/JCLI-D-16-0123.1

Manwell JF, McGowan JG, Rogers AL (2009) WIND ENERGY EXPLAINED Theory, Design and Application Second Edition. Wiley

Reddy BN, Venkatesan R, Osuri KK, Mathew S, Kadiyam J, Joseph KJ (2018) Comparison of AMSR-2 wind speed and sea surface temperature with moored buoy observations over the Northern Indian Ocean. J Earth Syst Sci 127:14. https://doi.org/10.1007/s12040-017-0902-3

Sudha AK, Prasada Rao CVK (2013) Comparison of Oceansat-2 scatterometer winds with buoy observations over the Indian Ocean and the Pacific Ocean. Remote Sens Lett. https://doi.org/10.1080/2150704X.2012.713140

Taylor KE, Stouffer RJ, Meehl GA (2012) An Overview of CMIP5 and the experiment design. Am Meteorol Soc. https://doi.org/10.1175/BAMS-D-11-00094.1

van Vuuren DP et al (2011) The representative concentration pathways: an overview. Climatic Change 109:5–31https://doi.org/10.1007/s10584-011-0148-z

Young IR, Zieger S, Babanin AV (2011) Global trends in wind speed and wave height. Science. New Series 332(6028):451–455. https://www.jstor.org/stable/29784125

Chapter 25
Trend Analysis of Temperature for Eastern Ganga Canal Command

Radha Krishan, Bhaskar R. Nikam, and Deepak Khare

Abstract Temperature and its analysis play a significant role in planning any irrigation project. However, expeditious climate change and shift have evolved with such footings, which were implausible earlier. On this respect, in the present study, analysis of temperature trend has been carried out over the command area of Eastern Ganga Canal (EGC) project for pre-climate shift period and post-climate shift period. Non-parametric Mann–Kendall and Sen's methods have been applied to study the trends in annual and seasonal temperature (*Kharif*, *Rabi* and *Summer*). Results showed significant decreasing trends in annual, *Kharif*, *Rabi* and *Summer* season temperature with −0.03 °C/year, −0.02 °C/year, −0.04 °C/year, and −0.02 °C/year respectively, during pre-climate shift period; whereas the significant increasing trend has been observed in all the four parameters during post-climate shift period with 0.02 °C/year. This rise in temperature is directly affecting the water requirement of crops, by rise in reference crop evapotranspiration (ET_0). The ET_0 showed decreasing trends for above-mentioned period during pre-climate shift whereas significant increasing trends have been observed during post-climate shift. These results will help project managers and farmers in understanding the climate shift and manage them to the development of alternative water management strategies.

Keywords Climate shift · Temperature trend · Mann–Kendall test · Sen slope estimator · Eastern Ganga Canal (EGC) command

25.1 Introduction

Sound scrutiny of water resource potential is very valuable for profitable management of water resources, for which study of temperature is the starting point. Data referring spatio-temporal variability in temperature are quite significant from the scientific

R. Krishan (✉) · D. Khare
Department of WRD&M, Indian Institute of Technology (IIT) Roorkee, Roorkee, India
e-mail: rkrishan@wr.iitr.ac.in

B. R. Nikam
Water Resources Department, Indian Institute of Remote Sensing (IIRS-ISRO), Dehradun, India

R. Jha et al. (eds.), *Climate Change Impacts on Water Resources*, Water Science and Technology Library 98, https://doi.org/10.1007/978-3-030-64202-0_25

perspective (Basistha et al. 2009). Analysis has demonstrated that global warming has occurred at the pace of 0.74 ± 0.18 °C, over 1906–2005 (IPCC WG II 2007; Basistha et al. 2009). Air temperature is noticed as a positive signal for assessing the prominence of global climate (Jhajharia and Singh 2011). A rise of 0.8 °C/decade in the upper Danube basin, Europe, during summertime temperature was observed by Reiter et al. (2012). Vose et al. (2005) have reported that boost in the heat of the Earth's surface is because of surging minimal temperature than the maximal temperature. Sen and Balling (2005) observed rise in temperature in the Deccan Plateau region of India, however, the periodic temperature series observed in the country was not momentous except for Kashmir region. Arora et al. (2005), Kothawale and Rupa (2005), Dash et al. (2007), Pal and Al-Tabbaa (2011) and Mondal et al. (2015) have observed an increase in temperature over India and the subcontinent.

Intergovernmental Panel for Climate Change (IPCC WG II 2007) has described spatial, interseasonal and interannual instability in precipitation trends over Asia. It was observed that varying precipitation pattern has a remarkable brunt on water and agricultural sector of the Asia–Pacific region (Kundu et al. 2015). The decadal change in precipitation pattern has turned feeble since the late 1970s over East Asia (Zhou et al. 2009; Li et al. 2010; Kundu et al. 2015). Decrease in rainfall with increase in temperature and evapotranspiration have induced drying and added droughts in many regions. The tropics are getting extremely afflicted from the droughts (Mondal et al. 2015).

Rising global surface temperatures are expected to affect the rainfall due to variations in atmospheric circulation (Dore 2005). Nitta and Yamada (1989) observed that the tropical sea surface temperature (SST) in the Pacific and Indian Ocean has been increasing since the late 1970s. They have recorded 0.3–0.4 °C rise in average SST of tropical region during 1980s in comparison to 1970s. Folland et al. (1984) scrutinised inter-annual fluctuations of global SST and near-surface temperature of marine air for the period 1856–1981 and got various number of inter-decadal variations. Folland et al. (1986) have demonstrated that wet and dry periods in the Sahel region of Africa are linked to the SST anomalies on a global scale. Hsiung and Newell (1983) have analysed the global SST for the period 30 years (1949–1979) and obtained the El Ninõ mode as the important non-seasonal pattern. 'A pattern of basin wide SST anomalies involved with a transition to the positive phase of the Inter-decadal Pacific Oscillation (IPO), occurred in the mid-1970s with effects that extended globally' (Power et al. 1999; Mantua et al. 1997). This integral decadal capriciousness linked with the IPO existed till 1970s and must have forced climate shift in the 1970s from a negative to the positive phase of the IPO. People have viewed this change as natural phenomena caused by internally generated decadal variability of the Pacific climate system. However, during the mid-1970, there has been an increase in the global temperature due to changes in external forcing in the Pacific Ocean and other metrological parameters have also undergone changes. It was argued that the increase in the temperature is caused by the increase of greenhouse gases from the burning of fossil fuel along with external forcing. But later on, IPCC report in 2011 revealed that the change in climate in the twentieth century was because of the shift of climate in the mid of 1970s rather than the burning fossil fuels. Varying effects of this climate shift are observed globally (Power et al. 1999).

A very important cycle that draws the attention of the researchers is the deep study of the atmosphere and how it contributes to climate variability. This climate variability also affects the food chain and food web causing severe threats to various valuable species of plants and animals. In the recent years, there has been a lot of research going on related to temperature variability but at local scales (Singh 2013; Suhas and Goswami 2008; Trenberth et al. 2003; Konwar et al. 2012). India's climate is generally outlined by temperature. Therefore, change in temperature pattern between pre-climate and post-climate shift periods may incomparably affect the economy of the nation. The analysis of impact of climate shift on water resources is mandatory in all agricultural zones of India. Agriculture being the backbone of Indian economy, it is extremely important to study the distribution of climatic parameters that govern the output from the agriculture sector. Much of research has been carried out in context of climate change but very few works are done regarding climate shift. To analyse the impact of the climate shift on status and trends of climatic parameters, an attempt is being made in the present study, to quantify the long-term trend of air temperature for both pre-climate and post-climate shift periods across the agricultural dominated command area of Eastern Ganga Canal project. Kothyari and Singh (1996) observed rising annual maximum temperature over the Ganga basin. In this regards, the present study has been carried out to check the shifts in climate during pre-implementation and post-implementation of EGC command.

Incidentally, the Eastern Ganga Canal (EGC) project was operationalised during 1970s and the data used for designing the project belong to pre-climate shift era and actual implementation of this project belongs to post-climate shift period. This is the reason that makes EGC project an ideal location for analysing the impact of climate shift on climatic availability and demand for water. Keeping this in mind, an effort has been made in the present study to analyse and discuss the trends in annual and seasonal temperature during pre-climate shift and post-climate shift periods over Eastern Ganga Canal (EGC) command area, India.

25.2 Study Area and Data Used

Eastern Ganga Canal (EGC) system is a part of Ganga Canal system, which is one of the Asia's longest canal systems. EGC originates from the left bank of Bhimghorda Barrage at Haridwar, India. Command area of EGC project covers major part of two districts, i.e. Haridwar (in Uttarakhand) and Bijnor (in Uttar Pradesh), which extends from 29° to 30° N and from 78° to 78°45′ E. EGC project was implemented 1970 onwards to utilise the surplus water of River Ganga during monsoon period for the development of *Kharif* cultivation in the districts of Bijnor, Haridwar and a very small part of Moradabad. The project has a gross command area (GCA) of 3.01 lakh ha, among which 2.33 lakh ha is culturable command area (CCA) and ultimate irrigation potential (IP) of the project is around 1.05 lakh ha. In the period other than

Kharif season, the command area is mainly depended on precipitation occurred in the region, as the EGC project is designed to supply water only in *Kharif* season. The location map of the study area is shown in Fig. 25.1.

Daily gridded temperature data of 1° × 1° spatial resolution were collected from Indian Meteorological Department (IMD), Pune for the EGC command area. The daily temperature data (1951–2012) were extracted for the EGC command and surrounding area, which covers two grids of IMD data. The daily mean temperature data were converted into seasonal (*Kharif*, *Rabi* and Summer) and annual scale in two time periods as discussed above, however, the time span of pre-climate and post-climate shift periods is modified to 1951–1976 and 1977–2012, respectively due to non-availability of gridded temperature data prior to 1951 and after 2012. The conversion of daily temperature data into seasonal temperature data is done by using the weighted temporal average of 4 months i.e. July to October for *Kharif* season; November to March for *Rabi* season, and April to June for summer season.

25.3 Analysis of Trend

Attempts for the disclosure of significant trends in climatologic time series can be formulated as parametric and non-parametric approaches. Parametric trend tests require data to be independent and normally distributed, while non-parametric trend tests work with the assumption of data is sovereign (Gocic and Trajkovic 2013). In this study, two non-parametric techniques, e.g. Mann–Kendall (MK) test and Sen's slope estimator technique, were used to detect and quantify the trends in the annual temperature and seasonal temperature (*Kharif*, *Rabi* and Summer season). Before employing the MK test, all the data series were proven for serial correlation using Lag-1 autocorrelation (Anderson 1941; Duhan and Pandey 2013) at 1, 5, and 10% significance level to eliminate the effect of serial correlation. All the series were found to be serially absolute, hence, MK test was directly enforced to original data series to disclose the trend using the two-sided hypothesis at 5% significant level. The value of statistically significant trend was calculated using Sen's estimator. The statistical and qualitative tests/methods used in the present study to analyse the trend are briefly described below:

(i) Mann–Kendall Test

The Mann–Kendall test (Mann 1945; Kendall 1975) searches for a trend in a time series without specifying whether the trend is linear or non-linear. The trend test is applied to a time series x_i ranked from $i = 1, 2\ldots n-1$, and x_j ranked from $j = i + 1, 2, \ldots n$. Each data point x_i is used as a reference point and is compared with all other data points x_j such that

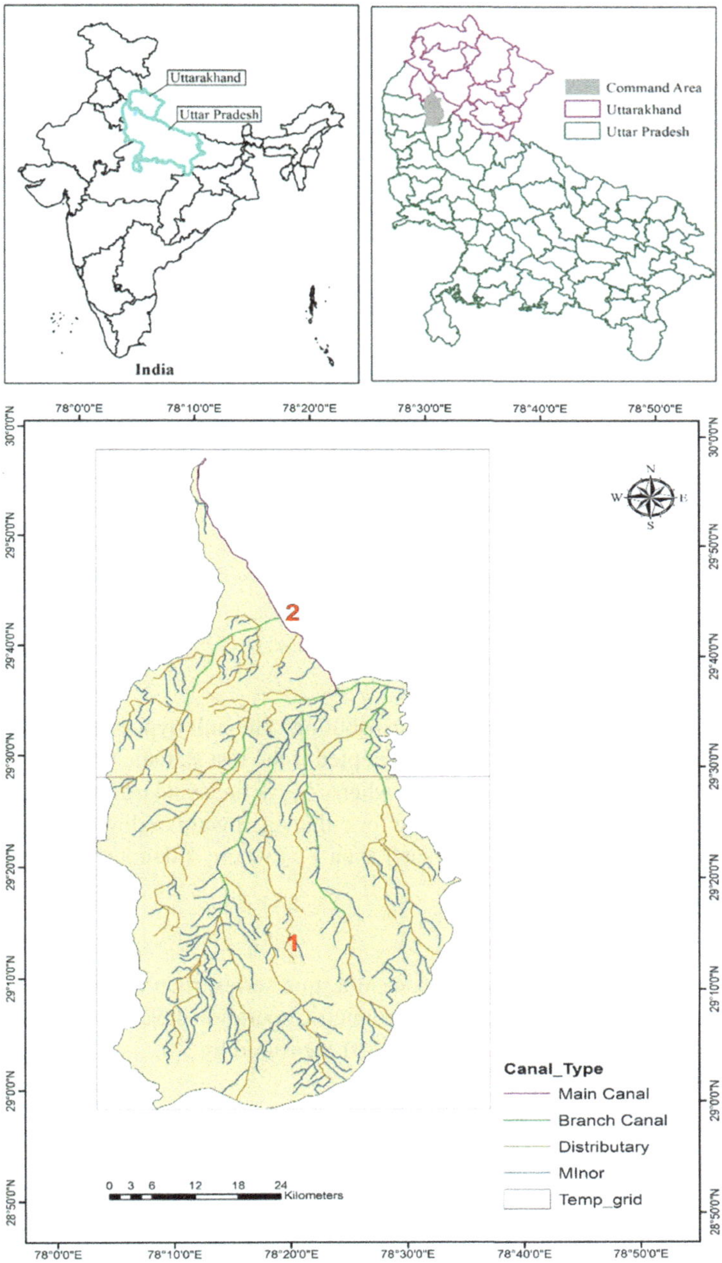

Fig. 25.1 Location map of the study area with the coverage of IMD rainfall data grids

$$\text{Sgn}(x_j - x_i) = \begin{cases} +1, > (x_j - x_i) \\ 0, = (x_j - x_i) \\ -1, < (x_j - x_i) \end{cases} \tag{25.1}$$

The Kendall statistics S is estimated as

$$S = \sum_{i=1}^{n-1} \sum_{j=i+1}^{n} \text{sgn}(x_j - x_i) \tag{25.2}$$

The variance of the statistic S is defined by

$$\text{Var}(S) = \frac{n(n-1)(2n+5) - \sum_{i=1}^{n} ti(i)(i-1)(2i+5)}{18} \tag{25.3}$$

where, t_i denotes the number of ties up to sample i.

The test statistics Z_c is estimated as

$$Z_c = \begin{cases} \frac{S-1}{\sqrt{\text{Var}(S)}}, S > 0 \\ 0, S = 0 \\ \frac{S+1}{\sqrt{\text{Var}(S)}}, S < 0 \end{cases} \tag{25.4}$$

where, Z_c follows a standard normal distribution. The null hypothesis, H_0, meaning that no significant trend is present, is accepted if the test statistic (Z_c) is not statistically significant, i.e. $-Z_{a/2} < Z < Z_{a/2}$, where $Z_{a/2}$ is the standard normal deviate. A significance level (α) is also used to test for either an upward or downward monotone trend (a two-tailed test). If Z_c is greater than $Z_{\alpha/2}$ where α denotes the significance level, then the trend is significant.

(ii) Sen's Slope Estimator Test

In cases where a linear trend is present in a time series, then the true slope can be predicted by employing a simple non-parametric operation developed by Sen (1968). The slope estimate of N pairs of data is first measured by

$$Q_i = \frac{x_j - x_k}{j - k} \text{ for } i = 1 \ldots . N \tag{25.5}$$

where, x_j and x_k are data values at times j and k ($j > k$), respectively. The meridian of these N values of Q_i is Sen's estimator of slope. If N is odd, then Sen's estimator is calculated by $Q_{\text{med}} = Q_{(N+1)/2}$ and if N is even, then Sen's estimator is calculated by $Q_{\text{med}} = [Q_{N/2} + Q_{(N+2)/2}]/2$ (Partal and Kahya 2006). Finally, Q_{med} is tested by a two-sided test at $100(1 - \alpha)\%$ confidence interval and true slope may be achieved by the non-parametric test.

25.4 Results and Discussion

The trend in temperature at various time scales has been studied for critical irrigation command area of EGC project with the objective of quantifying the change in trends of temperature during pre and post climate shift periods, which coincides with pre- and post-implementation periods of the project. Trends in annual mean temperature, seasonal mean temperature, annual ET_0 and Seasonal ET_0 for *Kharif*, *Rabi* and Summer seasons were estimated for both the time periods (e.g. pre-climate shift and post-climate shift) using MK test and Sen's slope estimator. For the qualitative analysis, the spatial pattern before and after the climate shift periods has been assessed for temperature time series. The results obtained in the present study are deliberated subsequently.

Temperature analysis

Analysis of trend in average of annual temperature and average of seasonal (*Kharif*, *Rabi* and summer) temperature has been carried out for 27 years data (1951–1976) for pre-climate shift period and 36 years data (1977–2012) for post-climate shift period over EGC command area. Trends are plotted for average annual and seasonal temperature for whole EGC command area in Fig. 25.2a–h.

Annual average temperature in the region varies from 22.07 to 23.66 °C during pre-climate shift period while it reduces to 21.83–23.64 °C during post-climate shift period. *Kharif* season temperature varies from 25.66–27.84 °C for pre-climate shift while 25.63–28.07 °C for post-climate shift period in the EGC command area. Pre- and post-climate shift period temperature in *Rabi* season varies from 15.40 °C to 17.58 °C and 15.29 °C to 17.74 °C, respectively. During summer season temperature varies from 27.11 °C to 30.39 °C during pre-shift and 26.56 °C to 30.43 °C in post-shift periods, respectively.

Trends result at 5% significance level for annual temperature, *Kharif* season temperature, *Rabi* season temperature and summer season temperature for pre- and post-climate shift within EGC command are shown in Fig. 25.2a–h, respectively. The results of trend analysis for the pre-climate shift period (1951–1976) showed that average annual temperature, average *Kharif* season temperature and average *Rabi* season temperature have significant decreasing trend at 95% confidence level with Sen's slope value of −0.03 °C, −0.02 °C and −0.04 °C respectively. Average summer season temperature shows decreasing trend with Sen's slope value of −0.02 °C, however, this trend appears to be statistically insignificant at 95% confidence level. On the contrary to the trends observed during pre-climate shift period, the average annual and seasonal (*Kharif* and *Rabi*) temperature during post-climate shift period (1977–2012), exhibits increasing trend of 0.02 °C in each case respectively, which is statistically significant at 95% confidence level as listed in Tables 25.1 and 25.2. The trend in summer temperature is increasing with Sen's slope value of 0.02 °C, however, here too this slope value is statistically insignificant at 95% confidence level, which may be due to high variability in summer temperature data.

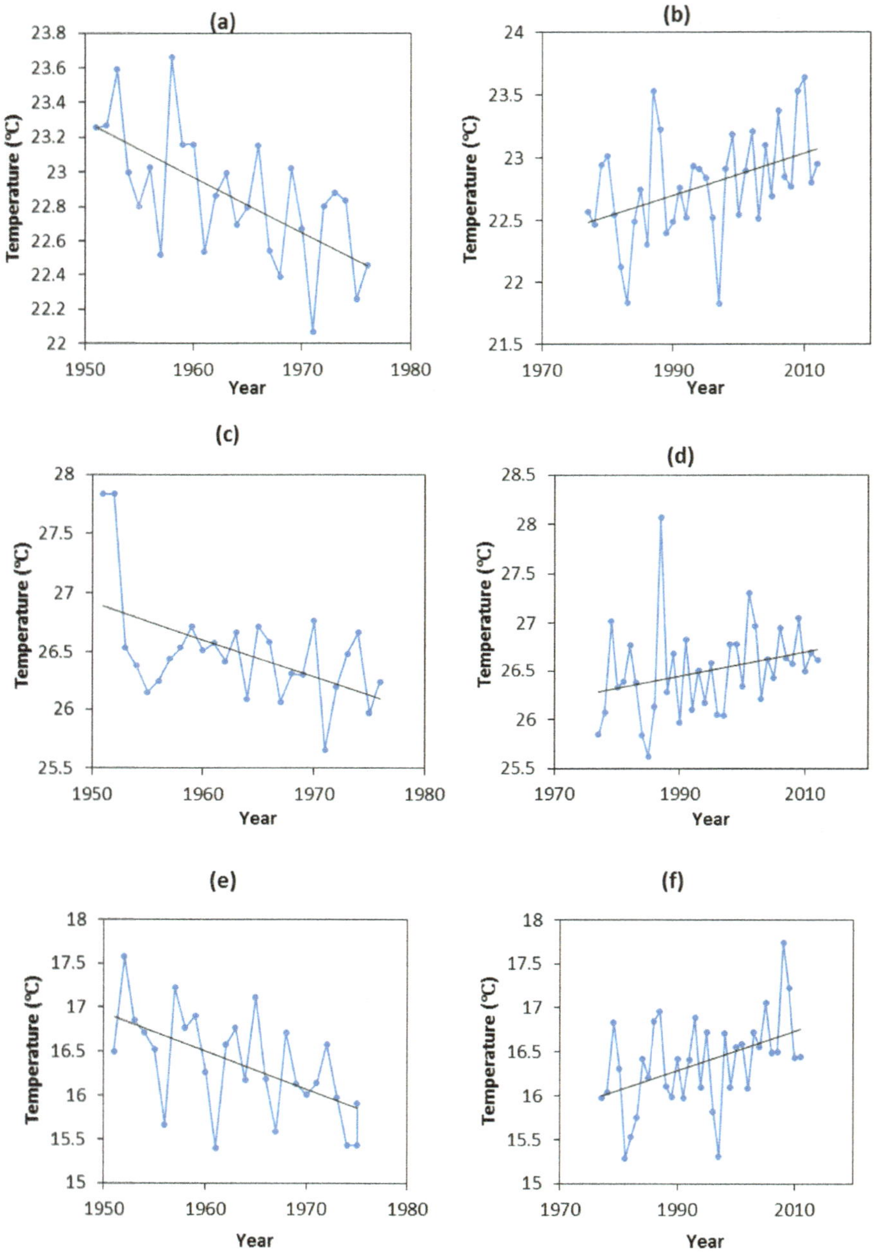

Fig. 25.2 Trend in pre- and post-climate shift period in annual temperature (**a** and **b**), *Kharif* season temperature (**c** and **d**), *Rabi* season temperature (**e** and **f**), summer season temperature (**g** and **h**) of EGC

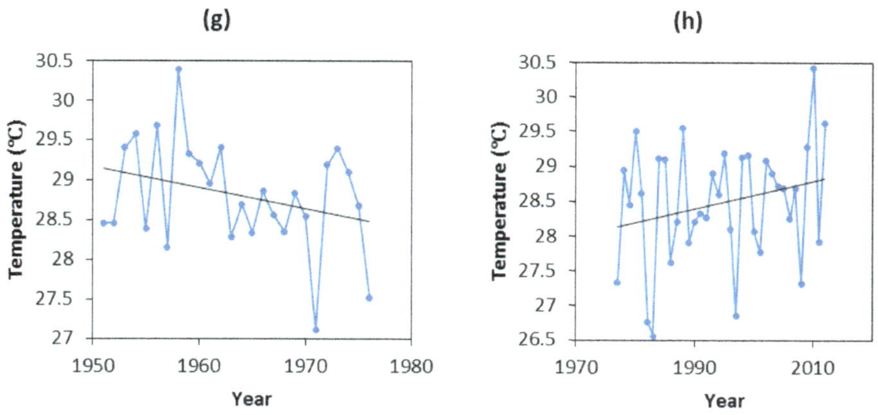

Fig. 25.2 (continued)

Table 25.1 Sen's estimator test values for pre- and post-climate shift for temperature

Shift	Average annual temperature	Average *Kharif* temperature	Average *Rabi* temperature	Average summer temperature
Pre-climate shift	−0.03	−0.02	−0.04	−0.02
Post-climate shift	0.02	0.02	0.02	0.02

Table 25.2 Trend result using Mann–Kendall and Sen's estimator test at 5% significance level for temperature

Shift	Average annual temperature	Average *Kharif* temperature	Average *Rabi* temperature	Average summer temperature
Pre-climate shift	Yes	Yes	Yes	No
Post-climate shift	Yes	Yes	Yes	No

ET_0 Analysis

From the above-mentioned tables, it is clear that there is decreasing trend for all the four parameters considered (i.e. annual temperature, *Kharif* season temperature, *Rabi* season temperature and summer season temperature) during pre-climate shift period while these parameters exhibit increasing trend during post-climate shift period; which may lead to increased evapotranspiration loss. So, trend for ET_0 is analysed for annual and seasonal ET_0.

Results showed that annual average ET_0 in the region varies from 235.90 to 290.95 mm during pre-climate shift period while it increases from 237.41 to 293.13 mm during post-climate shift period. *Kharif* season ET_0 varies from 103.98 to 132.89 mm for pre-climate shift while 106.46 to 138.24 mm for post-climate shift period in the EGC command area. Pre- and post-climate shift period ET_0 in *Rabi* season varies from 33.46 mm to 47.76 mm and 32.78 mm to 45.75 mm, respectively. During summer season, ET_0 varies from 91.53 mm to 120.92 mm during pre-shift and 89.06 mm to 133.15 mm in post-shift periods, respectively.

Trends results at 5% significance level for annual ET_0, *Kharif* season ET_0, *Rabi* season ET_0 and summer season ET_0 for pre- and post-climate shift within EGC command is carried out. The results of trend analysis for the pre-climate shift period (1951–1976) showed that average annual ET_0, average *Kharif* season ET_0 and average *Rabi* season ET_0 have significant decreasing trend at 95% confidence level with Sen's slope value of −3.33, −2.03 and −1.83 respectively. Average summer season ET_0 shows decreasing trend with Sen's slope value of −1.83, however, this trend appears to be statistically insignificant at 95% confidence level. While significant increasing trends observed during post-climate shift period at 95% confidence level with Sen's slope value of 2.41, 2.67 and 2.22 for average annual ET_0, average *Kharif* season ET_0 and average *Rabi* season ET_0. Average summer season ET_0 for post-climate shift shows increasing trend with Sen's slope value of 0.18, however, this trend appears to be statistically insignificant at 95% confidence level which may be due to high variability in summer temperature data (Tables 25.3 and 25.4).

Increase in annual and seasonal evapotranspiration loss is leading in reducing water availability in the spatial or temporal domain in the command area. On one side, water availability is decreasing, simultaneously population is increasing, resulting in increased water demand leading to adverse cumulative effect. In such situation, the knowledge of annual and seasonal assured water availability (rainfall) and conjunctive use may help decision-makers in proper planning of optimum utilisation of the existing water resources. Proper operation of canal system should be done to get the maximum benefit of the canal water considering these inputs.

Table 25.3 Sen's estimator test values for pre- and post-climate shift for ET_0

Shift	Average annual ET_0	Average *Kharif* ET_0	Average *Rabi* ET_0	Average summer ET_0
Pre-climate shift	−3.33	−2.03	−1.83	−1.83
Post-climate shift	2.41	2.67	2.22	0.18

Table 25.4 Trend result using Mann–Kendall and Sen's estimator test at 5% significance level for ET_0

Shift	Average annual ET_0	Average *Kharif* ET_0	Average *Rabi* ET_0	Average summer ET_0
Pre-climate shift	Yes	Yes	Yes	Yes
Post-climate shift	Yes	Yes	Yes	No

25.5 Conclusions

The trend analysis for temperature showed opposite trends between pre-climate shift and post-climate shift periods for annual average temperature, average *Kharif* season temperature, average *Rabi* season temperature, and average summer season temperature. During pre-climate shift period (1951–1976), the temperature is showing significant decreasing trends at 95% confidence level in case of annual average temperature, average *Kharif* season temperature and average *Rabi* season temperature. Whereas, in post-climate shift period (1977–2012), there are significant increasing trends in these variables. In a case of average summer season temperature, same opposite trends are observed during pre- and post-climate shift periods, however, both the trends were found to the statistically insignificant at 95% confidence level.

A significant increasing trend in post-climate shift period (also known as post-implementation phase of EGC), poses questions on the irrigation distribution strategies. These irrigation distribution strategies have been designed during project design phase, which coincides with pre-climate shift phase. The data and trends (trends in data) used to design these irrigation water distribution strategies have now changed in post-implementation phase of the project (post-climate shift period), especially the trends. The increasing trend in annual and seasonal temperature with increased losses through evapotranspiration from the command area will reduce natural water availability for agriculture in the command area. The reduction in natural water availability has to be compensated by increased irrigation supply in the area. To plan the increase/change in irrigation supply spatially or temporally, the knowledge of existing and future natural water availability and losses is the prerequisite. Based on this study, various policies can be derived for the better management of water resource in the command area. Furthermore, there is a need of an unified governmental plan for change detection, impact assessment, adaptation and mitigation pertaining to changes in climatic parameters. This particular study highlights the climatic variable along with their trends were changed from pre- to post-climate shift periods with in the EGC command region, so for better implementation of long-term water resources management, proper planning should be incorporated to overcome the effect of the climate shift in the EGC command area.

References

Anderson RL (1941) Distribution of the serial correlation coefficients. Ann Math Stat 8(1):1–13

Arora M, Goel NK, Singh P (2005) Evaluation of temperature trends over India. Hydrol Sci J 50(1):81–93

Basistha A, Arya DS, Goel NK (2009) Analysis of historical changes in rainfall in the Indian Himalayas. Int J Climatol 29:555–572

Dash SK, Jenamani RK, Kalsi SR, Panda SK (2007) Some evidence of climate change in twentieth-century India. Clim Change 85:299–321

Dore MHI (2005) Climate change and changes in global precipitation patterns: what do we know. Environ Int 31:1167–1181

Duhan D, Pandey A (2013) Statistical analysis of long term spatial and temporal trends of precipitation during 1901–2012 at Madhya Pradesh, India. Atmos Res 122:136–149
Folland CK, Parker DE, Kates FE (1984) Worldwide marine temperature fluctuations 1856–1981. Nature 310:6710–7673
Folland CK, Palmer TN, Parker DE (1986) Sahel rainfall and worldwide sea temperature, 1902–85. Nature 320:602–607
Jhajharia D, Singh VP (2011) Trends in temperature, diurnal temperature range and sunshine duration in Northeast India. Int J Climatol 31:1353–1367
Gocic M, Trajkovic S (2013) Analysis of changes in meteorological variables using Mann–Kendall and Sen's slope estimator statistical tests in Serbia. Global Planet Change 100:172–182
Hsiung J, Newell RE (1983) The principal non-seasonal modes of variation of global sea surface temperature. J Phys Oceangr 13:1957–1967
IPCC WG II (2007) Climate Change 2007: impacts, adaptation and vulnerability, working group II contribution to the intergovernmental panel on climate change, fourth assessment report, summary for policymakers
IPCC (2011) IPCC special report on renewable energy sources and climate change mitigation. Prepared by working group III of the intergovernmental panel on climate change, Cambridge University Press, Cambridge, UK
Kendall MG (1975) Rank correlation methods. Griffin, London
Konwar M, Parekh A, Goswami BN (2012) Dynamics of east–west asymmetry of Indian summer monsoon rainfall trends in recent decades. Geophys Res Lett 39:L10708
Kothawale DR, Rupa KK (2005) On the recent changes in surface temperature trends over India. Geophys Res Lett 32:L18714
Kothyari UC, Singh VP (1996) Rainfall and temperature trends in India. Hydrol Process 10:357–372
Kundu S, Khare D, Mondal A, Mishra PK (2015) Analysis of spatial and temporal variation in rainfall trend of Madhya Pradesh, India (1901–2011). Environ Earth Sci 73(12):8197–8216
Li H, Dai A, Zhou TJ, Lu J (2010) Response of East Asian summer monsoon to historical SST and atmospheric forcing during 1950–2000. Clim Dyn 34:501–514
Mann HB (1945) Non-parametric tests against trend. Econometrica 13:245–259
Mantua NJ, Hare SR, Zhang Y, Wallace JM, Francis RC (1997) A pacific inter-decadal climate oscillation with impacts on salmon production. Bull Am Meteorol Soc 78:1069–1079
Mondal A, Khare D, Kundu S (2015) Spatial and temporal analysis of rainfall and temperature trend in India. Theor Appl Climatol 122:143–158
Nitta T, Yamada S (1989) Recent warming of tropical sea surface temperature and its relationship to the northern hemisphere circulation. Meteorol Soc Jpn 67(3):375–383
Pal I, Al-Tabbaa A (2011) Assessing seasonal precipitation trends in India using parametric and non-parametric statistical techniques. Theor Appl Climatol. 103:1–11
Partal T, Kahya E (2006) Trend analysis in Turkish precipitation data. Hydrol Process: Int J 20(9):2011–2026
Power S, Casey T, Folland C, Colman A, Mehta V (1999) Inter-decadal modulation of the impact of ENSO on Australia. Clim Dyn 15(5):319–324
Reiter A, Weidinger R, Mauser W (2012) Recent climate change at the upper Danube—a temporal and spatial analysis of temperature and precipitation time series. Clim Change 111:665–696
Sen RS, Balling RC Jr (2005) Analysis of trends in maximum and minimum temperature, diurnal temperature range, and cloud cover over India. Geophys Res Lett 32:L12702
Sen PK (1968) Estimates of the regression coefficient based on Kendall's tau. J Am Stat Assoc 63:1379–1389
Singh C (2013) Characteristics of monsoon breaks and intraseasonal oscillations over central India during the last half century. Atmos Res 128:120–128
Suhas E, Goswami BN (2008) Regime shift in Indian summer monsoon climatological intraseasonal oscillations. Geophys Res Lett 35
Trenberth K, Dai A, Rasmussen M, Parsons DB (2003) The changing character of precipitation. Bull Am Meteorol Soc 84:1205–1217

Vose RS, Easterling DR, Gleason B (2005) Maximum and minimum temperature trends for the globe: an update through 2004. Geophys Res Lett 32:L23822
Zhou TJ, Gong DY, Li J, Li B (2009) Detecting and understanding the multi-decadal variability of the East Asian Summer Monsoon-recent progress and state of affairs. Meteorol Z 18:455–467

Chapter 26
Analysis of Long-Term Rainfall Trends in Rajasthan, India

Darshan Mehta and S. M. Yadav

Abstract The climatic variability for an area is referred to the long-term change in precipitation, temperature, humidity, evaporation, wind speed and other meteorological parameters. In order to identify the change, quantification of environmental change is essential that has occurred already and will be further helpful to make forecast for future. This will result into a better awareness for natural disasters. The objective of the study is to examine the rainfall variability in east as well as in west Rajasthan state in India. This will give an understanding about rainfall trends or changes. In this study, trend analysis has been carried out on monthly, seasonal and annual scale for the 33 districts of the arid as well as semiarid east and west Rajasthan state, India. Mann–Kendall test and Sen's slope estimator (statistical trend analysis techniques) used to detect trends at the 5% significance level on time series data of the east and west Rajasthan state for the time period, 1871–2016. This test was applied to identify the change in magnitude and direction of existing trend over the time. Trend detection of rainfall using techniques over 146 years shows increasing trend in premonsoon, southwest monsoon and annual precipitation in west Rajasthan while in east Rajasthan, it shows increasing trend in premonsoon, postmonsoon and annual precipitation. For proper water resource management and its planning, the analysis of climatic variables like heavy rainfall, temperature and humidity is helpful in adverse climatic conditions.

Keywords Climate change · Precipitation · Rajasthan · Trend analysis · Water resources

D. Mehta (✉) · S. M. Yadav
Department of Civil Engineering, Sardar Vallabhbhai National Institute of Technology, Surat 395010, India
e-mail: darshanmehta2490@gmail.com

S. M. Yadav
e-mail: shivnam27@gmail.com

R. Jha et al. (eds.), *Climate Change Impacts on Water Resources*, Water Science and Technology Library 98, https://doi.org/10.1007/978-3-030-64202-0_26

26.1 Introduction

Water resource has come to be a major concern for any change and development including food manufacture, controlling of flood and effective management. Universal change in climate might affect rainfall trend, which affects water obtainability alongside the drought risk as well as floods increases. Panel of intergovernmental on climate change demonstrates that the temperature has expanded by 0.74 °C ± 0.18 °C throughout the most recent 100 years (1905–2005), also rainfall is probably going to raise by 0.2–0.3% every period over zones of land in twenty-first century (IPCC 2007). To detect environmental changes and its effects on the various parts, the impact of climate change on agriculture, increased water shortage, rapid melting of glaciers and decrease in stream flows should be properly justified. Rainfall is the most important variable among other climatic variables, which involves spatial and temporal water availability patterns. The trend detection of rainfall is a vast topic in the field of water resource management, which deserves vital and methodical attention, as it affects freshwater availability and the production of food (Jain et al. 2012). Trend magnitude of India among the most recent century is equivalent to the worldwide situation (Kumar et al. 2010). Severe hydrological parameter changes such as rainfall, humidity, temperature and stream-flow are impelling the stream regimes significantly (Wani et al. 2017). Hydrologic parameters such as rainfall are reliant on factors of atmospheric. It is important to set up the associations among climatic and hydrological factors, which gives valuable intuitions into the conceivable changes in the area of hydrology and furthermore can help in the issues related to water resources management (Basistha et al. 2009). Detection of trends of rainfall in various scales will give an improved understanding to issues related with flood inundations and scarcities of water for utilizing as for climatic conditions. Temperature of air demonstrates a decent sign of the climatic condition as a result of its capacity to characterize the process in exchange of energy above the surface of world with sensible precision (Roy et al. 2015). Assessment of rainfall and analysis of annual maximum daily precipitation would upgrade the management of water resource applications as well as the viable usage of water resources (Pingale et al. 2016). Probability and occurrence investigation of precipitation data enables us to evaluate the expected rainfall at various chances (Mondal et al. 2012). Several scientists contributed to the study of environmental change (Hajani et al. 2014) with rainfall data of long term. Various series data are studied and have demonstrated that there is either decreasing or increasing in trend, both if there should be an occurrence of temperature and precipitation. Human interfering is also prompting to environmental change with changing area usage from the effect of rural and water system practices (Kalnay et al. 2003). To detect the variations in any series of time, it is basically important to do quality check furthermore series length due to their substantial impact in the analysis. For detecting trends, wrong and missing data records may lead to unauthentic interpretations. High-quality and informatory data consideration is basic for trend analysis correlated to perception of changes in series of time. In this paper, various methodologies are utilized to check all the changes, commencement of trend and trend changes over the series of time (Sonali and kumar 2013). The precipitation dispersion in Rajasthan is most uneven and differs substantially after region to region. The proper understanding and efficient utilization of the natural resources particularly precipitation are therefore, of great concern for the upgrading and manageability of agriculture in rain nourished region. The objective of this study is to explore the changeability of the rainfall in east and west Rajasthan state, India. Uncertainties associated with rainfall pattern will offer an information for better water resource management, farming, generation of hydropower and activities related to the water in the study area.

26.2 Objective of Study

The objective of study is to carry out long-term trend analysis of rainfall on monthly, seasonal and annual scale for the 33 districts of the parched and semidry east and west Rajasthan state, India.

26.3 Study Area

Rajasthan is a state in the north western region of India covering a widespread area of 342,239 km^2 (10.4% of the country). According to census 2011, population is about 68 million and 75% of it lives in provincial zone and reliant on availability of precipitations for farming furthermore for other utilizes. Out of 33 districts, 4 can be clearly delineated in the state, west with desolate hills, north east with sandy plains, Aravali slopes extending from north to south in the center, and south eastern plateau. The Aravalli goes equivalent to the cloud bearing breezes and neglects to make some orographic precipitation throughout the months of monsoon in Rajasthan state. High temperature when tropic of cancer goes through the southern regions throughout the late spring (Qin et al. 2007). The arid and semiarid regions of the state experience high diurnal and seasonal differences in temperature. The temperature of summer is around 26 °C to 46 °C and 8 °C to 28 °C in winter. The annual rainfall totals to just 100 mm in the western parts and around 600 mm in the whole eastern parts. Variation in climate from semiarid to damp in the part of south east. Less as well as irregular rainfall describes, as one of the parched state in India, it gets around 574 mm normal yearly precipitation which diverges from 100 mm in Jaisalmer which is lowest to 550 mm in Ajmer and highest (1638 mm) in Mount Abu (Sirohi district) located on southeast districts.

Mean annual rainfall indicates extensive variation, as the outrageous western parts of Rajasthan, i.e. Jaisalmer region get precipitation under 100 mm (low) in compared with 900 mm (high) in the eastern parts of Jhalawar and Banswara (Pingale et al. 2014, 2016). East and west Rajasthan have mean yearly rainfall of about 64.9 cm and 66.7 cm. South-eastern region receives extreme rainfall in the state (Singh et al.

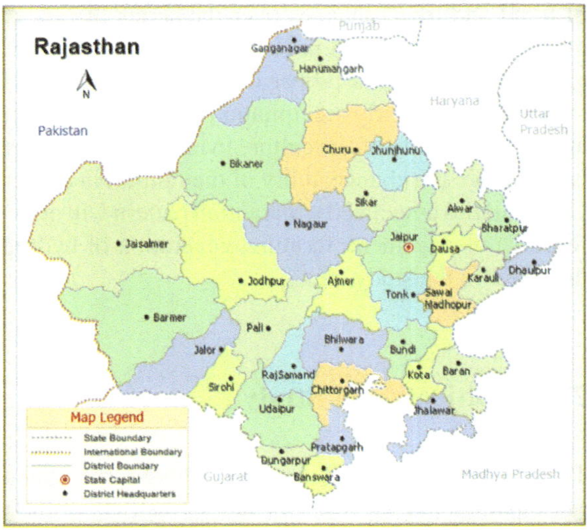

Fig. 26.1 Administrative Map of Rajasthan

2016). Whereas, Pali and Jalore areas of west part get highest volume of rain of about 50 cm and 43 cm. Annual rainfall of Bikaner, Ganganagar, Jaisalmer is about 26 cm, 24 cm and 17 cm respectively. The adjacent parts of these regions found the parched region. Figure 26.1 shows the administrative map of Rajasthan.

Based on the variations in climatic, soil and physiographic conditions, Rajasthan has additionally been classified into different agro climatic regions: arid western plain, irrigated north western plains, transitional plain of inland drainage, semiarid eastern plain, flood-prone eastern plains, subhumid southern plains and the Aravalli hills which govern the agro-pastoral livelihood of the people, the cropping pattern and water demand conditions.

26.4 Data Collection

Subdivisional monthly rainfall data of east and west Rajasthan were used in the study, which is prepared by the tropical meteorology of Indian Institute. To see the trends of rainfall in recent decades only 1871–2016, 146 years are considered in the study. Analysis was assessed on month wise, season wise and on annual basis. Seasons are considered as per classification of Indian Meteorological Department (IMD) as winter (January–February), pre-monsoon (March–May), southwest-monsoon (June–September) and post-monsoon (October–December).

26.5 Methodology

In the present study, rainfall trend analysis for 33 centers of east and west Rajasthan were aggregated to prepare month wise, season wise and on annual series of time. Statistical parameters like mean, standard deviation and coefficient of variation were computed for monthly, seasonal and annual time series of rainfall data. To analyze the trend of rainfall, rank-based nonparametric Mann–Kendall (MK) test and slope-based Sen's slope (SS) estimator were conducted on rainfall time series of 146 years using the XLSTAT 2016 software.

26.5.1 Nonparametric Mann–Kendall Test

For trend analysis statistical test, i.e. Mann–Kendall (M-K) test is commonly used for time series of meteorological and hydrological data. Two important benefits are using for this test. First, for distributing normally, it does not require any data. Second, the test has little affectability to sudden disruptions because of in homogeneous series of time. Various records described as "nondetects" are involved by assigning them a generalized significance which is less than the minimum estimated in the data series. From this test, if the hypothesis is null, i.e. H_0 then it shows "no trend." It is again tested with another hypothesis h_1, which shows that there is a trend. Mann–Kendall (MK) test statistics is calculated by the given below equations:

$$S = \sum_{i=1}^{n-1} \sum_{j=i+1}^{n} \text{sign}\left(x_j - X_i\right)$$

$$\text{Sign}\left(x_i - X_j\right) = \begin{cases} -1;\ \left(x_j - X_i\right) < 0 \\ 0;\ \left(x_j - X_i\right) = 00;\ \left(x_j - X_i\right) = 0 \\ 1;\ \left(x_j - X_i\right) > 0 \end{cases}$$

The MK tests adopt the two hypotheses; null hypothesis (h_0) and alternative hypothesis (h_a), in which null hypothesis assumes that there is no trend in time series or in other words the data values are independent and randomly ordered while alternative hypothesis assumes that there is a trend in time series. By performing trend analysis by XLSTAT 2016, the value of MK statistic is denoted by value p. Two-tailed test was performed at 95% level of confidence interval for rainfall series of time.

Performing the Mann–Kendall test by XLSTAT one more statistic Kendall's tau is obtained, which shows the correlation between the two variables in time series. Same as the Mann–Kendall test and Spearman rank correlation test Kendall's tau is also a rank-based correlation test. Values of Kendall's tau lies between the -1 and $+1$. In which positive correlation indicates that the ranks of two variables increase together while negative values of correlation imply that the rank of one variable increases and other one decreases.

26.5.2 Sen's Slope Estimator Test

Sen (1968), a method for robustly fitting a line to sample points in the plane, magnitude of the slope using this method can be obtained by using as follows:

$$\text{Sen's slope} = \text{med}\left[\frac{y_i - Y_j}{(i - j)}\right]; j < i$$

Y_i and y_j are the values at period of time i and j. Positive value shows increasing trend while negative value recommends a decreasing trend. If data points having total number in the series is n, then there will be $\frac{n(n-1)}{2}$ slope and the test statistics b_{sen} is the intermediate of all estimated slope. Value obtained positive as well as negative of statistics test shows increasing and decreasing trends.

26.6 Result and Discussion

26.6.1 Rainfall Statistics Characteristics

Statistical characteristics of rainfall over the east and west Rajasthan state for 146 years (1871–2016) were carried out monthly, seasonal and annual time series. Statistical parameters like mean, standard deviation, coefficient of variation, maximum and minimum values are described in Table 26.1 for east Rajasthan and Table 26.2 for west Rajasthan.

The statistical analysis of rainfall characteristics for the period of 146 years (1871–2016) indicated that the mean annual rainfall for east Rajasthan is 693.70 mm with a standard deviation of 168.7 mm and a coefficient of variation of 24.3% whereas for west Rajasthan is 298.9 mm with a standard deviation of 107.9 mm and a coefficient of variation of 36.1%. The high CV denotes that the variability of rainfall of that area is not equally distributed and the amount of rainfall is lowest.

Table 26.1 Rainfall statistics over east Rajasthan (1871–2016) for monthly, seasonal and annual rainfall

Month	Rainfall (mm) for east Rajasthan					
	Mean	SD	CV (%)	% contribution to annual	Maximum (mm)	Minimum (mm)
January	6.8	8.2	120.6	1.0	37.1	0.0
February	5.98	8.3	138.8	0.9	38.8	0.0
March	4.74	9.2	194.1	0.7	68.1	0.0
April	3.54	6.3	178.0	0.5	44.7	0.0
May	13.06	13.4	102.6	1.9	87.7	0.0
June	74.33	48.4	65.1	10.7	227.5	3.1
July	231.81	84.7	36.5	33.4	428.8	20.6
August	223.80	101.5	45.4	32.3	465.3	5.8
September	104.83	76.2	72.7	15.1	344.3	2.5
October	14.39	25.1	174.4	2.1	179.6	0.0
November	5.21	12.4	238.0	0.8	79.1	0.0
December	5.19	12.2	235.1	0.7	120.6	0.0
Annual	693.70	168.7	24.3	100.0	1336.8	275.7
Pre-monsoon	21.33	17.9	83.9	3.1	101	0.5
S-W monsoon	634.77	163.2	25.7	91.5	1118	194.2
Post-monsoon	24.79	30.3	122.2	3.6	185.2	0.0
Winter	12.80	12	93.8	1.8	56.8	0.0

26.6.2 Trend Analysis of Rainfall

26.6.2.1 Long-Term Rainfall Trends for East Rajasthan

To identify trend detection Mann–Kendall (nonparametric rank) two-tailed assessment and slope-based Sen's slope estimator (SS) were performed over Rajasthan state on rainfall time series. Test is carried out on monthly seasonal and annual time scale for 1871–2016 (146 years). Rainfall trend detection is carried out on month wise, season wise and on annual basis. Time series of rainfall plots on annual basis and season wise of east Rajasthan are presented in Figs. 26.2 and 26.3. Trend analysis for individual month was tested at 95% significance level (Table 26.3). Slope (s) positive value shows an increasing trend while negative value shows a decreasing trend. By performing trend analysis, results show that the month wise rainfall of June shows an insignificant decreasing trend ($p = 0.78$), while July month shows increasing trend ($p = 0.874$). Annual rainfall shows an insignificant increasing trend ($p = 0.895$). Trend analysis for seasonal rainfall shows insignificant decreasing trend for winter rainfall and southwest monsoon ($p = 0.465$ and 0.66) while for premonsoon and

Table 26.2 Rainfall statistics over west Rajasthan (1871–2016) for monthly, seasonal and annual rainfall

Month	Rainfall (mm) for west Rajasthan					
	Mean	SD	CV (%)	% contribution to annual	Maximum (mm)	Minimum (mm)
January	4.1	5.5	134.1	1.4	26.4	0.0
February	5.5	8.6	156.4	1.8	43.7	0.0
March	4.3	8	186.0	1.4	64.7	0.0
April	3.6	6.5	180.6	1.2	45.7	0.0
May	11.4	12.9	113.2	3.8	61.2	0.0
June	30.7	25.1	81.8	10.3	172.7	0.0
July	93.4	49.5	53.0	31.2	220.2	0.2
August	94.9	64.3	67.8	31.7	370.6	0.7
September	40.4	39.1	96.8	13.5	185	0.0
October	5.5	11.9	216.4	1.8	74.9	0.0
November	2.0	5.4	270.0	0.7	37.7	0.0
December	2.5	5.2	208.0	0.8	30.6	0.0
Annual	298.9	107.9	36.1	100.0	722.3	36.6
Pre-monsoon	19.4	18.2	93.8	6.5	123.7	0.1
S-W monsoon	259.7	98.3	37.9	86.9	572.8	33.4
Post-monsoon	10.1	14	138.6	3.4	81.1	0.0
Winter	9.6	10.1	105.2	3.2	43.7	0.0

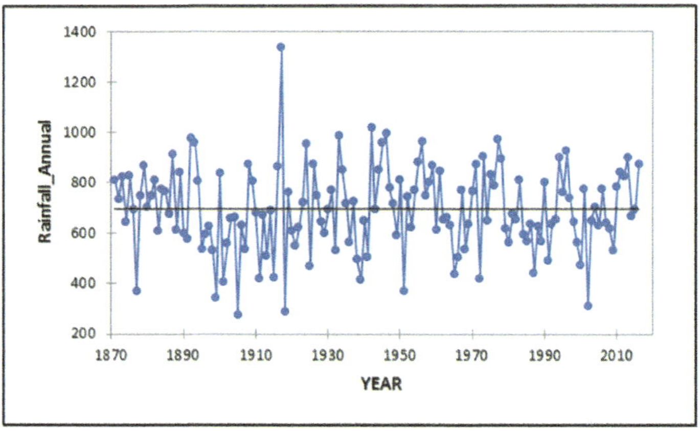

Fig. 26.2 Time series of annual rainfall (1871–72 to 2015–16) over East Rajasthan

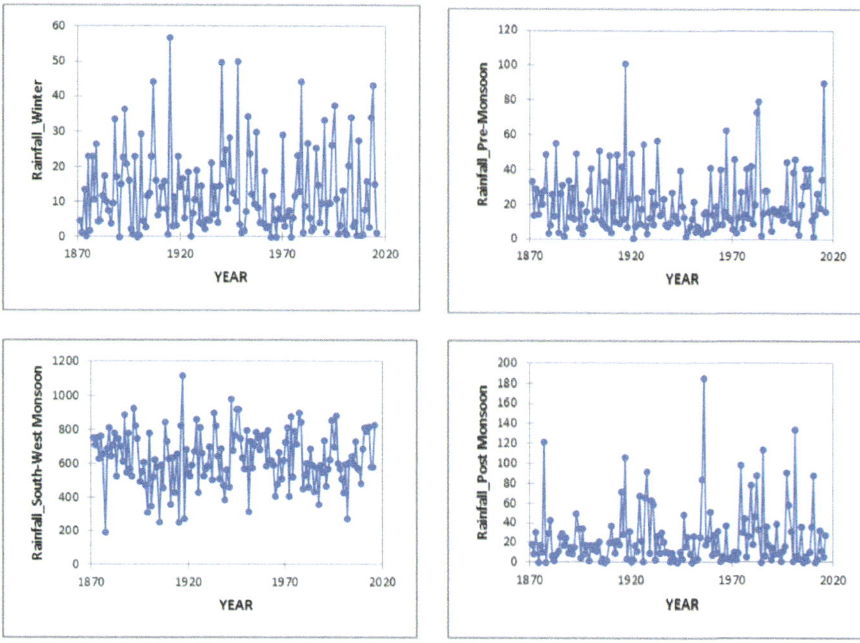

Fig. 26.3 Rainfall series of time during **a** winter **b** pre monsoon **c** southwest monsoon **d** post monsoon for period of 146 years (1871–72 to 2015–16) over East Rajasthan

postmonsoon, shows an insignificant increasing trend ($p = 0.586$ and $p = 0.832$) which is shown in Table 26.3 respectively.

Rejection of h_0, if value of p is less than the level of significance i.e. $A = 0.05$ and shows that there is a trend, while accepting h_0 shows no trend was identified. Rejection of null hypothesis results to be statistically significant. The analysis using Mann–Kendall test shows annual basis increasing trend of rainfall in east Rajasthan with Mann–Kendall tau coefficient shows rising (0.007) with Sen's slope (0.895) respectively. The decreasing trend was found in winter and south west monsoon with Mann–Kendall tau coefficient (−0.041) with Sen's slope (−0.011) and (−0.025) with Sen's slope (−0.151). Value of Sen's slope shows zero value defines no trend in the time series of the data value.

26.6.2.2 Long-Term Rainfall Trends for East Rajasthan

Trend analysis of rainfall time series is assessed on month wise, season wise and on annual basis. Time-series data plots on annual and seasonal scales of west Rajasthan are presented in Figs. 26.4 and 26.5. Trend analysis for individual month was carried out by Mann–Kendall two-tailed test and Sen's slope estimator at 95% significance level (Table 26.4). By performing trend analysis, results show that the monthly rainfall

Table 26.3 Results of the Mann–Kendall test for trends in rainfall data (1871–72 to 2015–16) over east Rajasthan

Rainfall period	Kendall's tau	Trend interpretation	P-value (Mann–Kendall test)	Sen's slope	Test interpretation
January	−0.053	Falling	0.345	−0.005	Insignificant decreasing trend
February	0.014	Rising	0.802	0	No trend
March	0.031	Rising	0.59	0	No trend
April	0.103	Rising	0.069	0.004	Insignificant increasing trend
May	−0.023	Falling	0.688	−0.006	Insignificant decreasing trend
June	−0.016	Falling	0.78	−0.026	Insignificant decreasing trend
July	0.009	Rising	0.874	0.022	Insignificant increasing trend
August	0.066	Rising	0.241	0.23	Insignificant increasing trend
September	−0.027	Falling	0.631	−0.053	Insignificant decreasing trend
October	0.039	Rising	0.496	0.004	Insignificant increasing trend
November	0.034	Rising	0.573	0	No trend
December	−0.103	Falling	0.077	0	No trend
Annual	0.007	Rising	0.895	0.038	Insignificant increasing trend
Winter	−0.041	Falling	0.465	−0.011	Insignificant decreasing trend
Premonsoon	0.031	Rising	0.586	0.012	Insignificant increasing trend
Southwest monsoon	−0.025	Falling	0.66	−0.151	Insignificant decreasing trend

(continued)

Table 26.3 (continued)

Rainfall period	Kendall's tau	Trend interpretation	P-value (Mann–Kendall test)	Sen's slope	Test interpretation
Postmonsoon	0.012	Rising	0.832	0.005	Insignificant increasing trend

Significant at 0.05 level
Note if $p < 0.05$ (statistically significant); $p > 0.05$ (statistically insignificant); (ii) Sen's slope = + ve increasing trend, Sen's slope = −ve decreasing trend, Sen's slope = 0 no trend

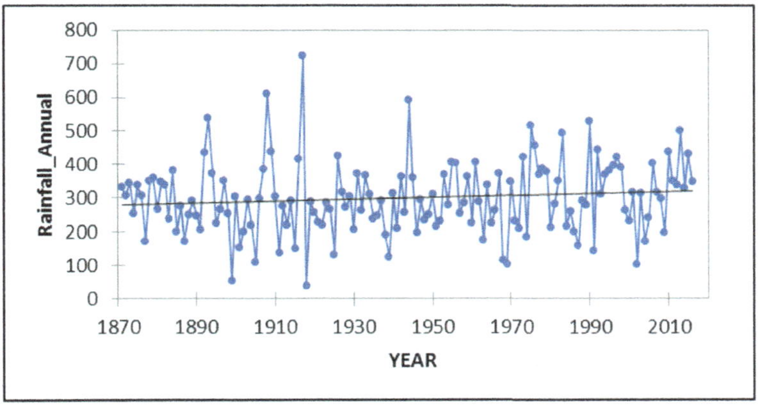

Fig. 26.4 Time series of annual rainfall (1871–72 to 2015–16) over West Rajasthan

of January indicates an insignificant decreasing trend ($p = 0.331$), while April month shows increasing trend ($p = 0.016$). Significant increasing trend obtained in annual rainfall ($p = 0.12$). Trend analysis for seasonal rainfall shows no trend for postmonsoon, while in winter rainfall shows an insignificant decreasing trending ($p = 0.359$) and for premonsoon and southwest monsoon, shows an insignificant increasing trend ($p = 0.139$ and $p = 0.282$) which is shown in Table 26.4 respectively.

Rejection of h_0, if value of p is less than the level of significance i.e. A = 0.05 which shows that there is a trend, while accepting h_0 shows no trend was identified. Rejection of null hypothesis results to be statistically significant. The analysis using Mann–Kendall test shows annual basis increasing trend of rainfall in west Rajasthan with Mann–Kendall tau coefficient shows rising (0.12) with Sen's slope (0.324) respectively. The decreasing trend was found in winter and postmonsoon with Mann–Kendall tau coefficient (−0.051) with Sen's slope (−0.011) and (−0.005) with Sen's slope (0).

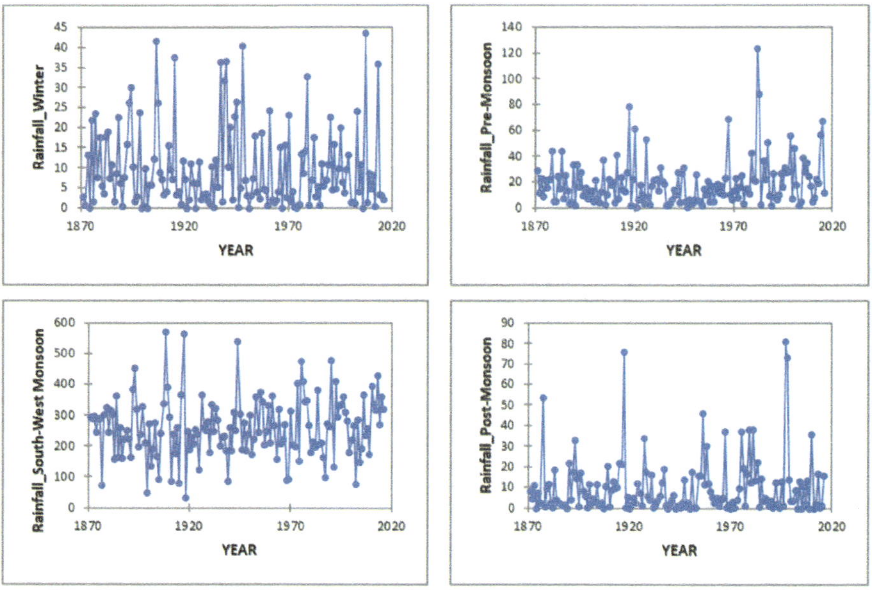

Fig. 26.5 Rainfall series of time during **a** winter **b** premonsoon **c** southwest monsoon and **d** postmonsoon of 146 years (1871–72 to 2015–16) over West Rajasthan

26.7 Summary and Conclusion

Trend analysis of rainfall during the winter, premonsoon season, and south-west monsoon season, postmonsoon and on annual basis was carried out for period of 1871–2016 over Rajasthan. Estimation of Sen's slope and percentage change in variables of climate were assessed for the period of 146 years. There is a variation of rainfall between dissimilar months with the difference ranging through decades. It reveals slight decreasing rainfall trend over Rajasthan. From the rainfall analysis of trend, it tends to be inferred that despite the fact that insignificant variation in rainfall pattern occurs over the couple of decades in many portions of the Rajasthan, there is suggestion of about changes in trend of rainfall in monsoon months. More investigation might be taken up for this area to examine the existing farming practices, patterns of land use, erosion as well as sedimentation levels and levels of reaction to existing hydrological system parameters in the area. Increasing rate of rainfall may help in improving productivity of farming and satisfy the demand of irrigation. At last, this will assist policy manufacturers and technologists to concentrate on district-wise planning with some methods for impact of climate change as well as environmental change variation and alleviation, by considering regional and local scale unpredictable trends as compared with worldwide climatic patterns.

Table 26.4 Results of the Mann–Kendall test for trends in rainfall data (1871–72 to 2015–16) over West Rajasthan

Rainfall period	Kendall's tau	Trend interpretation	Mann–Kendall test p-value (two tailed test)	Sen's slope	Test interpretation
January	−0.055	Falling	0.331	−0.002	Insignificant decreasing trend
February	0.025	Rising	0.664	0	No trend
March	0.099	Rising	0.086	0.003	Insignificant increasing trend
April	0.138	Rising	0.016	0.006	Significant increasing trend
May	−0.001	Falling	0.984	0	No trend
June	0.065	Rising	0.249	0.044	Insignificant increasing trend
July	0.068	Rising	0.222	0.131	Insignificant increasing trend
August	0.029	Rising	0.601	0.054	Insignificant increasing trend
September	−0.006	Falling	0.919	−0.005	Insignificant decreasing trend
October	0.05	Rising	0.388	0	No trend
November	−0.018	Falling	0.779	0	No trend
December	−0.102	Falling	0.088	0	No trend
Annual	0.087	Rising	0.12	0.324	Insignificant increasing trend
Winter	−0.051	Falling	0.359	−0.011	Insignificant decreasing trend
Premonsoon	0.083	Rising	0.139	0.034	Insignificant increasing trend
Southwest monsoon	0.06	Rising	0.282	0.212	Insignificant increasing trend

(continued)

Table 26.4 (continued)

Rainfall period	Kendall's tau	Trend interpretation	Mann–Kendall test p-value (two tailed test)	Sen's slope	Test interpretation
Postmonsoon	−0.005	Falling	0.931	0	No trend

Significant at 0.05 level
Note if $p < 0.05$ (statistically significant); $p > 0.05$ (statistically insignificant); (ii) Sen's slope = + ve increasing trend, Sen's slope = −ve decreasing trend, Sen's slope = 0 no trend

References

Basistha A, Arya DS, Goel NK (2009) Analysis of historical changes in rainfall in the indian himalayas. Int J Climatol 29(4):555–572

Hajani E, Rahman A, Haddad K (2014) Trend analysis for extreme rainfall events in new south wales, australia. Int J Environ Chem Ecol Geol Geophys Eng 8(12):834–839

Ipcc (2001) Climate change 2001: the scientific basis. Contribution of working group to the third assessment report of the intergovernmental panel on climate change. Vol. 1. Cambridge univ. Press. Isbn 0521014956 pp 8

Jain SK, Kumar V (2012) Trend analysis of rainfall and temperature data for india. Current Sci, 37–49

Kalnay E, Cai M (2003) Impact of urbanization and land-use change on climate. Nature 423:528–531

Kumar V, Jain SK, Singh Y (2010) Analysis of long-term rainfall trends in india. Hydrol Sci J–Journal des sciences hydrologiques 55(4):484–496

Mondal A, Kundu S, Mukhopadhyay A (2012) Rainfall trend analysis by mann-kendall test: a case study of north-eastern part of cuttack district, orissa. Int J Geol Earth Environ Sci 2(1):70–78

Pingale SM, Khare D, Jat MK, Adamowski J (2016) Trend analysis of climatic variables in an arid and semi-arid region of the ajmer district, rajasthan, india. J Water Land Develop 28(1):3–18

Pingale SM, Khare D, Jat MK, Adamowski J (2014) Spatial and temporal trends of mean and extreme rainfall and temperature for the 33 urban centers of the arid and semi-arid state of rajasthan, india. Atmosph Res 138:73–90

Qin D, Chen Z, Averyt KB, Miller HL, Solomon S, Manning M, Tignor M (2007) Ipcc, 2007: "summary for policymakers."

Roy AD (2015) Trend detection in temperature and rainfall over rajasthan during the last century. Asian J Res Soc Sci Humanities 5(2):12–26

Sen PK (1968) Estimates of the regression coefficient based on Kendall's tau. J Am Stat Assoc 63(324):1379–1389

Singh B (2016) Variability and trend analysis of rainfall data of jhalawar district of rajasthan, india. J Appl Natural Sci 8(1):116–121

Sonali P, Kumar DN (2013) Review of trend detection methods and their application to detect temperature changes in india. J Hydrol 476:212–227

Wani JM, Sarda VK, Jain SK (2017) Assessment of trends and variability of rainfall and temperature for the district of mandi in himachal pradesh, india. Slovak J Civil Eng 25(3):15–22

Chapter 27
Statistical Downscaling of GCM Output and Simulation of Rainfall Scenarios for Brahmani Basin

Lasyamayee Lopamudra Sahoo and Kanhu Charan Patra

Abstract The change in climate threatens the abundance of usable water across the globe. Most of the river basins are unable to cope up with the impact of climate change. Hence, assessing the future scenario has become the need of today. General Circulation Model (GCM) provides information at a course grid resolution. Downscaling can help in getting the information at a local scale level from GCM data, which helps the researchers to work on a regional level. Statistical downscaling method is preferred over dynamic downscaling method due to its less complex calculations. Statistical downscaling model (SDSM) is widely used in prediction of future climate scenarios. Here Brahmani–Baitarani river basin is selected as a case study for the downscaling of precipitation in the monthly time scale. SDSM version 4.2 is used as the model and precipitation is taken as the predictand. Predictors are chosen from the NCEP global variables like air temperature, geopotential height, specific humidity, zonal and meridional wind velocities, precipitable water and surface pressure data. The outcome of the study shows an increasing trend in the rainy season of the year. The mean rainfall increases significantly in 2040s than other epoches.

Keywords Climate change · GCMs · Statistical downscaling model (SDSM) · Brahmani–baitarani river system

27.1 Introduction

Climate change has adverse impact on the surface of earth starting from forest ecosystem to flood plain of rivers. Surface hydrology, forestry, floods, soil erosion, land use changes, ground water, environment, living beings and their ecosystems; all are affected by the climate change. Its adverse effect on water resources threatens the abundance of usable water availability. Water is the basis of lifeline on earth.

L. L. Sahoo (✉) · K. C. Patra
Department of Civil Engineering, NIT Rourkela, Rourkela 769008, Odisha, India
e-mail: lasyakunmun2009@gmail.com

K. C. Patra
e-mail: kcpatra@nitrkl.ac.in

R. Jha et al. (eds.), *Climate Change Impacts on Water Resources*, Water Science and Technology Library 98, https://doi.org/10.1007/978-3-030-64202-0_27

Population explosion associated with various anthropogenic activities like per capita use, industrialization and others require more water in coming decades. Increase in demand of water with population explosion but possible decrease in availability of usable water creates a critical situation for the water resources planners (Chiew et al. 2010).

Study by researchers indicates that "the central India has been found to be the most vulnerable to climate change. Parts of north western, north eastern and southern India appear to be resilient to cope with droughts while the rest of the country is non resilient" (Sharma et al. 2017). Therefore, a proper assessment of past and prediction of probable future precipitation and the resulting run off over time is necessary for hydrologist (Anandhi et al. 2008).

General circulation models (GCMs) are considered as the most effective tools to simulate climatic conditions on earth. They provide information at a coarse grid resolution (usually 1°–2°). But data at a finer grid are required to work on a smaller study area like a smaller catchment. To manage the gap between the lower resolution and higher resolution, downscaling is used. It tries to link between the GCM information and information needed by the hydrologists (Walsh 2011).

Downscaling methods can be broadly classified into two groups; dynamical and statistical. Statistical downscaling method is preferred to dynamic downscaling method because of less computational work. Again it is classified into three subgroups; regression methods, weather generators and weather typing schemes. All the methods deal with the basic concept that regional climates (predictand) are the function of the large scale atmospheric state (predictor). This relationship between predictor and predict and can be deterministic or probabilistic function.

SDSM (statistical downscaling model) is the most commonly used model for this purpose. SDSM combine uses a conceptual water balance model and a mass-balance water quality model to investigate climate change impact assessment (Wilby and Harris 2006). Many authors have compared SDSM with other statistical downscaling models. Harpham and Wilby (2005) concluded that SDSM yields better daily precipitation quantiles and intersite correlation when compared with artificial neural networks (ANNs). Khan et al (2006) also concluded that SDSM is very efficient in reproducing various statistical parameters of data set in the downscaled results with a confidence level of 95%. Rath et al. concluded that future trends in precipitation for annual and seasonal period from the SDSM indicates a decrease in precipitation pattern for the time period 2020s and 2080s while an increase in 2050s for both A2 and B2 scenarios (Rath et al. 2016).

The present work focuses on the application of SDSM to the Brahmani–Baitari river basin in India to simulate the future scenarios of the precipitation and other parameters.

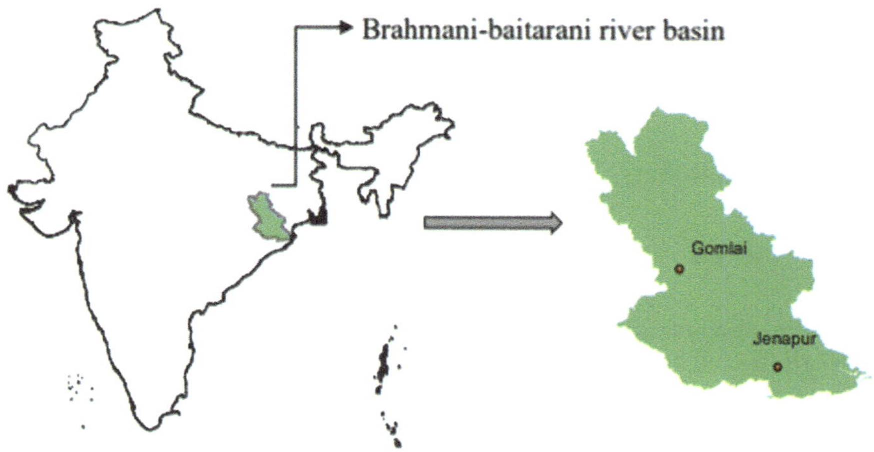

Fig. 27.1 Schematic diagram of Brahmani–Baitarani river basin

27.2 Study Area

Brahmani and Baitarani river basin is situated in the central-east India between latitude 20° 28′ to 23° 35′ N and longitude 83° 52′ to 87° 30′ E. The basin extends over the states of Odisha, Jharkhand and Chhattisgarh draining an area of 51, 822 Sq.km which is 1.7% of total geographical area of the country. Major part of its catchment area is situated in the state of Odisha. Both the rivers are seasonal in nature. They are rejuvenated at the onset of monsoon as they are fed by rain. At the time of summer, their discharge is significantly decreased. Though 90% of the basin receives an average annual rainfall of between 1400 and 1600 mm, some places like Dhenkanal and Jashpur district are drought prone areas. Two hydro-observation stations Jenapur and Gomlai are taken for this study. Jenapur station has a drainage area of 33,955 km^2 and Gomlai has a drainage area of 21,950 km^2. Figure 27.1 shows the two stations in the considered river basin.

27.3 Data

The observed large-scale predictors have been derived from the NCEP reanalysis data sets that contain 41 years of daily observed predictor data normalized over the period 1961–1990. These data have been interpolated into the grid size of 2.5 latitude *3.75 longitude before the normalization is implemented. The HadCM3 (Hadley center for climate and prediction and research, UK) model output both for A2 and B2 scenarios are directly downloaded from website http://climate-scenarios.canada.ca. The long-term meteorological data from the period 1981–2016 are obtained from the central water commission (CWC) India on daily basis at the Gomlai and Jenapur in the state of Odisha. A total 36 years of data are taken as baseline period, out of which 25 years are used for calibration and 11 years are needed for validation of the model.

27.4 Methodology

27.4.1 SDSM

Among the handful models that are available for downscaling, the SDSM is very popularly used. Statistical downscaling model (SDSM) is a statistical weather generator based on linear multiple regression. It is used to predict the climate parameters such as the precipitation or temperature in long time duration. It uses large-scale atmospheric variables to condition the local scale weather generators. It also uses stochastic techniques in variance of daily time series. In fact it is the combination of transfer function and stochastic weather generator methods.

27.4.2 Multiple Linear Regressions

Multiple linear regressions are used to explain the relationship between one continuous dependent variable (predictand) with one or more than one independent variables (GCM outputs). For a given data set, a linear regression model assumes a linear relationship between the variables. The equation used for the multiple linear regressions is written as:

$$(y/x) = a + b_1x_1 + b_2x_2 + \cdots b_nx_n. \tag{1}$$

where, y is the dependent predictant variable with respect to x, x_1, x_2, x_3.... x_n the independent predictor variables, a, b_1, b_2...b_n the intercepts or parameters of the equation. Figure 27.2 shows the flowchart of climate scenario generation in SDSM.

27.5 Results and Discussion

Any error in the observed data may result the model to fail in the prediction of future climate scenario. Hence, before the simulations, the observed data are subjected to quality control in order to check for any missing data codes or gross data errors. For selecting a set of predictors, it is necessary to access the effect of predictors on the precipitation at that particular station. A correlation matrix is generated among the large scale global predictors and user specified predictand. Predictors are selected based on the highest correlation value. Partial r, p values and scatter plots are also

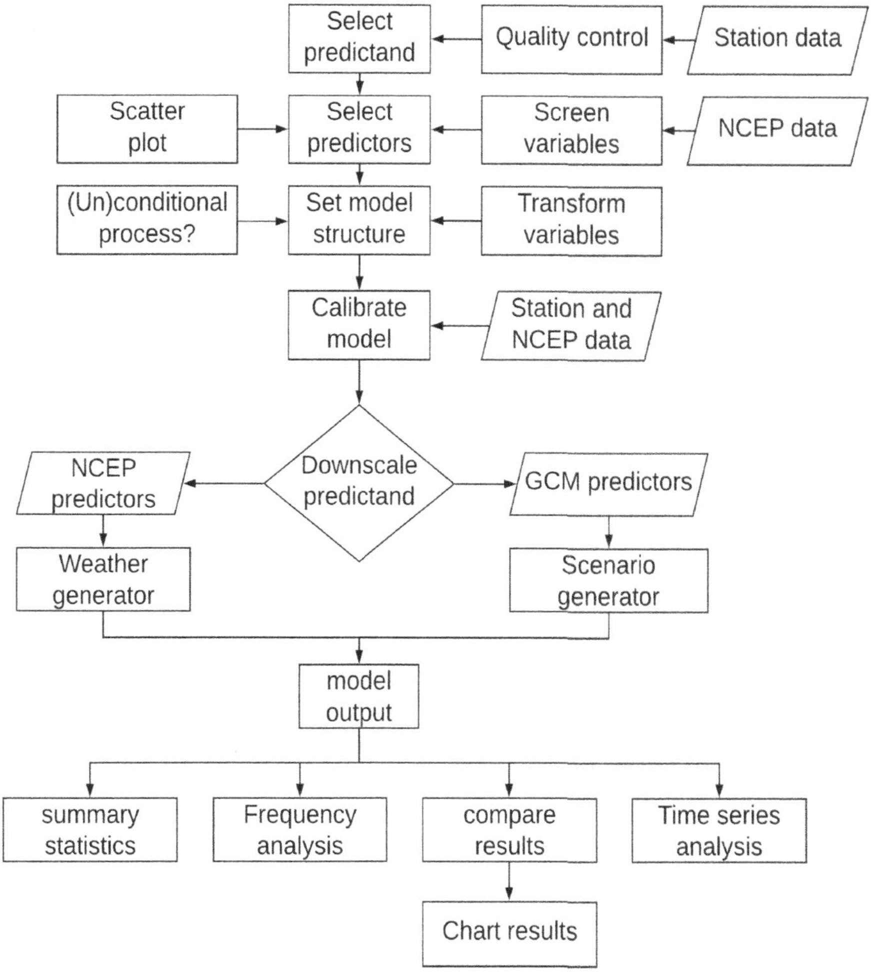

Fig. 27.2 Climate scenario generation in SDSM (*Source* SDSM user manual)

considered in screening of the predictors. Predictor variables selected for each station after the screening test are given in Table 27.1.

For calibration of the model, 25 years of precipitation data (collected from CWC), between 1980 and 2005, are considered. A multiple linear regression is established between the NCEP variables and precipitation at that particular station. The intercepts of the regression equation are calculated by the forced entry method. Synthetic daily weather series is created by weather generation using the observed predictors. When a calibrated model is selected, SDSM automatically relates all necessary predictors to regression model weights. In the present study, monthly time scale is taken for analysis and no conditional factors are added. An ensemble size of 20 is used for the analysis. Results of model calibration for Gomlai station are shown in Fig. 27.3.

Table 27.1 Predictors selected for each station and their correlation values

Station name	NCEP codes	Predictor variable	Correlation value	Partial r	P value
Jenapur	Shumas	Surface specific Humidity	0.265	0.056	0.0000
Jenapur	rhumas	Near surface relative humidity	0.229	0.011	0.2729
Jenapur	r850as	Relative humidity at 850hpa	0.212	– 0.015	0.1412
Jenapur	p5thas	500hpa wind direction	0.177	0.018	0.0703
Jenapur	p-uas	Surface zonal velocity	0.174	– 0.001	0.5632
Gomlai	shumas	Surface specific humidity	0.323	– 0.092	0.0000
Gomlai	r850as	Relative humidity at 850hpa	0.277	0.021	0.0274
Gomlai	rhumas	Near surface relative humidity	0.287	– 0.029	0.0023

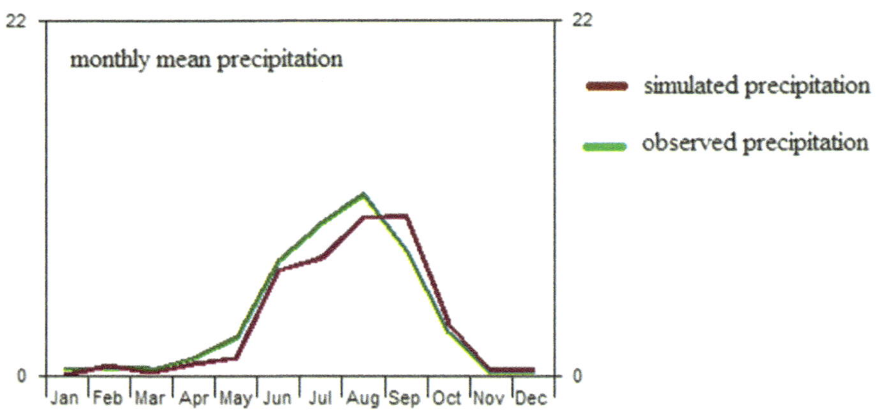

Fig. 27.3 Calibration of model

Remaining 11 years data (2006–2016) are used for validation of the model. The statistical plot between observed and simulated value is plotted to compare the model output. Figure 27.4 shows the plot for validation of the model.

It is quite clear from Fig. 27.4 that the model operates efficiently for calibration and validation. Hence future scenario is generated using this calibrated model. Scenario A2 experiment results show a very heterogeneous world with continuously increasing global population and regionally oriented economic growth that is more fragmented and slower than in other experiments; while that of B2 scenario experiment represents a world in which the emphasis is on local solutions to economic,

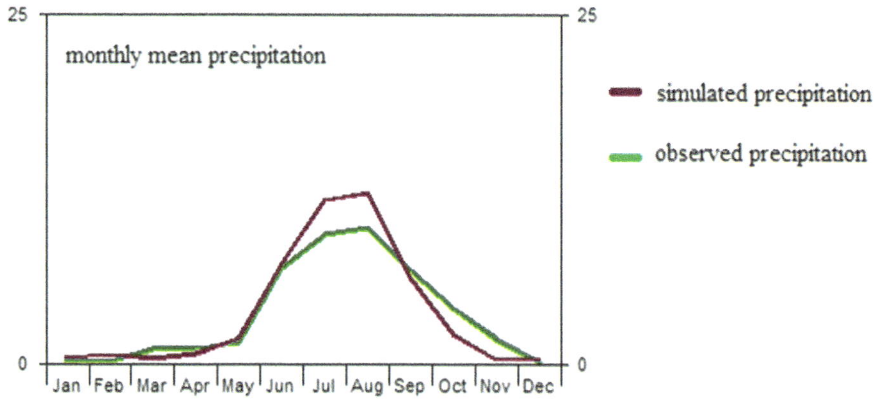

Fig. 27.4 Validation of the model

social, and environmental sustainability, with continuously increasing population (lower than A2) and intermediate economic development. Both the scenarios are considered for the present study.

Outcome of climate model HadCM3 is used for scenario generation for six decades (2030–2090) in two subgroups as 2030–2060 and 2060–2090. These GCM outputs are normalized to 360 days, 12 months with 30 days of each duration. Results are presented in Tables 27.2 and 27.3.

Table 27.2 Simulated future precipitation for A2 and B2 scenarios at Jenapur station

Months	Observed values (1981–2016)	Predicted values 2030-60-A2	Predicted values 2030-60 B2	Predicted values 2060-90-A2	Predicted values 2060-90 B2
January	15.218	2.465	2.431	2.416	2.440
February	16.126	4.894	4.864	4.907	4.858
March	11.978	4.211	4.293	4.213	4.266
April	12.219	6.787	6.733	6.798	6.736
May	10.165	12.441	12.455	12.475	12.524
June	12.459	19.477	19.548	19.249	19.361
July	13.828	20.513	20.361	20.447	20.561
August	15.225	22.139	21.972	22.093	21.966
September	15.670	18.176	17.777	17.901	18.042
October	15.742	16.768	16.925	16.901	16.963
November	16.787	6.376	6.432	6.401	6.323
December	14.468	1.809	1.809	1.812	1.807
Annual total	169.615	136.056	135.6	135.613	135.847

Table 27.3 Simulated future precipitation for A2 and B2 scenarios at Gomlai station

Months	Observed values (1981–2016)	Predicted values 2030-60-A2	Predicted values 2030-60 B2	Predicted values 2060-90-A2	Predicted values 2060-90 B2
January	19.742	2.969	2.980	2.966	2.933
February	15.15	3.220	3.159	3.187	3.143
March	10.96	1.843	1.855	1.871	1.844
April	11.88	3.045	3.043	3.041	3.033
May	9.29	6.065	6.1081	6.145	6.197
June	10.03	16.630	16.533	16.654	16.454
July	12.295	21.510	21.532	21.201	21.446
August	16.436	22.650	22.380	22.490	22.288
September	17.537	14.708	14.765	14.612	14.896
October	15.758	9.062	9.047	9.040	9.049
November	15.940	3.286	3.236	3.294	3.262
December	17.83	2.571	2.584	2.589	2.577
Annual total	172.848	107.559	107.221	107.093	110.384

Graphical representation of the mean monthly precipitation of baseline period (observed period) and forecasted period are shown in Fig. 27.5, (a) Jenapur station, (b) Gomlai station; which provides a comparative analysis of change in precipitation in future time period.

a) At Jenapur station

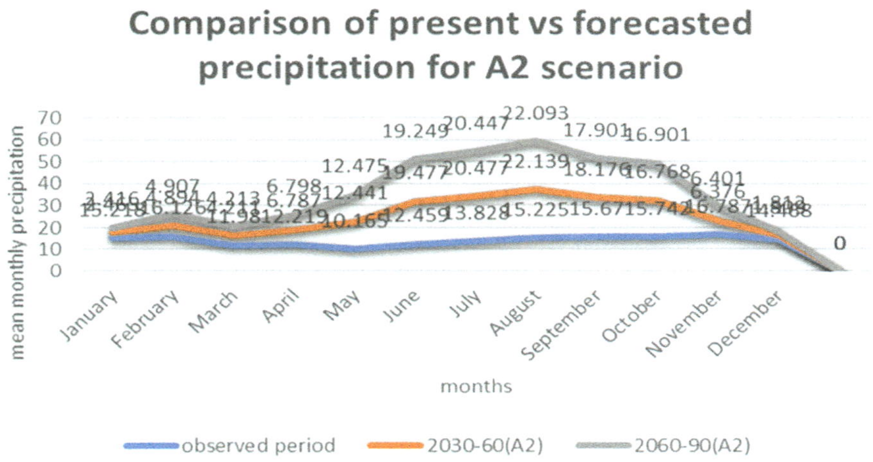

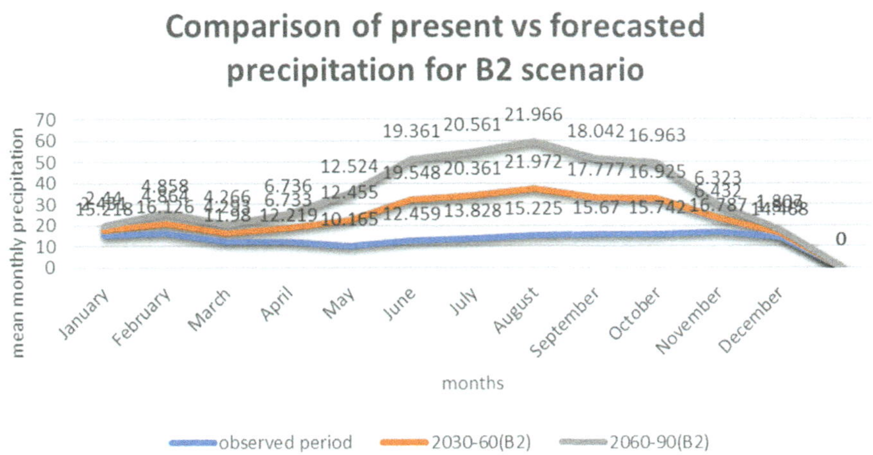

Fig. 27.5 Comparison of monthly mean precipitation in forecasted period with that of baseline period

b) At Gomlai station

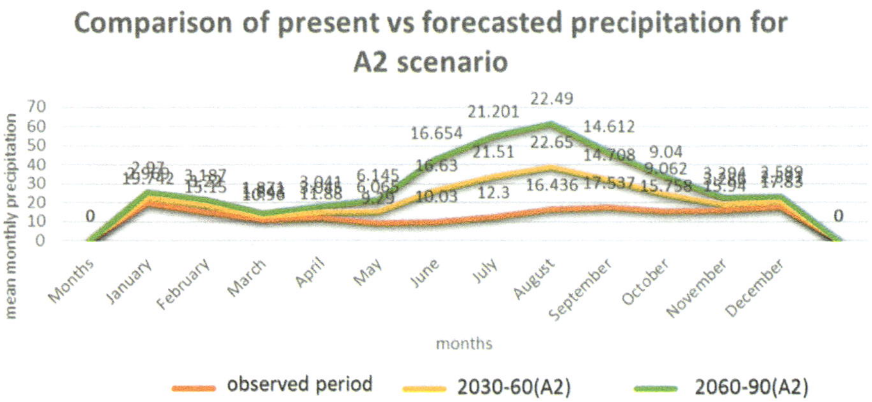

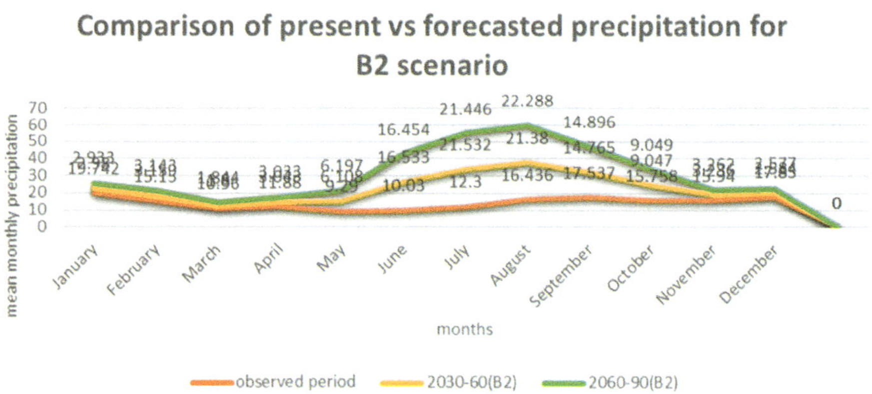

Fig. 27.5 (continued)

27.6 Conclusions

As the model is statistical in nature, the accuracy is highly dependent on the selection of predictors and user's expertise. The predictor selection process is based on the correlation values, for which, sometimes the model underperforms for conditional predictands like precipitation. The performance of model was satisfactory for the two stations considered in this study; however analysis at more number of stations is required to evaluate model performance in complete river basin. Based on the present study on the statistical downscaling of GCM outputs and simulation of rainfall scenarios for Brahmni–Baitarani River basins in Odisha, the quantity of mean monthly precipitation shows an increasing graph with time for all epoches. It can also be noted that the increase is more in the rainy season (June, July, August) with respect to other time of the year, which might be indicating increase in runoff as well leading to flood events. But no conclusion can be drawn regarding increase in discharge without considering other factors like temperature, land-use, land-cover change etc.

References

Anandhi A, Srinivas VV, Nanjundiah RS, Nagesh Kumar D (2008) Downscaling precipitation to river basin in India for IPCC SRES scenarios using support vector machine. Int J Climatol 28:401–420. https://doi.org/10.1002/joc.1529

Chiew FHS, Young WJ, Cai W, Teng J (2010). Current drought and future hydroclimate projections in southeast Australia and implications for water resources management. Stochastic Environ Res Risk Assess, 602–612

Khan MS, Coulibaly P, Dibike Y (2006) Uncertainty analysis of statistical downscaling methods. J Hydrol 319:357–382

Pervez MS, Henebry GM (2014) Projections of the ganges-brahmaputra precipitation-downscaled from GCM predictors. J Hydrol 517:120–134

Prayas R, Patra KC (2016) Assessment of the impact of climate change on the hydrological parameters using statistical downscaling. Thesis submitted to the NIT, Rourkela in partial fulfillment of the requirements for the dual degree of Bachelor and Master of Technology in Civil Engineering

Sharma A, Goyal MK (2017) Assessment of ecosystem resilience to hydroclimatic disturbances in India. Global Change Biol

Walsh J (2011) Statistical downscaling. In: NOAA Climate Services Meeting

Wilby RL, Harris I (2006) "A framework for assessing uncertainties in climate change impacts: low-flow scenarios for the river thames UK. Water Resour Res 42:W02419. https://doi.org/10.1029/2005WR004065

Wilby RL, Dawson CW, Barrow EM (2002) SDSM—a decision support tool for the assessment of regional climate change impacts. Environ Model Softw 17(2):145–157

Chapter 28
Impact of Land Use–Land Cover Changes on the Streamflow of the Kolab River Basin Using SWAT Model

Partha Sarathi Bhunia and Kanhu Charan Patra

Abstract Hydrological parameters are affected by many factors, including long-term effects such as climate change that alter rainfall–runoff relationships, and short-term effects related to human intervention (e.g., dam construction, land-use and land-cover change (LUCC)). The consequential impacts of human-induced climate changes and land-use changes on hydrological parameters have become a big challenge and convinced to give great attention of many researchers. The Kolab river watershed, Odisha is an important watershed supporting drinking water and recreational activities. In this paper, it is assessed the long-term impacts of LULC change and climate changes on hydrological parameters using the Soil and Water Assessment Tool (SWAT) and a detailed LULC record from 1995–2013.

Keywords Arc GIS · LUCC pattern · SWAT modelling · SWAT-CUP · SUFI2 algorithm

28.1 Introduction

Land use refers to the purpose the land serves, for example, recreation, wildlife habitat, or agriculture. Land cover refers to the surface cover on the ground, whether vegetation, urban infrastructure, water, bare soil or other. Human activities and few natural processes are the principal influences of land use–land cover changes. As agriculture is the prime sector of India's economic system and the population of this country is growing up abruptly, the existing land cover has to be processed. Land-use changes are often non-linear and might trigger feedbacks to the system, stress living conditions, and threaten people with vulnerability. A comprehensive knowledge is required for effective watershed management and ecological restoration of the hydrological process of the watershed. Land use–land cover changes are one

P. S. Bhunia (✉) · K. C. Patra
Department of Civil Engineering, NIT Rourkela, Rourkela 769008, Odisha, India
e-mail: sarathipartha94@gmail.com

K. C. Patra
e-mail: kcpatra@nitrkl.ac.in

R. Jha et al. (eds.), *Climate Change Impacts on Water Resources*, Water Science and Technology Library 98, https://doi.org/10.1007/978-3-030-64202-0_28

of predominant influences on hydrological parameters and soil erosion including climate change.

Many studies have taken into consideration the impact of land use–land cover changes on streamflow (Matheussen and Kirschbaum 2000; Hurkmans et al. 2009). Few studies have carried out to investigate the impact of land use–land cover changes due to urbanization on surface microclimate and hydrology: a satellite perspective (Carlson and Arthur 2000). Though a lot of hydrological models are available, like Water Erosion Prediction Project (WEPP), Hydrologic Simulation Program Fortran (HSPF), the Soil and Water Assessment Tool (SWAT) and the physically-based distributed hydrological model Systeme Hydrologique Européen TRANsport (SHETRAN), that could be used in simulating the runoff. But we have taken the SWAT for the current study because it is widely used and it is user friendly in terms of handling input data (Arnold et al. 1998).

The objective this study was to assess the impact of past land-use change on streamflow of the Kolab river basin. The results obtained from this study will be helpful for understanding the interactions between the streamflow and land-use change which is required for water resource planning.

28.2 Study Area

The present study was conducted for the Kolab river basin (See Fig. 28.1). The Kolab river originates from the western slopes of Eastern Ghats in Odisha state from Sinkaram hill ranges at 1370 m MSL. It is also known as Sabari river and it is one of the tributaries of Godavari river. Sileru river and Pateru river are the tributaries of this river. The annual average rainfall over this entire basin is approximately 1250 mm. The basin extends over states of Chhattisgarh, Odisha and Andhra Pradesh having a basin area of 20,427 Sq. km. which is approximately 1.6% of the total area of India with a maximum length of 418 km. It lies between 81°15′37″ E and 83°E longitudes and 17°34′13″ N and 19°°N latitudes. The topography of this region is undulating and highly dissecting. The climate of the region is tropical. Late December and early February are the coldest months. Mainly four seasons are there in the region throughout a single year. They are Hot Weather, South-west monsoon, post-monsoon and Cold Weather. December and February are the coldest months with minimum average temperature of 11.5 °C. The average annual maximum temperature of the basin is 30.57 °C. while the average annual minimum it is 18 °C. April and May months are the hottest months with average temperature of 34.5 °C. Location of the study area is given below (see Fig. 28.1).

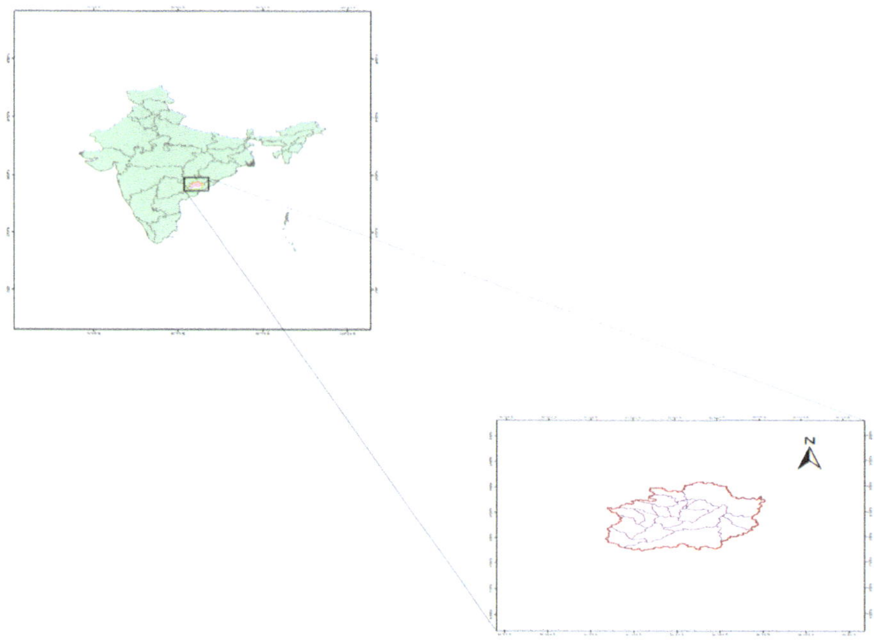

Fig. 28.1 Location map of Kolab river Basin, India

28.3 Materials and Methodology

28.3.1 SWAT Model

SWAT, a semi-distributed, encyclopaedic, physical-based continuous hydrological model, which is developed to predict the effects of land management on the hydrology, agricultural chemical yields and sediment yields in different watersheds under varying land use and soil conditions (Neitsch et al. 2009). To achieve this, the model needs specific climatic input parameters such as temperature, rainfall, solar radiation, relative humidity and wind speed. Along with these weather data, the model requires some spatial data such as DEM, soil map, land-use map and hydrological data such as streamflow.

The framework of this model is to project the digital elevation model, land use/land cover and soil map into a standard projection system. The delineation process takes place in the model with the help of DEM data of the basin and the whole basin can be divided into several number of sub-basins. The grid formats of the soil map, land use- land cover map and the slope categories were overlaid and reclassified into hydrological response units (HRUs). HRUs are used for simplification of the model run by uniting all identical soils and land-use areas into a single response unit. The model evaluates different hydrological parameters such as evapotranspiration, peak rate of runoff, surface runoff and other components by individual HRU using water balance equation. The SCS curve number procedure (USDA-SCS 1972) and Green and Ampt. infiltration method (Green and Ampt. 1911) are two basic methods of the model for estimating the surface runoff and modified rational method for estimating peak runoff rate. The estimation of hydrologic cycle at individual HRU was carried out using water balance equation as follows.

$$SW_t = SW_\circ + \sum_{i=1}^{t}\left(R_{day} - Q_{surf} - E_a - w_{seep} - Q_{gw}\right) \quad (28.1)$$

where SW_t is the final soil water content in mm H_2O, SW_O is the initial soil water content on day i in mm H_2O, t is the time of days, R_{day} the amount of precipitation on day i in mm H_2O, Q_{surf} is the amount of surface runoff on day i in mm H_2O, E_a is the amount of evapotranspiration on day i in mm H_2O, w_{seep} is the amount of water entering the vadose zone from soil profile on day i in mm H_2O, Q is the amount of return flow on the day i in mm.

28.3.2 SWAT Model Input Data

The major inputs for the model required for simulation are land-use map, soil cover map, digital elevation model, rainfall and temperature data (Table 28.2). Shuttle radar topographic mission (SRTM) provides the digital elevation model data (DEMs) having a 90 m resolution. These DEM data are available in the form of square-shaped tiles with a specific dimension. The DEM downloaded from SRTM website has projection system of WGS 1984 UTM, zone 44 N at 90 m resolution. These DEMs were used for watershed delineation and those were also used to depict the streamflow, drainage pattern, flow accumulation and networks of the basin area.

Landsat 8 and Landsat 5 satellite images (spatial resolution 30 m) for 2013 and 1995 were obtained from earth explorer in the form of tiles. Figure 28.2c and d shows, the land-use maps, used in SWAT simulation. Soil map was prepared using FAO digital soil map of the world having a scale of 1:5,000,000. The meteorological parameters (precipitation, maximum and minimum temperature, relative humidity, solar radiation, wind speed etc.) control the hydrological cycle. The meteorological

Table 28.1 Different land use and cover area in Kolab basin

ID	Land cover specification	1995 Area(km^2)	2013 Area(km^2)	Change (%)
1	Built up	8460.21	10463.36	(+) 0.09
2	Forest	5820.38	4129.23	(−) 0.08
3	Barren Land	4314.67	3450.04	(−) 0.04
4	Agriculture	1387.56	1929.89	(+) 0.02
5	Water	239.48	250.04	(+) 0.0005

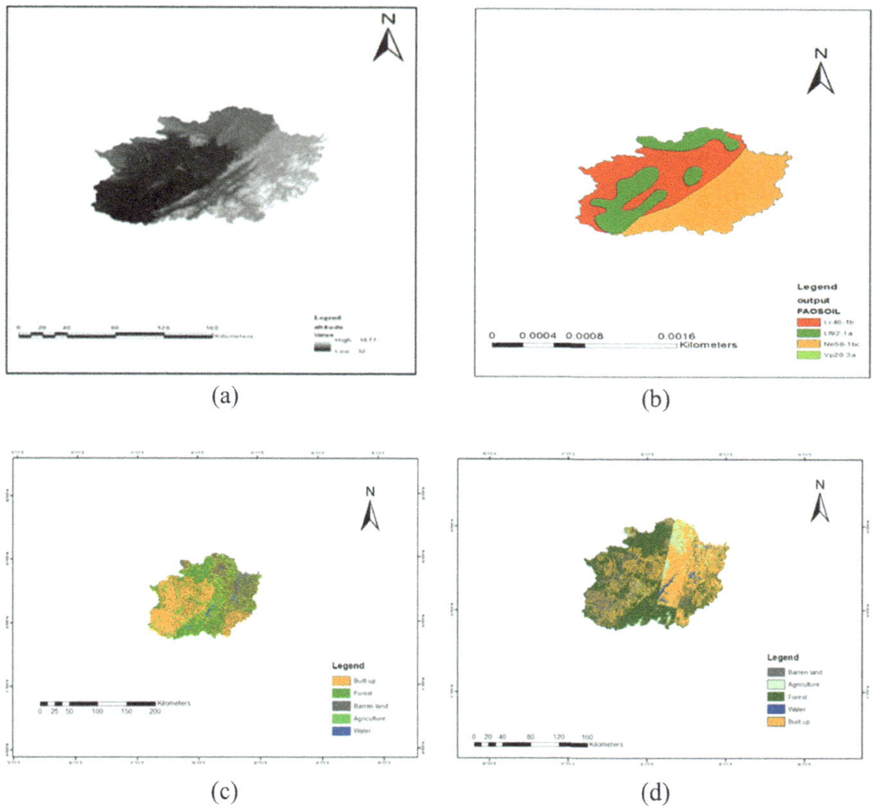

Fig. 28.2 **a** Digital Elevation Model of the Basin. **b** Soil map of the basin. **c** Landuse land cover map (1995). **d** Landuse land cover map (2013)

data for 1979 to 2013 were collected from The National Centers for Environmental Prediction (NCEP).

28.3.3 SWAT Model Setup and Simulation

Arc SWAT 12, an interface of SWAT model in ArcGIS 10.1, was used to simulate the streamflow of the Kolab basin. To develop the model, the following five steps are needed: (1) sub-basin discretization; (2) HRU definition; (3) climate station formation; (4) parameter sensitivity analysis; and (5) calibration and validation.

The digital elevation model data were used to delineate the area of Kolab basin along with the stream networks using Arc SWAT 12. The Kolab basin had produced 20 HRUs based on soil and land use–land cover data. The simulation of the model was done for the year of 1979–2013 on daily time step with a warm-up period of

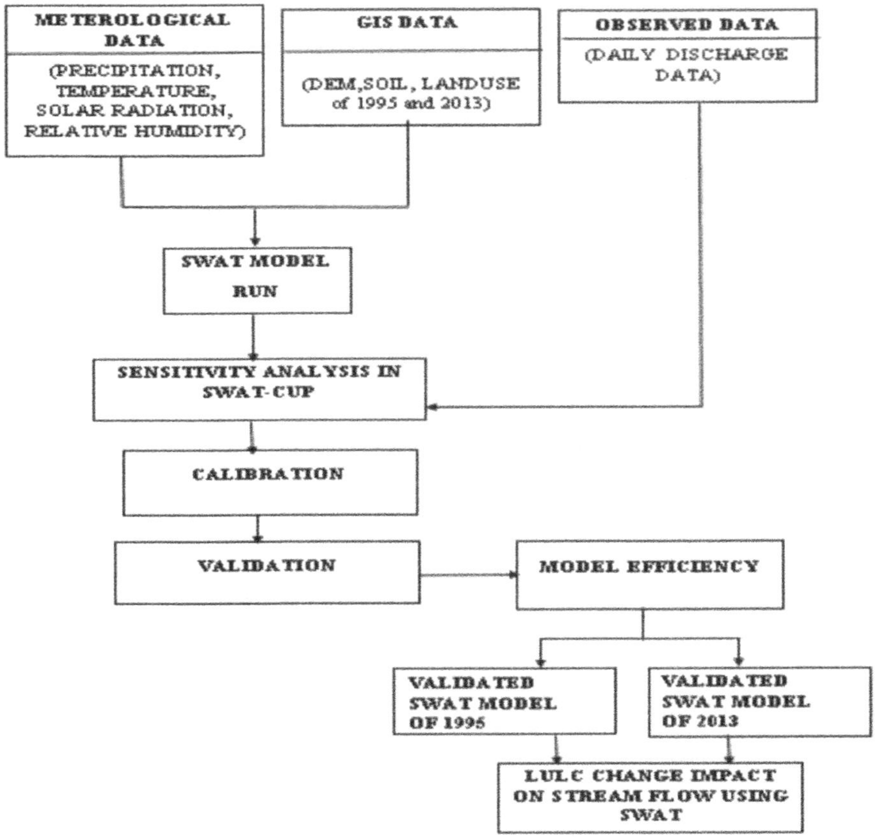

Fig. 28.3 Schematic diagram of methodology

5 years. Figure 28.3 shows the schematic diagram of complete methodology of this study.

28.3.4 Sensitivity Analysis, Calibration and Validation

Sensitivity analysis can identify the dominant parameters, which affect the SWAT output using global sensitivity approach of Sequential Uncertainty Fitting (SUFI2) algorithm in SWAT-CUP (Calibration and Uncertainty Procedure) (Abbaspour 2007). In this study, 14 parameters were taken initially to investigate their sensitiveness, ranges were set from SWAT-CUP user manual. After an initial iteration run, Global sensitivity approach checks the sensitivity of one parameter relative to another and arrange the parameters by ranks according to their t-stat and p-values. Sensitivity

Table 28.2 Spatial model input data for the Kolab River catchment

Data type	Description	Resolution	Sources
Topographic map	Digital elevation map (DEM)	90 m	SRTM
Land-use map	Land-use classification	30 km	Landsat 8 & 5
Types	10 km	FAO	(USGS) Soil map
Weather	Daily precipitation	–	NCEP
Hydrological data	Discharge	–	WRD, Koraput

analysis gives the allowable ranges and the best-fitted values. In this study, 14 streamflow influencing parameters were tested for sensitivity (Table 28.5).

Calibration is examining the precision of a measurement instrument by comparing it to reference standards. In this present study, only selected sensitive parameters were used for calibration and validation process. For 1995, land-use scenario, the model was calibrated for the period of 1986–1992 and validated for the period of 1993–1995. Similarly, 2013 land-use scenario model was calibrated for the period of 2004–2010 and validated for the period of 2011–2013. To evaluate the model performance, four statistical standards, Nash-Sutcliffe efficiency (NSE), Coefficient of determination (R^2) and the percent bias (PBIAS) were used (Moriasi et al. 2007).

$$\text{NSE} = 1 - \frac{\sum_{i=1}^{n}(O_i - S_i)^2}{\sum_{i=1}^{n}(O_i - O)^2} \tag{28.2}$$

$$\text{PBIAS} = \frac{\sum_{i=1}^{n}(O_i - S_i) * 100}{\sum_{i=1}^{n}(O_i)} \tag{28.4}$$

$$\text{RSR} = \frac{\sqrt{\sum_{i=1}^{n}(O_i - S_i)^2}}{\sqrt{\sum_{i=1}^{n}(O_i - O_i)^2}} \tag{28.5}$$

$$R^2 = \frac{\left[\sum_{i=1}^{n}(O_i - O)(S_i - S)\right]^2}{\left[\sum_{i=1}^{n}(O_i - O)^2\right] * \left[\sum_{i=1}^{n}(S_i - S)^2\right]} \tag{28.6}$$

where O_i is the observed daily discharge, S_i is the simulated daily discharge, O is the average measured discharge, S is the average simulated discharge, n is the number of observations.

NSE illustrates the prediction ability of hydrological model by comparing the simulated output to the observed data. It ranges from $-\infty$ to 1. NSE value close to 1, indicates the model is accurate. R^2 is used to understand how well-observed output is replicated by the model, on the basis of the proportion of total variance of the simulated output data. The range of R^2 varies from 0 to 1. An R^2 value close to 1 specifies the model is precise. RSR is the ratio of the root mean square error to the standard deviation of measured data. It standardizes RMSE (root-mean-square error) using the observation standard deviation. RSR varies from 0 to large positive values. The lower value of RSR indicates better the model fit. A model simulation can be accepted satisfactory if NSE is greater than o.4, PBIAS is $\pm$ 25% and R^2 is greater than 0.5 for streamflow (Arnold et al. 2012).

28.4 Result and Discussion

Using the Maximum Likelihood Classification tool in Arc-GIS the change in the LULC area for 1995 and 2013 were estimated (Table 28.1). It showed that the built-up area and agricultural field were increased by 0.09% and 0.02% respectively. Therefore, the forest area and barren land were reduced by 0.08% and 0.04% respectively. The calibration and validation of the SWAT-CUP model based on two different land-use scenarios showed that only five parameters out of fourteen were very sensitive in Global sensitivity analysis (Tables 28.4 and 28.6). Based on the sensitiveness, the sensitive parameters are presented for 1995 land use (LU) scenario model (Table 28.3). The sensitivity analysis indicates the most sensitive parameter for the streamflow was Base-flow alpha factor for bank storage (ALPHA_BNK). The other parameters that were sensitive were SCS runoff curve number (CN2), available water capacity of the soil layer (SOL_AWC), effective hydraulic conductivity in main channel alluvium (CH_K2), saturated hydraulic conductivity (SOL_K), soil evaporation compensation factor (ESCO). Rest of all the parameters were found not sensitive to streamflow in the catchment as their p-values were greater than 5% (Anaba et al. 2016). The best-fitted values and ranges obtained after sensitivity analysis of these five most sensitive parameters are presented in Table 28.4.

In the second case, the result was shown with the help of land-use scenario model (2013). The sensitive parameters for 2013 land use (LU) scenario model are listed below (see Table 28.3). Here in this study, only five parameters were sensitive to streamflow. The most sensitive parameter for the streamflow was effective hydraulic conductivity in main channel alluvium (CH_K2). The other parameters which were sensitive were SCS runoff curve number (CN2), Base-flow alpha factor for bank storage (ALPHA_BNK), Saturated hydraulic conductivity (SOL_K), Groundwater revap coefficient (GW_REVAP). Rest of all were found very less sensitive to the streamflow in the catchment as their p-values were greater than 5% (Anaba et al. 2016). The best-fitted values and ranges obtained after sensitivity analysis of these five most sensitive parameters are presented in Table 28.6.

Table 28.3 Sensitivity Analysis of streamflow parameters for 1995 scenario model

SL. No.	Parameter name	t-stat	p-value
1	CH_K2 (Effective hydraulic conductivity in main channel alluvium)	9.205	0.000
2	CN2 (SCS runoff curve number)	5.921	0.000
3	ALPHA_BNK (Base-flow alpha factor for bank storage)	5.266	0.000
4	SOL_K (Saturated hydraulic conductivity)	2.320	0.000
5	GW_REVAP (Groundwater "revap" coefficient)	−1.922	0.044
6	REVAPMN (Threshold depth of water for revap or percolation to occur)	1.877	0.061
7	GWQMN (Threshold depth of water in the shallow aquifer required for return flow to occur)	−1.583	0.114
8	HRU_SLP (Average slope steepness)	1.347	0.179
9	CH_N2 (Manning's "n" value for the main channel)	1.273	0.204
10	SOL_AWC (Available water capacity of the soil layer)	0.963	0.334
11	ESCO (Soil evaporation compensation factor)	−0.912	0.362

Table 28.4 Ranges and best-fitted values of flow calibration parameters

Rank	Parameter name	Fitted value	Minimum value	Maximum value
1	v_GW_DELAY.gw	171.539	30.000	440.000
2	r_CN2.mgt	0.070	−0.200	0.200
3	r_SOL_K.sol	0.097	−0.800	0.800
4	v_ALPHA_BF.gw	0.463	0.000	1.000
5	v_GW_REVAP.gw	0.016	0.000	0.200

28.4.1 SWAT Model Calibration and Validation Result

The calibrated (Fig. 28.4a) and validated (Fig. 28.4b) results for 1995 LU model show the graphical representations of the streamflow. Figure 28.5a and b shows the graphical representation of calibrated and validated streamflow for 2013 LU scenario respectively. It also depicts the calibration results showed a good match compare to the validated streamflow results in both cases (Table 28.7). Besides that, the statistical objective functions in SWAT-CUP evaluated the model performance was good (Moriasi et al. 2007).

Table 28.5 Sensitivity Analysis of streamflow parameters for 2013 scenario mode

SL. No.	Parameter name	t-stat	p-value
1	K2 (Effective hydraulic conductivity in main channel alluvium	−2.986	0.001
2	CN2 (SCS runoff curve number)	11.001	0.000
3	ALPHA_BNK (Base-flow alpha factor for bank storage)	13.536	0.000
4	SOL_K (Saturated hydraulic conductivity)	3.020	0.002
5	GW_REVAP (Groundwater "revap" coefficie	−1.556	0.120
6	REVAPMN (Threshold depth of water for revap or percolation to occur)	−1.129	0.259
7	GWQMN (Threshold depth of water in the shallow aquifer required for return flow to occur)	0.983	0.326
8	HRU_SLP (Average slope steepness)	0.805	0.414
9	CH_N2 (Manning's "n" value for the main channel)	−1.301	0.193
10	SOL_AWC (Available water capacity of the soil layer)	−3.129	0.000
11	ESCO (Soil evaporation compensation factor)	−2.035	0.042
12	ALPHA_BF (Base-flow alpha factor)	0.707	0.479
13	GW_DELAY (Groundwater delay)	0.397	0.691
14	SLSUBBSN (Average slope length)	1.668	0.095

Table 28.6 Ranges and best-fitted values of flow calibration parameters

Rank value	Parameter name	Fitted value	Minimum value	Maximum
1	v_ALPHA_BNK.rte	0.949	0.000	1.000
2	r_CN2.mgt	0.150	−0.200	0.200
3	r_SOL_K.sol	0.063	−0.200	0.400
4	v_CH_K2.r	117.625	5.000	130.000
5	v_ESCO.hru	0.875	0.800	0.100

Table 28.7 SWAT model calibration and validation statistical objective functio

Stages of model	Statistical parameters			
	R^2	NSE	RSR	PBIAS
Calibration (1986–1992)	0.700	0.661	0.512	−15.0
Validation (1993–1995)	0.677	0.623	0.551	−17.0
Calibration (2004–2010)	0.685	0.621	0.560	−14.0
Validation (2011–2013)	0.637	0.611	0.574	−17.0

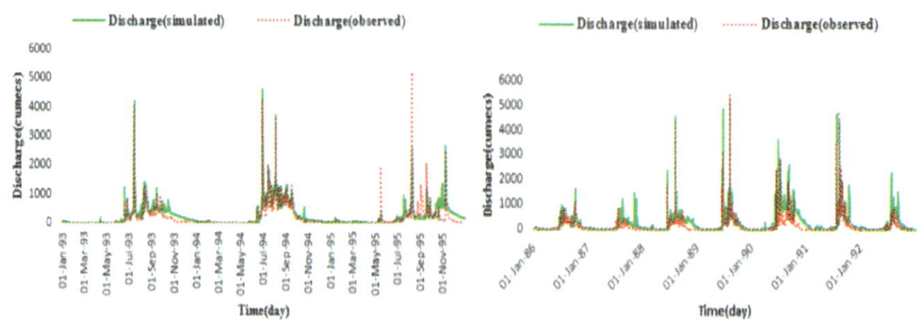

(a) (b)

Fig. 28.4 **a** Observed and simulated daily discharge during calibration period (1986–1992) for 1995 LU model. **b** Observed and simulated daily discharge during validation period (1993–1995) for 1995

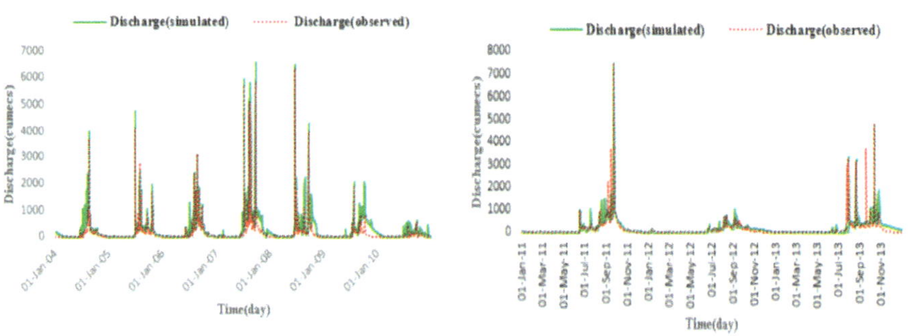

(a) (b)

Fig. 28.5 **a** Observed and simulated daily discharge during calibration period (2004–2010) for 2013 LU model 2013) **b** Observed and simulated daily discharge during calibration period (2011–2013) for 2013 LU model

28.4.2 Modelling Streamflow Response to Land-Use Dynamics

Establishing scenarios to assess the impacts of land use.

The impact assessment of land-use changes on streamflow under different scenarios was done by using SWAT 2012. The one factor at a time approach (Li et al. 2009) was taken into consideration. In this study, we have taken two scenarios based on different land use and land cover condition (for 1995 & 2013) and showed the difference in mean annual streamflow. It shows that the mean annual streamflow increased by 3.75% due to change in land use and land cover area (Table 28.8).

Scenario I: Climate of 2008–2013 and land use of 2013.

Scenario II: Climate of 2008–2013 and land use of 1995.

Table 28.8 Difference in mean annual stramflow based on scenario I and II

Scenario	Mean annual Streamflow (m3/s)
I	229.69
II	221.33
Change (I–II) with respect to scenario II	+3.75%

28.5 Conclusion

The present study analyzed the LULC changes and their impacts on the streamflow. It showed that the Kolab river basin had experienced a remarkable change in eighteen years (1995–2013). With substantial changes in land-use/land-cover patterns in different areas of the Kolab basin, the effect of land-use/land-cover change deserves more attention when evaluating the impacts of different factors on the streamflow. Therefore, to estimate the changes in streamflow we should not eliminate the effects of land use and land cover changes along with climate change. Key findings from the study can be summarized as follows: i) This study showed that an increment in the built-up area (0.09%), agricultural area (0.02%), and decrement of the area covered by the forests (0.08%), barren land (0.04%) altered the overall mean annual streamflow by 3.75%. ii) The calibration of the SWAT model was good and it produced a convenient result for the streamflow simulation in this region, and this model can be very useful to predict the streamflow in future scenarios. iii) Since the land use and land cover changes are explicitly related to the streamflow, this study will be helpful for better management of this watershed.

References

Abbaspour KC (2007) User manual for SWAT-CUP, SWAT calibration and uncertainty analysis programs. Swiss Federal Institute of Aquatic Science and Technology, Eawag, Duebendorf, Switzerland, 93

Anaba LA., Banadda N, Kiggundu N, Wanyama J, Engel B, & Moriasi D (2016) Application of SWAT to assess the effects of land use change in the Murchison Bay catchment in Uganda

Arnold JG, Moriasi DN, Gassman PW, Abbaspour KC, White MJ, Srinivasan R, ... & Jha MK (2012) SWAT: model use, calibration, and validation. Trans ASABE 55(4):1491–1508

Arnold JG, Srinivasan R, Muttiah RS, Williams JR (1998) Large area hydrologic modeling and assessment part I: model development 1. JAWRA J Am Water Resour Assoc 34(1):73–89

Carlson TN, Arthur ST (2000) The impact of land use—land cover changes due to urbanization on surface microclimate and hydrology: a satellite perspective. Global Planet Change 25(1–2):49–65. Ellis E, Pontius R (2007) Land-use and land-cover change. Encyclopedia Earth, 1–4

Green WH, Ampt GA (1911) Studies on soil phyics. J Agric Sci 4(1):1–24

Hurkmans RTWL, Terink W, Uijlenhoet R, Moors EJ, Troch PA, Verburg PH (2009) Effects of land use changes on streamflow generation in the Rhine basin. Water Resour Res 45(6)

Li Z, Liu WZ, Zhang XC, Zheng FL (2009) Impacts of land use change and climate variability on hydrology in an agricultural catchment on the Loess Plateau of China. J Hydrol 377(1–2):35–42.

Matheussen B, Kirschbaum RL, Goodman IA, O'Donnell GM, Lettenmaier DP (2000) Effects of land cover change on streamflow in the interior Columbia River Basin (USA and Canada). Hydrol Process 14(5):867–885

Matheussen B, Kirschbaum RL, Goodman IA., O'Donnell GM, Lettenmaier DP (2000) Effects of land cover change on streamflow in the interior Columbia River Basin (USA and Canada). Hydrol Proces 14(5):867–885

Moriasi DN, Arnold JG, Van Liew MW, Bingner RL, Harmel RD, Veith TL (2007) Model evaluation guidelines for systematic quantification of accuracy in watershed simulations. Trans ASABE 50(3):885–900

Neitsch SL, Arnold JG, Kiniry JR, Williams JR (2011) Soil and water assessment tool theoretical documentation version 2009. Texas Water Resources Institute

Chapter 29
Statistical Downscaling of Climatic Variables in Indo-Gangatic Alluvial Plain

Prabhakar Shukla and Raj Mohan Singh

Abstract Climate change influences events such as droughts, floods, extreme temperatures and rainfall. The changes in climatic conditions may adversely affect food production, energy generation, and water resources. Rising greenhouse gas (GHG) concentrations in the atmosphere are considered the dominant cause of climate change. General circulation models (GCMs) are used for the projection of climate into future, accounting for the GHG concentrations. However, the coarse spatial resolution of GCM outputs does not permit their direct use in catchment-scale studies. Therefore, either dynamic or statistical downscaling techniques are used for linking GCM outputs to catchment scale hydroclimatic variables. In the present study, beta regression-based statistical downscaling of daily scale climatic variables (precipitation, minimum temperature, and maximum temperature) is performed. The daily frequency climatic variables obtained from Hadley global environment model 2—earth system (HADGEM2-ES) of CMIP 5 project is statistically downscaled for future period (2016–2095) using NCEP potential variables in MATLAB. The methodology is performed for Sai–Gomti interfluve region, Uttar Pradesh, India. The statistical analysis of observed and simulated hydro-climatic variables shows that the average maximum temperature may rise by 4 °C while average minimum temperature may fall by 4.5 °C during 2016–2095.

Keywords Statistical downscaling · HADGEM2-ES · CMIP 5 · NCEP · Climate change

---

P. Shukla (✉)
Scientist, Water Security and Sustainable Development Hub, IIT Delhi, Hauz Khas, New Delhi 110016, India
e-mail: prabhakar@fitt.iitd.ac.in

R. M. Singh
Professor, Department of Civil Engineering, MNNIT Allahabad, Prayagraj 211004, Uttar Pradesh, India
e-mail: rajm@mnnit.ac.in

R. Jha et al. (eds.), *Climate Change Impacts on Water Resources*, Water Science and Technology Library 98, https://doi.org/10.1007/978-3-030-64202-0_29

29.1 Introduction

The variations in the distribution of surface and groundwater resources are governed by the changes in the climatic variables such as precipitation, temperature and evaporation (Chew et al. 2008). The global climate is expected to change throughout the twenty-first century (Dessai et al. 2005). Impacts of climate change on health of humans (Thomas et al. 2012), agricultural food production (Ziska et al. 2012), floods (Prudhomme et al. 2013) and water resources (Arnell and Gosling 2013) are major aspects discussed by researchers. Since the climate is highly influential on the water resources within a catchment in many different ways as described earlier, knowledge of changing climate into the future is imperative for the effective management of water resources in a catchment.

General circulation models (GCMs) are used to simulate the effect of rising greenhouse gas (GHG) concentrations on the global climate. The simulations produced by GCMs are often in the order of a few hundred kilometers and hence cannot explain the local-scale variability of climate in a catchment (Tripathi et al. 2006). Downscaling is implemented to produce fine-scale hydro-climatic information derived from GCM simulations at coarse resolution via the application of various statistical numerical models. There are two types of downscaling techniques: (i) dynamical downscaling (DD) and (ii) statistical downscaling (SD). In dynamic downscaling, an atmospheric physics-based model called a regional climate model (RCM) is nested in a GCM for the simulation of regional climate (murphy 1998). The high computational cost associated with the implementation of dynamic downscaling is considered to be a major drawback (Haas And Pinto 2012). Statistical downscaling techniques attempt to develop empirical relationships between the GCM outputs and the catchment scale hydroclimatic variables (Sachindra 2014). The SD models are considerably dependent on the predictors that are used as inputs to these models (Fowler et al. 2007). The simulations produced by a statistical downscaling model are sensitive to the atmospheric domain for which the inputs are extracted (Wilby and Wigly 2000).

The prime objective of the present study is to perform beta regression-based statistical downscaling of daily scale GCM climatic variables (precipitation, minimum temperature, and maximum temperature) for future period (2016–2095) using national centers for environmental prediction (NCEP) predictors.

29.2 Study Area

The study area is shown in Fig. 29.1. The study area is part of Uttar Pradesh in India comprising districts of Barabanki, Raebareli, Sultanpur, Pratapgarh and Jaunpur, India. The study area lies between latitude 25^0 41′6″ n and 26^0 45′36″ n, and longitude 81^0 6′36″ e to 82^0 49′12″ e is estimated to be 8287.50 km^2. The area is representative of the whole Sarda Sahayak canal command. It is expected that the methodology adopted and conclusions arrived, would be applicable elsewhere

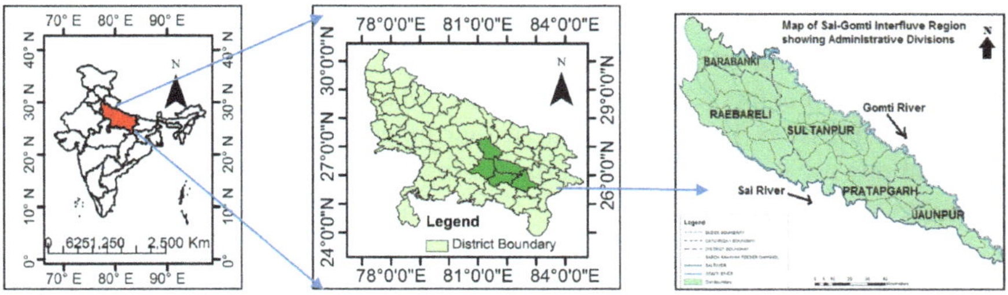

(a) (b) (c)

Fig. 29.1 Location map of study area **a** India, **b** Uttar Pradesh and **c** Sai–Gomti Interfluve

in the canal command area. The study area is interfluve bounded by Gomti and Sai rivers in north and west directions respectively. The confluence of Sai–Gomti River in Jaunpur district forms extremity of the area.

29.3 Methodology

29.3.1 Selection of Potential Predictors

The predictor data are extracted from the national centers for environmental prediction/national center for atmospheric research (NCEP/NCAR) reanalysis dataset (Kalnay et al. 1996). The eight climate variables (predictors) namely air temperature, precipitable water, relative humidity, sea level pressure, u wind, v wind, precipitation rate, and pressure obtained from NCEP reanalysis are analyzed over the period 1965–2012 (47 years) on daily frequency. Table 29.1 shows summary of correlation analysis performed between NCEP variables and observed rainfall data obtained from India meteorological department (IMD), Pune.

Table 29.1 Correlation analysis of NCEP variables and observed rainfall data

	Air temperature	Precipitable water	Relative humidity	Sea level pressure	U wind	V wind	Precipitation rate	Pressure
Raebareli	−0.31	−0.201	0.211	0.328	0.016	0.044	−0.239	0.333
Barabanki	−0.275	−0.177	0.189	0.301	0.017	0.058	−0.219	0.309
Sultanpur	−0.318	−0.199	0.213	0.326	0.017	0.057	−0.240	0.332
Thiessen polygon method	−0.378	−0.017	0.288	0.335	0.017	0.053	−0.218	0.308
Average of stations	−0.357	−0.214	0.262	0.395	0.016	0.052	−0.283	0.405

The threshold value for selection of potential variables (predictors) is kept as 0.3 (Bhowmik 2012). Based upon analysis, following four NCEP variables have been selected as potential predictors: air temperature, relative humidity, sea level pressure, and surface pressure. Furthermore, these four potential predictors are applied for statistical downscaling of climatic variables (precipitation, minimum temperature, maximum temperature).

29.3.2 Beta Regression-Based Statistical Downscaling

In the present study, daily scale climatic variables such as precipitation, minimum temperature and maximum temperature are downscaled using multisite downscaling method based on the beta regression (BR) developed by Mandal et al. (2016). This regression method based on the beta distribution has proven to be very versatile and flexible to model exogenous variables (Ferrari And Cribari-Neto 2004) and is novel in its application as a statistical downscaling technique. The BR model is developed in MATLAB. Both, the historical observed values of climatic variables, as well as historical and future GCM output datasets are utilized for the BR model. The BR model is developed based on a statistical relationship between potential NCEP predictors (air temperature, relative humidity, sea level pressure, and surface pressure) and a predictand (precipitation, minimum temperature, and maximum temperature). The climatic data from Hadley center global environment model version 2 [HADGEM-2] GCM for representative concentration pathway-2.6 (RCP2.6) are extracted and applied for both historical period. HADGEM2-ES is a coupled atmosphere–ocean GCM with interactive land and ocean carbon cycles and dynamic vegetation (Martin et al. 2011). The observed values of daily scale precipitation, minimum temperature, and maximum temperature are obtained from India meteorological department (IMD), Pune. The BR model is calibrated for the duration 1965–2012. Further, the calibrated model is validated with observed values of climatic variables for the period 2013–2015 (Fig. 29.2).

The comparison between observed and simulated climatic variables showed satisfactory results at all stations in study area. The Sultanpur district ($R^2 = 0.82$) showed best correlation for maximum temperature while Raebareli ($R^2 = 0.84$) showed high correlation between observed and simulated minimum temperatures. Also, the best correlation between observed and simulated rainfall is found at Raebareli ($R^2 = 0.73$). The statistics of comparative results between observed and simulated climatic parameters is shown in Table 29.2.

Further, the calibrated and validated BR model is applied for statistical downscaling of daily scale climatic variables (rainfall, maximum temperature, minimum temperature) for the future period up to 2095 (Fig. 29.3a–i)

The results show that the average maximum temperature may rise by 4 °C while average minimum temperature may fall by 4.5 °C during 2016–2095. The high frequency of peak events in rainfall is observed. This is possible due to over estimation of rainfall events using NCEP/NCAR reanalysis and GCM datasets (Sachindra 2014).

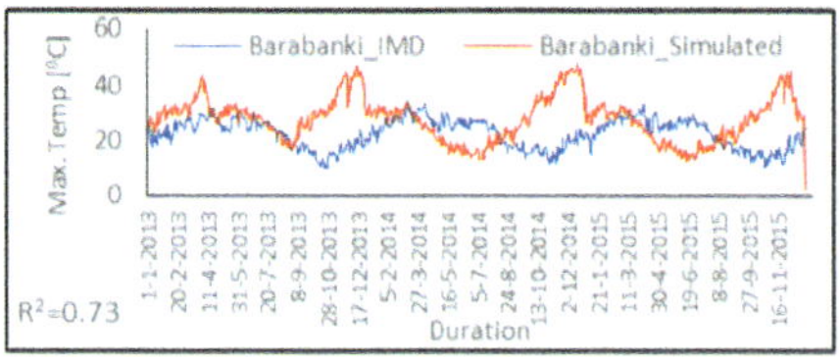

(a): validation of max. Temp. At Barabanki

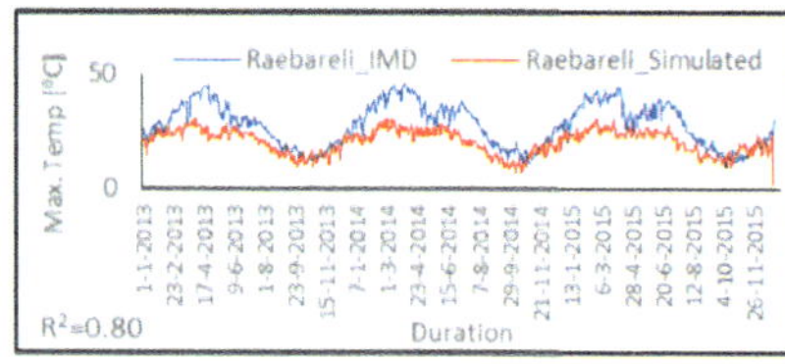

(b): validation of max. Temp. At Raebareli

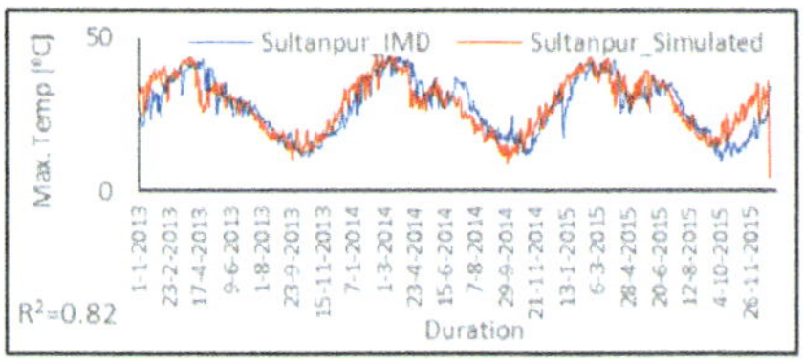

(c): validation of max temp. At sultanpur

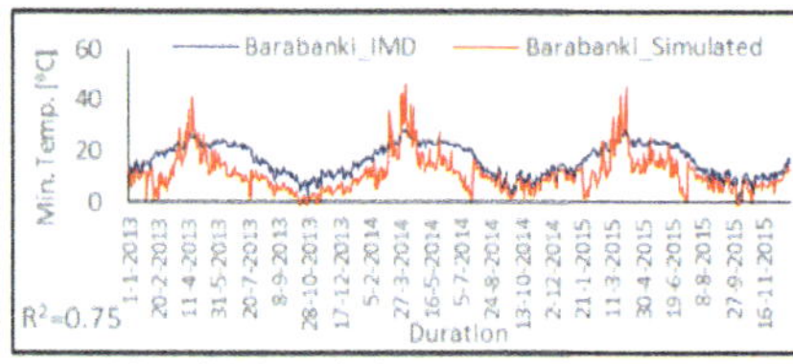

(d): validation of min. Temp. At barabanki

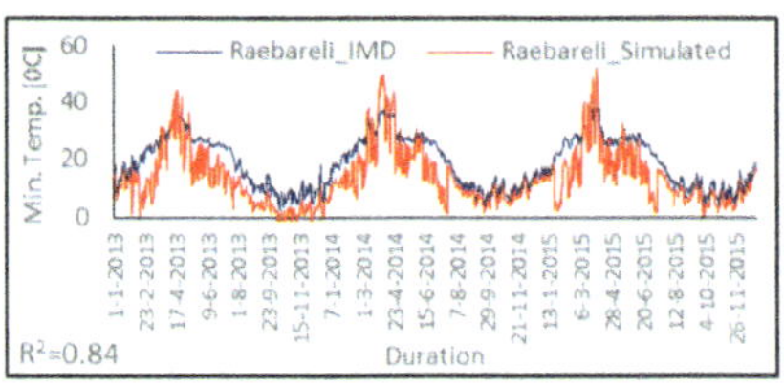

(e): validation of min. Temp. At Raebareli

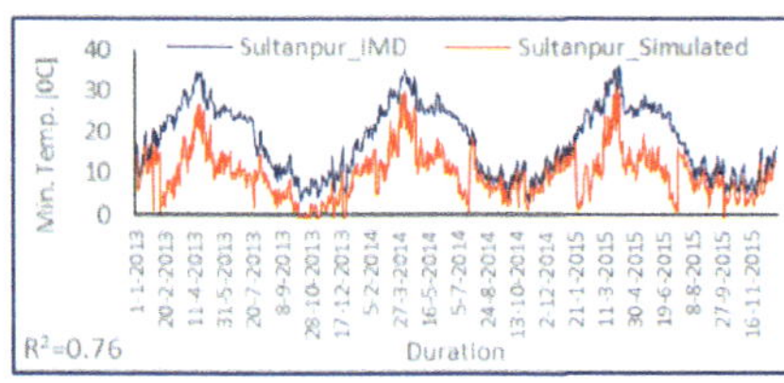

(f): validation of min. Temp. At Sultanpur

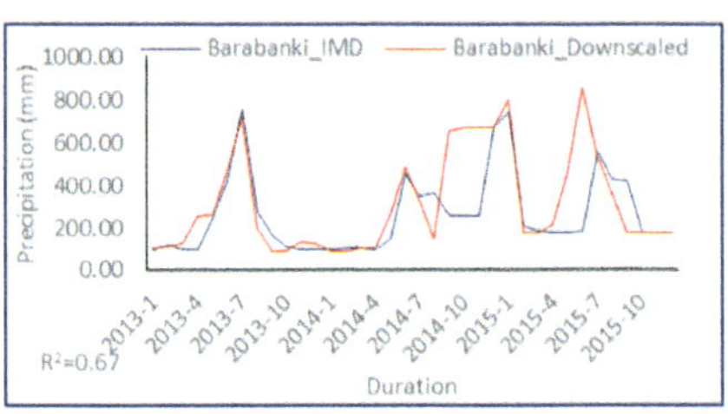

(g): validation of rainfall at Barabanki

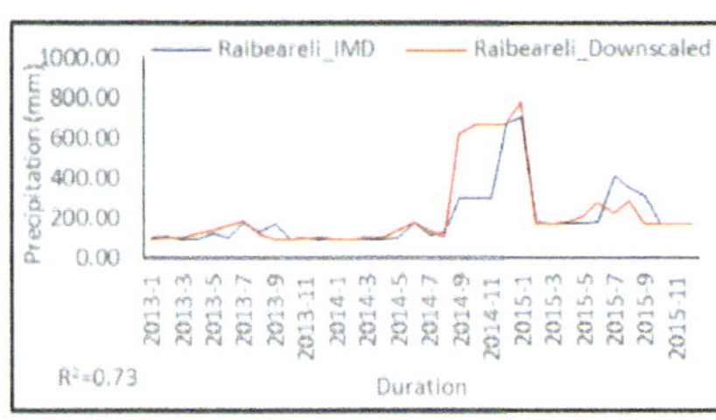

(h): validation of rainfall at Raebareli

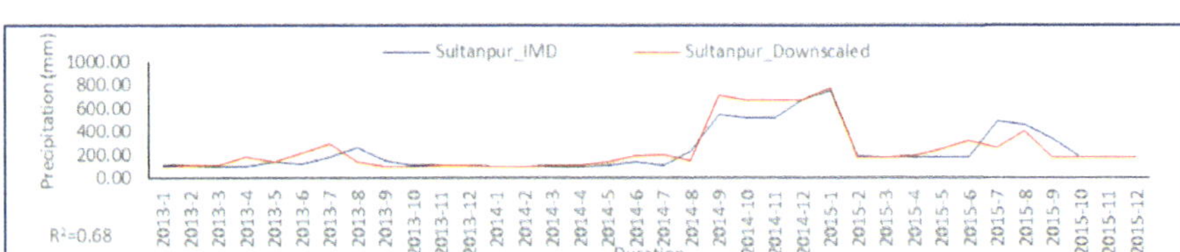

(i): validation of rainfall at Sultanpur

Fig. 29.2 Validation of downscaled climatic variables

Table 29.2 Statistics of observed and simulated climatic variables

statistical parameters	Climatic variables	Barabanki	Raebareli	Sultanpur
R^2	Maximum temperature	0.73	0.80	0.82
	Minimum temperature	0.75	0.84	0.76
	Precipitation	0.67	0.63	0.78
Average % error	Maximum temperature	13.88	27.75	12.89
	Minimum temperature	11.71	22.45	25.85
	Precipitation	23.77	29.36	39.5
Nash–Sutcliffe efficiency	Maximum temperature	0.60	0.73	0.72
	Minimum temperature	0.63	0.67	0.54
	Precipitation	0.39	0.35	0.35

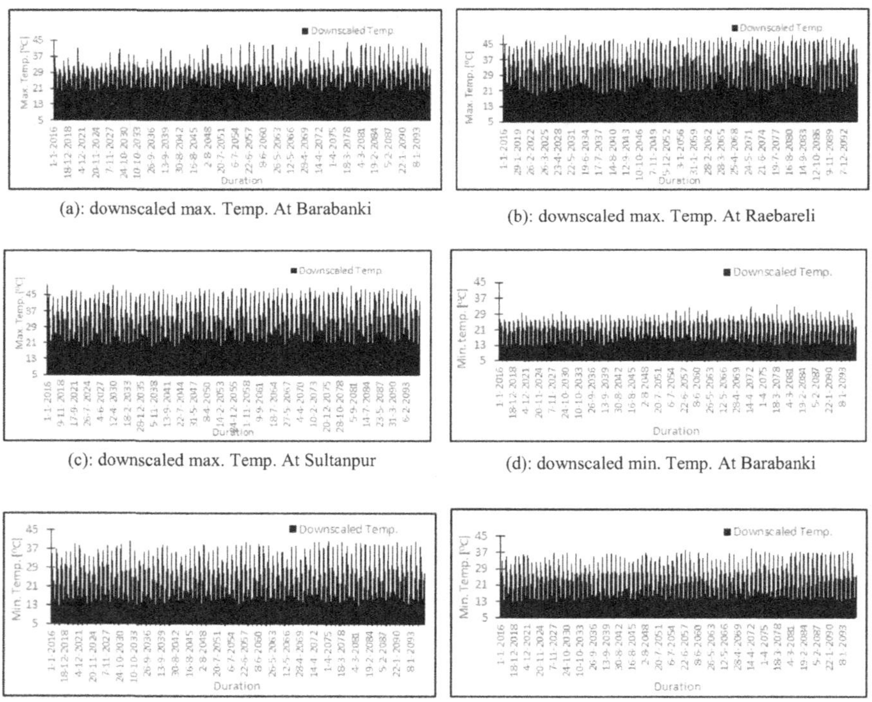

(a): downscaled max. Temp. At Barabanki

(b): downscaled max. Temp. At Raebareli

(c): downscaled max. Temp. At Sultanpur

(d): downscaled min. Temp. At Barabanki

(e): downscaled min. Temp. At Raebareli

(f): downscaled min. Temp. At Sultanpur

Fig. 29.3 Downscaled climatic variables for future period

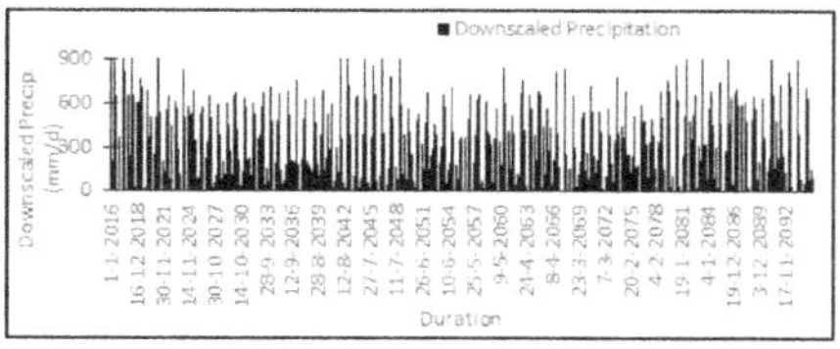

(g): downscaled precipitation at Barabanki

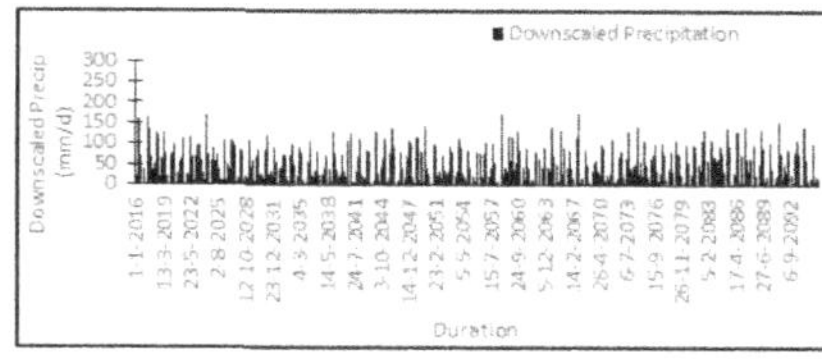

(h): downscaled precipitation at Raebareli

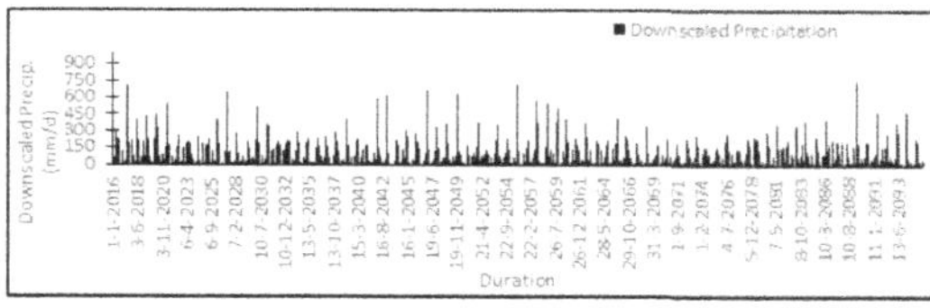

(i): downscaled precipitation at Sultanpur

Fig. 29.3 (continued)

29.4 Conclusion

In the present study, maximum temperature, minimum temperature, and rainfall are statistically downscaled using beta-regression [BR] technique for future period [up to 2095] by incorporating potential NCEP predictors [air temperature, relative humidity, sea level pressure, and surface pressure]. The BR technique is employed in MATLAB platform using future HADGEM2-ES GCM model outputs for Sai–Gomti interfluve region, Uttar Pradesh, India. The model performance statistics (Table 29.2) shows that developed model is satisfactorily able to capture the maximum temperature, minimum temperature, and rainfall. The calibrated and validated BR model is further applied for statistical downscaling of aforementioned climatic variables for future period (up to 2095). During the period 2016–2095, increase of 4 °C and 4.5 °C is observed in maximum temperature and minimum temperature, respectively.

Acknowledgements This work was supported by Science and Engineering Research Board (SERB), Department of Science and Technology (DST), Government of India, New Delhi under grant number Sb/s3/cee/0012/2014. The grant awarded for this study was used for financial support to the first author.

References

Aarnell NW, Gosling SN (2013) The impacts of climate change on river flow regimes at the global scale. J Hydrol 486:351-364

Bhowmik RD (2012) Statistical downscaling of global climatic variables to daily rainfall in a small spatial window in south-central India using neural networks & sdsm4.2 m.tech. Thesis, iit kanpur

Chiew FHS, Vaze J, Viney NR et al (2008) Rainfall-runoff modelling across the murray-darling basin. A report to the australian government from the csiro murray-darling basin sustainable yields project
Dessai S, Lu X, Hulme M (2005) Limited sensitivity analysis of regional climate change probabilities for the 21st century. J Geophys Res-atmos 110:1–17
Ferrari S, Cribari-neto F (2004) Beta regression for modelling rates and proportions. J Appl Stat 31:799–815
Fowler HJ, Blenkinsop S, Tebaldi C (2007) Linking climate change modelling to impacts studies: recent advances in downscaling techniques for hydrological modelling. Int J Climatol 27:1547–1578
Haas R, Apinto JG (2012) A combined statistical and dynamical approach for downscaling large-scale footprints of european windstorms. Geophys Res Lett 39:123804
Kalnay E, Kanamitsu M, Kistler R et al (1996) The NCEP/ncar reanalysis project. Bullet Amer Meteorol Soc 77:437–471
Mandal S, Srivastav RK, Simonovic SP (2016) use of beta regression for statistical downscaling of precipitation in the campbell river basin, british columbia, canada. J. Hyd 538:49–62
Martin E, Timbal B, Brun E (1997) Downscaling of general circulation model outputs: simulation of the snow climatology of the french alps and sensitivity to climate change. Clim Dyn 13:45–56
Prudhomme C, Crooks S, Kay AL, Reynard N (2013) Climate change and river flooding: part 1 classifying the sensitivity of british catchments. Clim Change119:933–948
Sachindra DA (2014) Catchment scale downscaling of hydroclimatic variables from general circulation model outputs. Phd thesis, victoria university, melbourne australia
Thomas P, Swaminathan A, Lucas RM (2012) Climate change and health with an emphasis on interactions with ultraviolet radiation: a review. Glob Chang Biol 18:2392–2405
Tripathi S, Srinivas VV, Nanjundiah RS (2006) Downscaling of precipitation for climate change scenarios: a support vector machine approach. J hydrol 330:621–640
Wilby RL, Wigley TML (2000) Precipitation predictors for downscaling: observed and general circulation model relationships. Int J Climatol 20:641–661
Ziska LH, Bunce JA, Shimono H et al (2012) Food security and climate change: on the potential to adapt global crop production by active selection to rising atmospheric carbon dioxide. Proc R Soc Biol Sci 279:4097–4105

Chapter 30
Comparing Global High-Resolution Precipitation Data with Rain Gauge Data in Assam, India

Pulendra Dutta, Dipsikha Devi, and Arup Kr. Sarma

Abstract Rainfall is the main driving variable to impact the hydrologic model results. In the context of unavailability of actual rainfall data, researchers use global data, derived from the satellite for hydrologic assessment of a watershed. In this study, Climate Forecast System Reanalysis (CFSR) precipitation data are tested for its reliability in regard to Assam, India; which is characterized by a wide spatial variation of rainfall. Based on the observed wet days' records, the CFSR precipitation datasets are compared with two sets of actual gauge rainfall data, viz. Indian Meteorological Department (IMD) data and Tea Garden data. From the results of this study, the CFSR data are found to be 34.94% deficient in annual rainfall volume than the actual gauge measurements.

Keywords Hydrologic model · Global data · Gauge data · Watershed

30.1 Introduction

In the present era, there are numerous tools available for assessment of hydrologic response from a watershed. With the advent of computer-based software, the hydrological models are found to be much capable to provide necessary information about the ground response. However, it is difficult to generate accurate water resource availability information without adequately accurate climate (temperature, rainfall, evaporation, etc.) input information.

The ground observations for weather variables are bare, especially in developing and under-developed countries due to lack of available resources (Hughes 2006). The observed rainfall records are often inadequate, irregular and frequently contain discrepancies (Koutsouris et al. 2016). It is, therefore, the researchers undergo difficulties in watershed modeling due to the data scarcity situations. However, these

P. Dutta · D. Devi
IIT Guwahati, Guwahati, Assam, India

A. Kr. Sarma (✉)
Civil Engg, IIT Guwahati, Guwahati, Assam, India
e-mail: aks@iitg.ac.in

R. Jha et al. (eds.), *Climate Change Impacts on Water Resources*, Water Science and Technology Library 98, https://doi.org/10.1007/978-3-030-64202-0_30

difficulties are overcome with the advancement of the remote-sensing techniques through which monitoring rainfall over large land areas including mountainous and oceanic regions has become possible. These data provide more homogeneous quality, although the time series accuracy is lower than the existing and the corresponding rain gauge-based observations datasets (Schulz et al. 2009).

In the event of sufficient available data, it is necessary to statistically analyze the datasets (Nyatuame et al. 2014), especially for variability and trend characteristics before using for water resources studies. In a study, Bora and Borah (2017) carried out some statistical analysis on 12 IMD stations rainfall data over Northeast India. Using l-moment test, they found low values of discordancy measure for all the stations' data than a critical value of 2.757 (Hosking and Wallis 1993); and concluded that development of regional maximum rainfall estimation was required for all the IMD stations data. Besides, the use of statistical analysis has been forwarded in some region-specific studies (Elamir and Sheult 2003; Trefry et al. 2005) for investigating the extreme rainfall events. Whereas, Shabri et al. (2011) investigated the extreme value distribution of rainfall data over Malayasia using two different approaches viz. L-Moment and TL-Moment methods, and they found the second method as more efficient in the estimation of rainfall statistics.

In the absence of actual measured data, the researchers are bound to adopt synthetic weather data, which need analysis in terms of its evaluation and inter-comparison to test its applicability of use for a particular watershed (Koutsouris et al. 2016). There have been numerous region-specific statistical analyses in order to check the applicability of synthetic weather data (Todd et al. 1999; Thorne et al. 2001; Grimes and Diop 2003). The application of different statistical methods was described in many kinds of literatures (Sapiano and Arkin 2008; Dinku et al. 2011; Romilly and Gebremichael 2011; Liechti et al. 2012; Guo and Liu 2016; Rossi et al. 2017) to evaluate and compare the global, gridded rainfall with respect to the actually measured data at gauge level. So, it is important to analyze the global rainfall datasets prior to applying for a watershed modeling since it may not possess a good agreement to the gauge observations. As for example, the radar estimated rainfall significantly differs from the gauge rainfall (Barnston and Thomas 1983; Masunga et al. 2002). In another study by Espinosa et al. (2015), a large amount of disagreement was observed in the radar estimated data, while compared with the actually observed rainfall over South California, USA.

Rainfall is the main influencing factor for the hydrologic assessment from a basin. The hydrologic models cannot provide desired water availability information without the use of adequately correct rainfall values as input data. As there are insufficient actual observed rainfall data to represent temporal and spatial variations across Assam, India; the use of global precipitation data becomes popular. In this study, the global CFSR (Climate Forecast System Reanalysis) precipitation data are analyzed by comparing with IMD as well as Tea-Garden data to justify whether these data are reasonably applicable towards hydrologic assessment over Assam, India.

30.2 Study Area, Materials, and Methods

Assam, one of the eight Northeast India's states is situated toward the south of the eastern Himalayas covering an area of 78,438 km^2 (Fig. 30.1). It is located toward the west of 96 °E and toward the south of 28 °N and lies in the middle reach of the Brahmaputra and Barak rivers. The maximum temperature of Assam is 35–38 °C during summer, whereas the minimum temperature during winter falls to 6–8 °C. Moreover, the state of Assam experiences heavy rainfall that is more than many parts of India.

The present study aims to test acceptance of CFSR precipitation data over Assam, India so as to help the hydrologic community in studying watershed response. The three sources of data used in this analysis are: (i) Global precipitation provided by https://globalweather.tamu.edu, as the CFSR data, (ii) Gauge rainfall data (station codes: s1, s2, s3, s4) collected from Indian Meteorological Department (IMD), and (iii) Gauge rainfall data (station codes: s5, s6, s7, s8) collected from various tea estates across Assam. The eight global weather stations are chosen at the most proximity to the corresponding gauge weather stations.

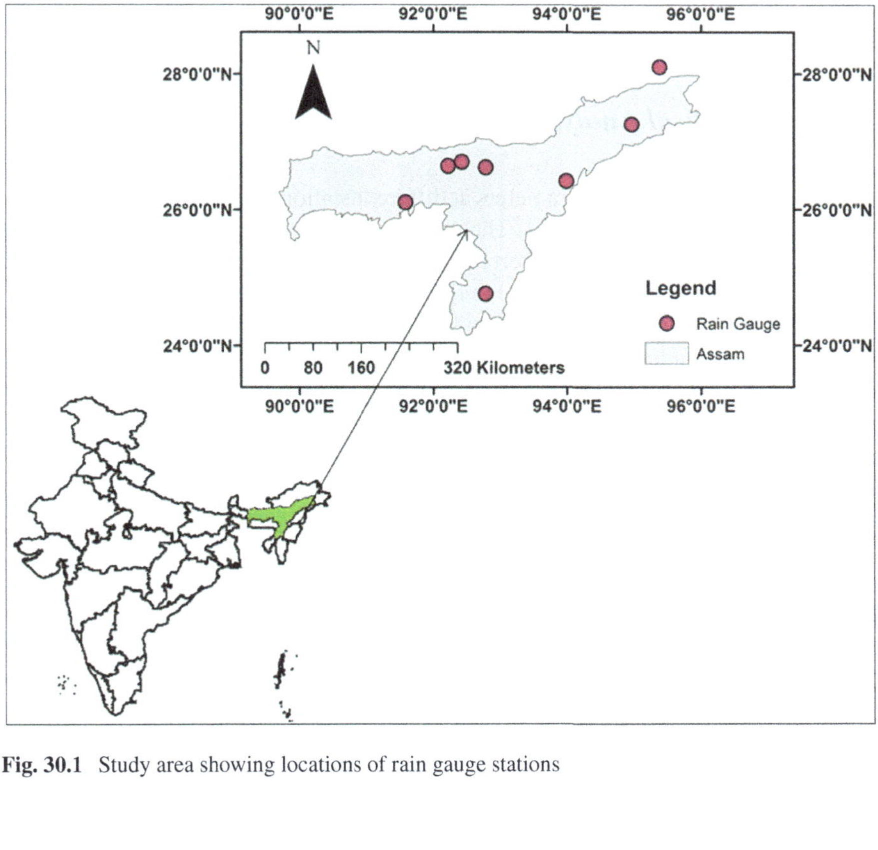

Fig. 30.1 Study area showing locations of rain gauge stations

For our analysis, we used daily rainfall values of the global datasets and then compared with respect to the corresponding values of the gauge datasets during 2006–2011. Due to the lack of continuous data in the gauge records, longer periods could not be taken up for this study. No gap filling was required for the gridded global data, however very few days if found with missing values in gauge records are ignored and omitted in this study. Since the observed stations suffer to record continuous data over a long period, due to malfunctioning of rain measuring instrument or due to some other reasons, more stations could not be taken up in this study, though one station is considered outside the boundary of the study area.

The statistical parameters of rainfall values are calculated separately for each year and then averaged over 6 years during 2006–2011, for both the gauge and global datasets. The rainfall values of the wet days only are taken up in this study, since global datasets contain more homogeneous and continuous daily records. Wet days are considered as those days which actually experience rainfall events according to the gauge observations. The total annual rainfall at a station is obtained by summing up the wet days' rainfall values in a year, and then averaged over the entire duration of the study period.

30.3 Results and Discussion

30.3.1 Statistical Analysis

Table 30.1 shows statistical parameters at different stations for wet days rainfall over 6 years of data, during 2006–2011. The amount of annual precipitation at the global stations widely varies from the corresponding gauge stations' values. Dhekiajuli station suffers the highest difference in the annual average rainfall between the two data sets, the global one being less than half of the observed values. The significant variation in the standard deviation values indicates wide disagreement between the two datasets. Moreover, different skewness values represent large temporal variation between the duos.

30.3.2 Scatter Plot

Scatter plot of global rainfall values with respect to the corresponding rainfall values that are actually measured at rain gauges is shown for only two stations for the year 2006 (Fig. 30.2) as for example. Whereas similar trends were also observed for the other years at these two stations, and for the other six stations during 2006–2011. A perfect match between the values of the two datasets forms a straight line. But, it is clear from the figure that the values are highly scattered, and hardly concentrate along the 1:1 straight line. This implies that the CFSR values for precipitation are

Table 30.1 Summary of Rainfall statistics indicating average values over 6 years Daily Rainfall data during 2006–2011

Station Name (code)	Source	Annual (mm)	Mean (mm)	Maximum (mm)	Standard Deviation (mm)	Variance	Skewness
Guwahati (S1)	Gauge	1854	13.26	113.62	18.69	349.31	3.44
	Global	2492.4	17.82	116.41	23.52	547.75	1.72
Passighat (S2)	Gauge	3895.3	22.69	186.34	30.72	943.72	2.87
	Global	3577.8	19.13	131.56	21.65	468.93	1.81
Silchar (S3)	Gauge	3328.3	17.81	127.40	21.20	449.44	3.63
	Global	3263.2	17.46	152.93	24.67	608.44	3.61
Tezpur (S4)	Gauge	1738.3	11.86	91.73	16.45	270.60	3.04
	Global	1721.2	11.74	109.42	17.85	318.62	3.85
Dhekiajuli (S5)	Gauge	1563.1	15.16	70.25	14.49	209.96	2.55
	Global	709.6	6.88	80.67	12.09	146.17	2.91
Golaghat (S6)	Gauge	1398.5	13.74	78.84	15.20	231.04	2.08
	Global	831.2	8.17	76.36	13.67	186.86	1.97
Mangaldoi (S7)	Gauge	1612.9	12.99	95.73	15.86	251.22	3.03
	Global	1436.3	11.56	123.67	17.30	299.29	1.45
Dibrugarh (S8)	Gauge	1409	10.61	63.75	11.31	127.91	3.02
	Global	725	5.45	53.29	9.06	82.01	3.34

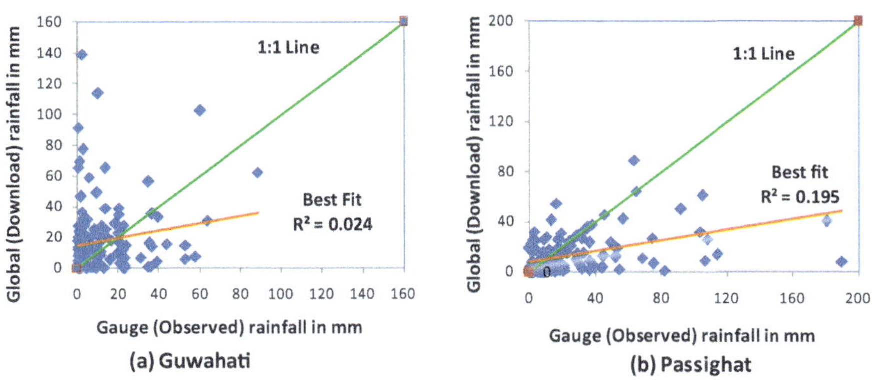

Fig. 30.2 Scatter plot of global precipitation versus gauge precipitation during 2006 at weather stations: **a** Guwahati, and **b** Passighat

different from the corresponding observed values. Since the daily records are highly scattered, hence the comparison and the analysis is further forwarded on the basis of annual rainfall volumes.

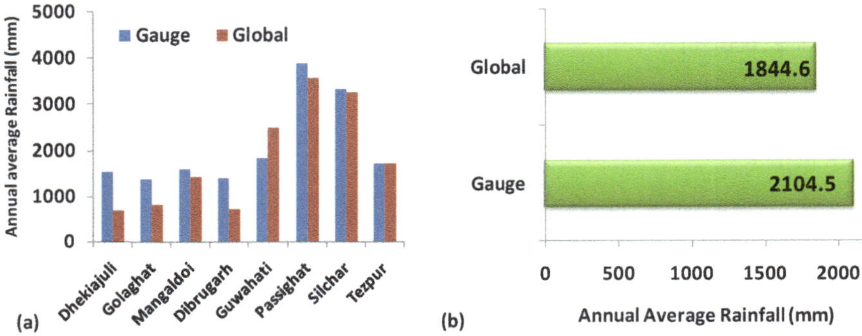

Fig. 30.3 Comparison of annual average rainfalls **a** station wise, **b** all stations

30.3.3 Comparison of Annual Rainfall

The annual average rainfall at each weather station is shown in Fig. 30.3a. Here, the wet days' rainfall is summed up during the individual years and then averaged over 6 years of data at the corresponding stations. It is evident from the results that the annual average rainfall as recorded by the CFSR global stations falls below the corresponding records at the actual gauge observations, except Guwahati station. While considering all the stations as a lump (Fig. 30.3b), the average annual rainfall is only 1844.6 mm for the global data sets against 2104.5 mm at the gauge level.

Table 30.2 shows the calculation for percent variation in annual rainfall of the global datasets over the gauge measurements by considering all the stations over the entire duration of this study. The annual rainfall values at all the global stations are found to be lesser than the actual measurements, except the Guwahati station where the gauge observation is found 25.61% deficient than the global annual rainfall. The actual gauge values fall higher than the global rainfall, even more than cent percent for Dhekiajuli station. While averaging over all the stations' values, it is evident that the global measurement is sufficiently lower than the gauge measurements.

30.4 Conclusions

The climate forecast system reanalysis (CFSR) data, designed by the National Centers for Environmental Prediction (NCEP), include all conventional and satellite observations. There are only few IMD weather stations across Assam, a majority of which lack continuous daily records over a long period. Tea garden stations are therefore added for the present study to cover better spatial representations. It is obvious from the results that the CFSR precipitation values fall below the actual gauge precipitation measurements. While considering all stations across the present study area, the average annual precipitation values at the CFSR global stations are found to be

Table 30.2 Comparison of datasets based on annual rainfall variations

Particulars	Guwahati	Passighat	Silchar	Tezpur	Dhekiajuli	Golaghat	Mangaldoi	Dibrugarh
Gauge stations' annual rainfall (mm) i	1854	3895.3	3328.3	1738.3	1563.1	1398.5	1612.9	1409
Global stations' annual rainfall (mm) ii	2492.4	3577.8	3263.2	1721.2	709.6	831.2	1436.3	725
Difference. (mm) (i–ii)	−638.4	+317.5	+65.3	+26.1	+853.5	+567.3	+176.6	+684
Variation of gauge data over global data (%)	−25.61	+8.87	+2.00	+1.52	+120.23	+65.87	+12.30	+94.34
Average variation of gauge data over global data (%)	+34.94							

34.94% deficient than the gauge observations. However, the tea garden datasets are seen to provide more disagreement to the global rainfall values than that of the IMD stations datasets. This may be due to the fact that references to IMD datasets were probably made while generating the global datasets. This analysis may be helpful to the research community while assessing hydrologic response from a watershed over the present study area.

References

Barnston AG, Thomas JL (1983) Rainfall measurement accuracy in face: a comparison of gauge and radar rainfall. J Climate Appl Meteorol 22:2038–2052

Bora DJ, Borah M (2017) A statistical study on rainfall frequency analysis of north east india. phd thesis, dept. Of mathematical sciences, tezpur university, assam, india

Dinku T, Connor S, Ceccato (2011) Evaluation of satellite rainfall estimates and gridded gauge products over the upper blue nile region. In: nile river basin, melesse am (ed). Springer: dordrecht, 109–127. https://doi.org/10.1007/978-94-007-0689-7

Elamir EA, Seheult AH (2003) Trimmed l-moments. J Comput Statist Data Anal 43:299–314

Espinosa B, Hromadka TV, Perez R (2015) Comparison of radar data versus rainfall data. J methodsx (sciencedirect), methodsx 2:423–431. https://doi.org/10.1016/j.mex.2015.10.007

Grimes D, Diop M (2003) Satellite-based rainfall estimation for river flow forecasting in africa. I: rainfall estimates and hydrological forecasts. J Hydrol Sci 48(4)

Guo R, Liu Y (2016) Evaluation of satellite precipitation products with rain gauge data at different scales: implications for hydrological applications. www.mdpi.com/journal/water, https://doi.org/10.3390/w8070281

Hosking JRM, Wallis JR (1993) Some statistics useful in regional frequency analysis. J Water Resour Res 29(2):271–281

https://globalweather.tamu.edu

Hughes DA (2006) Comparison of satellite rainfall data with observations from gauging station networks. J Hydrol (2006) 327:399– 410

Koutsouris, a. J., chen, d., and lyon, s. W. 2016."comparing global precipitation data sets in eastern africa: a case study of kilombero valley, tanzania. Int J Climatol 36:2000–2014

Liechti CT, Matos JP, Boillat JL, Schleiss AJ (2012) Comparison and evaluation of satellite derived precipitation products for hydrological modeling of the zambezi river basin. J Hydrol Earth Syst Sci 16:489–500. https://doi.org/10.5194/hess-16-489-2012

Masunaga H, Iguchi T, Oki R, Kachi M (2002) Comparison of rainfall products derived from trmm microwave imager and precipitation radar. Amer Meteorol Soc 41

Nyatuame M, Gyimah VO, Ampiaw F (2014) Statistical analysis of rainfall trend for volta region in ghana. Int J Atmosph Sci. https://doi.org/10.1155/2014/203245

Romilly TG, Gebremichae M (2011) Ealuation of satellite rainfall estimates over ethiopian river basins. J Hydrol. Earth Syst Sci 15:1505–1514. https://doi.org/10.5194/hess-15-1505-2011

Rossi M, Kirschbaum D, Valigi D, Mondini AC, Guzzetti F (2017) Cmparison of satellite rainfall estimates and rain gauge measurements in italy, and impact on landslide modeling. J Climate 5:90. https://doi.org/10.3390/cli5040090

Sapiano MRP, Arkin PA (2008) Anintercomparison and validation of high-resolution satellite precipitation estimates with 3-hourly gauge data. J Amer Soc. https://doi.org/10.1175/2008jhm1052.1

Schulz J, Albert P, Behr HD, Caprion D, Deneke H, Dewitte S, Durr B et al (2009) Operational climate monitoring from space: the eumetsat satellite application facility on climate monitoring (cm-saf). Atmos Chem Phys 9:1687–1709. https://doi.org/10.5194/acp-9-1687

Shabri AB, Doud ZM, Ariff NM (2011) Regional analysis of annual maximum rainfall using tl-moments methods. J Theor Appl Climatol 104:561–570

Thorne V, Coakley P, Grimes D, Dugdale G (2001) Comparison of tamsat and cpc rainfall estimates with rainfall, for southern africa. Int J Remote Sens 22(10):1951–1974

Todd MC, Barret EC, Beaumont MJ, Bellerby TT (1999) Estimation of daily rainfall over the upper nile river basin using a continuously calibrated satellite infrared technique. J Meteorol Appl 6(3):201–210

Trefry C, Watkins D, Johnson D (2005) Regional rainfall frequency analysis for the state of michigan. J Hydrol Eng 10(6):437–449

Chapter 31
Variability of Rainfall, Temperature and Potential Evapotranspiration at Annual Time Scale Over Tapi to Tadri River Basin, India

Prem Mahyavanshi, V. D. Loliyana, and Priyank J. Sharma

Abstract The change in the meteorological variables is one of the most important factors that affect crop water requirements, and subsequently, water allocation for food production in agriculture-based countries like India. Present study evaluates the application of statistical trend detection tests and examines the magnitude slope of trends in climatic variables viz., rainfall, potential evapotranspiration (PET) and temperature over the west-flowing rivers in the Tapi to Tadri basin, India. In the present study, high-resolution daily gridded rainfall dataset of India Meteorological Department (IMD) at $0.25^\circ \times 0.25^\circ$ resolution, while the PET and temperature data of Climate Research Unit (CRU) at $0.5^\circ \times 0.5^\circ$ resolution have been analyzed for period of 116 years (1901–2016) at annual scale. The trends in aforesaid climate variables have been detected using nonparametric Mann–Kendall (MK) and Modified Mann–Kendall (MMK) tests, and the slope of trend magnitude is computed from Sen's Slope (SS) test. The results indicated increasing trend in annual rainfall across 61% grids, while decreasing across 37% grids and no trends were observed at remaining grids out of 119 grids. Further, increasing trend in potential evapotranspiration, maximum, mean and minimum temperatures were observed at all the 40 grids. Thus, increase in temperature was greatly responsible for increasing trends in potential evapotranspiration across the study region. The outcomes of the present study provide insight of the climate variability and interaction among the meteorological variables across the Tapi to Tadri basin for the study period.

P. Mahyavanshi
Department of Civil and Construction Engineering, University of Wolverhampton, Rustomjee Academy for Global Careers, Dahanu Road (East), Dahanu 401602, Maharashtra, India
e-mail: mahyavanshiprem@gmail.com

V. D. Loliyana (✉)
Department of Civil and Construction Engineering, University of Wolverhampton, Rustomjee Academy for Global Careers, Dahanu Road (East), Dahanu 401602, Maharashtra, India
e-mail: viraj.nishi@gmail.com

P. J. Sharma
DST INSPIRE Fellow, Department of Civil Engineering, Sardar Vallabhbhai National Institute of Technology, Surat 395007, Gujarat, India
e-mail: pjs230688@gmail.com

R. Jha et al. (eds.), *Climate Change Impacts on Water Resources*, Water Science and Technology Library 98, https://doi.org/10.1007/978-3-030-64202-0_31

Keywords Climatic variability · Meteorological variables · Nonparametric tests · Tapi to tadri basin

31.1 Introduction

The climate variability and its dynamics are the basis of several researches on climate change and their impacts. The variability in precipitation and temperature results in climate changes that require consideration owing to their impact on the water–energy–nexus (Dore 2005). Numerous trends and indices of climatic variables obtained from the empirical data indicate upward trend of average surface global temperature since the 1950s, and the highest rate of changes was detected since the middle of 1970s (Kundu et al. 2016; Quirk 2012; Alexander et al. 2006; Frich et al. 2002). Besides the basic meteorological variables viz., precipitation and temperature, evapotranspiration (ET) is an important climate variable and essential element of the hydrological cycle that has substantial implications in the agricultural fields. The determination of evapotranspiration is necessary for effective water resources management in an agriculturally based country like India. The changing patterns of potential evapotranspiration (PET) have consequences on crop production and different hydrological processes (Hobbins et al. 2004). As per Doorenbosand and Pruitt's (1977), "the PET takes place when the ground is completely covered by actively growing uniform green grass of 8 to 15 cm tall in abundance of soil moisture." PET could rise in most parts of the globe in near future as a result of water-retaining capacity of atmosphere maximizes with higher temperature such as 7% increase per 1 °C warming (Trenberth 2011). Subsequently, evapotranspiration rate is expected to intensify due to water vapor deficit. The factors affecting PET are the meteorological parameters, crop factor and soil moisture content. The meteorological parameters are humidity, windspeed, rainfall, temperature, duration of sunshine hours, vapor pressure, station level pressure, etc. Therefore, long-term analysis of historical data of climatic variables is substantially required in understanding the basin behavior from hydro-climatological perspective.

In recent past, number of researchers analyzed trends in mean and extreme precipitation as well as temperature (Zhai et al. 2005; Alexander et al. 2006; Martinez et al. 2012; Zhang et al. 2012; Boccolari and Malmusi 2013; Song et al. 2014; Sharma and Babel 2014; Soltani et al. 2016; Soro et al. 2016; Liu et al. 2016; Xiao et al. 2016; He et al. 2017; Caloiero 2017; Ye and Li 2017; Sharma et al. 2017), and largely reported their increasing trends across the globe. Few researchers reported trends in PET across various parts of world (Goyal 2004; Helfer et al. 2012; Huo et al. 2013; Liu et al. 2013, 2014; Haijun et al. 2014; Sonali and Kumar 2016). The Indian summer monsoon rainfall (ISMR) exhibits wide spatial and temporal variability across India because of influence of numerous local scale factors like variation in growth rate of population, rapid urbanization, deforestation, etc. Their study analyzed data at finer scale, demonstrating high spatial variability in climate variables due to local level changes (in terms of population and urbanization), which contradicts in results analyzed at larger scales (Ghosh et al. 2009; Sharma et al. 2017). Despite changing climate conditions, it is observed that very few studies are focusing on change detection in meteorological variables other than rainfall, viz., PET and temperature, at basin scale in India. The present study focuses on detection of long-term spatio-temporal changes in meteorological variables at finer spatial scales for west-flowing river basins from Tapi to Tadri. Further, short length of meteorological data may exhibit to specious inferences in trend analysis, hence, data set of 116 years (1901–2016) has been used in the present study. In this paper, the following objectives were studied out for Tapi to Tadri river basin, India using high resolution daily gridded rainfall data (119 grids) at $0.25° \times 0.25°$ resolution, while the PET (40 grids) and temperature (40 grids) data at $0.5° \times 0.5°$ resolution: (a) computation of the primary statistical parameters (mean, standard deviation, coefficient of variation, skewness, kurtosis) of the annual time series for 116 years (1901–2016), (b) trend detection of annual series by application of non-parametric methods namely Mann-Kendall, Modified Mann-Kendall and Sen's slope tests.

31.2 Study Area

The Tapi to Tadribasin, see Fig. 31.1 (area $\approx$55,940 km^2), exhibits an array of topographic elements ranging from laterite platforms, alluvial tracts along lagoons or estuary or rivers, coastal sand dunes or flats of mud, sandy beaches and erosional surfaces in the residual hills or hard basement rock. With respect to physiographical nature, entire basin is distinct corridor of low land, having hills whose elevations rise up to 150 m at places to an excess of 300 m. The steep face of Sahyadris runs parallel to the basin and unveils an elevation between 760 m and 1220 m. Along the western coast, plains of eroded marines and elevated beaches which were formed in earlier times are found at height varying from 30 to 90 m respectively. It can be observed that the highest recorded elevation of the basin is at 1402 m (Jain et al. 2007). The watershed area showcases a humid and moist atmosphere climate along the coastal belt which is relatively hot with higher degree of humidity reaching an apex of 90%. The south-west and north-east monsoons control the summer and winter climate, with the winter beginning by December. The rainfall in the region is directly influenced by south-west monsoon, which is usually heavy and assured between June and August periods. Between the months of June and November, 90% of the rainfall is received. The peak rainfall received around Vasisthi was 2539.76 mm in 1975, whereas the minimum annual rainfall in entire basin was 1601.7 mm in year 2002. Out of all, Tillari dam, an interstate project between Goa and Maharashtra; Damanganga dam, an interstate project between Gujarat, Daman and Diu, and Dadra and Nagar Haveli; Surya Dam in Thane district of Maharashtra and Anjunem Irrigation Project, located over Costi Nadi, Goa are some of the important projects in the basin. Also, the Koynaand Kalinadi hydroelectric projects are two of the most important hydroelectric projects in the Tapi to Tadri basin. The agricultural land (44%) is dominant land use in the basin followed by forest land (35%) (India-WRIS report 2014).

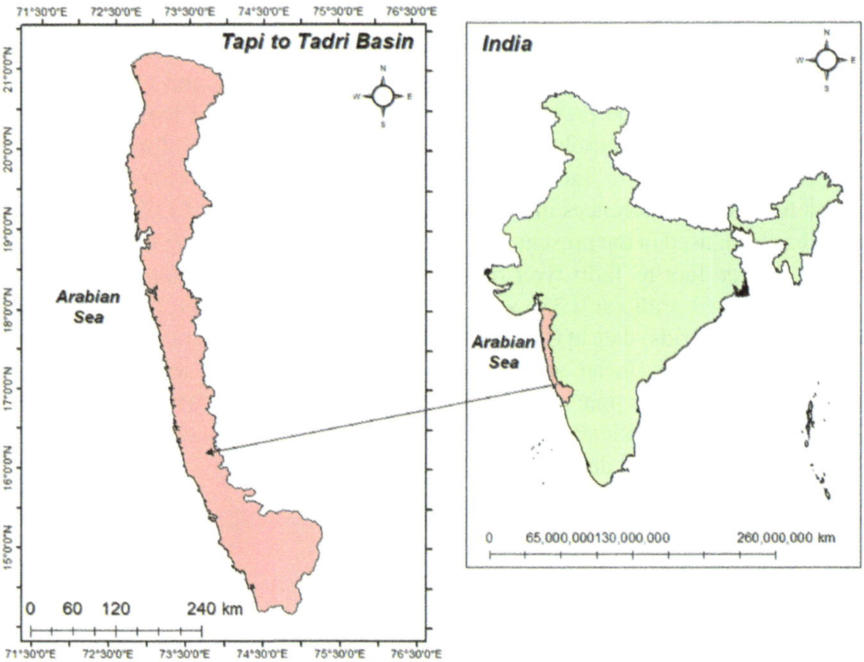

Fig. 31.1 Index map of study area

31.3 Materials and Methods

The daily rainfall data procured from India Meteorological Department (IMD), distributed in a gridded format (0.25° × 0.25°), over the basin were used as input data for the study. The daily temperature and potential evapotranspiration data downloaded from Climate Research Unit (CRU) website (https://climatedataguide.ucar.edu/ accessed on June 13, 2018), distributed in a gridded format (0.5° × 0.5°), over the basin were also considered for the analysis, see Fig. 31.2.

31.3.1 Preliminary Statistical Analysis

The primary statistical parameters such as maximum, minimum, mean, standard deviation, coefficient of variation, skewness, and kurtosis were computed for annual rainfall, temperature, and PET time series at each grid for period 1901–2016. It is observed that overall mean and standard deviation of total annual rainfall in the Tapi to Tadri basin are 2396.8 mm and 654.3 mm, respectively, indicating it as a zone of high rainfall compared with other parts of the country.

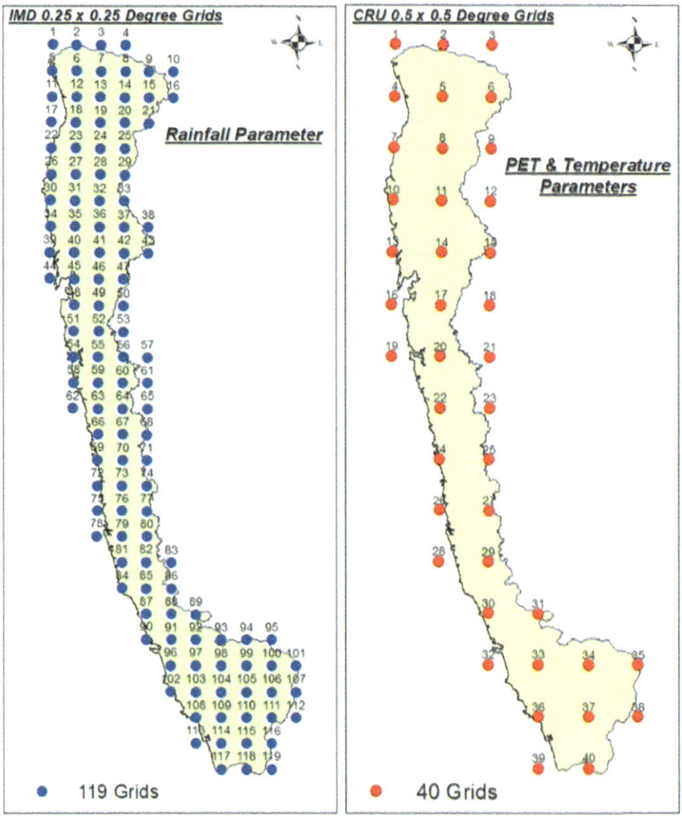

Fig. 31.2 IMD and CRU data grids for Tapi to Tadri river basin

The Tapi to Tadri basin receives heavy rainfall due to orographic influence of the Western Ghats which bring heavy rainfall in the region. The average temperature in the basin varies from 21.9 to 32.5 °C during the entire year, while the highest and lowest temperatures recorded in the basin have been in May and January months respectively. The mean PET is about 1568 mm within the Tapi to Tadri basin. As per Koppen-Geigger climate classification (Peel et al. 2007), the Tapi to Tadri basin falls in Monsoon climate (Am). This region receives rainfall principally during the monsoon season and has short dry winter season.

31.3.2 Methodology

The stepwise procedure adopted to ascertain the long-term temporal and spatial variability in rainfall, temperature and PET data is included in Fig. 31.3. Before applying the trend test, the data were examined for the presence of serial correlation,

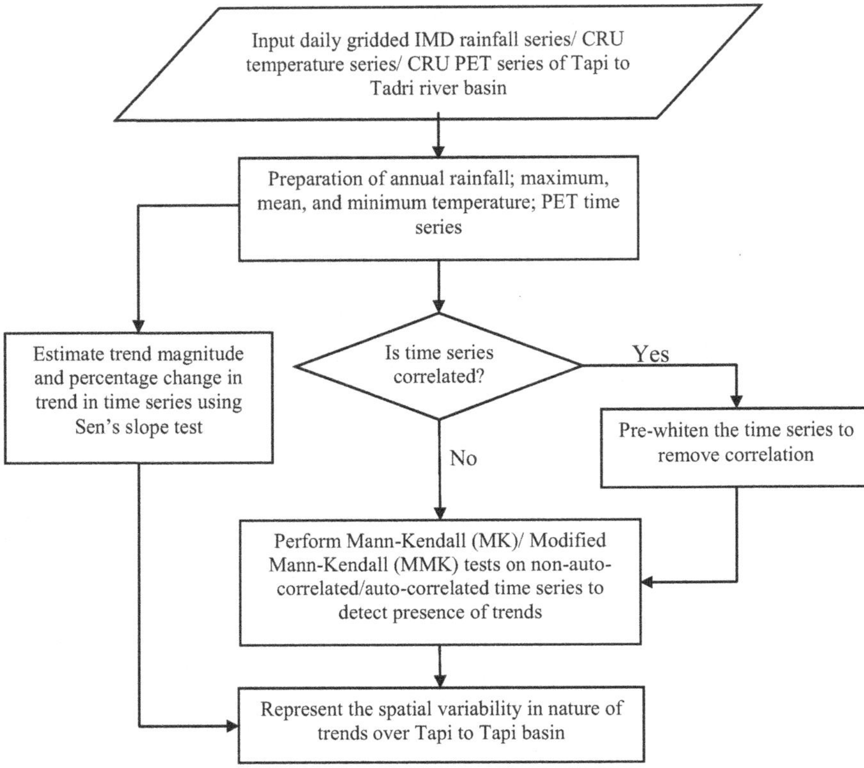

Fig. 31.3 Methodology adopted in the present study

which needs to be removed before applying Mann–Kendall test (Mann 1945; Kendall 1975). Autocorrelation is a common problem in time-series analysis and can cause the data to be biased to an extent. If there exists significant serial correlation, prewhitening process has to be performed on time series to remove autocorrelation. The modified Mann–Kendall test (Hamed and Rao 1998) has a distinct advantage over the Mann–Kendall test that it can be applied to correlated data also. After detecting the nature of trends, the Sen's slope estimator method (Hirsch et al. 1982) is applied to estimate the slope of trend magnitude.

31.4 Results and Discussions

31.4.1 Trends in Rainfall

The nonparametric tests such as MK, MMK and Sen's slope estimator tests have been applied on total annual rainfall series at each grid. The results indicated increasing trend in annual rainfall across 61% grids (72 grids), while decreasing across 37% grids (44 grids) and no trends were observed at remaining three grids out of 119 grids. Further, from MMK Z-statistic values, 49 grids exhibited significant increasing trends at 5% significance level, of which 43 grids exhibited significant increasing trend at 1% significance level also, see Table 31.1. On other hand, 17 grids exhibited significant decreasing trend at 5% significance level, of which 13 grids showed significant decreasing trend at 1% significance level also, see Table 31.1. Moreover, the interior grids (i.e., away from the Arabian Sea and close to the Western Ghats) exhibited larger decreasing trends vis-à-vis the exterior grids (i.e., near to the Arabian Sea), see Fig. 31.4a. These might have resulted due to deforestation activities in the forest area along the Western Ghats and rapid pace of urbanization in the basin. The Tapi to Tadri basin being an agriculture dominated basin depends largely on monsoon rainfall for its water needs. The decrease in rainfall would affect the water requirements for the Kharif crops being cultivated in the basin. The maximum and minimum values of Sen's slope for total annual rainfall are found to be 35.36 mm/year and −20.50 mm/year at Grid-80 and Grid-50 respectively, see Table 31.1.

31.4.2 Trends in Temperature

The maximum mean and minimum average temperature series of each grids at annual time scales were investigated for the presence of trends. The trend detection carried out using MK, MMK and, Sen's slope estimator tests yielded results which agreed with each other. From the trend analyses of maximum, mean, minimum annual average temperature, increasing all 40 grids exhibited increasing trend across the basin (except at Grid-11 for mean temperature which exhibited no trend), see Table 31.2 and Fig. 31.4c–e. The increasing trend in maximum, mean and minimum temperature at all the grids has been found to be significant at 1% significance level. Thus, the temperature in the basin is found to severely rise over the past 116 years. The Tapi to Tadri river basin, located nearer to the ocean in the west–south core of India, this increase in temperature would affect the water availability in the basin. The overall increase in temperature implies increase in water vapor and its circulation pattern in the atmosphere and, which could increase the frequency of occurrence of extreme events in the basin. However, the linkages of increase in temperature and their impact on frequency of extreme events in Tapi to Tadri basin are not discussed in present study and would form an interesting hypothesis for exploration by the research community. The maximum and minimum values of Sen's slope for maximum, mean and minimum temperatures are found to be 0.0076°C/year and 0.0050°C/year at Grid-40 and Grid-18;0.0071°C/year and 0.0052°C/year at Grids-38 &40 and Grid-18; 0.0073°C/year and 0.0054°C/year at Grid-6 and Grid-18 respectively, see Table 31.2.

Table 31.1 Results of trend analysis of total annual rainfall series

Grid	MKZ	MMKZ	Sen β (mm/year)	Grid	MKZ	MMKZ	Sen β (mm/year)	Grid	MKZ	MMKZ	Sen β (mm/year)
1	2.04	2.04	2.43	25	−1.44	−3.00	−2.24	49	0.16	0.15	0.48
2	0.18	0.16	0.24	26	3.87	3.87	6.08	50	−5.45	−6.31	−20.90
3	0.89	0.89	1.19	27	3.91	4.13	6.28	51	2.26	2.29	4.41
4	−0.35	−0.89	−0.46	28	0.17	0.31	0.23	52	5.10	5.04	11.27
5	2.50	2.50	3.38	29	−2.09	−2.20	−3.16	53	−4.33	−3.26	−13.20
6	1.67	1.99	2.14	30	3.56	3.67	5.44	54	1.28	1.66	2.43
7	1.12	1.05	1.52	31	1.02	0.80	1.82	55	4.07	4.07	7.61
8	−0.96	−0.89	−1.28	32	0.53	0.76	0.86	56	−3.82	−3.27	−9.92
9	0.57	0.57	0.52	33	−0.25	−0.26	−0.51	57	−4.57	−3.99	−6.64
10	3.03	2.72	1.49	34	2.41	2.97	3.47	58	−2.07	−2.45	−3.93
11	1.62	1.95	2.38	35	−3.09	−3.09	−5.64	59	1.76	1.76	3.56
12	−0.04	−0.05	−0.10	36	−3.66	−7.36	−6.92	60	−1.46	−1.77	−3.84
13	−0.17	−0.17	−0.33	37	1.93	1.82	3.16	61	0.94	0.81	0.92
14	2.56	2.94	4.00	38	−0.10	−0.17	−0.06	62	−2.27	−2.27	−5.06
15	1.37	1.46	1.82	39	2.78	3.97	4.11	63	1.60	1.55	3.24
16	3.72	3.72	1.86	40	−3.86	−3.86	−6.95	64	0.00	0.00	−0.02
17	1.48	1.77	2.11	41	−5.69	−6.71	−13.83	65	−1.03	−1.03	−1.11
18	−0.08	−0.07	−0.12	42	−5.29	−5.73	−17.52	66	2.13	3.92	3.76
19	−0.64	−0.64	−1.16	43	−2.57	−2.09	−2.04	67	1.03	1.03	1.79
20	1.83	1.79	2.69	44	2.72	2.72	3.72	68	−1.54	−1.54	−1.86
21	4.97	4.13	4.30	45	0.51	0.00	0.88	69	2.41	3.10	4.07
22	3.66	3.66	5.14	46	−4.10	−4.10	−10.33	70	3.33	8.51	5.14
23	2.94	2.94	5.02	47	−5.35	−8.97	−19.22	71	−1.40	−1.40	−1.39
24	−0.76	−0.82	−1.27	48	0.22	0.23	0.33	72	4.87	5.14	8.88

(continued)

Table 31.1 (continued)

Grid	MKZ	MMKZ	Sen β (mm/year)	Grid	MKZ	MMKZ	Sen β (mm/year)
73	4.61	4.61	8.04	97	7.08	8.69	16.03
74	6.66	6.20	9.08	98	−1.70	−1.43	−2.35
75	4.40	3.78	8.04	99	0.42	0.42	0.24
76	3.99	4.36	8.29	100	−0.10	−0.09	−0.08
77	3.72	2.81	8.42	101	0.37	0.32	0.16
78	3.85	3.85	5.51	102	4.41	2.63	10.50
79	5.36	5.36	13.70	103	6.97	3.06	16.54
80	6.07	3.52	35.36	104	−0.98	−1.27	−1.72
81	5.74	5.74	11.66	105	−1.62	−1.16	−2.09
82	4.94	5.70	10.78	106	3.03	2.31	1.77
83	−0.83	−1.11	−1.06	107	−0.09	−0.09	−0.06
84	2.87	3.27	4.25	108	2.00	0.00	3.20
85	−0.48	−0.58	−0.97	109	4.40	3.92	8.72
86	−4.91	−4.26	−10.79	110	−0.47	−0.47	−0.70
87	0.93	0.99	1.54	111	3.77	2.86	2.52
88	4.75	3.49	12.74	112	−0.27	−0.25	−0.14
89	6.25	3.89	13.72	113	−0.05	−0.07	−0.14
90	3.18	3.92	4.47	114	6.36	3.69	18.67
91	6.43	4.92	19.87	115	2.20	2.53	3.71
92	7.71	8.64	22.24	116	−1.88	−1.44	−1.93
93	−0.45	−0.39	−0.58	117	5.07	5.07	10.11
94	1.30	1.30	0.76	118	4.80	4.80	8.76
95	−1.55	−1.67	−1.03	119	−0.19	−0.15	−0.23
96	4.31	7.54	10.91	–			

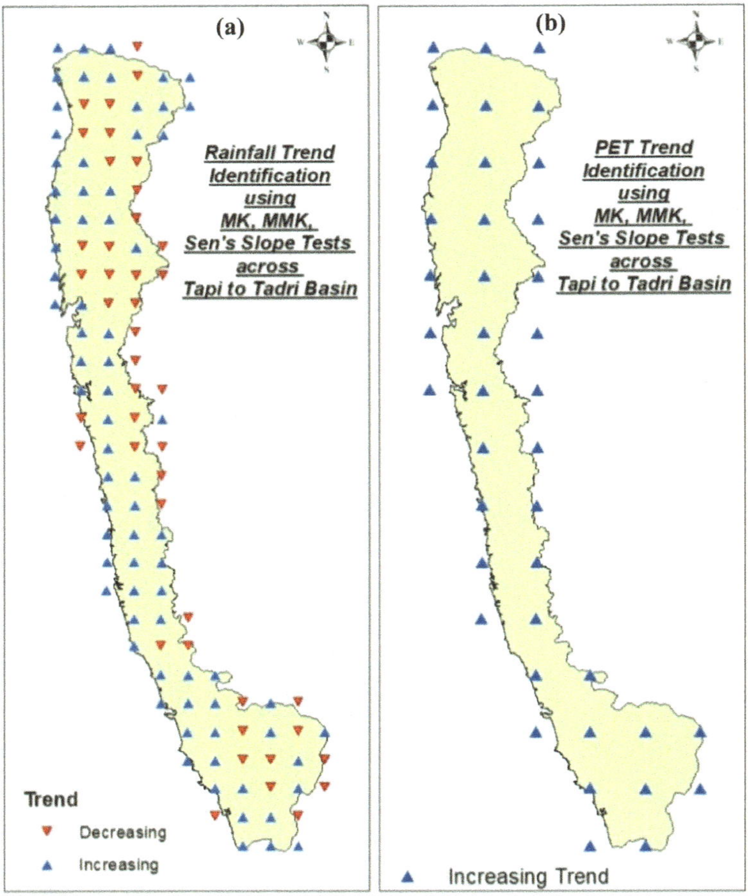

Fig. 31.4 Trends in annual time series of **a** total rainfall, **b** PET, **c** maximum temperature, **d** mean temperature and **e** minimum temperature over Tapi to Tadri river basin

31.4.3 Trends in PET

The MK, MMK, and Sen's slope estimator tests have been applied on PET series of each grids at annual time scales. The trends in annual PET exhibited increasing trend at all 40 grids across the basin, see Fig. 31.4b. Further, from MMK Z-statistic values, 23 out of 40 exhibited significant increasing trend at 5% significance level, of which 17 grids showed significant increasing trend at 1% significance level also. The increase in PET is directly influenced by significant increase in temperature across the basin over the period of 116 years. The maximum and minimum values of Sen's slope for annual average PET are found to be 0.279 mm/year and 0.018 mm/year at Grids-36 & 40 and Grid-18 respectively, see Table 31.2. The highest increase in PET is observed in the southern part of the basin, like maximum and mean temperature, with

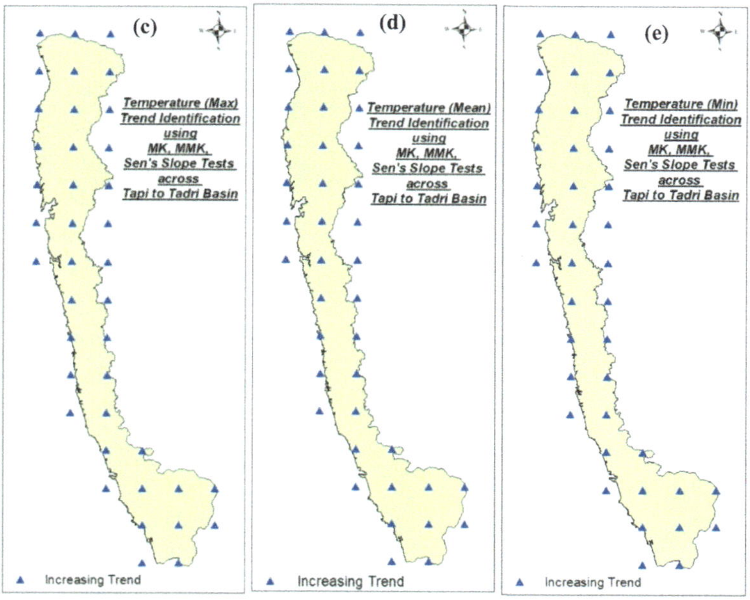

Fig. 31.4 (continued)

agricultural and forest land. The interrelationship of rainfall, PET and temperature indicates that in areas of lower rainfall away from ocean, higher PET and temperature are reported across the basin, which might require attention in terms of the supply and demand of available water resources.

31.5 Conclusions

The trends and long-term variability in rainfall, PET and temperature have been analyzed for Tapi to Tadri basin, India for period 1901–2016. The key conclusions resulting from present study are outlined underneath:

(1) The Tapi to Tadri basin experiences Monsoon climate and the average annual rainfall over the basin for period 1901–2016 is found to be 2400 mm. The maximum and minimum rainfall was found to be 5039.8 mm and 626.5 mm at Gird-60 and Grid-16 respectively. Thus, the basin experiences wide spatial and temporal rainfall variability.
(2) The annual total rainfall has been found to increase and decrease at, respectively, 72 and 44 grids out of 119 grids across Tapi to Tadri basin. This nonuniformity could exhibit to complications in management of the water resources within the basin.

Table 31.2 Results of trend analysis of annual PET and temperature series

Potential evapotranspiration				Maximum temperature			
Grid	MKZ	MMKZ	Sen β (mm/year)	Grid	MKZ	MMKZ	Sen β (°C/year)
1	2.58	3.51	0.141	1	6.36	5.86	0.0069
2	1.62	1.49	0.091	2	5.91	8.56	0.0063
3	1.44	1.70	0.065	3	5.83	5.45	0.0061
4	1.64	1.64	0.093	4	6.33	9.87	0.0066
5	1.24	1.12	0.067	5	6.08	11.69	0.0063
6	1.38	1.24	0.074	6	6.56	9.17	0.0069
7	0.99	0.84	0.052	7	6.45	7.54	0.0066
8	1.76	2.20	0.097	8	6.71	6.71	0.0069
9	1.47	1.47	0.079	9	6.38	6.38	0.0066
10	2.45	3.23	0.145	10	6.78	9.17	0.0069
11	1.29	1.24	0.073	11	6.46	6.97	0.0065
12	0.98	1.92	0.045	12	6.39	9.36	0.0064
13	2.04	2.37	0.117	13	6.58	6.58	0.0067
14	1.76	1.72	0.087	14	6.23	6.23	0.0062
15	1.49	1.54	0.074	15	5.92	7.33	0.0059
16	1.31	1.31	0.075	16	6.22	4.98	0.0061
17	1.00	1.64	0.055	17	5.92	5.61	0.0056
18	0.42	0.39	0.018	18	5.25	4.74	0.0050
19	1.53	1.33	0.082	19	6.19	8.73	0.0060
20	1.68	2.10	0.079	20	6.06	5.21	0.0057
21	1.86	1.63	0.077	21	6.15	7.51	0.0057
22	1.99	1.99	0.102	22	6.43	6.85	0.0059
23	1.83	1.83	0.070	23	6.11	6.11	0.0057
24	2.60	2.40	0.136	24	6.45	6.45	0.0060
25	2.15	2.15	0.100	25	6.52	5.86	0.0060
26	3.45	3.82	0.174	26	6.74	7.78	0.0062
27	3.25	2.63	0.161	27	6.71	6.71	0.0061
28	4.66	3.91	0.261	28	6.85	6.85	0.0062
29	4.81	4.34	0.256	29	6.86	11.42	0.0062
30	4.62	4.98	0.254	30	7.21	11.59	0.0065
31	4.70	4.70	0.242	31	7.31	7.80	0.0065
32	4.93	4.33	0.258	32	7.52	9.29	0.0068
33	5.03	5.31	0.267	33	7.58	7.23	0.0068
34	5.52	7.13	0.268	34	7.67	7.67	0.0070
35	4.91	5.74	0.224	35	7.78	7.32	0.0071

(continued)

Table 31.2 (continued)

Potential evapotranspiration				Maximum temperature			
Grid	MKZ	MMKZ	Sen β (mm/year)	Grid	MKZ	MMKZ	Sen β (°C/year)
36	5.25	6.45	0.279	36	7.76	7.76	0.0071
37	5.51	5.63	0.267	37	7.80	9.44	0.0072
38	5.43	6.02	0.257	38	7.84	7.33	0.0075
39	5.03	4.15	0.247	39	7.70	7.05	0.0074
40	5.46	5.96	0.279	40	7.87	24.21	0.0076
Mean temperature				Minimum temperature			
Grid	MKZ	MMKZ	Sen β (°C/year)	Grid	MKZ	MMKZ	Sen β (°C/year)
1	7.03	9.13	0.0070	1	6.77	6.77	0.0072
2	6.68	7.38	0.0065	2	6.38	6.38	0.0068
3	6.51	6.51	0.0063	3	6.27	10.79	0.0066
4	7.07	6.87	0.0066	4	6.79	6.79	0.0070
5	6.81	8.21	0.0063	5	6.45	6.45	0.0066
6	7.24	12.04	0.0070	6	6.95	6.95	0.0073
7	6.99	6.99	0.0065	7	6.70	5.95	0.0069
8	7.28	0.00	0.0069	8	7.01	8.67	0.0071
9	7.15	7.15	0.0067	9	6.74	10.93	0.0069
10	7.33	8.97	0.0070	10	7.06	7.06	0.0072
11	7.21	9.10	0.0067	11	6.74	8.13	0.0069
12	6.98	10.23	0.0065	12	6.65	9.96	0.0067
13	7.27	7.27	0.0068	13	6.89	6.73	0.0071
14	6.94	6.94	0.0063	14	6.54	6.05	0.0066
15	6.74	6.74	0.0061	15	6.19	8.27	0.0063
16	6.86	5.90	0.0061	16	6.57	10.26	0.0064
17	6.44	6.44	0.0057	17	6.28	7.41	0.0060
18	5.79	8.07	0.0052	18	5.47	4.79	0.0054
19	6.95	9.81	0.0062	19	6.47	7.28	0.0063
20	6.70	8.18	0.0059	20	6.23	8.53	0.0061
21	6.71	8.30	0.0058	21	6.24	5.58	0.0061
22	7.08	8.18	0.0061	22	6.53	6.71	0.0064
23	6.89	8.45	0.0060	23	6.21	7.23	0.0062
24	7.19	7.19	0.0062	24	6.50	13.66	0.0064
25	7.17	7.17	0.0063	25	6.43	6.43	0.0064
26	7.45	7.45	0.0065	26	6.65	6.65	0.0067
27	7.41	7.41	0.0065	27	6.55	7.15	0.0066

(continued)

Table 31.2 (continued)

Mean temperature				Minimum temperature			
Grid	MKZ	MMKZ	Sen β (°C/year)	Grid	MKZ	MMKZ	Sen β (°C/year)
28	7.61	6.92	0.0066	28	6.64	9.19	0.0067
29	7.44	7.97	0.0065	29	6.50	9.18	0.0067
30	7.43	9.96	0.0066	30	6.42	7.48	0.0065
31	7.46	6.92	0.0065	31	6.38	7.00	0.0065
32	7.64	10.96	0.0068	32	6.31	6.31	0.0067
33	7.51	9.35	0.0067	33	6.24	5.36	0.0065
34	7.59	7.59	0.0068	34	6.31	7.13	0.0067
35	7.59	8.58	0.0069	35	6.33	6.59	0.0066
36	7.64	11.22	0.0070	36	6.31	5.46	0.0066
37	7.73	9.50	0.0069	37	6.34	5.66	0.0065
38	7.75	9.14	0.0071	38	6.42	6.42	0.0065
39	7.75	9.30	0.0070	39	6.28	7.19	0.0066
40	7.85	8.28	0.0071	40	6.45	8.67	0.0066

(3) From the results, it is clear that increase in PET in the study area is mainly due to a significant increase in the air temperature. A marginal increase in PET due to climate change would put enormous pressure on the existing water resources within the catchment. However, analyses at seasonal, monthly and extreme indices may give better understating of interrelationship of climatic variables.

(4) For planning and management of agriculture and water resources is important to understand the distribution and changing trend of rainfall, temperature and potential evapotranspiration under climate variability. Therefore, the agro scientists, planners and water resources managers shall formulate strategies to counter this effect, and, implement them to ensure better water management practices in the basin.

Acknowledgments The authors are also thankful to *India Meteorological Department (IMD)* for providing necessary data to conduct the present study. The third author gratefully acknowledges the financial support received from *Department of Science and Technology (DST), Ministry of Science and Technology, Government of India* vide their letter no. DST/INSPIRE Fellowship/2015/IF150634 dated January 11, 2016.

References

Alexander LV, Zhang X, Peterson TC, Caesar J, Gleason B, Klein Tank AMG, Haylock M, Collins D, Trewin B, Rahimzadeh F, Tagipour A, Rupa Kumar K, Revadekar J, Griffiths G, Vincent L, Stephenson DB, Burn J, Aguilar E, Brunet M, Taylor M, New M, Zhai P, Rusticucci M, Vazquez-Aguirre JL (2006) Global observed changes in daily climate extremes of temperature and precipitation. J Geo Res 111(5):D05109-1-22
Boccolari M, Malmusi S (2013) Changes in temperature and precipitation extremes observed in Modena, Italy. Atmos. Res. 122:16–31
Caloiero T (2017) Trend of monthly temperature and daily extreme temperature during 1951–2012 in New Zealand. Theor Appl Climtol 129(1–2):111–127
Doorenbos J, Pruitt WO (1977) Guidelines for predicting crop water requirements. Irrigation and Drainage Paper No 24, 2nd edn., Food and Agriculture Organization, Rome, p 156
Dore MHI (2005) Climate change and changes in global precipitation patterns: what do we know? Environ Int 31:1167–1181
Frich P, Alexander LV, Gleason B, Haylock M, Tank AMGK, Peterson T (2002) Observed coherent changes in climatic extremes during the second half of the twentieth century. Clim Res 19:193–212
Ghosh S, Luniya V, Gupta A (2009) Trend analysis of Indian summer monsoon rainfall at different spatial scales. Atmos Sci Lett 10(4):285–290
Goyal RK (2004) Sensitivity of evapotranspiration to global warming: a case study of arid zone of Rajasthan (India). Agric Water Manage 69:1–11
Haijun L, Yan L, Tanny J, Ruihao Z, Guanhua H (2014) Quantitative estimation of climate change effects on potential evapotranspiration in Beijing during 1951–2010. J Geogr Sci 24:93–112
Hamed KH, Rao AR (1998) A modified Mann-Kendall trend test for autocorrelated data. J Hydrol 204:182–196
He Y, Tian P, Mu X, Gao P, Zhao G, Wang F, Li P (2017) Changes in daily and monthly rainfall in the Middle Yellow River, China. Theor Appl Climtol 129(1–2):139–148
Helfer F, Lemckert C, Zhang H (2012) Impacts of climate change on temperature and evaporation from a large reservoir in Australia. J Hydrol 475:365–378
Hirsch RM, Slack JR, Smith RA (1982) Techniques of trend analysis for monthly water quality data. Water Resour Res 20(6):727–732
Hobbins MT, Ramírez JA, Brown TC (2004) Trends in pan evaporation and actual evaporation across the conterminous U.S.: paradoxical or complementary? Geophys Res Lett 31:L13503
Huo Z, Dai X, Feng S, Kang S, Huang G (2013) Effect of climate change on reference evapotranspiration and aridity index in arid region of China. J Hydrol 492:24–34
India-WRIS report (2014) West flowing rivers from Tapi to Tadri basin. Central Water Commission, Government of India
Jain SK, Agarwal PK, Singh VP (2007) Hydrology and water resources of India, vol 57. Springer Science & Business Media
Kendall MG (1975) Rank correlation methods. Charles Griffin, London
Kundu S, Khare D, Mondal A (2016) Interrelationship of rainfall, temperature and reference evapotranspiration trends and their net response to the climate change in Central India. Theor Appl Climtol 1–22
Liu M, Tian H, Yang Q, Yang J, Song X, Lohrenz SE, Cai WJ (2013) Long-term trends in evapotranspiration and runoff over the drainage basins of the Gulf of Mexico during 1901–2008. Water Resour Res 49:1988–2012
Liu Y, Zhuang Q, Pan Z, Miralles D, Tchebakova N, Kicklighter D, Chen J, Sirin A, He Y, Zhou G, Melillo J (2014) Response of evapotranspiration and water availability to the changing climate in Northern Eurasia. Clim Change 1–15
Liu B, Chen J, Lu W, Chen X, Lian Y (2016) Spatiotemporal characteristics of precipitation changes in the Pearl River Basin, China. Theor Appl Climtol 123(3–4):537–550
Mann HB (1945) Nonparametric tests against trend. Econometrica J Econom Soc 245–259

Martinez CJ, Maleski JJ, Miller MF (2012) Trends in precipitation and temperature in Florida, USA. J Hydrol 452–453:259–281
Peel MC, Finlayson BL, McMahon TA (2007) Updated world map of the Köppen-Geiger climate classification. Hydrol Earth Syst Sci Dis 4(2):439–473
Quirk T (2012) Did the global temperature trend change at the end of the 1990s? Asia-Pac J Atmos Sci 48:339–344
Sharma D, Babel MS (2014) Trends in extreme rainfall and temperature indices in the western Thailand. Int J Climatol 34(7):2393–2407
Sharma PJ, Loliyana VD, Resmi SR, Timbadiya PV, Patel PL (2017) Spatio-temporal trends in extreme rainfall and temperature indices over Upper Tapi Basin, India. Theor Appl Climatol 1–26
Soltani M, Laux P, Kunstmann H, Stan K, Sohrabi MM, Molanejad M, Zawar-Reza P (2016) Assessment of climate variations in temperature and precipitation extreme events over Iran. Theor Appl Climatol 126(3–4):775–795
Sonali P, Kumar DN (2016) Spatio-temporal variability of temperature and potential evapotranspiration over India. J Water Climate Change 7(4):810–822
Song X, Zhang J, AghaKouchak A, Roy SS, Xuan Y, Wang G, He R, Wang X, Liu C (2014) Rapid urbanization and changes in spatiotemporal characteristics of precipitation in Beijing metropolitan area. J Geophy Res: Atmos 119(19)
Soro GE, Noufé D, Goula Bi TA, Shorohou B (2016) Trend analysis for extreme rainfall at sub-daily fand daily timescales in Côte d'Ivoire. Climate 4(3):37
Trenberth KE (2011) Changes in precipitation with climate change. Climate Res 47(1/2):123–138
Xiao C, Wu P, Zhang L, Song L (2016) Robust increase in extreme summer rainfall intensity during the past four decades observed in China. Sci Rep 6
Ye Z, Li Z (2017) Spatiotemporal variability and trends of extreme precipitation in the Huaihe River basin, a climatic transitional zone in East China. Adv. Met. Article ID 3197435
Zhai P, Zhang X, Wan H, Pan X (2005) Trends in total precipitation and frequency of daily precipitation extremes over China. J. Climate 18(7):1096–1108
Zhang Q, Singh VP, Peng J, Chen YD, Li J (2012) Spatial–temporal changes of precipitation structure across the Pearl River basin, China. J Hydrol 440:113–122

Chapter 32
Climate Change and Water Resources: Emerging Challenges, Vulnerability and Adaptation in Indian Scenario

Y. Shiva Shankar, Abhishek Kumar, and Devendra Mohan

Abstract Global warming has adversely affected the climatic systems on the earth. Long-term alterations in global weather patterns, particularly rise in global temperature have been significantly influencing the hydrological patterns leading to climate change. One of the biggest challenges posed by the global warming-induced climate change is its direct effect on the well-being of humans through the changes in hydrological cycle affecting the sectors such as agriculture, industrial growth, hydropower generation, domestic water supply etc. Subsequently, the researchers have predicted that impacts of climate change such as the melting of glaciers, increased frequency of floods and droughts, changes in evapotranspiration, variations in surface runoff and rising sea level have significant implications to water resources. Developing economies like India are highly vulnerable to these changes as the majority of its population is dependent on agriculture. In addition to the above the increasing water needs for domestic and industrial purposes, depleting groundwater levels and surface water pollution have been stressing the policymakers for sustainable water resource management. The present research exhaustively reviews the influence of climate change on hydrological components, its qualitative and quantitative implications to surface and groundwater resources in Indian scenario through various case studies. The research also proposes the corrective measures for the impact minimization on the available water resources in the present context.

Keywords Global warming · Climate change · Water resources · Hydrologic cycle and temperature rise

Y. Shiva Shankar
Civil Engineering Department (Formerly), Ujjain Engineering College, Ujjain 456010, India
e-mail: shivjuet@gmail.com

A. Kumar
Central Mine Planning and Design Institute, Coal India Limited, Ranchi 834008, Jharkhand, India
e-mail: abmishra16@gmail.com

D. Mohan (✉)
Civil Engineering Department, Indian Institute of Technology, Banaras Hindu University, Varanasi 221005, Uttar Pradesh, India
e-mail: devendra.civil.iitbhu@gmail.com

R. Jha et al. (eds.), *Climate Change Impacts on Water Resources*, Water Science and Technology Library 98, https://doi.org/10.1007/978-3-030-64202-0_32

32.1 Introduction

Accelerated increase in the greenhouse gases (GHG) like carbon dioxide, nitrous oxide, methane, Chlorofluoro carbon's etc. in the atmosphere is the major cause for global warming resulting in climate change. Though both the natural and anthropogenic factors control the climate change, anthropogenic factors have been important due to the rapid alterations in average weather. The impacts of climate change were significant in this century due to the rapid industrialization, increased use of fossil fuels, urbanization, excessive agriculture and land-use changes. As per the IPCC report, the maximum growth in the emission of greenhouse gases (GHG) has occurred between 1970 and 2004. The sectorial contribution for the growth of GHG's is as follows: 145% increase from energy supply sector, 120% from transport, 65% from industry, 40% from change in land-use patterns and during this period global population increased by 69%. Enhanced concentration of GHG due to these activities into the earth's atmosphere is responsible for global warming and consequent climate change ultimately affecting the natural cycles (Panwar and Chakrapani 2013; Misra 2014; Ministry of Statistics and Programme Implementation (MSPI) 2013).

Water resources and food security are the key areas that are highly vulnerable due to climate change. These two sectors have a significant impact on economic and social development of the country. Water is essential for sustaining all forms of life, food production, economic development, and for general wellbeing. The surface water and groundwater resources of the country play a major role in agriculture, hydropower generation, livestock production, industrial activities, forestry, fisheries, navigation, recreational activities, etc. (Kumar et al. 2005, 2013; National Academy of Agricultural Sciences (NAAS) 2013). For agrarian nation like India, agriculture contributes to 18.5% of GDP and two-thirds of population depends upon agriculture, changing climatic scenarios pose a bigger threat to this sector affecting the livelihood of major population (Khan et al. 2009; Singh et al. 2009).

32.1.1 Impacts on Water Resources

Water remains the key factor globally in attainment of Sustainable Development Goals (SDG's). Attainment of SDG's aiming at elimination of poverty, supplying clean energy, health and sanitation, clean water supply, sustainable cities etc. are directly or indirectly associated with water. Hence water remains the vital component in planning of framework for achieving SDG's. Nations have been focusing on broader perspective, i.e., water nexus constituting the interactions between food, water, cities, energy and environment; that have been detrimental for their sustainable development. In the Indian scenario, the issues are much more complex and vulnerable due to rapid economic progress in the last two decades exerting pressure on existing water resources. In this context, forecasting the future water requirements in the era of climate change for devising the adaptation strategies without compromising the present needs is of prime importance. Sustainable water resources planning and management is very critical for combating global warming-induced climate change (World Bank Group (WBG) 2016; Kumar et al. 2013; Nath and Behera 2011; Madhusoodhanan et al. 2016).

Water resource sector has been facing several issues due to the excessive utilization and human interventions, for satisfying the growing water demands. There are 12 major river basins in the country. Major perennial rivers like Ganga, Brahmaputra, Indus, etc. flow through the northern part of the country originating from Himalayas, whereas the river basins flowing in southern part of India are Krishna, Godavari, Penna etc. Ganga–Brahamputra–Meghna river basin is the largest in the country with catchment area of about 11.0 lakh km^2 (more than 43% of the total catchment area of all the major rivers in the country). Ganges basin has been supporting population more than 650 millions living in countries such as China, Nepal, India and Bangladesh. Other major river basins with catchment area more than one lakh km^2 are Indus, Mahanadi, Godavari and Krishna. On the other hand, India has been severely dependent on its potential of annually replenishable groundwater sources for satisfying its water requirements. Surface runoff is the major component that is responsible for groundwater recharge (Ministry of Water Resources (MOWR) 2008; Jeuland et al. 2013).

The reports of Inter Governmental Panel on Climate Change (IPCC) indicated that the average global surface air temperature has increased by 0.2–0.6 °C since the late nineteenth century and it is projected to increase by 1.4–5.8 °C over the period 1990–2100 (Indian Meteorological Department (IMD) 2009; Inter Governmental Panel on Climate Change (IPCC) 2008). All India mean annual temperature has shown warming trend of 0.05 °C/10 years during 1901–2003, but warming trend was observed to be on rise during 1971–2003 with a rise of 0.22 °C/ 10 years. The temperature data between 2001 and 2016 suggest that 12 of its 15 hottest years were experienced during this period. Increased temperature coupled with air pollution due to aerosols, enhancing the humidity has been aggravating the health effects even causing the deaths due to hot waves. These hot days also have an economic impact due to loss of human productivity (Kothawale and Kumar 2005; Oldenborgh et al. 2018; Burgess et al. 2017). Temperature rise has direct influence on meteorological disasters (hurricanes, storms etc.), hydrological events (floods) and climatological events (heat and cold waves, drought, wild fire etc.) (Visser et al. 2014).

The impacts of global warming induced climate change on water resources could be either qualitative or quantitative. Quantitative effects are melting of glaciers, increased frequency of floods and droughts, effect on groundwater quantity due to decreased runoff and rising sea level. Quantitative effects are important for agricultural output as major freshwater resources in our country are glacier fed, non perennial rivers are dependent on runoff and country is surrounded long coastal line. The qualitative effects include increase in sediment load due to the flash floods, variation in dissolved oxygen levels, growth of algal blooms, saltwater intrusion, increase in concentration of pollutants, etc. (National Water Mission (NWM) 2011).

The present work discusses about the impacts of climate change on water resources, emerging vulnerabilities and adaptive measures for Indian scenario.

32.2 Quantitative Effects of Climate Change

32.2.1 Melting of Glaciers in the Himalayas

Himalayas constitute the largest sources of freshwater with major perennial rivers of the subcontinent like Ganga, Brahmaputra and Indus; originating from these glaciers. Glacial retreat in the Himalayan region poses multiple threats such as increasing magnitude of floods and droughts, decreasing the water quality, changes in hydropower generation, influence on biodiversity etc.; that has a strong hold on lives of millions of people in that region. Glacial lakes in Himalayas causing the outburst floods in this region have been increasing in the last few decades resulting in catastrophic damages and fatalities. It has been observed that there has been change in the nature of glacial lakes in the region, with significant numbers of them have disappeared and certain proglacial lakes have expanded. But the emerging threat was observed to be due to the formation of glacial lakes at higher elevation due to the temperature rise that could increase the probability of flash floods in the location. Additionally, these retreating glaciers have negative effects such as soil erosion and contamination in the downstream rivers. Increased frequency of extreme rainfall events acts as a medium for transportation of nutrients and chemicals leading to the formation of biomass blooms affecting the nearby habitat's (Nie et al. 2017; Pritchard 2017; Zhang et al. 2017; Kumar et al. 2005; Wheatley et al. 2017).

32.2.2 Modifications in Rainfall Pattern

Monsoon rainfall in India displayed annual variability considerably lower and higher than normal rainfall over large areas of Indian subcontinent during a period of several years leading to wide spread of droughts and floods. The year-to-year variability in monsoon rainfall leads to extreme hydrological events (large-scale droughts and floods) resulting in serious reduction in agricultural output causing food security and economic loss. Rainfall fluctuations in India have been largely random over a century, with no systematic change detectable on either annual or seasonal scale but significant changes were observed on regional scale. However, areas of increasing trend in the seasonal rainfall have been found along the West Coast, North Andhra Pradesh and Northwest India and those of decreasing trend over East Madhya Pradesh, Orissa and Northeast India during recent years. The randomness of the rainfall pattern is presented in Fig. 32.2 (Auffhammer et al. 2012; Mall et al. 2006b, 2007; Guhathakurta et al. 2011).

The changes observed in temperature and precipitation during 2000–2015 were significant in comparison to previous century. The last two decades have witnessed a devastating extreme weather events, drought of 2016 spread over 10 states has affected population of 330 million (Goyal and Rao 2018). Changes in the precipitation could result in different type of droughts such as meteorological drought (due to rainfall shortage), agricultural drought (due to lesser soil moisture), hydrological drought (due to water shortage in the water bodies) and socioeconomic drought (due to reduction in water for societal utility) (Mukherjee et al. 2018).

Utilization of groundwater resource for irrigation and drinking water needs has been increasing due to the improper water management in the past combined with population growth, industrialization and urbanization. Groundwater accounts for about 80% of domestic water requirement and more than 45% of the total irrigation in the country (Chakraborti et al. 2011; Kumar et al. 2005). It is a replenishable resource and has inherent advantages over the surface water as it is comparatively pure, less evaporation loss and wide distribution (Senthilkumar and Elango 2013). Increased utilization for domestic, irrigation and industrial water supply has resulted in overexploitation of groundwater affecting the groundwater in terms of the quality and quantity (Raju et al. 2009). Most of the groundwater recharge occurs due to the rainfall only. Recharge of an aquifer depends mostly on the type of soil, vegetation, amount of precipitation, surface temperature, wind speed, evapotranspiration, runoff and discharge patterns. The increase of global temperature affects the major component of groundwater recharge, i.e. by rainfall and seepage, through various sources (Panwar and Chakrapani 2013; Russell et al. 2013).

32.2.3 Change in Evapotranspiration

Evapotranspiration is the water lost to the atmosphere by two processes evaporation and transpiration. Evaporation is the loss of water from surface water bodies, wetlands, bare soil, and snow cover while transpiration is the loss from living-plant surfaces. As evapotranspiration is dependent on the vegetation, soils types and on the amount of water available. With the rising surface temperature, rate of evapotranspiration is expected to show spatio-temporal changes due to solar radiation and wind speed. The major reason behind this influence was observed to be due to its effect on hydrologic cycle due to the evaporation and consumption of solar radiation. In the current scenario of temperature rise, the effects were predicted to aggravate that would result in changes in precipitation affecting the water management in future. Groundwater is directly affected by changes in the rate of precipitation and evapotranspiration. Higher temperature leads to higher evapotranspiration rates, resulting in lesser moisture content and increases dryness in the soils. The high precipitation in wet climate will increase surface runoff causing soil erosion (Pan et al. 2015; King et al. 2015; Chattopadhyay and Hulme 1997; Panwar and Chakrapani 2013; Mall et al. 2007).

32.2.4 Sea-Level Rise

Sea-level rise due to climate change has been a serious global threat in recent days. The rate of global sea-level rise was faster from 1993 to 2003, about 3.1 mm per year, as compared with the average rate of 1.8 mm per year from 1961 to 2003 as reported in IPCC report and significantly higher than the average rate of 0.1–0.2 mm/yr increase recorded by geological data over the last 3000 years (Dasgupta and Meisner 2009). Surface water at prevailing temperature expands when the water temperature rises, with raising surface temperature due to global warming, these phenomena may to lead to floods in coastal area. This will cause potentially direct consequences for many highly populated areas located near sea level. Melting of glaciers and icecaps will accelerate this phenomenon further. With rising deglaciation at Antarctica, Intergovernmental Panel on Climate Change suggests that average sea level may rise by 15–95 cm by the year 2100. Decrease in levels of water table in surface and subsurface waters can cause seawater intrusion into the surface and groundwater, which increases the salinity of the water, making it unsuitable for drinking purpose. In the Indian context, this problem could be more evident due to the long coastline supporting millions of population and metropolitan cities such as Mumbai, Chennai and Kolkata (Panwar and Chakrapani 2013; Mall et al. 2007; IPCC 2008). Sea-level rise has been detrimental to biodiversity on coastal islands housing more than 20% of the terrestrial plant and vertebrate species in the world, within 5% of the global terrestrial area. Climate change causing tidal ranges has been leading to periodic floods affecting the nonsaline habitats. Sea-level rise also contributes to coastal erosion and salt water intrusion affecting the aquifers (Courchamp et al. 2004).

32.2.5 Challenges to Indian Agriculture

Climate change impacts agriculture both directly and indirectly. The type and magnitude of impact will vary depending on the degree of change in climate, geographical region and nature of crop. The following are the areas vulnerable to climate change that affect the agricultural productivity: variation in meteorological parameters effecting hydrologic cycle, effect on soil composition (fertility, erosion, moisture content, salinity, biological process), change in the water use and application of fertilizers and growth of pests (NAAS 2013; Kang et al. 2009; Khan et al. 2009). On the other hand, population is increasing at an unprecedented rate, with increasing glacier melting, changing rainfall patterns, seawater rise, decreased surface runoff and groundwater recharge; will have serious implications on water availability in the future. Studies have predicted that water is likely to be scarce by 2050. The water requirement in India by 2050 will be in the order of 1450 km^3, which is significantly higher than the estimated water resources of 1122 km^3 per year (Misra 2014).

There are two major crop growing season in India, summer or kharif crop growing season (June–September) and winter or rabi crop growing season starts after the summer monsoon, and continues till early summer. The major kharif crops are rice, maize, sugarcane, cotton, jute, groundnut, soybean and Bajra etc. The major rabi crops are wheat mustard, Barley, potato, onion and gram etc. Depending on crop duration, kharif crops can be harvested during the autumn or winter months also. The southwest monsoon is critical to the kharif crop, which accounts for more than 50% of the food-grain production and 65% of the oilseeds production in the country. The variations in the monsoon rainfall annually in India lead to large-scale droughts and floods, resulting in a major effect on Indian food grain production. Rainfall occurring at the end of the monsoon season provides stored soil moisture and often irrigation water for the rabicrop. Climate change is projected to have serious implications for these major crops. The greater losses in rabi season have a significant impact on wheat yield and interannual variability of monsoon affects the kharif crops like rice (Kumar et al. 2014; Mall et al. 2006a; Ranuzzi and Srivastava 2012; NAAS 2013).

Statistical analyses of state-level data confirm that drought and extreme rainfall adversely affected rice yield in predominantly rainfed areas during 1966–2002. The drought has been found to have much greater impact than extreme rainfall events. On an aggregated scale, the mean of all emission scenarios indicates that climate change is likely to reduce irrigated rice yields by ~4% in 2020 (2010–2039), ~7% in 2050 (2040–2069), and by ~10% in 2080 (2070–2099) climate scenarios. Modeling studies on wheat indicated that with every rise of 1 °C in mean temperature, India could lose 4–5 million tons of wheat. Maize production has also been projected to decrease due to variations in climate (NAAS 2013; Kang et al. 2009). Similarly increase in temperature may reduce yields of soybean, mustard, groundnut, and potato by 3–7%. Losses were also significant in other crops, such as mustard, peas, tomatoes, onion, garlic and other vegetable and fruit crops. Climate change is predicted to have a major impact on horticulture also melting of ice cap in the Himalayan regions affecting the production of apple, saffron, rhododendron, orchid, etc. Commercial production of horticultural plants particularly grown under open field conditions will be severely affected. High temperature and air pollution have significant effect on their growth (Datta 2013). With the varying meteorological conditions due to climate change, impact on food security is expected to be higher in future.

32.3 Qualitative Effects of Climate Change

Quality of water is affected adversely by the Climate Change, changes in amount or pattern of precipitation will change the residence time of water in the watershed, thereby changing its quality (Eckhardta and Ulbrich 2003). These impacts depend on magnitude of changes created in the environment and the consequences can be different according to type of water body (rivers, lakes, dams, ponds, wetlands, etc.) and characteristics (water residence times, size, shape, depth, etc.). In case of streams, main parameters affected are dissolved organic matter and nutrients.

Pathogens and cyanobacteria/cyanotoxins are more important in case of lakes. In addition, micropollutants concentration, both inorganic and organic, may also be frequently affected (Jackson et al. 2007).

All the physico-chemical, biochemical and biological reactions are controlled by temperature. According to Arrhenius equation, kinetics of a given chemical reaction may be doubled for a temperature increase of 10 °C (Wolfenden et al. 1999). This phenomenon globally has increased the concentrations of dissolved substances in water, affecting the content of dissolved gases like oxygen. Increase in surface water temperature combined with reduced dissolved oxygen content, changes mixing patterns affect the self-purification capacity and favors the growth of algal blooms (Komatsu et al. 2007; Jöhnk et al. 2008). Solar irradiation can also alter water quality and especially characteristics of natural organic matter in freshwater systems both by warming and Ultraviolet (UV) radiation (increasing photolysis) (Delpla et al. 2009; Tu 2009). In case of droughts, groundwater recharge decreases significantly. Hence over-extraction of water will lead to degradation in water quality leading to problems like saltwater intrusion (Ranjan et al. 2006). The climate change impacts on surface water quality can be summarized as shown in Fig. 32.1, which is the modified version of figure taken from Delpla et al. 2009. Figure 32.1 considers the effects droughts and floods of the two main factors (temperature and rainfall). These impacts depend on natural or artificially created environment and the consequences can be different according to type of water body (rivers, lakes, dams, ponds, wetlands, etc.) and characteristics (water residence times, size, shape, depth, etc.). In case of streams, main parameters affected are dissolved organic matter and nutrients. Pathogens and cyanobacteria/cyanotoxins are more important in case of lakes. In addition, micropollutants, both inorganic and organic, may also be frequently affected (Jackson et al. 2007). These changes could lead to the outbreak of various infectious water borne and vector-borne diseases such as yellow fever, diarrhoea, malaria, cholera etc. (Moors et al. 2013).

32.4 Vulnerability

The following have been the major vulnerabilities in the Indian scenario due to the global warming-induced climate causing changes in economical, ecological and social scenarios in the country, enhancing the stress on available water resources:

- Proportion of urban and rural populations is 31.2% and 68.8%, respectively, (as per 2011 census) projected to rise to 55% by 2050 (Lohani et al. 2016).
- Rapid alterations in the ecologically sensitive Himalayan glaciers due to large scale deforestation, mining and quarrying, cultivation and grazing, urbanization and tourism (Tiwari and Joshi 2012).
- Larger number of marginal and small-scale farmers in the agricultural sector (Poddar et al. 2014).
- Larger threat to biodiversity (Pecl et al. 2017).

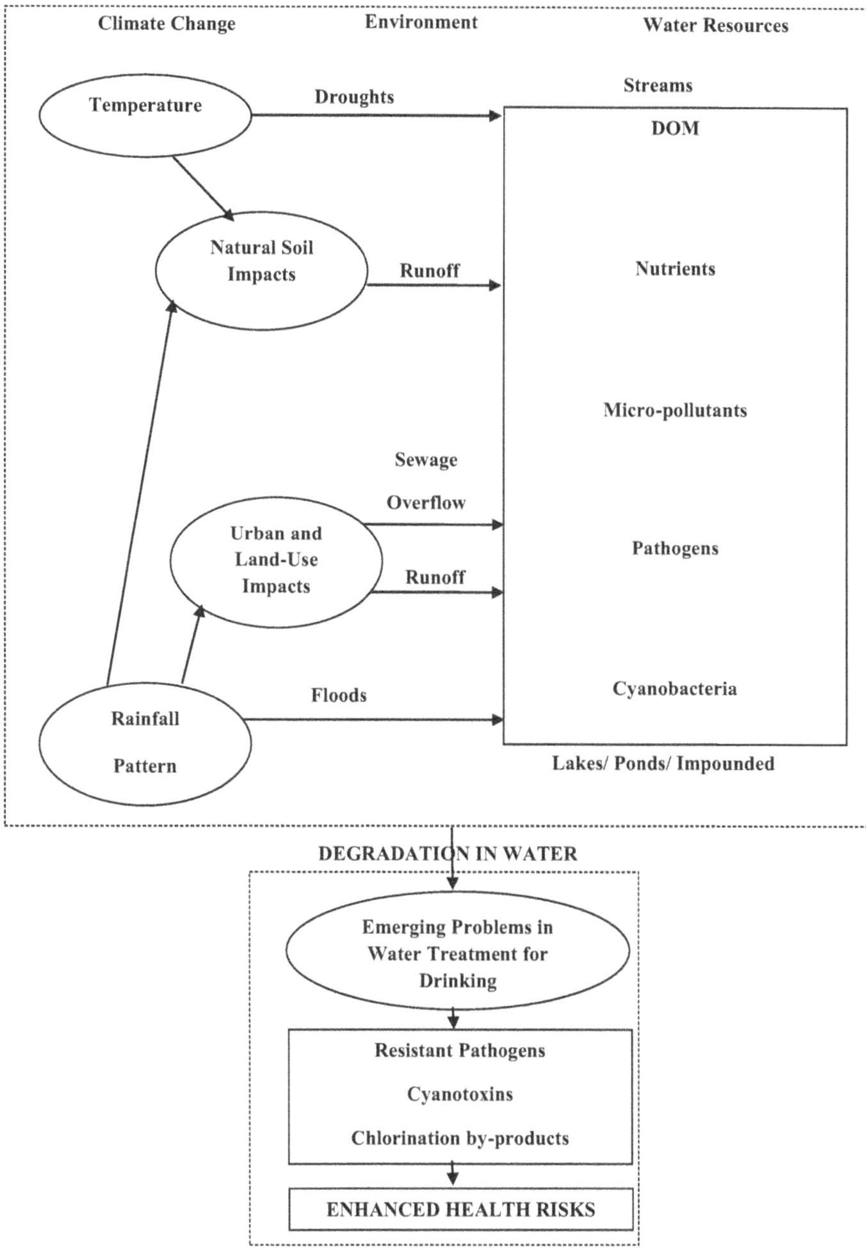

Fig. 32.1 Impacts of climate change on water resources and water quality

- Increase of crimes in the society due to the competition over the available resources fostering the conflicts, migrations, economic downfall etc. aggravating the poverty increasing the crimes in the society (Agnew 2011).
- Migrations and larger populations in coastal areas of the country (Bhagat 2014).

32.5 Adaptation in the Indian Scenario

The major initiative in the Indian scenario for combating the effects of climate change is enhancing the adaptive capacity and building of climate-resilient infrastructure. The implementation should be enhanced with proper good governance, coordination among stakeholders, smart infrastructure in the urban areas, adaptation of clean technologies for pollution prevention, capacity building for translating the ideas into reality, sustainable use of resources, incentives for green practices, vulnerability and risk analysis, education and enhanced research for analyzing the future threats.

References

Agnew R (2011) Dire forecast: a theoretical model of the impact of climate change on crime. Theor Criminol 16(1):21–42

Auffhammer M, Ramanathan V, Vincent JR (2012) Climate change, the monsoon, and rice yield in India. Clim Change 111(2):411–424

Bhagat RB (2014) Change vulnerability and migration in India: overlapping of hot spots. National Workshop on Migration and Global Environmental Change in India, organized by UNESCO and Foresight, 4–5 March 2014, New Delhi

Burgess R, Deschenes O, Donaldson D, Greenstone M (2017) Weather, climate change and death in India. http://www.lse.ac.uk/economics/Assets/Documents/personal-pages/robin-burgess/weather-climate-change-and-death.pdf

Chakraborti D, Das B, Murill TM (2011) Examining India's groundwater quality management. Environ Sci Technol 45:27–33

Chattopadhyay N, Hulme M (1997) Evaporation and potential evapotranspiration in India under conditions of recent and future climate change. Agricult Forest Meteorol 87:55–73

Courchamp F, Hoffmann BD, Russell JC, Leclerc C, Bellard C (2004) Climate change, sea-level rise, and conservation: keeping island biodiversity afloat. Trends Ecol Evol 29(3):127–130

Dasgupta S, Meisner C (2009) Climate change and sea level rise a review of the scientific evidence. The World Bank Environment Department, pp 3–36. https://www.openknowledge.worldbank.org/bitstream/handle/10986/18382/485250REPLACEM1Change0and0Sea0Level.pdf?sequence=1

Datta S (2013) Impact of climate change in Indian horticulture—a review. Int. J. Sci Environ Technol 2(4):661–671

Delpla I, Jung AV, Baures E, Clement M, Thomas O (2009) Impacts of climate change on surface water quality in relation to drinking water production. Environ Int 35:1225–1233

Eckhardta K, Ulbrich U (2003) Potential impacts of climate change on groundwater recharge and streamflow in a central European low mountain range. J Hydrol 284:244–252

Goyal MK, Rao YS (2018) Impact of climate change on water resources in India. J Environ Eng 144(7):1–10

Guhathakurta P, Sreejith OP, Menon PA (2011) Impact of climate change on extreme rainfall events and flood risk in India. J Earth Syst Sci 120(3):359–373
Indian Meteorological Department (2009) The warmest year since 1901. http://www.imd.gov.in/doc/warm2009.pdf
Inter Governmental Panel on Climate Change (IPCC) (2008) Climate Change and water—IPCC Technical Paper VI. http://www.ipcc.ch/pdf/technical-papers/climate-change-water-en.pdf
Jackson LJ, Lauridsen TL, Sondergaard M, Jeppesen E (2007) A comparison of shallow Danish and Canadian lakes and implications of climate change. Freshw Biol 52:1782–1792
Jeuland M, Harshadeep N, Escurra J, Blackmore D, Sadoff C (2013) Implications of climate change for water resources development in the Ganges basin. Water Policy 15:26–50
Jöhnk KD, Huisman J, Sharples J, Sommeijer B, Visser PM, Stroom JM (2008) Summer heatwaves promote blooms of harmful Cyanobacteria. Glob Change Biol 14:495–512
Kang Y, Khan S, Ma X (2009) Climate change impacts on crop yield, crop water productivity and food security—a review. Progr Nat Sci 19:1665–1674
Khan SA, Kumar S, Hussain MZ, Kalra N (2009) Climate change, climate variability and Indian agriculture: impacts vulnerability and adaptation strategies. In: Chapter 2: Climate change and crops. Springer Nature, Switzerland. https://doi.org/10.1007/978-3-540-88246-6_2. ISBN 978-3-540-88245-9
King DA, Bachelet DM, Symstad AJ, Ferschweiler K, Hobbins M (2015) Estimation of potential evapotranspiration from extraterrestrial radiation, air temperature and humidity to assess future climate change effects on the vegetation of the northern great plains, USA. Ecol Model 297:86–97
Komatsu E, Fukushima T, Harasawa H (2007) A modeling approach to forecast the effect of long-term climate change on lake water quality. Ecol Model 209:351–366
Kothawale DR, Kumar KR (2005) On the recent changes in surface temperature trends over India. Geophys Res Lett 32(1–4):L18714
Kumar R, Singh RD, Sharma KD (2005) Water resources of India. Curr Sci 89(5):794–811
Kumar RN, Reddy SK, Nathawat MS, Patel N, Rahore VS (2013) Climate change implications on water resources in India-review. Environ Ecol 31(2C):1085–1091
Kumar SN, Aggarwal PK, Rani DNS, Saxena R, Chauhan N, Jain S (2014) Vulnerability of wheat production to climate change in India. Clim Res 59:173–187
Lohani B, Siddiqi Y, Kilroy G (2016) Economic growth and climate proofing Asia through sustainable water resources management. In: Biswas AK, Tortajada C (eds) Water security, climate change and sustainable development. Springer Nature, Switzerland, pp 83–105
Madhusoodhanan CG, Sreeja KG, Elhdo TI (2016) Climate change impact assessments on the water resources of India under extensive human interventions. Ambio 45(6):725–741
Mall RK, Gupta A, Singh R, Srinivasan G, Rathore LS (2006a) Impact of climate change on Indian agriculture: a review. Clim Change 78:445–478
Mall RK, Gupta A, Singh R, Singh RS, Rathore LS (2006b) Water resources and climate change, an Indian perspective. Curr Sci 90(12):1610–1626
Mall RK, Pandey SN, Bhatla R (2007) Water resources in India and impact of climate change. Jalvigyan Sameeksha 22:157–176
Ministry of Statistics and Programme Implementation (MSPI) (2013) Statistics related to climate change—India. http://www.mospi.nic.in/mospi_new/upload/climate_change_29nov13.pdf
Ministry of Water Resources (MOWR) (2008) Preliminary consolidated report on effect of climate change on water resources in India. http://www.cwc.gov.in/main/downloads/Preliminary_Report_final.pdf
Misra AK (2014) Climate change and challenges of water and food security. Int J Sustain Built Environ 3:153–165
Moors E, Singh T, Siderius C, Balakrishnan S, Mishra A (2013) Climate change and waterborne diarrhea in northern India: impacts and adaptation strategies. Sci Tot Environ 468–469(1):S139–S151
Mukherjee S, Mishra A, Kevin ET (2018) Climate change and drought: a perspective on drought indices. Curr Clim Change Rep 4(2):145–163

Nath PK, Behera B (2011) A critical review of impact of an adaptation to climate change in developed and developing economies. Environ Dev Sustain 13(1):141–162
National Academy of Agricultural Sciences (NAAS) (2013) Climate resilient agriculture in India. Policy Paper 65. http://www.naasindia.org/Policy%20Papers/Polic,y%2065.pdf
National Water Mission (NWM) (2011) National action plan on climate change. Comprehensive mission document, vol 1. http://wrmin.nic.in/writereaddata/nwm16606419934.pdf
Nie Y, Sheng Y, Liu Q, Liu L, Liu S, Zhang Y, Song C (2017) A regional-scale assessment of Himalayan glacial lake changes using satellite observations from 1990 to 2015. Remote Sens Environ 189:1–13
Oldenborgh GJV, Philip S, Kew S, Weele MV, Uhe P, Otto F, Singh R, Pai I, Cullen H, Rao KA (2018) Extreme heat in India and anthropogenic climate change. Nat Hazards Earth Syst Sci 18:365–381
Pan S, Tian H, Dangal SRS, Yang Q, Yang J, Lu C, Tao B, Ren W, Ouyang Z (2015) Responses of global terrestrial evapotranspiration to climate change and increasing atmospheric CO_2 in the 21st century. Earth's Fut 3:15–35
Panwar S, Chakrapani GJ (2013) Climate change and its influence on groundwater resources. Curr Sci 105(1):37–46
Pecl GT et al (2017) Biodiversity redistribution under climate change: impacts on ecosystems and human well-being. Sci Mag 355(6332):1–9
Poddar R, Qureshi ME, Shi T (2014) A comparison of water policies for sustainable irrigation management: the case of India and Australia. Water Resourc Manag 28:1079–1094
Pritchard HD (2017) Asia's glaciers are a regionally important buffer against drought. Nature 545:169–178
Raju NJ, Ram P, Dey S (2009) Groundwater quality in the lower Varuna river basin, Varanasi district, Uttar Pradesh. J Geol Soc India 73:178–192
Ranjan P, Kazama S, Sawamoto M (2006) Effects of climate change on coastal fresh groundwater resources. Glob Environ Change 16:388–399
Ranuzzi A, Srivastava R (2012) Impact of climate change on agriculture and food security. ICRIER Policy Series, 16. http://icrier.org/pdf/Policy_Series_No_16.pdf
Russell SC, Bridget RS, Freddie SM, Robert CR, John BG, Lu Z (2013) Potential climate change effects on groundwater recharge in the High Plains Aquifer, USA. Water Resourc Res 49:1–16
Senthilkumar M, Elango L (2013) Geochemical processes controlling the groundwater quality in lower Palar river basin, Southern India. J Ear Sys Sci 122(2):419–432
Singh SP, Kulkarni SA, Singh VK, Gupta RS (2009) Impact of climate change on water resources availability and crop productivity in India. In: 60th international executive council meeting & 5th Asian regional conference, pp 1–9. http://www.indiaenvironmentportal.org.in/files/ImpactofClimateChangeonWater_complete.pdf
Tiwari PC, Joshi B (2012) Environmental changes and sustainable development of water resources in the Himalayan headwaters of India. Water Resourc Manag 26:883–907
Tu J (2009) Combined impact of climate and land use changes on streamflow and water quality in eastern Massachusetts, USA. J Hydro 379:268–283
Visser H, Petersen AC, Ligtvoet W (2014) On the relation between weather-related disaster impacts, vulnerability and climate change. Clim Change 125:461–477
Wheatley CJ, Beale CM, Bradbury RB, Higgins JWP, Critchlow R, Thomas CD (2017) Climate change vulnerability for species—assessing the assessments. Glob Change Biol 23:3704–3715
Wolfenden R, Snider M, Ridgway C, Miller B (1999) The temperature dependence of enzyme rate enhancements. J Amer Chem Soc 121:7419–7742
World Bank Group (WBG) (2016) High and dry—climate change, water, and the economy. International Bank for Reconstruction and Development, The World Bank, Washington, DC. https://openknowledge.worldbank.org
Zhang Q, Zhang F, Kang S, Cong Z (2017) Melting glaciers: hidden hazards. Science 356(6337):495

Chapter 33
Observed Spatio-Temporal Trends of Precipitation and Temperature Over Afghanistan

S. Rehana, P. Krishna Reddy, N. Sai Bhaskar Reddy, Abdul Raheem Daud, Shoaib Saboory, Shoaib Khaksari, S. K. Tomer, and U. Sowjanya

Abstract Afghanistan is a semi-arid country and most vulnerable to climate extremes related hazards, including droughts and floods that have caused huge impact on the socioeconomic development of the country. The present study analyzed the observed precipitation and temperature trends for seven agro-climatic zones of Afghanistan over the period 1951–2007 with Asian Precipitation-Highly-Resolved Observational Data Integration towards Evaluation of Water Resources (APHRODITE). Change in the magnitude of precipitation and temperatures in recent years with reference to distant past was assessed by dividing the historical data into two parts as 1951–1990 and 1991–2006. Further, the trend analysis was performed

S. Rehana (✉) · U. Sowjanya
Spatial Informatics, International Institute of Information Technology, Hyderabad, India
e-mail: rehana.s@iiit.ac.in

U. Sowjanya
e-mail: sowjanyauppuluri878@gmail.com

P. Krishna Reddy
Data Sciences and Analytics Center, IT for Agricultural Research Centre, International Institute of Information Technology, Hyderabad, India
e-mail: pkreddy@iiit.ac.in

N. Sai Bhaskar Reddy
MgtWell, Kabul, Afghanistan
e-mail: saibhaskarnakka@gmail.com

A. R. Daud · S. Khaksari
Ministry of Agriculture, Irrigation and Livestock, Kabul, Afghanistan
e-mail: daud.rahimi@mail.gov.af

S. Khaksari
e-mail: shoaib.khaksari@mail.gov.af

S. Saboory
Agromet, Agrometeorological Unit, Ministry of Agriculture, Irrigation and Livestock, Kabul, Afghanistan
e-mail: shoaib.saboory@mail.gov.af

S. K. Tomer
Satyukt Analytics Private Limited, Bengaluru, Karnataka, India
e-mail: satkumartomer@gmail.com

R. Jha et al. (eds.), *Climate Change Impacts on Water Resources*, Water Science and Technology Library 98, https://doi.org/10.1007/978-3-030-64202-0_33

on daily data to test the increasing or decreasing rainfall and temperature trends using Mann–Kendall trend test for each zone of Afghanistan. The maximum precipitation occurrence months were observed as January–May for all zones of Afghanistan. Whereas, June–December and generally considered as dry months. The maximum temperature was observed in the months of May–August, with hottest month as July for all seven zones of Afghanistan. The annual total precipitation has shown an increasing trend for the zones of South, South–West, East and Central, whereas, a decreasing trend has been observed for the zones of North, North–East and West zones. The trend analysis of the precipitation with gridded data sets reveals for the most part of the Afghanistan region the rainfall has been observed as decreasing. Whereas, for all seven agro-climatic zones of Afghanistan, an increasing trend of temperature in recent years of 2004–2016 was observed. Overall, the North, North–East and West zones of Afghanistan are more vulnerable with decreasing precipitation and increasing temperatures indicating more dry and warm periods indicating increasing drought conditions. Whereas, the South, South–West, East, and Central zones are more vulnerable with increasing trends of both precipitation and temperatures indicating increase of more wet and warm climates.

Keywords Climate extremes · APHRODITE · Afghanistan · Trend analysis · Mann–Kendall test

33.1 Introduction

An increase in the number and magnitude of extreme climate events has been observed globally in recent years causing huge loss of lives, extensive damages to crops, properties and immeasurable misery to millions of people (Hartmann et al. 2013). The issue of management of risks of extreme events and disasters under climate change adaptation has been the main focus by the Intergovernmental Panel on Climate Change (IPCC) in the Special Report on Extremes (SREX) (IPCC 2012). Globally, understanding past changes in the characteristics of extreme climate events became critical for reliable projections of future changes (Donat et al. 2013; Panda et al. 2017). Hence, understanding the variability of precipitation and temperature extreme events in the historical data is essential. Moreover, the study of historical trends of climate extremes will serve as the basis for understanding the possible physical vulnerabilities and adaptive measures. To this end, the study of climate events has gained higher scientific and societal interest over the last few decades (e.g. Zhang et al. 2005; Donat et al. 2013; Curry et al. 2014; Razavi et al. 2016). The study of such climate trends may be critical for any arid to semiarid counties such as Afghanistan as it can directly impact the droughts, floods, heat waves etc. and can be more intense under climate change. Afghanistan is a landlocked semi-arid country and most vulnerable to precipitation extremes related hazards, including droughts, floods and heatwaves that cause huge losses in life and property impacting the socio-economic development of the country. Currently, Afghanistan is among the countries with low levels of greenhouse gas (GHG) emissions. The Global Adaptation Index ranked Afghanistan as the most vulnerable countries in the world under climate change. There is a limitation over the climate extremes studies over Afghanistan due to limited weather data availability. More than three decades of war in Afghanistan that started in the late 1970s caused a huge interruption in water-resources data collection and destroying many of the older records of meteorological and hydrological data (Campbell 2015). Specifically, only a few (e.g. Aich et al. 2017) climate change assessment studies have been carried out in the literature. In this context, the present study tried to use the open source data sets to analyze the current trends in the precipitation and temperatures. We report the changes and trends in precipitation and temperature extremes in the recent years for seven agro-climatic zones of Afghanistan.

33.2 Case Study and Data

Afghanistan is a landlocked country and geographically highly heterogeneous with the glaciated peaks of the Hindukush and arid deserts of the South, located between 29 and 39° N and 60–75° E (Fig. 33.1). Entire Afghanistan is divided into seven agro-climatic zones following to climate classification scheme of Köppen-Geiger system (Köppen and Wladimir 1884), Beekeeping survey report (2014) and Ministry of

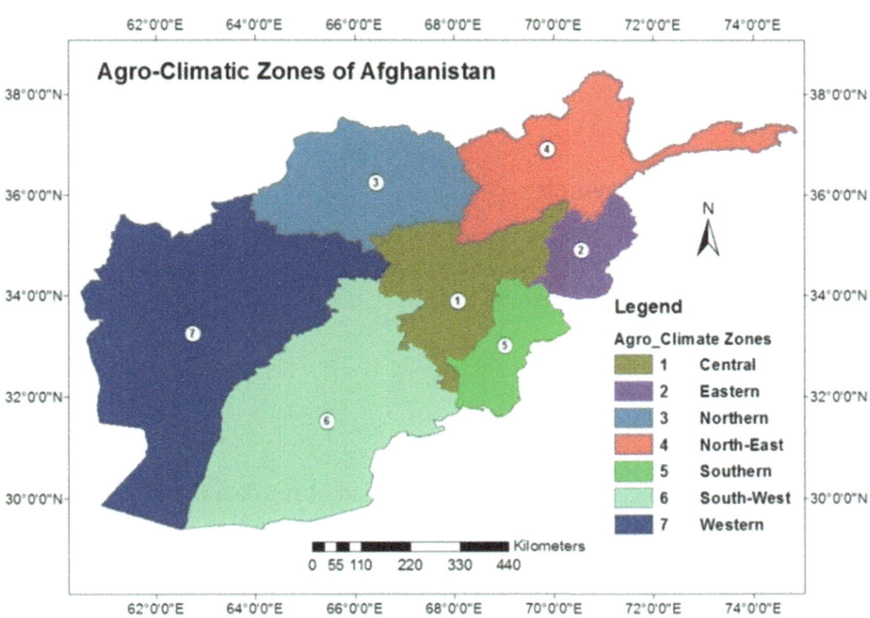

Fig. 33.1 Seven agro-climatic zones of Afghanistan with provinces

Agriculture, Irrigation and Livestock (MAIL) as Central, Eastern, Northern, North–East, Southern, South–West, and Western agro-climatic zones (Fig. 33.1). The parts of Central, North, North–East and South agro-climatic zones have mid-latitude steppe climate, desert climate and Mediterranean climate. The parts of East and South–West agro-climatic zones have Mediterranean climate; and tropical and subtropical—steppe and desert climates. The parts of South–West and West agro-climatic zones have mid-latitude steppe, and desert climate; tropical and subtropical desert climate; and Mediterranean climate.

One of the major challenges in the initiation of the present study is the availability of long time-series historical weather station meteorological data sets. Therefore, to understand the spatial variation of precipitation and temperature extremes, the study used a long time series fine-scale gridded dataset, Asian Precipitation-Highly-Resolved Observational Data Integration towards Evaluation of Water Resources (APHRODITE) datasets for a 48-year period from 1951 to 2007 for analysing the spatial trends in the precipitation over Afghanistan. APHRODITE daily gridded precipitation is the long-term continental-scale datasets which considered a dense network of daily rain-gauge data (about 5000–12,000) with 14 quality control processes, such as controlling for erroneous values, repetition, homogeneity etc. (Hamada et al. 2011). The APHRODITE data are mainly developed for Asia including Himalayas, South and Southeast Asia and mountainous areas in the Middle East at 0.5° resolution. The details of the data can be found at: https://climatedataguide.ucar.edu/climate-data/aphrodite-asian-precipitation-highly-resolved-observational-data-integration-towards. The APHRODITE data set is first extracted covering entire Afghanistan and further cropped to each agro-climatic zone to study the recent changes in the daily precipitation and temperature trends spatially.

33.3 Analysis Approach

The present study estimated the changes in each of the precipitation extreme statistics using linear and Mann–Kendall trend tests.

33.3.1 Linear Regression

One of the simplest methods to calculate the trend of the data is linear regression. The equation of linear regression line is given by:

$$Y = a + bX \tag{33.1}$$

where, X is the explanatory variable and Y is the dependent variable, b is the slope of the line and a is the intercept. The slope of the regression describes the trend, with positive as increasing and negative as decreasing trend. The observed trend study is conducted by considering the rainfall and temperatures as dependent variables and time as explanatory variable.

33.3.2 Mann–Kendall Trend Test

The Mann–Kendall (Mann 1945; Kendall 1975) trend test is a non-parametric trend test, which has been widely used for trend detection in hydrologic and climate data to assess if there is an upward (positive) or downward (negative) trend of a variable of interest over time. The test compares the relative magnitudes of sample data rather than the data values themselves (Gilbert 1987). The following procedure explains the Mann–Kendall trend test:

- The time series, x_i, of the variable, for which the trend test to be applied is considered as an ordered time series.
- Each of the data point, x_i, is compared with the all subsequent data values to estimate the Mann-Kendall statistic, S, as follows:

$$S_i = \sum_{i=2}^{n} \sum_{j=1}^{i-1} \text{sign}(x_i - x_j) \tag{33.2}$$

$$\text{where} \quad \text{sign}(x_i - x_j) = \begin{cases} 1 & \text{if } x_i > x_j \\ 0 & \text{if } x_i = x_j \\ -1 & \text{if } x_i < x_j \end{cases} \tag{33.3}$$

- A very high positive value of S is an indicator of an increasing trend, and a very low negative value indicates a decreasing trend.
- From the Mann–Kendall statistic, S, the normalized test statistics, Z, is computed as follows:

$$\text{if } S > 0, \quad Z = \frac{S - 1}{[\text{VAR}(S)]^{1/2}} \tag{33.4}$$

$$\text{if } S = 0, \quad Z = 0 \tag{33.5}$$

$$\text{if } S < 0, \quad Z = \frac{S + 1}{[\text{VAR}(S)]^{1/2}} \tag{33.6}$$

where VAR(S) is the variance of S. According to Kendall (1975) VAR(S) can be written as follows:

$$\mathrm{VAR}(S) = \frac{1}{18}\left[n(n-1)(2n+5) - \sum_{p=1}^{g} t_p(t_p-1)(2t_p+5)\right] \tag{33.7}$$

where n is the number of data points, g is the number of tied groups (a tied group is a set of sample data having the same value), and t_p is the number of data points in the Pth group. The Z-value follows a standard normal distribution. For testing the decreasing or increasing trend a significance level α is used. The probability associated with the computed test statistics, Z-value is estimated. The trend is identified as decreasing if Z-value is negative and the computed probability is less than the level of significance and the trend is identified as increasing if the Z-value is positive and the computed probability is less than the level of significance. If the computed probability is greater than the level of significance, there is no trend.

The trend analysis is performed on APHRODITE daily data to test the increasing or decreasing rainfall trends using Mann–Kendall trend test for each zone spatially. To estimate the rainfall trends in terms of increasing or decreasing, the Mann–Kendall trend test is carried out at each grid point. Changes in the daily precipitation trends in recent years with reference to distant past are assessed by dividing the historical data into two-time slices as 1951–1990 and 1991–2007. The period of 1951–1990 is considered as base period as followed by several studies in the literature (Sharma and Mujumdar 2017), which often depend on the availability of the climate data. Further, the period 1961 to 1990 is likely to have larger anthropogenic trends embedded in the climate data (https://www.ipcc.ch/ipccreports/tar/wg1/483.htm). Throughout the manuscript, the study considered a trend as being significant if it is statistically significant at the 5% level (p-value < 0.05).

33.4 Results and Discussion

The spatial average monthly precipitation from 1951 to 2007 for seven agro-climatic zones of Afghanistan is shown in Fig. 33.2. The maximum precipitation occurrence months are observed as January–May for all zones of Afghanistan. Whereas, June–December and generally considered as dry months. To study the precipitation trends at annual scale, the annual total precipitation for each zone is estimated for a period of 1951–2007 with APHRODITE data. Zone wise annual precipitation is calculated from the mean precipitation of all valid gridded points of APHRODITE data encompassing each zone. The annual total rainfall has decreased over North zone at a rate of 0.5 mm/decade, whereas for North–East zone as 17.3 mm/decade, the West zone at a rate of 6.3 mm/decade. However, the annual total rainfall over South zone has increased at a rate of 2.8 mm/decade, the South–West zone at a rate of 1.7 mm/decade, the East zone at a rate of 10.7 mm/decade, the Central zone at a rate of 6.3 mm/decade. Overall, the annual total rainfall has shown an increasing trend for the zones of South, South–West, East and Central, whereas, a decreasing trend has been observed for the

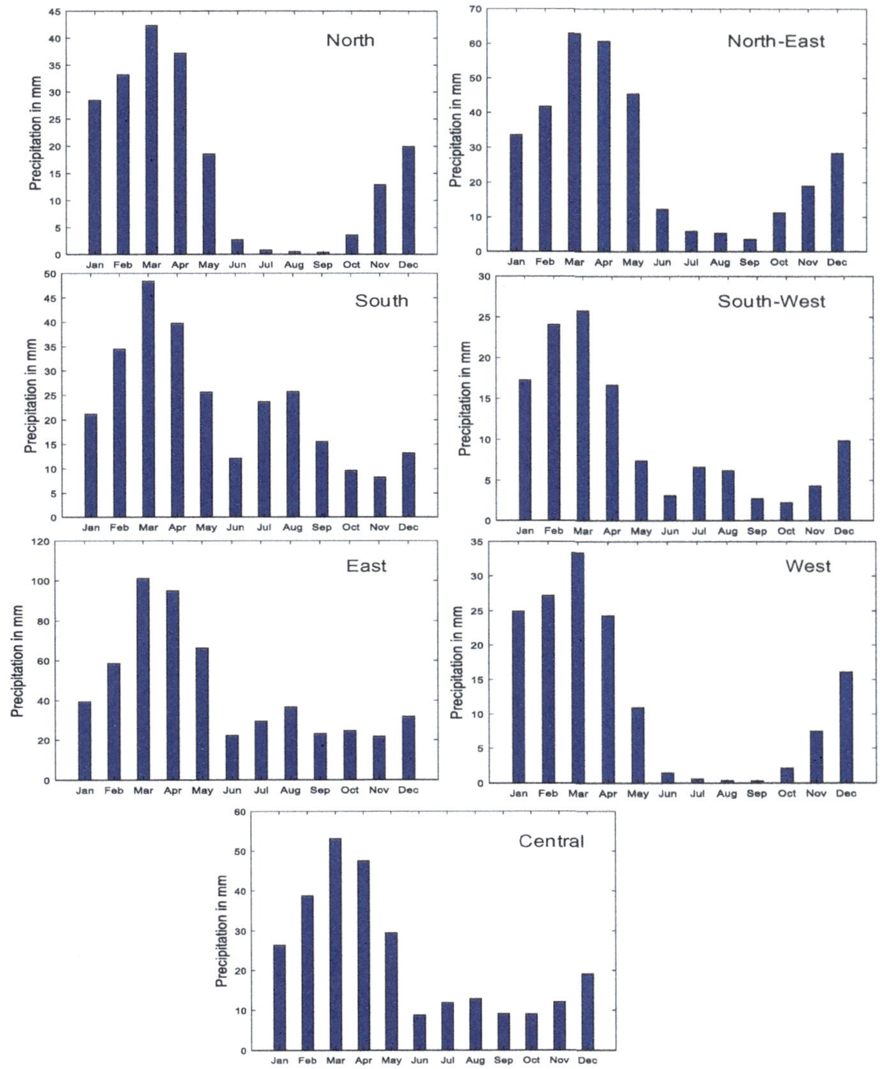

Fig. 33.2 The spatial average monthly precipitation from 1951 to 2007 for seven agro-climatic zones of zones of Afghanistan

zones of North, North–East and West zones based on linear regression trend test as shown in Fig. 33.3. Further, the trend analysis was performed on daily data to test the increasing or decreasing rainfall trends using Mann–Kendall trend test for each grid point for various zones of Afghanistan. The change in the rainfall variability in recent years with reference to distant past was assessed by dividing the historical data into two time slices as 1951–1990 and 1991–2007. Figure 33.4a–g shows the trend analysis results with Mann–Kendall trend test at 5% significance level for each zone

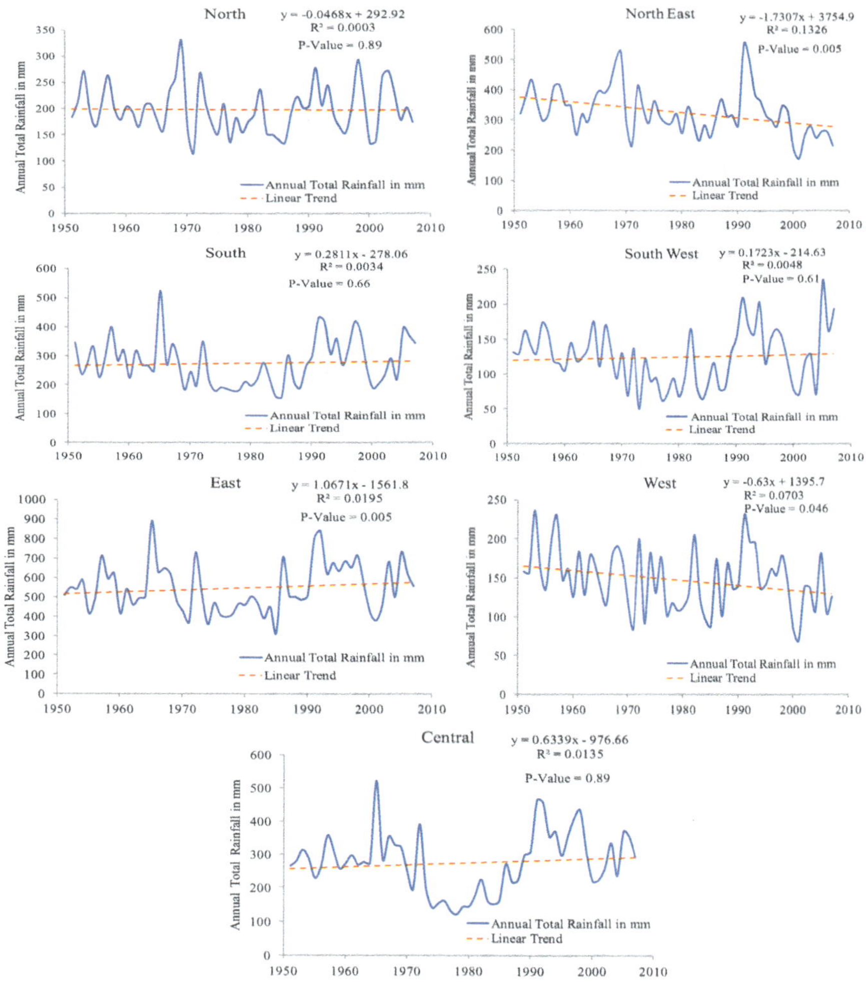

Fig. 33.3 The spatially averaged annual total rainfall trend over seven agro-climatic zones of Afghanistan for the observed periodof Afghanistan for the observed period of 1951 to 2007 with APHRODITE data

of Afghanistan for two time slices of 1951–1990 and 1991–2007. Figure 33.4a–g shows the spread of increasing (positive trend), decreasing (negative tend) and no trend at 5% significance level over North zone and most of the grid points shown a decreasing trend in rainfall in recent years over Afghanistan. The trend analysis of the rainfall with gridded data sets also reveals that for most part of the Afghanistan region the rainfall has been observed as decreasing.

The spatial average monthly temperature from 1961 to 2006 for seven agro-climatic zones of Afghanistan is shown in Fig. 33.5. The maximum temperature can be observed in the months of May–August, with hottest month as July for

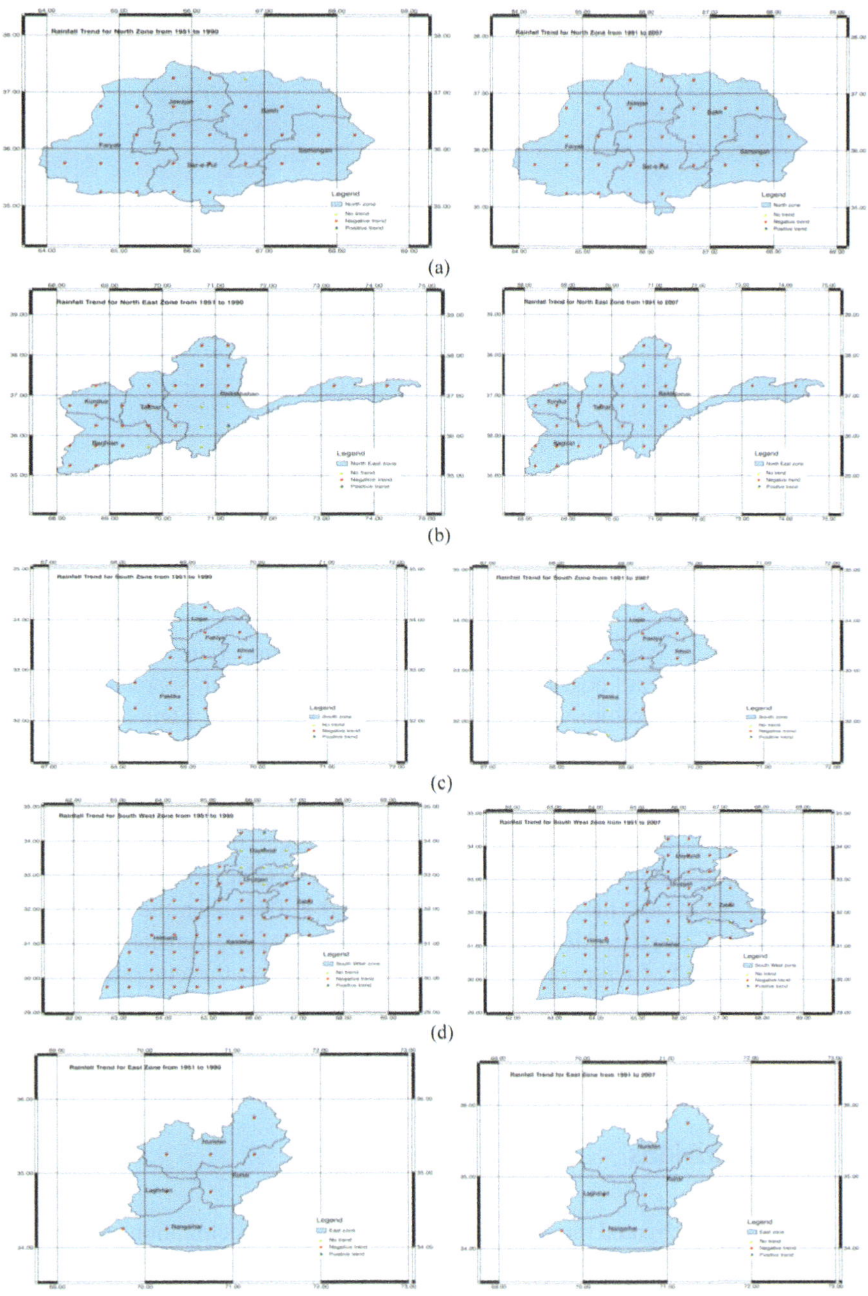

Fig. 33.4 Rainfall trend analysis for **a** North **b** North–East **c** South **d** South–West **e** East **f** West **g** Central zones with daily data of rainfall for two time slices of period of 1951–1990 and 1991–2007 with Mann–Kendall trend test. Positive Trend—Increasing rainfall trend; Negative Trend—Decreasing rainfall trend; No Trend—no trend at 5% significance level

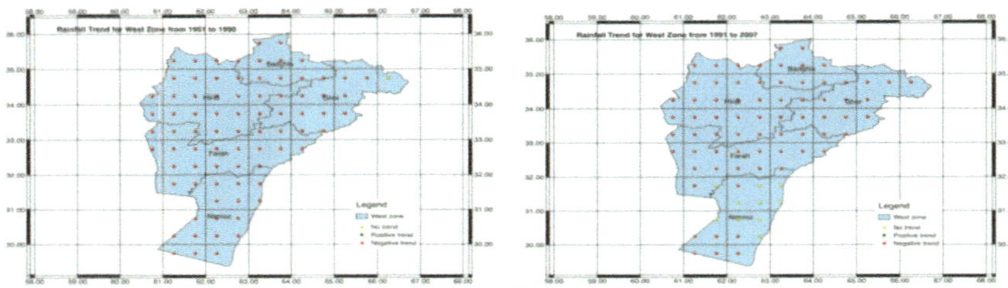

(f)

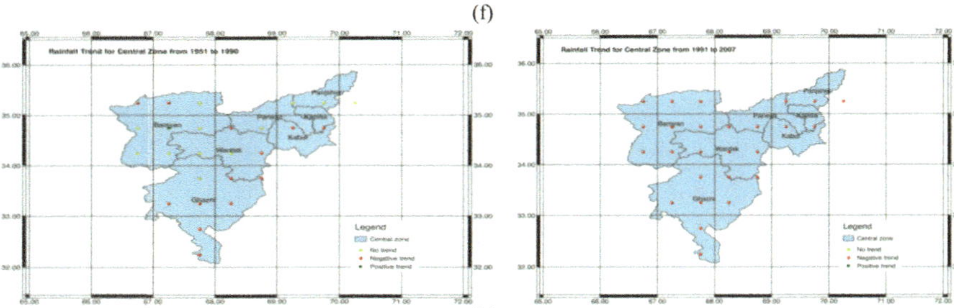

(g)

Fig. 33.4 (continued)

all seven zones of Afghanistan. The annual average temperatures from 1961 to 2006 are studied for the possible trend in the temperature with spatially averaged annual average temperature with APRODITE data (Fig. 33.6). Figure 33.6 shows the spatially averaged annual average temperature trends over seven agro-climatic zones of Afghanistan for the observed period of 1961 to 2006 with APRODITE data. The annual average air temperature has increased over the North zone at a rate of 0.4 °C/decade, the North–East zone at a rate of 0.3 °C/decade, the South zone at a rate of 0.02 °C/decade, the South–West zone at a rate of 0.4 °C/decade, the East zone at a rate of 0.1 °C/decade, the West zone at a rate of 0.4 °C/decade, the Central zone at a rate of 0.2 °C/decade. All seven agro-climatic zones of Afghanistan have shown an increasing trend of temperatures in recent years. The spatio-temporal trends of temperature for the time periods of 1961–1990 and 1991–2006 for each agro-climatic zone are shown in Fig. 33.7a–g.

33.5 Conclusions

The study has investigated the spatio-temporal analysis of precipitation and temperatures for seven agro-climatic zones of Afghanistan using gridded daily long time series datasets of APHRODITE from 1951 to 2007. The maximum precipitation occurrence months were observed as January–May for all zones of Afghanistan. Whereas, June–December and generally considered as dry months. The annual total precipitation has shown an increasing trend for the zones of South, South–West, East and Central, whereas, a decreasing trend has been observed for the zones of

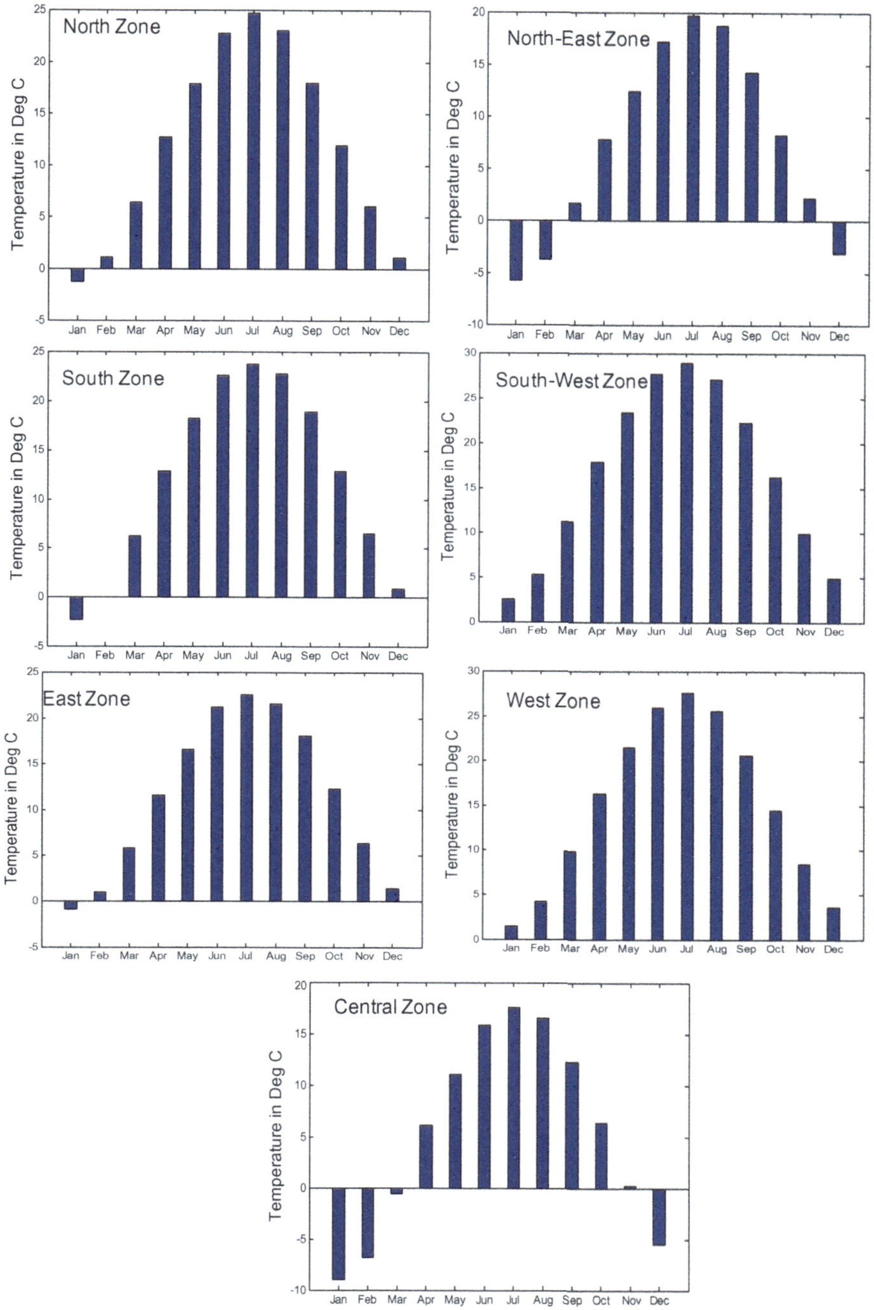

Fig. 33.5 The spatial average monthly temperature from 1961 to 2006 for seven agro-climatic zones of Afghanistan

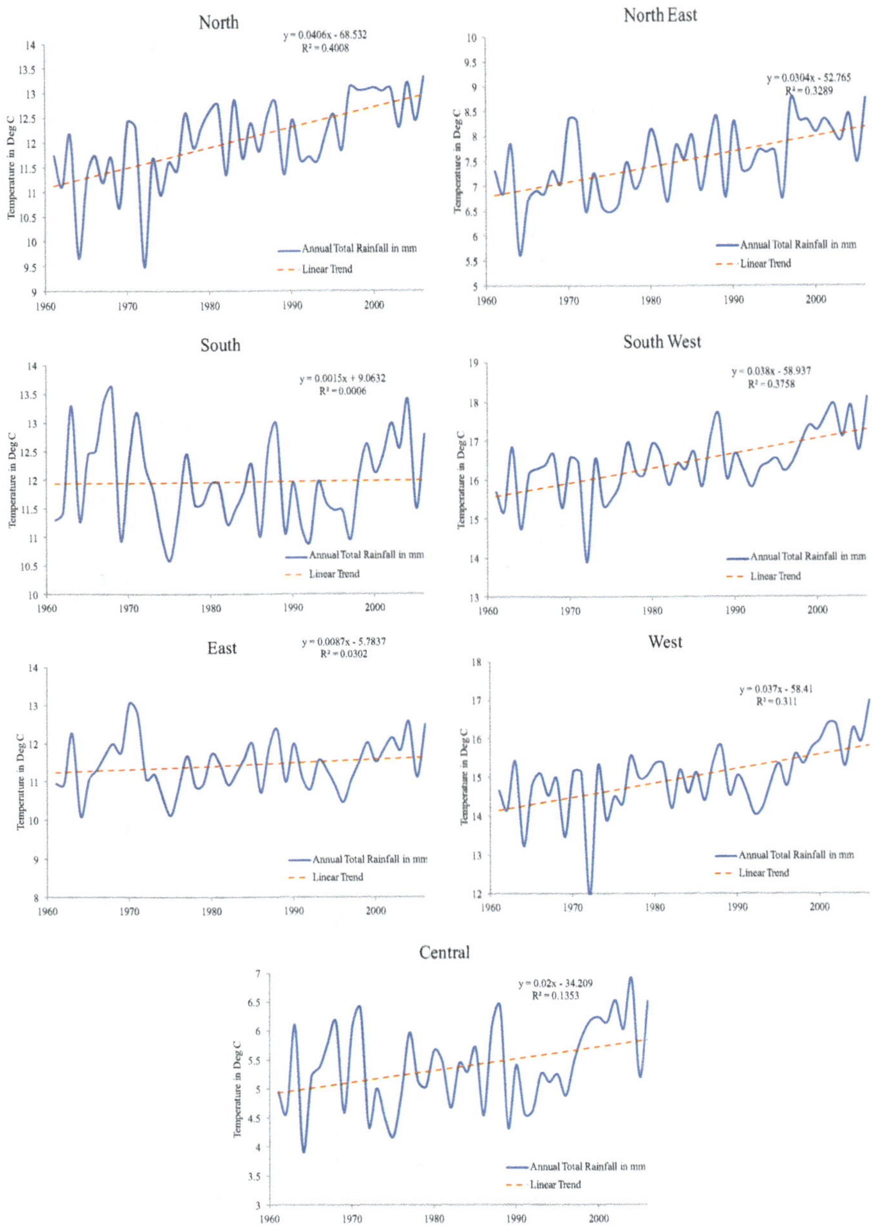

Fig. 33.6 The spatially averaged annual average temperature trend over seven agro-climatic zones of Afghanistan for the observed period of 1961 to 2006 with APHRODITE data

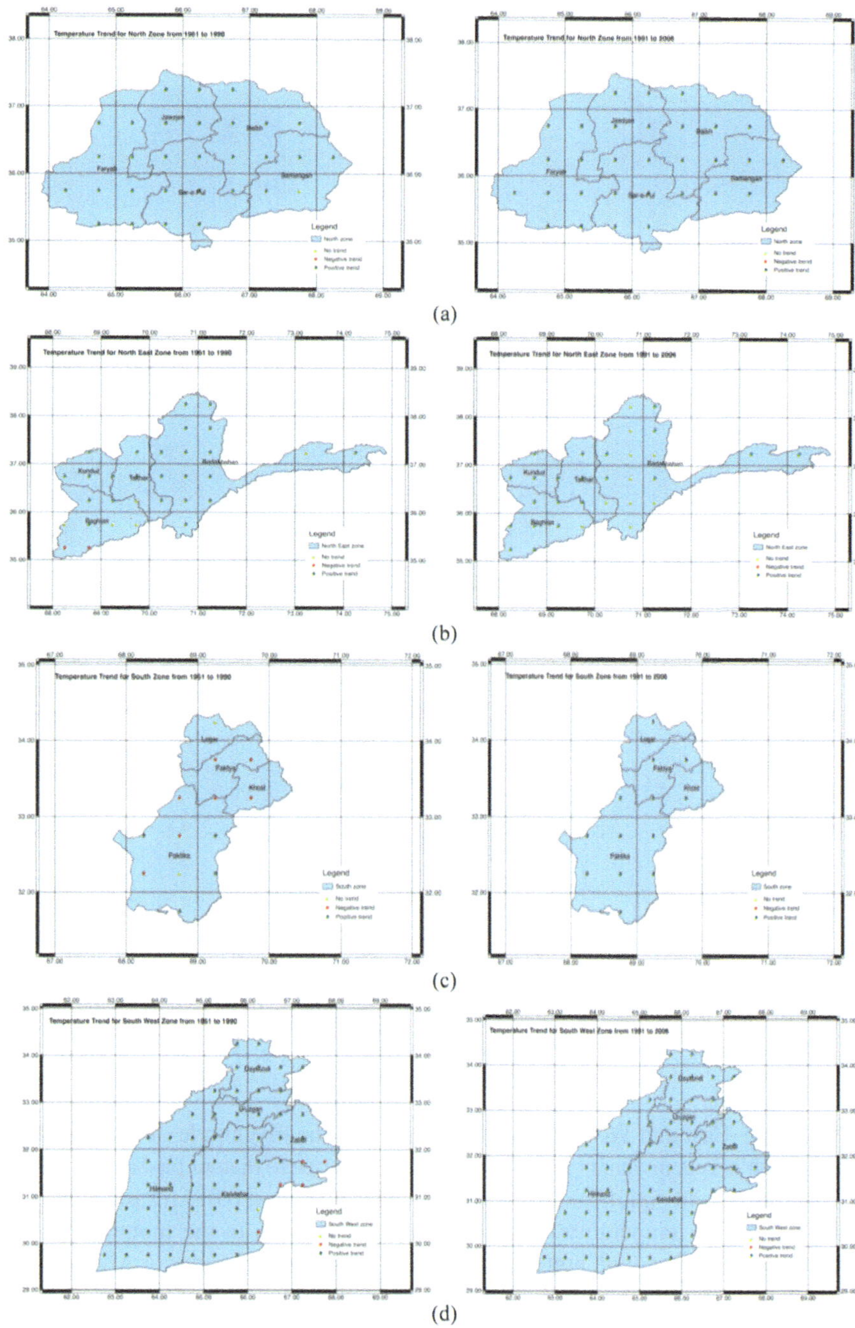

Fig. 33.7 Temperature trend analysis for **a** North **b** North–East **c** South **d** South–West **e** East **f** West **g** Central zones with daily data of rainfall for two time slices of period of 1951 to 1990 and 1991 to 2007 with Mann–Kendall trend test. Positive Trend—Increasing rainfall trend; Negative Trend—Decreasing rainfall trend; No Trend—no trend at 5% significance level

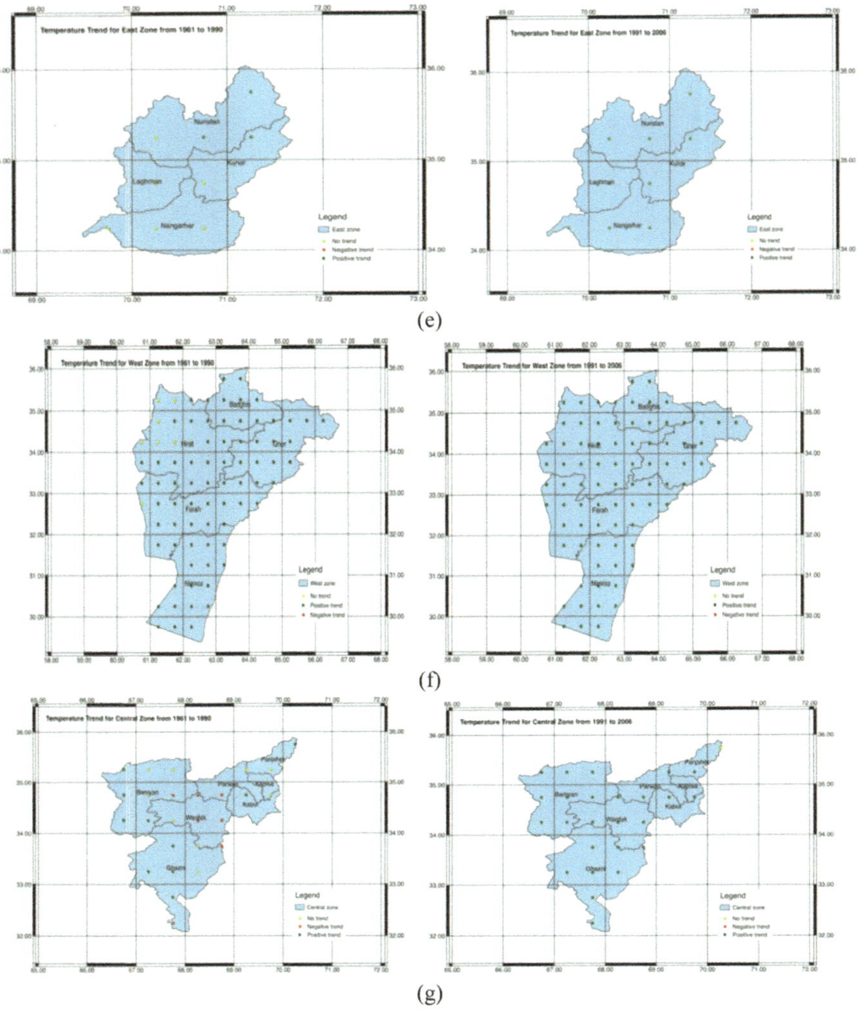

Fig. 33.7 (continued)

North, North–East and West zones. The changes in the magnitude of daily gridded precipitation in recent years (1991–2007) with reference to distant past (1951–1990) were analysed with significant decreasing trends (p-value < 0.05) for most parts of Afghanistan. The temperature was observed as maximum in the months of May, June, July and August, with hottest month as July for all seven zones of Afghanistan. The daily temperature trend analysis has revealed positive trends of temperatures for two time periods of 1951–1990 and 1991–2007 with Mann–Kendall trend test at 5% significance level. Overall, the North, North–East and West zones of Afghanistan are more vulnerable with decreasing precipitation and increasing temperatures indicating more dry and warm periods indicating increasing drought conditions. Whereas, the

South, South–West, East, and Central zones are more vulnerable with increasing trends of both precipitation and temperatures indicating increase of more wet and warm climates. The zonal level climate analysis carried in the present study has a benefit to the stakeholders and farmers in the agricultural water availability and demands perspective.

Acknowledgments The authors would like to thank the Ministry of Agriculture, Irrigation and Livestock (MAIL) and AGROMET, Afghanistan for the permission to use meteorological data. This publication is a part of the Climate Change Adaptation Project, MAIL, GoIRA by MgtWell Consulting Services through a project awarded ID no. 00087639 through the GEF/UNDP Grant No. 00076056 between the period of August 2016 and May 2017.

References

Afghanistan Statistical Organization (CSO) Ministry of Agriculture, Irrigation and Livestock, Agriculture data 2008–2016

Aich V, Akhundzadah NA, Knuerr A, Khoshbeen AJ, Hattermann F, Paeth H, Scanlon A, Paton EN (2017) Climate change in Afghanistan deduced from reanalysis and Coordinated Regional Climate Downscaling Experiment (CORDEX)—South Asia simulations. Clim 2017 5:38. https://doi.org/10.3390/cli5020038

Beekeeping survey report. Beekeeping and Animal Husbandry Development Project (BAHDP) and Beekeeping Development Project (BDP), Ministry of Agriculture, Irrigation and Livestock (MAIL), Conducted from September to November 2014

Campbell J (2015) A dry and ravaged land: investigating water resources in Afghanistan. Earth 60(1&2):48–55. http://www.earthmagazine.org/article/dry-and-ravaged-land-investigating-water-resources-afghanistan

Climate Change Scenarios for Agriculture of Afghanistan, Climate change adaptation project, Ministry of Agriculture Irrigation and Livestock, The Islamic Republic of Afghanistan, Mgtwell consulting Services, Kabul. GEF/UNDP Grant No: 00076056

Curry CL, Sillmann J, Bronaugh D, Alterskjaer K, Cole JNS, Ji D, Kravitz B, Kristjánsson JE, Moore JC, Muri H, Niemeier U, Robock A, Tilmes S, Yang S (2014) A multimodel examination of climate extremes in an idealized geoengineering experiment. J Geophys Res Atmos 119:3900–3923. https://doi.org/10.1002/2013JD020648

Donat MG, Alexander LV, Yang H, Durre I, Vose R, Dunn RJH, Willett KM, Aguilar E, Brunet M, Caesar J, Hewitson B, Jack C, Klein Tank AMG, Kruger AC, Marengo J, Peterson TC, Renom M, Oria Rojas C, Rusticucci M, Salinger J, Elrayah AS, Sekele SS, Srivastava AK, Trewin B, Villarroel C, Vincent LA, Zhai P, Zhang X, Kitching S (2013) Updated analyses of temperature and precipitation extreme indices since the beginning of the twentieth century: the HadEX2 dataset. J Geophys Res Atmos 118:2098–2118. https://doi.org/10.1002/jgrd.50150

Gilbert RO (1987) Statistical methods for environmental pollution monitoring. Wiley, New York

Hamada A, Arakawa O, Yatagai A (2011) An automated quality control method for daily rain-gauge data. Global Environ Res 15(2):183–192

Hartmann DL, Tank AMK, Rusticucci M, Alexander LV, Brönnimann S, Charabi YAR ... & Soden BJ (2013) Observations: atmosphere and surface. In: Climate Change 2013 the physical science basis: working group I Contribution to the fifth assessment report of the intergovernmental panel on climate change

IPCC (2012): Managing the risks of extreme events and disasters to advance climate change adaptation. A special report of working Groups I and II of the Intergovernmental panel on climate change [Field CB, Barros V, Stocker TF, Qin D, Dokken DJ, Ebi KL, Mastrandrea MD, Mach KJ,

Plattner G-K, Allen SK, Tignor M, Midgley PM (eds)]. Cambridge University Press, Cambridge, UK, and New York, NY, USA, 582 pp
Kendall MG (1975) Rank correlation method. Griffin, London, UK
Köppen and Wladimir (1884) Translated by Volken E, Brönnimann S. The thermal zones of the earth according to the duration of hot, moderate and cold periods and to the impact of heat on the organic world. Meteor. Zeits 20(3):351–360 (2011)
Mann HB (1945) Nonparametric tests against trend. Econometrica: J Econometric Soc 245–259
Panda DK, Aghakouchak A, Ambast SK (2017) Increasing heat waves and warm spells in India, Observed from a multiaspect framework. J. Geophys Res Atmos 122:3837–3858. https://doi.org/10.1002/2016JD026292
Razavi T, Switzman H, Arain A, Coulibaly P (2016) Regional climate change trends and uncertainty analysis using extreme indices: a case study of Hamilton, Canada. Climate Risk Manag 13:43–63. http://dx.doi.org/10.1016/j.crm.2016.06.002
Sharma S, Mujumdar P (2017) Increasing frequency and spatial extent of concurrent meteorological droughts and heatwaves in India. Sci Rep 7(1):15582. https://doi.org/10.1038/s41598-017-15896-3
Stocker TF et al (eds) Report of the intergovernmental panel on climate change. www.climatechange2013.org and www.ipcc.ch
Zhang X, Aguilar E, Sensoy S, Melkonyan H, Tagiyeva U, Ahmed N … & Albert P (2005) Trends in Middle East climate extreme indices from 1950 to 2003. J Geophys Res: Atmos 110(D22)

Chapter 34
Identification of Historical Shift, Dispersion, and Trend of the Monsoon Season for Guwahati City Using Fuzzy Segmentation and Trend Analyses

Amrutha Suresh and Pekkat Sreeja

Abstract Climate change has a potential impact on the water resources of a region as well as the occurrences of extreme events. Present study critically investigates the shift, dispersion, and trend of the monsoon season of urban Guwahati, India, using the historical climate data. The climatic variables considered in this study are the rainfall, maximum temperature, and the minimum temperature, collected over a period of 30 years (1980–2009). In this study, instead of considering the crisp boundaries for seasons, the time-series data were partitioned into internally homogeneous segments comprising comparable climatic conditions. The present study utilizes the fuzzy sets to segment the time-series climate data into different seasons. The homogeneity of the climatic variables within each segment/season was confirmed by simultaneous identification of the local probabilistic principal component analysis (PPCA) models. The algorithm facilitates a contiguous clustering in time by detecting changes in the hidden structure of multivariate time series. The present study defines and determines the shift or creep of the seasons as the movement of the center of the segment/season with time, whereas the dispersion as the variation in the length of the segment/season with time. The rainfall trend of the monsoon season was investigated by averaging the climatic variables for each segment of length $\mu \pm 1.5 \times \sigma$, where μ is the center of a season and σ is the standard deviation at the timescale. The present study noticed an early occurrence of the premonsoon and monsoon seasons and a late occurrence of the postmonsoon season that indicates the shift of the seasons. This study has observed a spread of the period of premonsoon, monsoon, and postmonsoon seasons after analyzing the dispersion characteristics. Similarly, the rainfall has shown a decreasing trend for all the seasons. Conclusively, this study has evidenced the historical variations in the monsoon season for urban Guwahati regarding shift, dispersion, and the reduction in the rainfall intensity.

A. Suresh (✉) · P. Sreeja
Civil Engineering Department, Indian Institute of Technology Guwahati, Guwahati 781039, Assam, India
e-mail: a.suresh@iitg.ac.in

P. Sreeja
e-mail: sreeja@iitg.ac.in

R. Jha et al. (eds.), *Climate Change Impacts on Water Resources*, Water Science and Technology Library 98, https://doi.org/10.1007/978-3-030-64202-0_34

Keywords Monsoon · Rainfall · Probabilistic Principal Component Analysis (PPCA) · Gath-Geva clustering

34.1 Introduction

Climate change poses a fundamental threat to the existence of life on the planet. Several researchers have shown much evidence of climate change. One of the evidence is the alteration of the climate pattern and the associated creep and dispersion of the seasons. Seasons can be defined as the time-series segments of similar climatic characteristics in a year. Generally, for the convenience of meteorological purposes, the seasons are defined with crisp boundaries based on the climatological normal. Indeed, the climate is not static hence the season may vary over a more extended period, called dispersion and creep of the seasons. Determination of the shift and dispersion of season is essential for appropriate decision-making to handle the climate change or to minimize its impact. The identification of the variations in the monsoon season is essential for the agronomic countries like India since the agricultural activities have a direct dependence on the monsoon season. Thus agriculture-related decision-making requires a detailed understanding of the seasons and their variations. Studying the shift and dispersion of seasons needs the knowledge of the temporal variations of the seasons. However, consideration of the crisp boundaries of the season may not be suitable for such analysis since it only implies the variations in the intensities of climatic variables. For studying the shift and dispersion of the seasons, it is important to segment the days in a year with similar climatic conditions as seasons, keeping a constant number of the season over the years. The present study emphasizes the time-series segmentation-based analysis of the historical climate data for a better understanding of the drift, dispersion, and the trend of the climatic variables over time.

Seasonality is a result of tilting of the Earth, while the monsoon weather systems are caused by the differences in land-sea temperature due to the solar radiation (Huffman et al. 1997). Earths' rotation and revolution around the Sun and the different land masses of the northern and southern hemispheres cause a difference in the climatic conditions over a year. The annual oscillation in the apparent position of the Sun between the tropics of Cancer and Capricorn causes the annual oscillation in the position of the thermal equator (region of maximum heating) on the Earth's surface. Such annual oscillation of the thermal equator causes the annual oscillation of temperature, pressure, wind, cloudiness, rain, and so on. The entire process results in the monsoon wind and regional variations in the intensity of monsoon. Figure 34.1 shows the monsoon map for India. It is to be noted that India receives 70% of its rainfall during the monsoon season. Table 34.1 shows the meteorological seasons considered in northeast India.

The present study emphasizes the importance of segmenting the time-series data to seasons based on a data-driven method instead of defining the crisp bounds. Time series segmentation is the clustering of the time series data with a time-ordered

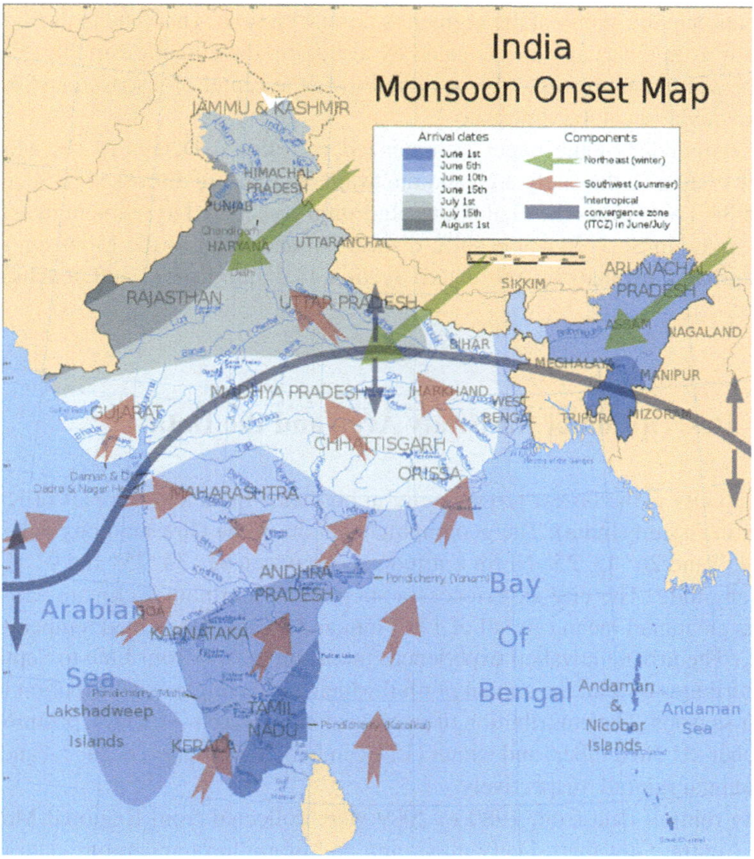

Fig. 34.1 India southwest summer monsoon onset map (*Source* Saravask 2007)

Table 34.1 The meteorological seasons of India

Season	Period
Winter	December–February
Premonsoon	March–May
Southwest Monsoon	June–September
Postmonsoon	October–November

structure. The clustering method considered in this study is the modified Gath-Geva clustering algorithm (Abonyi et al. 2005), which can also be referred to the Fuzzy Maximum Likelihood clustering with probabilistic principal component analysis. The clusters obtained from this method would be multivariate Gaussian functions. A Gaussian membership function is defined for homogeneous segmentation of the time-series data. The present study defines the shift or creep of the seasons as the movement of the center of the segment/season with time, whereas the dispersion as the variation in the length of the segment/season with time. The trend in the monsoon season over time was also investigated by averaging the climatic variables for each segment with a length $\mu \pm 1.5 \times \sigma$, where μ is the center of a season and σ is the standard deviation at the timescale.

The remainder of the paper is organized as follows. Details of the Monsoon weather system of the study area and the available data are discussed in Sect. 34.2. Section 34.3 overviews the Gath-Geva algorithm considered to segment the seasons in this study. In Sect. 34.4, the shift, dispersion, and trend of the Monsoon season for the urban Guwahati are presented. Section 34.5 summarizes and concludes the paper.

34.2 Description of the Study Area and the Data

Guwahati city is the fastest growing city in northeast India and is a part of Kamrup District in Assam (India). The geographic location of the Guwahati city is between 26° 4′ 45″ and 26° 13′ 25″ North Latitude and between 91° 34′ 25″ and 91° 52′ 00″ East Longitude. The city experiences a subtropical and humid climate. Guwahati receives an annual mean rainfall of 1729 mm and an average annual temperature of 23.5 °C. The urban Guwahati experiences monsoon season from June to September and contributes 63% of the annual rainfall which has the highest contribution among the four seasons. The contribution of premonsoon (1 March–31 May), postmonsoon (1 October–31 November) and winter (1 December–28 February) is 28, 7.4 and 2.4% of the annual rainfall, respectively.

Daily rainfall data from 1980 to 2009 were collected from Regional Meteorological Centre, Guwahati. Daily maximum and minimum temperature data for the period 1980–2009 were obtained from the SWAT database website (http://globalweather.tamu.edu/). Tables 34.2 and 34.3 shows the statistical description of rainfall and temperature data respectively for the period 1980–2009 for Guwahati city.

34.3 Segmentation of the Time-Series Data Using Gath-Geva (GG) Segmentation Algorithm

The fuzzy clustering algorithm was considered to segment the time-series climate data. The clustering algorithm performs a simultaneous identification of local probabilistic principal component analysis (PPCA) models and measures the homogeneity of the segments (Abonyi et al. 2005). The algorithm favors contiguous clusters in time and can detect changes in the hidden structure of multivariate time-series. The details of the GG segmentation algorithm are as follows.

Consider a time series of n samples denoted as $T = \{x_k | 1 \leq k \leq n\}$ where $k = 1, \ldots\ldots, n$ and $x_k = \left[x_{1,k}, x_{2,k}, \ldots\ldots\ldots\ldots\ldots\ldots, x_{n,k}\right]^T$ labeled by time points

Table 34.2 Statistical description of rainfall of Guwahati city for 1980–2009

Month/Season	Total mean rainfall (mm)	Average rainy days	Standard deviation (mm)	Coefficient of variation
January	11.7	1.13	15.1	129.0
February	23.1	2.33	21.5	92.8
March	57.9	4.67	41.6	71.8
April	179.2	10.17	111.9	62.4
May	242.2	13.37	128.9	53.2
June	290.8	14.33	121.1	41.6
July	323.5	15.83	112.3	34.7
August	273.5	13.07	136.8	50.0
September	196.7	9.8	84.8	43.1
October	109.5	4.67	90.3	82.5
November	15.3	0.97	21.8	142.6
December	6.0	0.6	8.4	139.6
Annual	1729.5	90.94	256.4	14.8
Premonsoon	479.4	28.21	150.3	31.3
Monsoon	1084.5	53.03	226.7	20.9
Postmonsoon	124.7	5.64	89.4	71.6
Winter	40.9	4.06	29.7	72.5

$t_1, t_2, \ldots\ldots\ldots\ldots, t_n$. The time series T is partitioned to c-segments represented as t_{n_i} where $1 \leq i \leq c$ which satisfies $1 = t_{n_0} < t_{n_1} < \cdots\cdots\cdots < t_{n_c} = t_n$. The time points $t_{n_0}, t_{n_1}, \ldots\ldots\ldots, t_{n_c}$ and the interval $\left[t_{n_{i-1}}, t_{n_i}\right]$ are the segment boundaries and segments respectively. The crisp segmentation cost is formulated as an optimization problem which minimizes the variance of the segment resulting in the following equation (Abonyi et al. 2005; Wang et al. 2012):

$$\text{cost}_{\text{crisp}}(t) = \sum_{k=1}^{N} \sum_{i=1}^{c} \beta_i(t_k) D^2\left(x_k, v_i^x\right) \tag{34.1}$$

where $D^2\left(x_k, v_i^x\right)$ is the distance between the center of the ith cluster (v_i^x) and x_k data point. $\beta_i(t_k) \in \{0, 1\}$ stands for the crisp membership of the kth data point in the ith segment.

$$\beta_i(t_k) = \begin{cases} 1 \ \ if \ \mathrm{t}_{n_{i-1}} < t_k < \mathrm{t}_{n_i}; \\ 0, \qquad \text{otherwise.} \end{cases} \tag{34.2}$$

Since defining a crisp boundary is not practical for most of the time series segmentation applications, Abonyi et al. (2005) have introduced a Gaussian membership

Table 34.3 Statistical description of the temperature of Guwahati city for 1980–2009

Month/Season	Minimum temperature (°C)	Maximum temperature (°C)	Mean temperature (°C)	Standard deviation (°C)	Coefficient of variation
January	3.59	30.21	22.23	4.24	19.05
February	12.73	35.18	26.07	3.18	12.20
March	19.92	38.07	30.89	2.98	9.64
April	20.83	41.40	32.70	3.70	11.33
May	20.74	40.62	32.70	3.56	10.90
June	22.92	41.14	32.26	3.46	10.71
July	22.51	39.19	31.26	2.72	8.70
August	23.77	38.81	31.50	2.74	8.71
September	20.55	37.69	30.45	2.68	8.81
October	17.66	34.84	28.65	3.02	10.56
November	10.84	31.84	26.13	2.52	9.65
December	1.96	29.79	22.20	5.21	23.48
Annual	1.08	33.52	23.55	4.92	20.89
Premonsoon	16.74	32.68	25.69	2.78	10.82
Monsoon	20.18	33.52	27.41	1.74	6.35
Postmonsoon	6.05	29.24	22.00	2.81	12.77
Winter	1.08	27.02	17.13	3.67	21.42

function for the time series data. Within the Gaussian membership function $A_i(t_k)$, the time coordinate is considered as an additional variable that divides the time series into fuzzy segments. Where,

$$A_i(t_k) = \exp\left(-\frac{1}{2}\frac{\left(t_k - v_i^t\right)^2}{\sigma_{i,t}^2}\right) \tag{34.3}$$

The new Fuzzy segmentation parameter $\beta_i(t_k)$ is defined as,

$$\beta_i(t_k) = \frac{A_i(t_k)}{\sum_{j=1}^{c} A_j(t_k)} \in [0, 1] \tag{34.4}$$

The objective function is modified as the sum of the weighted squared distances between z_k and η_i where $z_k = \left[t_k, x_k^T\right]^T$ and η_i is the cluster prototype. The objective function (J) is based on the assumption that the data points can be modeled as a mixture of multivariate Gaussian distribution (Abonyi et al. 2005; Wang et al. 2012). Where the objective function J is,

$$J = \sum_{k=1}^{n}\sum_{i=1}^{c}\left(\mu_{i,k}\right)^{m} D^2(z_k, \eta_i) \quad (34.5)$$

where $\mu_{i,k}$ represents the degree of membership of the observed data. "m" ($m \in [1, \infty)$) is the weighting component that decides the fuzziness of the cluster (usually $m = 2$). The distance $D^2(z_k, \eta_i)$ is inversely proportional to the probability of the data points z_k belongs to the ith cluster. As the time variable t_k is independent of x_k, the distance measure $D^2(z_k, \eta_i)$ is defined as (Abonyi et al. 2005; Wang et al. 2012).

$$p\left(z_k / \eta_i\right) = \frac{1}{D^2(z_k, \eta_i)} = \alpha_i \times p\left(t_k / \eta_i^t\right) \times p\left(x_k / \eta_i^x\right) \quad (34.6)$$

where,

α_i = Priori probability of the cluster, $p(\eta_i)$

$p\left(t_k / \eta_i^t\right)$ = Distance between kth data point and the center of the ith segment (v_i^t)

$$p\left(t_k / \eta_i^t\right) = G\left(t_k; v_i^t, \sigma_{i,t}^2\right) = \frac{1}{\sqrt{2\pi\sigma_{i,t}^2}} \exp\left(-\frac{1}{2}\frac{\left(t_k - v_i^t\right)^2}{\sigma_{i,t}^2}\right)$$

$\sigma_{i,t}^2$ = Variance of Gaussian function $G\left(t_k; v_i^t, \sigma_{i,t}^2\right)$

$p\left(x_k / \eta_i^x\right)$ = Distance between the cluster prototype (η_i) and the data in the feature space

$$\begin{aligned} p\left(x_k \eta_i^x\right) &= G\left(x_k; v_i^x, F_i^x\right) \\ &= \frac{1}{(2\pi)^{r/2}\sqrt{\det\left(F_i^x\right)}} \exp\left(-\frac{1}{2}\left(x_k - v_i^x\right)^T \left(F_i^x\right)^{-1}\left(x_k - v_i^x\right)\right) \end{aligned}$$

v_i^x = Center of Gaussian function $G\left(x_k; v_i^x, F_i^x\right)$

F_i^x = Covariance of Gaussian function $G\left(x_k; v_i^x, F_i^x\right)$

R = Rank of Covariance matrix, F_i^x.

Abonyi et al. (2005) developed the GG clustering algorithm based time series segmentation using Lagrange multipliers, and the objective function is given as

$$\bar{J} = \sum_{k=1}^{n}\sum_{i=1}^{c}\left(\mu_{i,k}\right)^{m} D^2(z_k, \eta_i) + \sum_{k=1}^{n}\lambda_k\left(\sum_{i=1}^{c}\mu_{i,k} - 1\right) \quad (34.7)$$

where λ_k is the Lagrange multiplier and $\mu_{i,k}$ is the membership degree of data point x_k to the ith cluster (i = 1,......, c) and $\mu_{i,k}$ is subjected to the following constraint

$$\mu_{i,k} \in [0, 1], \forall i, k;\ 0 < \sum_{k=1}^{n} \mu_{i,k}, \forall i;\ \sum_{i=1}^{c} \mu_{i,k} = 1, \forall k \tag{34.8}$$

34.3.1 Time-Series Segmentation Using Modified Gath-Geva Algorithm

The rainfall and temperature data were averaged for every 3 years from 1980 to 2009. The modified Gath-Geva algorithm was applied to the rainfall, maximum temperature, and minimum temperature such that to get the segments of homogeneous climatic conditions. The parameters of the Gath-Geva algorithm were adjusted in such a way that the final number of segments is five for each observation period. The final segments were assumed to be the non-rainy season, premonsoon season, monsoon season, postmonsoon season, and nonrainy season. Figure 34.2 shows the seasons obtained for every three years from 1980 to 2009. For further analyses, the center and standard deviation of each season on the time scale were determined. Corresponding to each season, the mean rainfall intensity was estimated over a period of $\mu \pm 1.5 \times \sigma$, where μ is the center of a season and σ is the standard deviation at the time scale, and performed the trend analyses. The detailed information of the shift, dispersion, and trend of the Monsoon season is given in the following section.

34.4 Determination of the Shift, Dispersion, and Trend of the Monsoon Season

This section analyses the shift, dispersion, and trend of the Monsoon season over past 30 years from 1980 to 2009 for urban Guwahati. The shift of the Monsoon season is determined by tracking the movement of the position of the central tendency of the seasons over the years. The dispersion is measured through the coefficient of variation of the segment length (period of the season) over the years. The trend of the rainfall during the monsoon season was also analyzed with the behavior of the average rainfall within the segment over the years. The average was determined for a period within the segment of length $\mu \pm 1.5 \times \sigma$, where μ is the center of a season and σ is the standard deviation at the timescale.

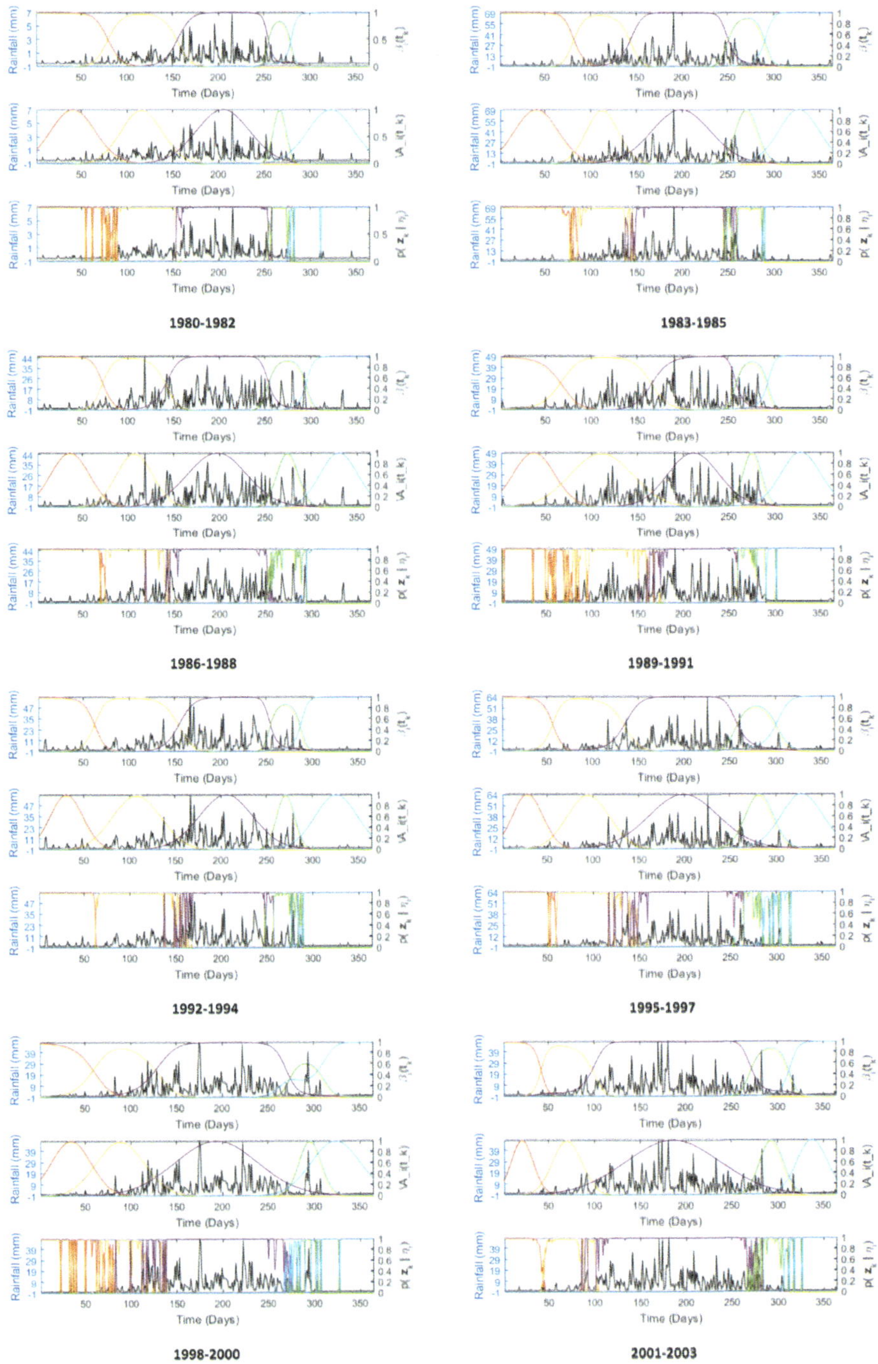

Fig. 34.2 The final segments after applying the modified Gath-Geva algorithm

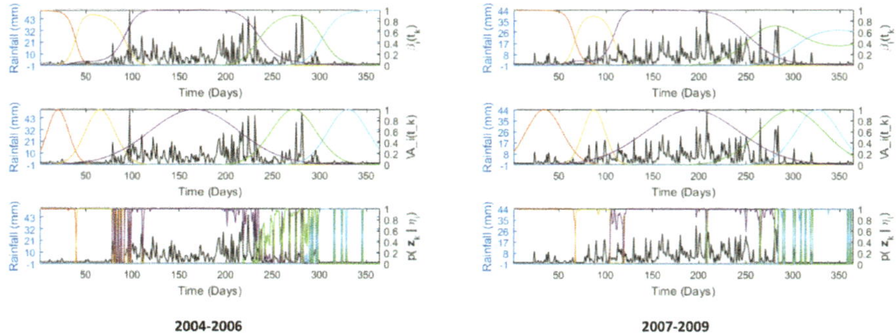

Fig. 34.2 (continued)

34.4.1 Analysis of the Shift/Creep of the Season

The analysis of the deviation of the center of the season with the time has shown a nonsignificant shift of the seasons over time. Figure 34.3 shows the shift of seasons corresponding to premonsoon, monsoon, and postmonsoon seasons. It is evident from the figures that the seasons are creeping to other seasons. Table 34.4 shows the statistics of the shift in the premonsoon, monsoon, and postmonsoon seasons. The analysis shows a nonsignificant shift in every season. The premonsoon and monsoon seasons shift backward (i.e., early occurrence of the premonsoon and monsoon seasons), whereas the postmonsoon seasons show a forward shift (i.e., late occurrence of the post-monsoon season). The premonsoon occurs early by 4.236 days/3-years whereas the Monsoon season occurs early by 2.306 days/3-years. The postmonsoon season occurs late by 2 days/3-years.

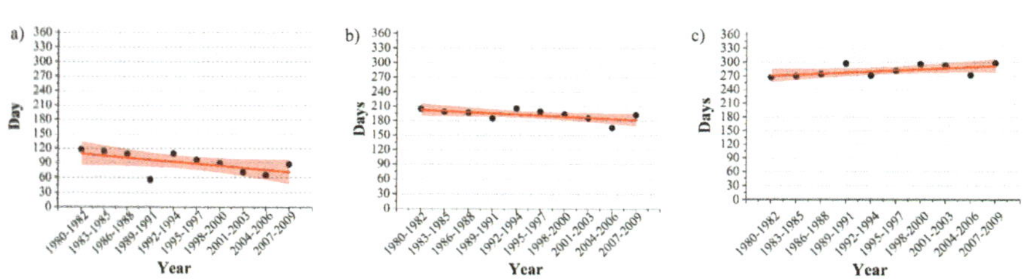

Fig. 34.3 Analysis of the shift of season; **a** shift of premonsoon, **b** shift of Monsoon, **c** shift of postmonsoon

Table 34.4 Statistics of the season shift analysis

Season	Estimated slope	Standard error
Premonsoon	−4.236	2.016
Monsoon	−2.306	1.101
Postmonsoon	2.382	1.206

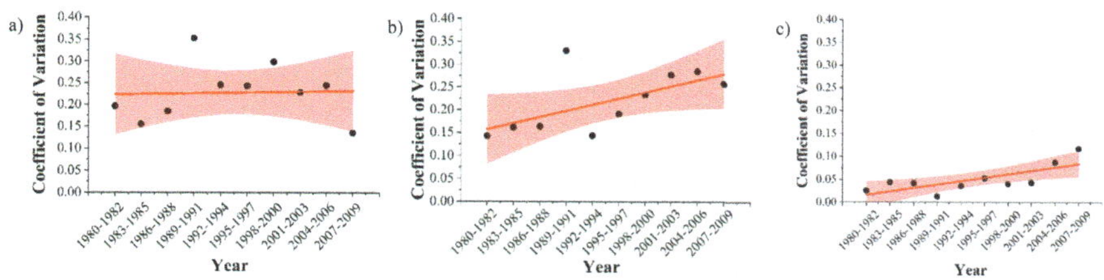

Fig. 34.4 Analysis of the dispersion of the seasons. **a** Dispersion of premonsoon. **b** Dispersion of Monsoon. **c** Dispersion of the postmonsoon

Table 34.5 Statistics of season dispersion analysis

Season	Estimated slope	Standard error
Premonsoon	0.001	0.008
Monsoon	0.014	0.006
Postmonson	0.007	0.002

34.4.2 Analysis of the Dispersion of the Season

The dispersion of the season is measured through the coefficient of variation of the length of the segment (duration of the season) over the years. Figure 34.4 shows the dispersion of the premonsoon, monsoon, and postmonsoon seasons. The monsoon and postmonsoon seasons show a nonsignificant dispersion of the seasons.

Table 34.5 shows the statistics of the dispersion analysis of the seasons. All the seasons have shown positive dispersion which indicates the spread of the seasons.

34.4.3 Trend Analysis

The trend of the rainfall during the premonsoon, monsoon, and the postmonsoon season was analyzed, and a nonsignificant decrease in rainfall was observed for all the seasons. The *rainfall during the premonsoon season shows a nonsignificant rainfall decrement by 0.237 mm/3-years. The monsoon rainfall decreases by 0.170 mm/3-years and the postmonsoon season have a rainfall decreasing rate of 0.077 mm/3-years. All these decrement rates are statistically insignificant.* Figure 34.5 shows the results corresponding to rainfall trend analysis results for different seasons.

The rainfall during the premonsoon season shows a nonsignificant rainfall decrement by 0.237 mm/3-years. The monsoon rainfall decreases by 0.170 mm/3-years and the postmonsoon season have a rainfall decreasing rate of 0.077 mm/3-years. All these decrement rates are statistically insignificant (Table 34.6)

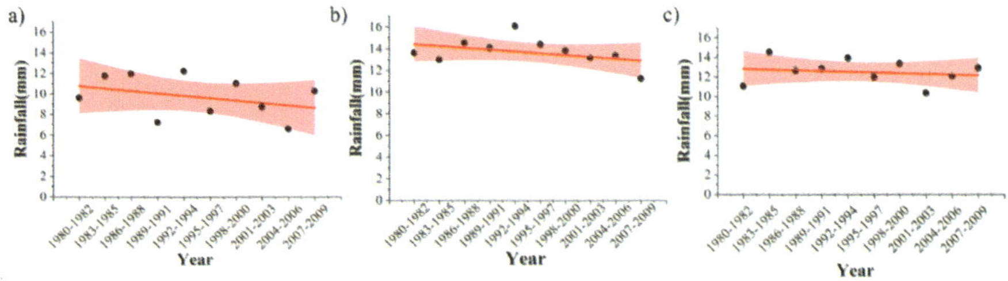

Fig. 34.5 Analysis of the trend of seasons. **a** Pre-Monsoon. **b** Monsoon. **c** Post-Monsoon

Table 34.6 Statistics of trend analysis

Season	Estimated slope	Standard error
Premonsoon	−0.237	0.220
Monsoon	−0.170	0.132
Postmonsoon	−0.077	0.145

34.5 Summary and Conclusion

Analysis of the shift and dispersion of the seasons with crisp boundaries for the season may not be appropriate since it may not provide a clear idea of the migration behavior of seasons over the timescale. The present study utilizes the fuzzy sets to segments the time-series climate data into different seasons instead of considering crip boundaries. The parameters of the GG algorithm were adjusted in such a way to obtain five segments after the segmentation process. The segments obtained from GG segmentation were assumed to be the nonrainy season, premonsoon season, monsoon season, postmonsoon season, and nonrainy season. From the creep analysis, it was noticed an early occurrence of the premonsoon and monsoon seasons and a late occurrence of the postmonsoon season. The rate of shift of seasons was observed as −4.236, −2.306, and 2.383 days/3-years for the premonsoon, monsoon, and postmonsoon seasons, respectively. The premonsoon, monsoon, and postmonsoon seasons have shown a positive dispersion that indicates spread of the seasons. A nonsignificant decreasing trend in rainfall was observed for the three seasons with a rate of 0.237, 0170, and 0.077 mm/3-years, respectively, for premonsoon, monsoon, and postmonsoon. The present study clearly shows the evidence of the shift and dispersion of the monsoon season for urban Guwahati.

References

Abonyi J, Feil B, Nemeth S, Arva P (2005) Modified Gath-Geva clustering for fuzzy segmentation of multivariate time-series. Fuzzy Sets Syst North-Holland 149(1):39–56

Huffman GJ, Adler RF, Arkin P, Chang A, Ferraro R, Gruber A, Janowiak J, McNab A, Rudolf B, Schneider U (1997) The Global Precipitation Climatology Project (GPCP) Combined Precipitation Dataset. Bull Amer Meteorol Soc Amer Meteorol Soc 78(1):5–20

Saravask (2007) India southwest summer monsoon onset map. Work by Planemad and Nichalp [CC BY-SA 3.0]. https://creativecommons.org/licenses/by-sa/3.0, https://commons.wikimedia.org/wiki/File:India_southwest_summer_monsoon_onset_map_en.svg

Wang N, Liu X, Yin J (2012) Improved Gath-Geva clustering for fuzzy segmentation of hydrometeorological time series. Stoch Env Res Risk Assess 26(1):139–155

Chapter 35
Analysis of Intensity–Duration–Frequency and Depth–Duration–Frequency Curve Projections Under Climate Variability

Manish Kumar Sinha, Klaus Baier, Rafig Azzam, M. K. Verma, and Ramakar Jha

Abstract This study focuses on the design of intensity–duration–frequency (IDF) curves under climate variability for the Raipur city in Chhattisgarh state, Central India. Design of IDF curve is the process of design of rainfall by the frequency analysis using historical rainfall data. The originality of this study comes with the attention of temporal rainfall variability consideration in design rainfall. To track the inherent rainfall events, shorter duration rainfall data have been used. As the shorter duration rainfall data will give the precise number of rainy events which may be ignored in larger duration rainfall events. The insight of the result gives the asses to check the vulnerability of the hydraulic structure as well as planning, designing and operation under climate uncertainty. The Gumbels extreme value distribution is used to find the design intensity for different durations and different intensities at different return periods. The annual maximum series of precipitation intensity for each duration is obtained and K-S & Chi-Square tests were performed to check the fitness of the rainfall data. To consider the nonstationarity of the rainfall, time series is divided into two parts from the change point. Trend of each series represents the change in intensity of the rainfall with time. These results are explaining the importance of the adaptation of changed climate.

Keywords Climate shift · Climate variability · Intensity–Duration–Frequency curve · Trend analysis · Raipur city

---

M. K. Sinha · K. Baier · R. Azzam
Department of Engineering Geology and Hydrogeology, RWTH Aachen University, Aachen, Germany

M. K. Verma
Chhattisgarh Swami Vivekanand Technical University, Bhilai, Chhattisgarh, India

R. Jha (✉)
Department of Civil Engineering, National Institute of Technology Patna, Patna, India
e-mail: rj@nitp.ac.in

R. Jha et al. (eds.), *Climate Change Impacts on Water Resources*, Water Science and Technology Library 98, https://doi.org/10.1007/978-3-030-64202-0_35

35.1 Introduction

The present study is a part of ongoing research "Impact of urbanization on water resources" at Department of Engineering Geology and Hydrogeology RWTH Aachen University Germany. The problem under consideration for this study is the anomaly in changing rainfall pattern in an urban area, which causes the urban hydraulic structures vulnerable to this changing climate. This study focuses particularly the impact of increasing urban heat which is caused by increasing urbanization and resulting imperviousness and built-up areas on rainfall extremes.

In developing countries like India, the present hydraulic systems (urban wastewater drains) are not adequate to drain off the storm water quickly under adverse climate conditions. It is not only caused by its capacity but also caused by the selection of design rainfall. This problem could be related to the change of area in respect of urban development and the change of climatic condition. The effect of change in landuse, vehicle exhaust, dust particle and many more related to urban development may have a considerable effect on the local climate. On the other hand, due to climate change effect, the increased rainfall intensity will make the area vulnerable to the flood anomaly. Hence, the combine effect of increased surface runoff and the high-intensity rainfall will make the area susceptible to flooding. Therefore, consideration of the importance of the city and vulnerability to flooding, for sound designing of the drainage system, knowledge of proper rainfall intensity for different return periods is necessary. The focus of this study was to provide the projection for design rainfall intensity under climate uncertainty.

To identify the design rainfall value for the hydraulic structure, frequency analysis has been done. The Intensity–Duration–Frequency (IDF) curve was used to perform frequency analysis to graphically represent the projected rainfall intensities. But the basic assumption of the no nonstationarity of the meteorological data is now having a big question mark in the frequency analysis. Because any time-series data may have a trend either increasing or decreasing due to the effect of climate change with in a similar return level. The sensitivity of the time series toward the overall trend in past rainfall data is ignored in IDF analysis. This study was the attempt to understand the influence of existence of trend/shift in rainfall data on account of climate uncertainty.

Shrestha et al. (2017) have been accepted the fact that those IDFs developed from past climatic scenarios cannot be valid for future climatic conditions unless the IDFs were updated to the future climate trends. There is a need to evaluate the climate-induced change in pattern, intensity and frequency of extreme rainfall events before embarking on any system performance analysis. A very popular Gumbels extreme value (GEV) distribution has been used in this study (Shrestha et al. 2017; Fadhel et al. 2017; Rasel and Islam 2015; DeGaetano and Castellano 2017; Huang et al. 2016; Madsen et al. 2009; Sivapalan and Blöschl 1998; Tfwala et al. 2017) and many authors used GEV for the preparation of IDF curves. Since the effect of climate uncertainty has been ignored due to its fundamental assumption of data no nonstationarity, IDF/design rainfall has to be updated. It is the demand of the situation that the process of development of design rainfall should have the component that considers the shift of the rainfall events there by modifying the IDF curves for the optimal designing and operation of urban infrastructures.

35.2 Study Area

This study has been carried out a developing city Raipur of Chhattisgarh state in India. Raipur area has a tropical wet and dry climate; temperatures remain moderate throughout the year (Figs. 35.1, 35.2, 35.3 and 35.4), except from March to June, which can be extremely hot. The temperature in April–May sometimes rises above 48 °C. These summer months also have dry and hot winds (Sinha et al., 2019). In summers, the temperature can also go up to 50 °C. Winters last from November to January and are mild, although lows can fall to 8 °C (Jaiswal et al. 2015; Meshram et al. 2017). The average (1980–2007) potential evapotranspiration varies from 104 mm in December to 258 mm in April. South west monsoon contributes more than 85% of the total rainfall. The annual normal rainfall of the area is ~1200 mm (CGWRD 2013; Khavse et al. 2015; Mukherjee et al. 2011).

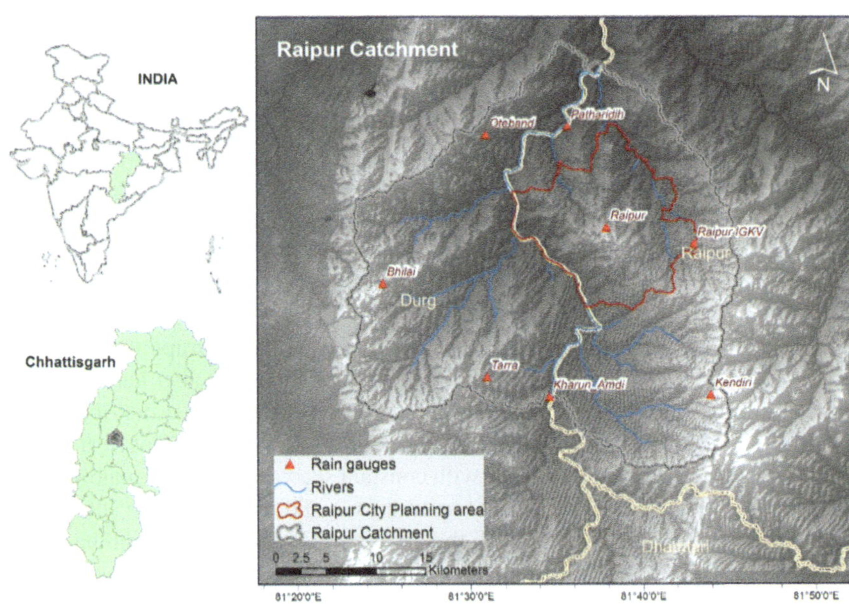

Fig. 35.1 Location map of the study area

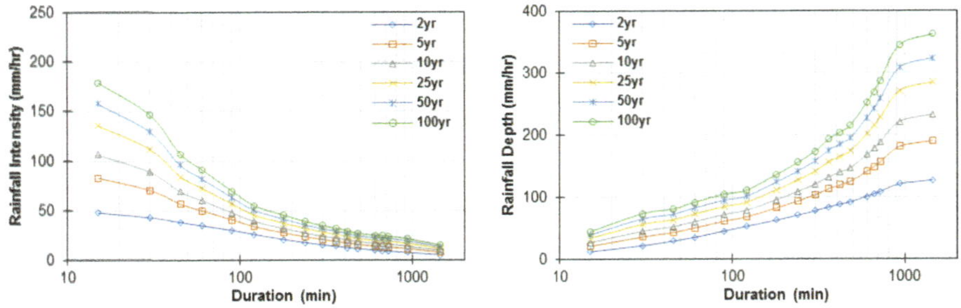

Fig. 35.2 IDF and DDF curves of Raipur city derived using GEV: with no nonstationary component (Data-base:1980–2016:15 min storm interval)

35.3 Materials and Methods

Data Collection

Figure 35.1 shows the location map of the study area indicating the Raipur urban planning area and Raipur catchment area. The rain gauge locations are also presented in Fig. 35.1. The rainfall data collection and gauging station information are given in Table 35.1. The rainfall data of 15 min interval was utilized to develop IDF relations were as all other rainfall data were used to check the data consistency and to evaluate climate shift and trend analysis.

Homogeneity test and trend analysis were performed in the time series of year 1980–2016. It was found that 1993 was the year of significant change point-based on applied hypothesis. The present paper only covers the impact of climate shift on IDF curves considering change point at year 1993. The data set was divided in two parts before and after year 1993; which was later used in IDF curve assessment.

Gumbel's Extreme Value Distribution

Gumbel (1958) showed that the n largest values of subsamples asymptotically follow an extreme value type 1 (EV1) distribution. This distribution, also known as the Gumbel Extreme Value distribution or double negative exponential distribution, is widely used for frequency analysis of floods, maximum rainfall, etc. This distribution is essentially a log-normal distribution with constant skewness (approximately 1.14). Let a series of large number of (N) observations of random variable be subdivided into n subsamples of size m each, such that N = nm. Each subseries shall have two extreme values: one maximum and one minimum. Its PDF and CDF are as follows:

$$f(x) = \alpha e^{\left(-\alpha(x-\beta) - e^{(-\alpha(x-\beta))}\right)}; \quad -\infty < x < \infty; \quad -\infty < \beta < \infty; \quad \alpha > 0 \tag{35.1}$$

$$F(x) = e^{\left(-e^{(-\alpha(x-\beta))}\right)} \tag{35.2}$$

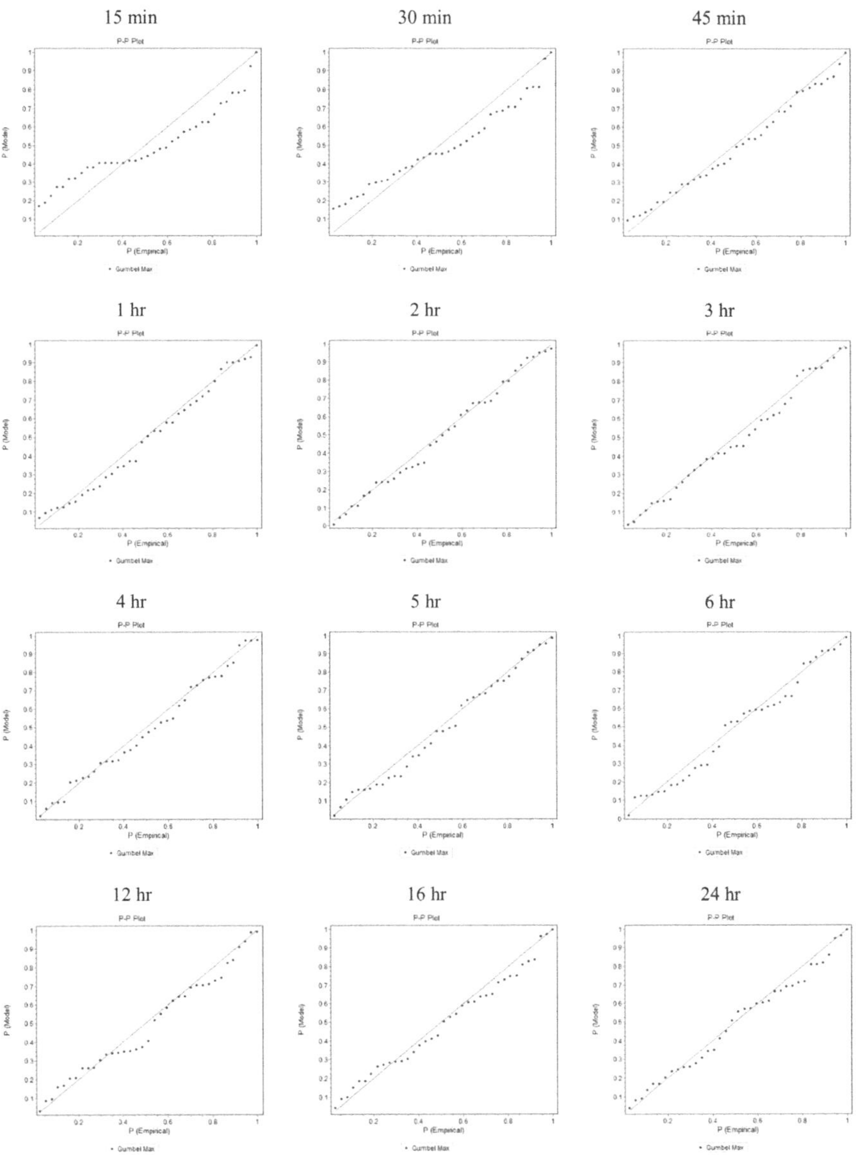

Fig. 35.3 Probability plot of modeled versus empirical frequencies of rainfall intensity for various durations of GEV

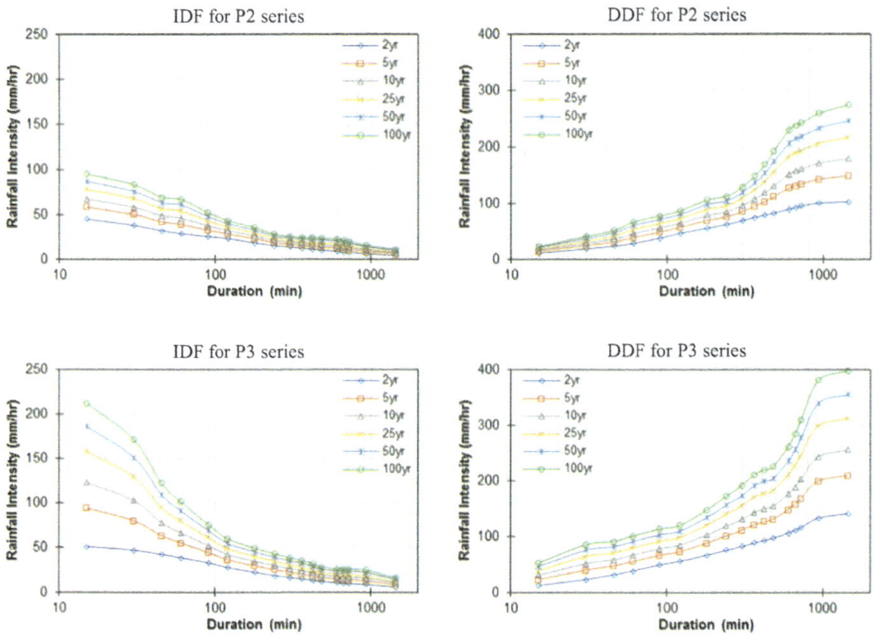

Fig. 35.4 Impact of climate variability on IDF and DDF curves of Raipur city derived using GEV

where α, and β are scale and location parameters. The estimates of parameters using the method of moments are: $\alpha = 1.283/\sigma_x$; $\beta = \bar{x} - 0.45\sigma_x$. According to Gumbel, the probability that an event with magnitude larger than x_0 occurs is (Subramanya 2013):

$$\text{Prob}(X \geq x_0) = 1 - e^{[-e^{(-y)}]}; \quad y = \alpha(x - \beta) \tag{35.3}$$

where, y is the reduced variate and substituting the values of β and α.

$$y = 1.283(x - \bar{x})/\sigma_x + 0.577 \tag{35.4}$$

The expression for the reduced variate y for return period T is:

$$y = -\left[0.834 + 2.303 \log\left(\log \frac{T}{T-1}\right)\right] \tag{35.5}$$

Now, variate x can compute with return period T and for the "n" finite length of series:

$$x_T = \bar{x} + k_n \sigma_{n-1} \tag{35.6}$$

Table 35.1 Gauging station location, gauge type and the climatic parameter with data scale

S no.	Station	Latitude	Longitude	Operated by	Gauging type	NRRG	RRG	Parameters	Data scale
1	Raipur	21°12′00″	81°39′00″	IMD	Tipping bucket Rain gauge		**X**	RF	15 min, Daily
2	Raipur_IGKV	21°14′00″	81°42′00″	IMD	Full Climatic Station		**X**	All	Daily
3	Kharun_Amdi	21°05′56″	81°34′37″	SWRD	–	**X**		RF	Daily
4	Patharidih	21°20′28″	81°35′45″	SWRD	–	**X**		RF	Daily
5	Otebandh	21°20′00″	81°31′00″	SWRD	–	**X**		RF	Daily
6	Kendri	21°06′00″	81°44′00″	SWRD	–	**X**		RF	Daily
7	Bhilai	21°12′00″	81°25′00″	SWRD	–	**X**		RF	Daily

Note IMD: Indian meteorological Department Nagpur, India
SWRD: State Water Resources Department Chhattisgarh, India
NRRG/RRG: (Non) Recording Rain gauge
All: climatic parameters including rainfall (RF), minimum temperature (MinT), maximum temperature (MaxT), relative humidity (RH), sunshine hour (SS), wind speed (WS), and evapotranspiration (EP)

$$k_n = \frac{(y_T - \bar{y}_n)}{s_n}$$

where σ_{n-1} is the standard deviation of the sample of size n. y_n and s_n are reduced mean and reduced standard deviations which are functions of n. Values of these can be obtained from standard tables that are widely available. Note that as $n \rightarrow \infty$, $y_n \rightarrow 0.577$ and $s_n \rightarrow 1.2825$ (Subramanya 2013).

Goodness-of-Fit Test

The goodness-of-fit test is used to test if sample data fit a distribution from a certain population, goodness-of-fit of a probability distribution can be tested by comparing the theoretical and sample values of the relative frequency or the cumulative frequency function. In this study, two testes were applied Chi-Square and Kolmogorov–Smirnov test to check the test fitness of GEV distribution.

Both tests are nonparametric test that is used to find out how the observed value of a given rainfall is significantly different from the expected value. In goodness of fit test, the sample data are divided into intervals, then the number of points that fall into the interval is compared, with the expected number of points in each interval. In the Chi-square and K–S tests, the null hypothesis makes a statement concerning how many cases are to be expected in each category if this hypothesis is correct. The Chi-square statistic is defined as

$$\chi^2 = \sum_i \frac{(O_i - E_i)^2}{E_i} \tag{35.8}$$

where, O_i is the observed number of cases in category i, and E_i is the expected number of cases in category i. This Chi-square statistic is obtained by calculating the difference between the observed number of cases and the expected number of cases in each category.

Similar to Chi-square test, K–S test is used to decide if a sample comes from a hypothesized continuous distribution. Assume that we have a random sample $x_1,\ldots, x_n$ from some distribution with CDF $F(x)$. The empirical CDF is denoted by

$$F_n(x) = \frac{1}{n}[\text{Number of observation } \leq x] \tag{35.9}$$

$$D = \max_{1\leq i\leq n}\left(F(x_i) - \frac{i-1}{n}, \frac{i}{n} - F(x_i)\right) \tag{35.10}$$

The KS statistic (D) is based on the largest vertical difference between the theoretical and the empirical cumulative distribution function.

Hypothesis Testing

The null and the alternative hypotheses are: H_0 is the data follow the GEV distribution and H_A is vice versa. The hypothesis regarding the distributional form is rejected at the chosen significance level (α) if the test statistic is greater than the critical value obtained from a table. The fixed values of α (0.2, 0.1 and 0.05) are used to evaluate the H_0 at various significance levels. A value of 0.05 is typically used in this study for making concluding remarks. The P-value, in contrast to fixed α values, is calculated based on the test statistic, and denotes the threshold value of the significance level in the sense that the H_0 will be accepted for all values of α less than the P-value.

35.4 Result

The result of GEV provides the projected intensities of annual rainfall maxima used for 15, 30, 45 min, 1–6, 12, 16 and 24 h duration at 2, 5, 10, 25, 50 and 100-year return period. From Fig. 35.2, it has been seen that the higher the return period higher will be the intensity of rainfall and this will start getting reduced when the duration gets increases. Therefore, it is required to have the knowledge of shorter duration rainfall (like 1 h, 30 min, 15 min, etc.) events so that the expected anomaly of high intensity rainfall could be track at the required return period.

It has been seen that the shorter duration rainfall (15 min) will give the precise number of rainfall events because in a day it is possible to have more than one events of same duration and also, possible to have the one events of different intensities. Therefore, rainfall data in daily time step may mislead the rainfall intensity and the distribution of rainfall throughout the day. Rainfall is a natural phenomenon and without knowing its actual distribution of duration and intensity it is highly difficult to access the reliable projection of the design rainfall. Due to the climate uncertainty intensification of rainfall in shorter duration is one of the major problems in Raipur city (Guha et al. 2017).

The fitness of the data was checked for the Gumble's extreme value distribution by performing K–S and Chi-Square Goodness of fit tests. These tests were done at the significance level of 0.20, 0.10 and 0.05. K–S test has failed the 15 min rainfall data at 0.2 significance level and Chi-square test failed the rainfall data at 0.20, 0.10 and 0.05 significance level, which represents that the 15 min data are not good to be fit in the Gumble's extreme value distribution. The 30 min rainfall data failed to fit in GEV as Chi-square test for 0.20 and 0.10 significance level rejects the hypothesis, and the 12 h rainfall not fit the Chi-square at 0.20 significance level. So, the results of these rejected data may give the ambiguous result (Table 35.2).

Figure 35.3 depicts the probability plot of modeled and empirical frequencies of rainfall intensity of the following duration used. These plots graphically represent the deviation of the modeled frequency. Probability plots help to easily understand the fitness of the data with the probability density function. It can be clearly seen that the plots for 15 min, 30 min and 12 h are not perfectly matched with the straight

Table 35.2 Results goodness of fit test of fitted GEV distribution (including values of location and shape parameters) using K–S and Chi-square test for 37 years at 0.2, 0.1 and 0.05 significance levels

			K–S test			Significance level for 37 sample size			
S no.	Duration	Parameters	Statistic	*p*-value	Rank	a	0.20	0.10	0.05
1	15 min	$\sigma = 7.7379$, $\mu = 9.1943$	0.1930	0.1110	6	Critical value	0.1719	0.1965	0.2183
						Reject?	Yes	No	No
2	30 min	$\sigma = 12.201$, $\mu = 17.071$	0.1560	0.2990	7	Critical value	0.1719	0.1965	0.2183
						Reject?	No	No	No
3	45 min	$\sigma = 12.265$, $\mu = 24.169$	0.0940	0.8730	6	Critical value	0.1719	0.1965	0.2183
						Reject?	No	No	No
4	1 h	$\sigma = 13.449$, $\mu = 29.797$	0.0860	0.9250	7	Critical value	0.1719	0.1965	0.2183
						Reject?	No	No	No
5	1.5 h	$\sigma = 14.054$, $\mu = 39.817$	0.0750	0.9740	3	Critical value	0.1719	0.1965	0.2183
						Reject?	No	No	No
6	2 h	$\sigma = 13.827$, $\mu = 47.067$	0.0840	0.9370	3	Critical value	0.1719	0.1965	0.2183
						Reject?	No	No	No
7	3 h	$\sigma = 17.378$, $\mu = 56.025$	0.0890	0.9050	5	Critical value	0.1719	0.1965	0.2183
						Reject?	No	No	No
8	4 h	$\sigma = 20.194$, $\mu = 62.894$	0.0720	0.9830	4	Critical value	0.1719	0.1965	0.2183
						Reject?	No	No	No
9	5 h	$\sigma = 22.709$, $\mu = 68.683$	0.0930	0.8740	6	Critical value	0.1719	0.1965	0.2183
						Reject?	No	No	No
10	6 h	$\sigma = 25.764$, $\mu = 73.845$	0.0910	0.8910	4	Critical value	0.1719	0.1965	0.2183
						Reject?	No	No	No
11	7 h	$\sigma = 27.394$, $\mu = 77.571$	0.0910	0.8900	6	Critical value	0.1719	0.1965	0.2183
						Reject?	No	No	No
12	8 h	$\sigma = 28.966$, $\mu = 81.087$	0.1410	0.4130	6	Critical value	0.1719	0.1965	0.2183
						Reject?	No	No	No

(continued)

Table 35.2 (continued)

			K–S test			Significance level for 37 sample size			
S no.	Duration	Parameters	Statistic	*p*-value	Rank	a	0.20	0.10	0.05
13	10 h	$\sigma = 35.475$, $\mu = 87.131$	0.0880	0.9120	5	Critical value	0.1719	0.1965	0.2183
						Reject?	No	No	No
14	11 h	$\sigma = 38.81$, $\mu = 90.02$	0.1000	0.8190	7	Critical value	0.1719	0.1965	0.2183
						Reject?	No	No	No
15	12 h	$\sigma = 42.172$, $\mu = 92.743$	0.1140	0.6820	6	Critical value	0.1719	0.1965	0.2183
						Reject?	No	No	No
16	16 h	$\sigma = 52.913$, $\mu = 101.06$	0.0880	0.9140	6	Critical value	0.1719	0.1965	0.2183
						Reject?	No	No	No
17	24 h	$\sigma = 55.66$, $\mu = 106.0$	0.0950	0.8620	6	Critical value	0.1719	0.1965	0.2183
						Reject?	No	No	No

chi^2 test			Significance level for 37 sample size			
Statistic	*p*-value	Rank	α	0.20	0.10	0.05
20.1900	0.0005	6	Critical value	5.989	7.779	9.488
			Reject?	Yes	Yes	Yes
9.0100	0.0609	5	Critical value	5.989	7.779	9.488
			Reject?	Yes	Yes	No
2.7940	0.5929	7	Critical value	5.989	7.779	9.488
			Reject?	No	No	No
1.2380	0.7440	7	Critical value	4.642	6.251	7.815
			Reject?	No	No	No
0.5150	0.9720	4	Critical value	5.989	7.779	9.488
			Reject?	No	No	No
2.6310	0.6214	5	Critical value	5.989	7.779	9.488
			Reject?	No	No	No
0.3310	0.9541	1	Critical value	4.642	6.251	7.815
			Reject?	No	No	No
2.3190	0.8035	5	Critical value	7.289	9.236	11.07

(continued)

Table 35.2 (continued)

chi² test			Significance level for 37 sample size			
Statistic	p-value	Rank	α	0.20	0.10	0.05
			Reject?	No	No	No
0.2890	0.9905	4	Critical value	5.989	7.779	9.488
			Reject?	No	No	No
1.4250	0.6997	7	Critical value	4.642	6.251	7.815
			Reject?	No	No	No
0.4080	0.9387	1	Critical value	4.642	6.251	7.815
			Reject?	No	No	No
1.1000	0.8943	6	Critical value	5.989	7.779	9.488
			Reject?	No	No	No
2.6040	0.7607	5	Critical value	7.289	9.236	11.07
			Reject?	No	No	No
2.2370	0.6922	5	Critical value	5.989	7.779	9.488
			Reject?	No	No	No
6.1570	0.1877	7	Critical value	5.989	7.779	9.488
			Reject?	Yes	No	No
2.2830	0.6839	7	Critical value	5.989	7.779	9.488
			Reject?	No	No	No
2.0130	0.5697	7	Critical value	4.642	6.251	7.815
			Reject?	No	No	No

line of probability density function. Hence the fitness of 15 min, 30 min and 12 h duration is not significant to use for Gumble's extreme value frequency analysis.

Figure 35.4 shows the IDF and DDF curves for the series Time 1 and Time 2. The variation of the projected intensities is different for both the series as it can be seen from the curves that the maximum value of intensity for Time 1 (1980–1993) series for 100 years return period is 96 mm/hr whereas for Time 2 (1994–2016) series is 212 mm/hr. The projected intensities were confined in the range between 45 mm/hr and 96 mm/hr for the 2, 5, 10, 25, 50 and 100 years return period in Time-1 series. On the other hand, Time-2 series expands its range on the higher side between 51 and 212 mm/h for the 2, 5, 10, 25, 50 and 100 years return period. Therefore, from Figs. 35.2 to 35.4, it can be clearly seen that the IDF and DDF curves are required to update with the new dataset at particular time interval.

The change in any component of a hydrological system will affect the other component of the system as the process is inherently dependent on each other. The physical change and the meteorological change of the area in a system were having temporal as well as spatial effect on the process/performance of the system. The temporal change in weather data will try to develop the nonstationary trend. In

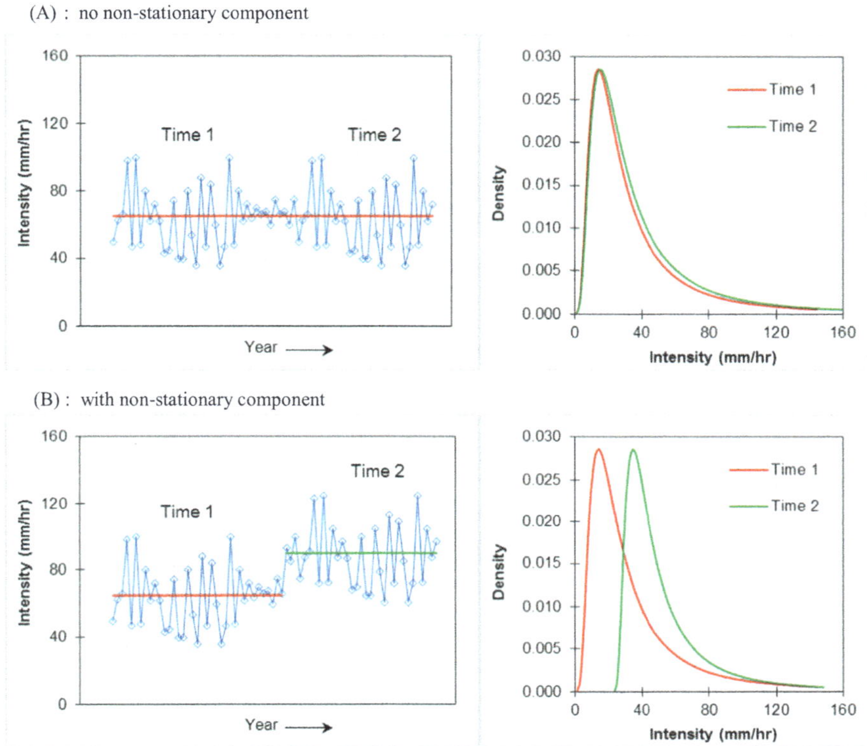

Fig. 35.5 Extreme Rainfall time series in left side; Graphical view of GEV distribution fitted with two parts of the time series namely: Time 1 and Time 2 in right side

this study to access the non-stationarity of the rainfall data, the time series (1980–2016) was broken into two parts in Time 1 (1980–1993) and Time 2 (1994–2016). In Fig. 35.4a, both the time series are analyzed without consideration of the non-stationarity. As depicted in the graph, the intensity remains same for both the series, and when nonstationarity takes into account both the series Time 1 and Time 2 are individually show the different trends in Fig. 35.5 as the intensities are getting on higher side. This result will provide the quantified change in the rainfall intensity. Therefore, it is necessary to consider the nonstationary behavior of the time series.

35.5 Conclusion

The effect of climate change in urban area can be seen as the change in urban landcover due to which the local changed temperature makes the area vulnerable to the effect of urban heat island. In this study, the increased frequency of high-intensity rainfall for smaller duration was utilized for frequency analysis. The time series of rainfall data (1980–2016) of 37 years was used to prepare the IDF and DDF curves. To check the nonstationarity of the rainfall data, time series was divided into two parts as Time 1 and Time 2. The trend of the time series provides the information of change in projection of IDF curves. This result will inform the change in design rainfall value, which can be directly used in the field for the design purpose. From this study, it can be concluded that the area under the study is needed to update the IDF curves in respect of present change in climate scenario.

Acknowledgments The authors would like to thank Deutscher Akademischer Austauschdienst (DAAD), as the research material is based upon work within the program "A New Passage to India." We would like to thank Indian Meteorological Department for allowing us to digitize and collect continuous climatic data of Raipur city at Regional center Nagpur.

References

Central Ground Water Resources Department (2013) Water Year Book Chhattisgarh. In: Divya RN, Sonkusale DK (eds) 1–99. Publication of Water Resource Department, Government of Chhattisgarh, India

DeGaetano AT, Castellano CM (2017) Future projections of extreme precipitation intensity-duration-frequency curves for climate adaptation planning in New York State. Climate Services 5:23–35

Fadhel S, Rico-Ramirez MA, Han D (2017) Uncertainty of Intensity–Duration–Frequency (IDF) curves due to varied climate baseline periods. J Hydrol 547:600–612

Guha S, Govil H, Mukherjee S (2017) Dynamic analysis and ecological evaluation of urban heat islands in Raipur city, India, vol 11. SPIE, 23

Gumbel EJ (1958) Statistical theory of floods and droughts. J Inst Water Eng Sci 12:157–184

Huang YF, Mirzaei M, Amin MZM (2016) Uncertainty quantification in rainfall intensity duration frequency curves based on historical extreme precipitation quantiles. Procedia Eng 154:426–432

Jaiswal RK, Lohani AK, Tiwari HL (2015) Statistical analysis for change detection and trend assessment in climatological parameters. Environ Processes 2(4):729–749

Khavse R, Deshmukh N, Manikandan N, Chaudhary J, Kaushik D (2015) Statistical analysis of temperature and rainfall trend in Raipur district of Chhattisgarh. Current World Environ 10(1):305–312

Madsen H, Arnbjerg-Nielsen K, Mikkelsen PS (2009) Update of regional intensity–duration–frequency curves in Denmark: tendency towards increased storm intensities. Atmos Res 92(3):343–349

Meshram SG, Singh VP, Meshram C (2017) Long-term trend and variability of precipitation in Chhattisgarh State, India. Theor Appl Climatol 129(3):729–744

Mukherjee R, Sahoo M, Naik KC (2011) Raipur City, Chhattisgarh. In: Groundwater scenario in major cities of India. Central Ground Water Board NCCR, Raipur, Chhattisgarh, India, pp 188–195

Rasel MM, Islam MM (2015) Generation of rainfall intensity-duration-frequency relationship for North–Western region in Bangladesh. IOSR J Environ Sci Toxicol Food Technol 9(9):41–47

Shrestha A, Babel M, Weesakul S, Vojinovic Z (2017) Developing Intensity–Duration–Frequency (IDF) curves under climate change uncertainty: the case of Bangkok, Thailand. Water 9(3):145

Sinha MK, Baier K, Azzam R, Baghel T, Verma MK (2019) Semi-distributed modelling of stormwater drains using integrated hydrodynamic EPA-SWM model. Springer, Singapore, pp 557–567

Sivapalan M, Blöschl G (1998) Transformation of point rainfall to areal rainfall: Intensity-Duration-Frequency curves. J Hydrol 204(1):150–167
Subramanya K (2013) Engineering hydrology, 4th edn. Tata McGraw-Hill Education
Tfwala CM, van Rensburg LD, Schall R, Mosia SM, Dlamini P (2017) Precipitation Intensity-Duration-Frequency curves and their uncertainties for Ghaap plateau. Climate Risk Manag 16:1–9

Chapter 36
Changes in Monthly Hydro-Climatic Indices for Middle Tapi Basin, India

Priyank J. Sharma, P. L. Patel, and V. Jothiprakash

Abstract The present study examines the changes in monthly streamflows and their linkages with rainfall variability in the Middle Tapi basin (MTB), India. The Middle Tapi basin (area $\approx$ 32,920 km^2) is part of Tapi basin (area $\approx$ 65,145 km^2) located in the western part of central India. The Girna, Bori and Panjhara rivers are the major gauged tributaries of the Tapi River in the Middle Tapi basin. The streamflow data of eight stream gauging stations on the Tapi River and its tributaries for the period 1973–2013 were collected and analyzed. The non-parametric Modified Mann–Kendall (MMK) and Spearman's Rho (SR) tests have been used to evaluate the trends in mean monthly streamflows and total monthly rainfall series for all the stream gauging stations. Further, the Sen's slope estimator test is employed to compute the slope of trend magnitude and percentage changes in the trend. The trends in mean streamflow exhibited spatial homogeneity, wherein, decreasing trends are observed at all stations in the MTB during June (except Savkheda stream gauging station) and August months; while, spatial heterogeneity is observed for July and September months. The total monthly rainfall exhibited dichotomic fluctuations in their trends, wherein, largely increasing trend has been reported for the August month, and decreasing trend is observed for the September month across the MTB. The analysis carried out in the present study would enhance the understanding of the hydro-climatic interactions and influence of climate variability on streamflows within the Middle Tapi basin.

Keywords Climatic variability · Non-parametric tests · Streamflow variability · Middle tapi basin

P. J. Sharma (✉) · P. L. Patel
Department of Civil Engineering, Sardar Vallabhbhai National Institute of Technology Surat, Surat 395007, India
e-mail: pjs230688@gmail.com

P. L. Patel
e-mail: premlalpatel1966@gmail.com

V. Jothiprakash
Department of Civil Engineering, Indian Institute of Technology Bombay, Mumbai 400076, India
e-mail: vprakash.iitb@gmail.com

R. Jha et al. (eds.), *Climate Change Impacts on Water Resources*, Water Science and Technology Library 98, https://doi.org/10.1007/978-3-030-64202-0_36

36.1 Introduction

The changes in hydro-climatic variables have impacts on the intensity and frequency of extreme events, which are vulnerable to the environment and human society (Thornton et al. 2014). The detection of trends in streamflows has received larger attention of the hydrologic community in the recent past (Lettenmaier et al. 1994; Zhang et al. 2001) due to the rising water demands for human consumption, irrigation, hydropower generation and sustenance of aquatic ecosystems. The streamflow or runoff, which is considered to be the response of a watershed, assimilates the impact of changes in different atmospheric variables over a basin. The observed changes in precipitation and temperature, if found consistent, then they will be reflected to some degree in the streamflow at the basin scale (Kliment et al. 2011). The analysis of trends in hydro-climatic variables such as precipitation, temperature, evapotranspiration, soil moisture, streamflow, etc. has been extensively reported by different researches over the past decade (Tian et al. 2010; Han et al. 2013; Abeysingha et al. 2016; Langat et al. 2017; Pan et al. 2018). Masih et al. (2011) analyzed the streamflow variables, viz., mean annual and monthly flows, 1- and 7-day maximum and minimum flows, timing of the 1-day maxima and minima, and the number and duration of high and low flow pulses, for the Karkheh basin in Iran and concluded that observed trends were spatially non-uniform, however, the decrease or increase in the streamflow trends was largely related to the precipitation changes in the basin. Tehrani et al. (2018) carried out the trend analysis of hydro-climate variables such as temperature, precipitation, and streamflow to derive useful information in understanding the hydrological changes associated with climate variability. Their study exhibited statistically significant relationships between precipitation and streamflow trends (p value <0.05) and showed that region was heading towards a more severe drought situation. The Middle Tapi basin has agriculture as predominant land use, wherein almost 76% of basin area is under seasonal/perennial irrigation. Further, the Middle Tapi basin receives lower rainfall compared to Upper and Lower Tapi basins. Thus, lesser water availability and greater water demand for agriculture and domestic uses often lead to water scarcity issues in the basin. The present study focuses upon the detection of trends and variability in the hydro-climatic indices in the Middle Tapi basin, India.

36.2 Study Area

The Tapi river is the second largest west-flowing river originating from the Indian Peninsular region. The Tapi basin (area $\approx$ 64,145 km^2) is further delineated into upper, middle and lower Tapi basins. The middle Tapi basin (MTB) extends from Hathnur dam to Ukai dam, covering a geographical area of 32,925 km^2, see Fig. 36.1. The Girna, Bori and Panjhara sub-catchments also fall under Middle Tapi basin. In MTB, the Tapi river covers a distance of 297 km and flows mostly through the

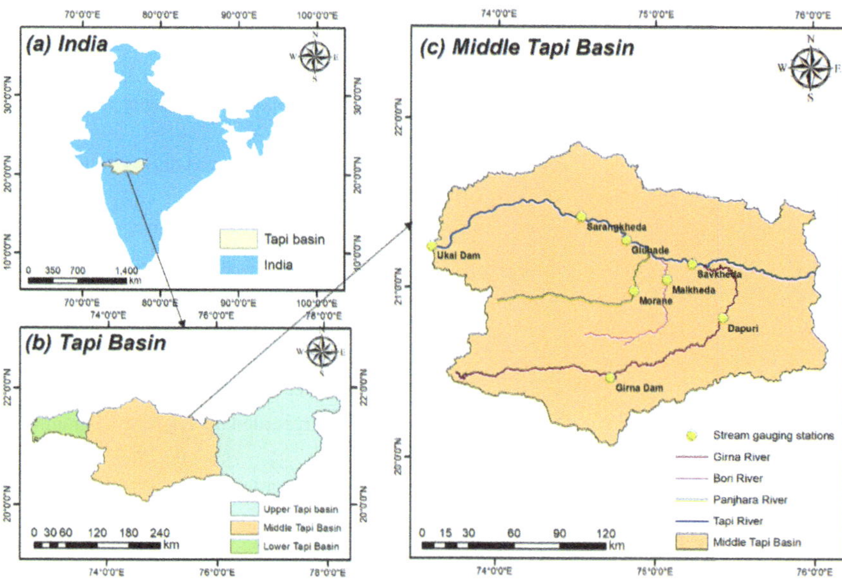

Fig. 36.1 Index map of Middle Tapi basin

plains of Khandesh region and contributes to the Ukai reservoir. The stream gauging stations along the Tapi River in MTB are Savkheda, Gidhade, Sarangkheda and Ukai dam, see Fig. 36.1. The Girna River is the second largest tributary of the Tapi River, which flows eastwards to join Tapi River, after covering a distance of 260 km. The Girna sub-catchment drains an area of 10,061 km^2, which is almost one-third of the MTB area. The stream gauging stations on the Girna River are Girna dam and Dapuri. The Bori and Panjhara Rivers are ephemeral in nature and carry very low discharge compared to the Girna River. The Bori and Panjhara sub-catchments drain 2850 km^2 and 3257 km^2 areas respectively (Jain et al. 2007). The Tapi basin receives about 90% of total rainfall during the monsoon months from June to September, of which around 50% is received in July and August months only (Sharma et al. 2018). The basin receives occasional rain during the post-monsoon period (October and November) and very scanty rainfall in rest of the periods. The Middle Tapi basin exhibits considerable spatial variability in rainfall, wherein the north western region (near Ukai dam and Navapur) receives highest rainfall around 1159.0 mm, while the south western region (around Kalvan and Malegaon) receives lowest rainfall of around 625.0 mm. The normal rainfall in the Middle Tapi basin is reported to be 766.7 mm.

The daily observed streamflow data for the stream gauging stations viz., Savkheda, Gidhade, Sarangkheda, Dapuri, Malkheda and Morane were collected from Central Water Commission (CWC), Tapi Division, Surat. The daily inflows into Ukai and Girna reservoirs were collected from Ukai Civil Circle, Ukai and Tapi Irrigation Development Corporation (TIDC), Jalgaon. Further, the daily rainfall data of 24

Table 36.1 Definition of hydro-climatic indices used in present study

Hydro-climatic indices	Notation	Definition
June mean streamflow	Q_{JUN}	The mean streamflow observed during June month (m^3/s)
July mean streamflow	Q_{JUL}	The mean streamflow observed during July month (m^3/s)
August mean streamflow	Q_{AUG}	The mean streamflow observed during August month (m^3/s)
September mean streamflow	Q_{SEP}	The mean streamflow observed during September month (m^3/s)
June total rainfall	R_{JUN}	The total rainfall recorded during June month (mm)
July total rainfall	R_{JUL}	The total rainfall recorded during July month (mm)
August total rainfall	R_{AUG}	The total rainfall recorded during August month (mm)
September total rainfall	R_{SEP}	The total rainfall recorded during September month (mm)

and 26 rain gauge stations in Upper and Middle Tapi basins, respectively, were collected from India Meteorological Department (IMD), Pune and CWC, Surat. The hydro-climatic indices adopted in the present study are defined in Table 36.1.

36.3 Methodology

The methodology adopted in the present study is discussed as under:

(i) Input the observed daily streamflows and rainfall data for the stream gauging stations and rain gauges used in present study.
(ii) Analyze the rainfall–runoff behaviour in the middle Tapi basin by computing annual values of runoff coefficient.
(iii) Derive the mean monthly streamflow series for all the stream gauging stations and total lumped rainfall for the sub-catchments represented by each stream gauging station using Thiessen's polygon method.
(iv) Analyze the nature and significance of trends in time series of hydro-climatic variables using Modified Mann-Kendall (MMK) and Spearman's Rho (SR) tests (Mann 1945; Kendall 1975; Lehmann 1975).
(v) Estimate the trend magnitude and percentage change in trends using the Sen's slope estimator test (Hirsch et al. 1982; Yue and Hashino 2003).

36.4 Results and Discussions

36.4.1 Rainfall–Runoff Relationship

The observed daily streamflow for the stream gauging stations, reported in Table 36.2, was used to derive the mean monthly streamflow time series. The lumped rainfall for the sub-catchments represented by each stream gauging station was computed from the daily station rainfall using Thiessen's Polygon approach for that period. The rainfall and runoff depths were then computed for each stream gauging station for the monsoon months viz., June, July, August and September months. The variability of monthly rainfall and runoff depths for different stream gauging stations are shown in Fig. 36.2. From Fig. 36.2, it can be seen that rainfall exhibits high sub-seasonal variability, wherein, the rainfall is higher during July and August months compared to the June and September months. Further, it is also seen that Girna, Bori and Panjhara sub-catchments receive lesser rainfall compared to the rest of MTB, see Table 36.2 and Fig. 36.2. The runoff depth has been found to increase gradually and attains highest value during the August month due to comparatively higher soil moisture levels than other months. The higher rainfall during the June and July months does not produce much runoff due to higher infiltration losses on account of lower soil moisture content within the catchment. Sharma et al. (2018) highlighted that in upper Tapi basin (UTB), the higher summer temperatures resulted into drying of the soil and thus much of the rainwater during the early phase of monsoon is infiltrated into the dry soil, thereby, resulting in reduced overland flow. The MTB has been found to exhibit similar climatic and topographic characteristics as that of UTB, hence, the above reasoning shall also be applicable for observed seasonal runoff variations in MTB. The annual runoff coefficients for the stream gauging stations on the main

Table 36.2 Statistical analysis of observed streamflows in Middle Tapi basin

Stream gauging station	River	Drainage area (km^2)	Data availability	Mean annual runoff depth (mm)	Mean annual rainfall depth (mm)	Mean annual runoff coefficient (mm/mm)
Savkheda	Tapi	48,136	1973–2004*	157.6	805.6	0.20
Gidhade	Tapi	54,750	1973–2013	147.9	786.7	0.19
Sarangkheda	Tapi	58,400	1978–2013	139.0	778.3	0.18
Ukai dam	Tapi	62,225	1975–2013	155.6	795.5	0.20
Girna dam	Girna	4729	1973–2013	106.4	669.4	0.16
Dapuri	Girna	8901	1973–2004*	48.7	681.6	0.07
Malkheda	Bori	1830	1978–2004*	43.9	644.1	0.07
Morane	Panjhara	1933	1978–2004*	63.4	574.6	0.11

*Data length available and reported in the present study

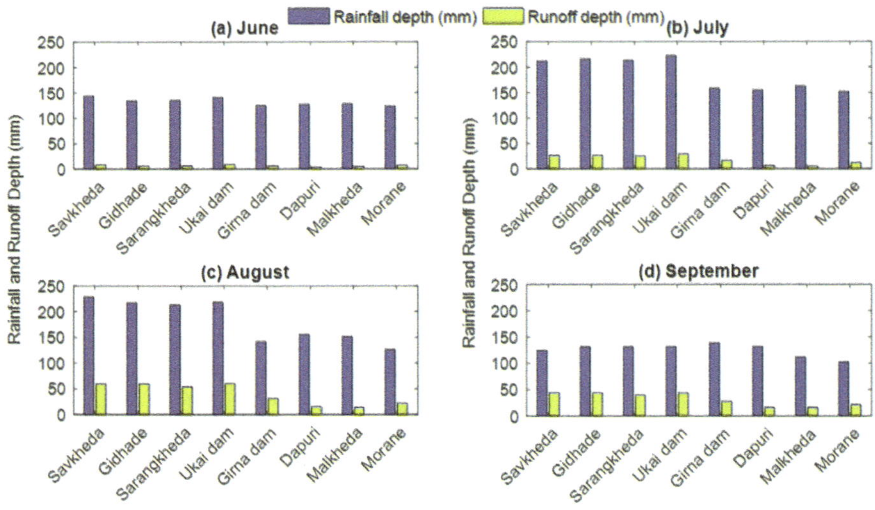

Fig. 36.2 Monthly rainfall and runoff depths for the stream gauging stations in Middle Tapi basin during **a** June, **b** July, **c** August and **d** September months

Tapi River were found to be higher than those on its tributaries, see Table 36.2. A sudden drop in the annual runoff coefficient for the most downstream gauging station on Girna River at Dapuri ($C = 0.07$) vis-à-vis Girna dam ($C = 0.16$) is due to the regulating effect of Girna reservoir on the streamflows of Girna River. The Girna reservoir exercises significant controlling effect on the downstream flows, wherein, the spillway gates were operated only on ten occasions since its operation in 1969.

36.4.2 Trends in Hydro-Climatic Indices

The monthly streamflows in the Girna river (at both stations) are reported to be decreasing during the monsoon months, viz., June, July, August and September, with significant decrease during the June month at Girna dam ($Z_{\text{MMK}} = -1.97$), see Fig. 36.3. The monthly rainfall, however, reported increasing trends at Girna dam during August and September months, while, at Dapuri during the June and September months. For the remaining period, the rainfall at both the stations exhibited decreasing trend. However, the runoff has been found to decrease at both the stations during all the 4 months. The decreasing trends in streamflows during all monsoon months, even with minor increasing trends of rainfall during August and September months in Girna sub-catchment, could be due to abstraction of streamflows from the minor storage structures, which have come up in the sub-catchment in the recent past.

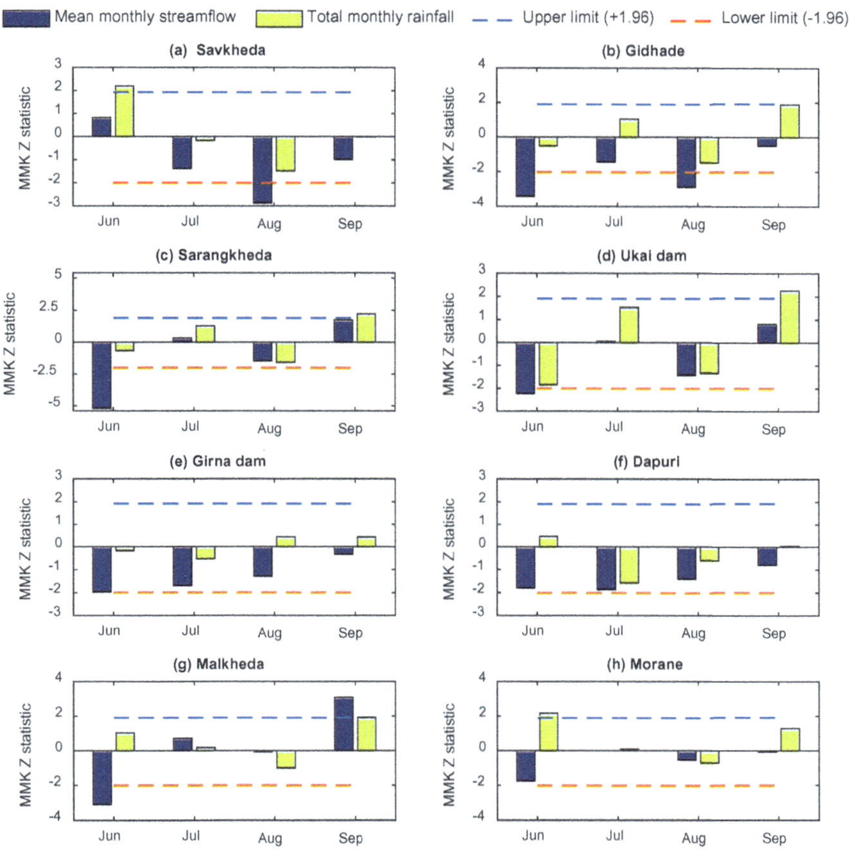

Fig. 36.3 MMK *Z*-statistic values of mean monthly streamflow and total monthly rainfall for **a** Savkheda, **b** Gidhade, **c** Sarangkheda, **d** Ukai dam, **e** Girna dam, **f** Dapuri, **g** Malkheda and **h** Morane stream gauging stations

The streamflow at Malkheda along the Bori River reported decreasing trend during June, August and September months, while significant decreasing trend (at 5% significance level) was reported during the June month ($Z_{MMK} = -3.09$). The July streamflow reported very slightly increasing trend having slope of trend magnitude (β) as 0.03 m^3s^{-1}/year. Further, the streamflow exhibited decreasing trend during June, August and September months for Morane along the Panjhara River, while no trend was reported during the July month. On other hand, the monthly rainfall for Bori and Panjhara sub-catchments were found to increase during all the months, except during August month. However, significant increasing trend in rainfall for Panjhara sub-catchment was observed during June month ($Z_{MMK} = 2.17$).

The streamflows in the Tapi River exhibited significant decreasing trend at Gidhade ($Z_{MMK} = -3.42$), Sarangkheda ($Z_{MMK} = -5.20$) and Ukai dam ($Z_{MMK} = -2.23$) during the June month, whereas marginal increasing trend was observed at Savkheda ($Z_{MMK} = 0.82$), see Fig. 36.3. The reported decreasing/increasing trends in streamflows at aforesaid stations are in line with respective decreasing/increasing trends in the rainfall in their sub-catchments. The nature of trends in other month streamflows, are invariably, at all stream gauging stations in line with trends in rainfall in their respective sub-catchments except during July and September months at Gidhade stream gauging station. The decreasing trend in streamflows at Gidhade stream gauging station during July month is in contradiction with increasing trends in rainfall at the same station. Such reverse trends are due to major decreasing trend in streamflows at upstream stream gauging stations at Savkheda on main Tapi River and Dapuri stream gauging station on Girna River. Similarly, marginal decreasing trends in streamflows during September month due to significant increasing trend in rainfall at Gidhade stream gauging station could be due to abstraction of streamflows on main Tapi River in Hathnur reservoir and minor hydraulic structures on ungauged tributaries upstream of Gidhade for the conservation purposes in the fag end of the monsoon.

36.4.3 Changes in Magnitude and Percentage Change of Trends

The magnitude of trend slope and percentage change in trends with respect to mean is computed for all the hydroclimatic indices analyzed in the present study. The Sen's slope value β is computed using the relationship given by Hirsch et al. (1982) and the percentage change in the trend has been calculated using β value as suggested by Yue and Hashino (2003). The percentage change in trend, either positive or negative, within 0–15% is quantified as 'nominal', between 15 and 30% as 'moderate', between 30 and 45% as 'severe', and >45% as 'very severe'. The results of Sen's slope estimator tests on mean monthly streamflow series are shown in Table 36.3. From Table 36.3, it is evident that the streamflows are consistently decreasing at all the stations during June and August months. The percentage change in streamflow trend during June and August months was found to be negatively severe and very severe for most stations. Compared with changes in streamflow trends, the changes in rainfall trends exhibited less severity across the study area during the period of analysis indicating the possibility of anthropogenic influences on the streamflows in the basin. The information of streamflow and rainfall changes and their magnitudes would be useful to the agricultural community in planning the appropriate crops within selected sub-catchments (Table 36.4).

Table 36.3 Results of Sen's slope estimator test for mean monthly streamflow series

Stream gauging station	Q_{JUN}		Q_{JUL}		Q_{AUG}		Q_{SEP}	
	β (m^3s^{-1} /year)	% change	β (m^3s^{-1} /year)	% change	β (m^3s^{-1} /year)	% change	β (m^3s^{-1} /year)	% change
Savkheda	0.42	8.7	-8.73	-57.7	-35.24	-102.6	-9.03	-35.3
Gidhade	-2.47	-88.6	-6.31	-46.1	-28.88	-94.4	-3.48	-15.4
Sarangkheda	-5.65	-153.6	2.10	13.4	-20.93	-62.7	10.09	40.4
Ukai dam	-4.53	-85.6	0.38	2.1	-10.95	-29.7	8.84	33.1
Girna dam	-0.23	-89.4	-0.30	-41.6	-0.61	-44.0	-0.17	-14.1
Dapuri	-0.13	-41.7	-0.50	-80.2	-0.97	-59.6	-0.45	-26.5
Malkheda	-0.09	-84.6	0.03	23.3	-0.01	-2.3	0.20	47.4
Morane	-0.08	-44.3	0.00	0.5	-0.12	-20.6	-0.01	-2.3

Legend: (% change in trend)	< -45%	-45 – (-30)%	-30 – (-15)%	0 – (-15)%	0 – 15%	15 – 30%	30 – 45%	> 45%

Table 36.4 Results of Sen's slope estimator test for total monthly rainfall series

Stream gauging station	R_{JUN}		R_{JUL}		R_{AUG}		R_{SEP}	
	β (mm /year)	% change	β (mm /year)	% change	β (mm /year)	% change	β (mm /year)	% change
Savkheda	1.5	33.7	-0.2	-3.4	-2.2	-30.6	-0.1	-1.4
Gidhade	-0.3	-9.6	0.8	15.6	-1.3	-24.5	1.4	42.3
Sarangkheda	-0.7	-19.7	1.5	24.6	-2.0	-34.2	2.2	59.2
Ukai dam	-1.0	-27.0	1.1	18.7	-1.1	-20.3	1.6	46.6
Girna dam	-0.2	-7.2	-0.3	-8.1	0.3	9.3	0.5	13.9
Dapuri	0.7	16.8	-1.7	-35.3	-1.0	-20.9	0.1	1.3
Malkheda	1.4	30.0	0.7	11.9	-1.3	-22.9	1.8	42.9
Morane	3.9	86.1	0.1	1.9	-1.4	-30.6	2.0	53.3

Legend: (% change in trend)	< -45%	-45 – (-30)%	-30 – (-15)%	0 – (-15)%	0 – 15%	15 – 30%	30 – 45%	> 45%

36.5 Conclusions

The following key conclusions derived from the foregoing study are as follows:

(i) The rainfall and runoff exhibited considerable spatial and temporal variability across the middle Tapi basin. The runoff response was found to be very low during the June and July months due to higher infiltration losses on account of lower soil moisture levels.

(ii) The annual runoff coefficient for the stream gauging stations on the main Tapi River has been found to be higher than those on the tributaries, wherein the

Girna reservoir was found to exhibit significant influence on downstream flows in the Girna River, i.e., at Dapuri stream gauging station.

(iii) The trends in mean monthly streamflows at all stream gauging stations are in line with the trends in rainfall in respective sub-catchments except at Gidhade stream gauging station. The reverse trends in streamflows vis-à-vis rainfall trends at Gidhade stream gauging station are due to possible anthropogenic influences in the upstream sub-catchments.

(iv) The percentage change in mean monthly streamflow trend during June and August months was found to be negatively severe and very severe for most stations across the MTB.

(v) The overall decreasing trends in streamflows in MTB on main Tapi River and its tributaries may have the severe consequences on the existence of aquatic systems in the river in future.

Acknowledgements The authors are also thankful to *Tapi Division, Central Water Commission CWC), Surat; Ukai Civil Circle, Ukai; Tapi Irrigation Development Corporation (TIDC), Jalgaon* and *India Meteorological Department (IMD), Pune* for providing necessary data to conduct the present study. The first author gratefully acknowledges the financial support received from *Department of Science and Technology (DST), Ministry of Science and Technology, Government of India* vide their letter number DST/INSPIRE Fellowship/2015/IF150634 dated January 11, 2016.

References

Abeysingha NS, Singh M, Sehgal VK, Khanna M, Pathak H (2016) Analysis of trends in streamflow and its linkages with rainfall and anthropogenic factors in Gomti River basin of North India. Theo Appl Climatol 123(3–4):785–799

Han JC, Huang GH, Zhang H, Li Z, Li YP (2013) Heterogeneous precipitation and streamflow trends in the Xiangxi River watershed, 1961–2010. J Hydrol Eng 19(6):1247–1258

Hirsch RM, Slack JR, Smith RA (1982) Techniques of trend analysis for monthly water quality data. Water Resour Res 18(1):107–121

Jain SK, Agarwal PK, Singh VP (2007) Hydrology and water resources of India, vol. 57. Springer, Dordrecht

Kendall MG (1975) Rank correlation methods. Charles Griffin, London

Kliment Z, Matoušková M, Ledvinka O, Královec V (2011) Trend analysis of rainfall-runoff regimes in selected headwater areas of the Czech Republic. J Hydrol Hydromech 59(1):36–50

Langat PK, Kumar L, Koech R (2017) Temporal variability and trends of rainfall and streamflow in Tana River Basin, Kenya. Sustain 9(11):1963

Lehmann EL (1975) Nonparametrics: statistical methods based on ranks. Holden-Day, San Francisco

Lettenmaier DP, Wood EF, Wallis JR (1994) Hydro-climatological trends in the continental United States, 1948–88. J Clim 7(4):586–607

Mann HB (1945) Nonparametric tests against trend. Econometrica 13:245–259

Masih I, Uhlenbrook S, Maskey S, Smakhtin V (2011) Streamflow trends and climate linkages in the Zagros Mountains, Iran. Clim Chang 104(2):317–338

Pan Z, Ruan X, Qian M, Hua J, Shan N, Xu J (2018) Spatio-temporal variability of streamflow in the Huaihe River Basin, China: climate variability or human activities? Hydrol Res 49(1):177–193

Sharma PJ, Loliyana VD, Resmi SR, Timbadiya PV, Patel PL (2018) Spatiotemporal trends in extreme rainfall and temperature indices over Upper Tapi Basin, India. Theo Appl Climatol 134(3-4):1329–1354
Tehrani EN, Sahour H, Booij MJ (2018) Trend analysis of hydro-climatic variables in the north of Iran. Theo Appl Climatol 1–13
Thornton PK, Ericksen PJ, Herrero M, Challinor AJ (2014) Climate variability and vulnerability to climate change: a review. Glob Chang Bio 20(11):3313–3328
Tian F, Yang Y, Han S, Moiwo JP, Qiu G (2010) Determination of the period of major runoff decline and related driving factors in Ye River Basin, North China. J Water Clim Change 1(2):154–163
Yue S, Hashino M (2003) Temperature trends in Japan: 1900–1996. Theo Appl Climatol 75(1–2):15–27
Zhang X, Harvey KD, Hogg WD, Yuzyk TR (2001) Trends in Canadian streamflow. Water Resour Res 37(4):987–998

Chapter 37
Multiobjective Automatic Calibration of a Physically Based Hydrologic Model Using Multiobjective Self-Adaptive Differential Evolution Algorithm

Saswata Nandi and M. Janga Reddy

Abstract The physically based hydrological models require the estimation of various model parameters through calibration. Several past studies that focused on parameter estimation of hydrological models have found that no single objective performance criterion is adequate for matching different essential characteristics of the observation data. Since physically based hydrological models simulate many of the catchment hydrological processes, it needs to define multiple performance criteria to effectively use the information from various datasets and application of multiobjective optimization for attaining Pareto optimal solutions. In the present study, a Multiobjective Self-adaptive Differential Evolution algorithm (MOSaDE) is applied to perform multiobjective calibration of hydrological models. MOSaDE is an advancement of well-known Differential Evolution (DE) algorithm, using the notion of Pareto dominance, fast nondominated sorting approach, diversity preservation using crowding distance and elitist strategy of joining parent and offspring population. The parameter self-adaptation strategy in the MOSaDE also increases the robustness of the algorithm and alleviate the needs of computationally demanding sensitivity analysis of the algorithm parameters. The methodology is verified for calibration of Variable Infiltration Capacity (VIC) model, which is a popular physically based hydrological model, for a case study in Krishna basin, in India and the results are found to be promising.

Keywords Hydrological modeling · VIC · MOSaDE · Calibration · Multiobjective optimization

37.1 Introduction

Over the last few decades, the application of hydrological models in water resource management and hydrological studies has increased tremendously. With the advancement of computational facilities, hydrological models have evolved from simple

S. Nandi (✉) · M. J. Reddy
Department of Civil Engineering, Indian Institute of Technology Bombay, Mumbai 400076, India
e-mail: ce.saswata@gmail.com

R. Jha et al. (eds.), *Climate Change Impacts on Water Resources*, Water Science and Technology Library 98, https://doi.org/10.1007/978-3-030-64202-0_37

conceptual models to complex process based fully distributed models, which can simulate the various catchment processes using physical laws and also can consider the spatial distribution of such processes. Past studies have highlighted the advantages of such physically based distributed hydrological models and their utilization of spatially varying model parameters and input forcing variables (Beven 1985). Reliability of these model predictions is often dependent on their parameterization scheme and the structure of the hydrological model. However, generally few of the model parameters are always associated with difficulty regarding their direct measurement from the field and need to be estimated. This initiates the needs to implement some parameter estimation schemes or calibration procedures, which can help finding the appropriate values of those unknown parameters to increase the model performance. Traditionally, a manual calibration approach is applied to estimate the unknown parameter values, where trial and error parameter adjustments are implemented for matching the model behavior with observations. Moreover, the performance of the calibrated model is also dependent on understanding of the modelers and behavior of various model parameters, which makes this calibration approach tedious and subjective in nature. Large number of model parameters in modern hydrological models and their highly nonlinear behaviors for simulating the hydrological processes possess further difficulty for the efficient application of manual calibration practices.

To overcome the discussed difficulties of manual calibration approach, automated calibration methods of hydrological models are introduced, which are found to be effective when a model has multiple parameters that are needed to be estimated and different parameters of the model are strongly inter-related (Ryu 2009). For automatic calibration, typically a stochastic optimization algorithm is coupled with hydrological models to identify the optimal parameter values enabling the model to simulate outputs that are closely associated with the observations. The evolutionary algorithms, for example, genetic algorithms (GA), particle swarm optimization (PSO), and shuffled complex evolution method (SCE-UA) are frequently applied for hydrological model calibration (Duan et al. 1992; Jha et al. 2006; Jung and Karney 2006). For implementation of such optimization techniques, various performance measures have been used for identifying the various aspects of model behaviors. For instance, Root Mean Square Error (RMSE) is a popular measure for evaluating the accuracy of a model in a way to minimize the disagreement between the model simulations and observations for the high flow values, while mean squared logarithmic error (MLSE) is used to give more focus on matching low flows from the model. Therefore, there exist different performance measures to match the different characteristics of the observation data sets, which are competitive among themselves. For this reason, increasing a model accuracy considering one measure may result into degrading model accuracy for other performance measures. This led to the beginning of multiobjective calibration approach for hydrological model application, where several competitive model performance measures and their trade-off relationships are explored.

For solving multiobjective problems, there exist two broad approaches are: aggregation-based approach and Pareto-based approach. In aggregation-based approaches, multiple objectives are combines into a single objective using some weighting rules, whereas in Pareto-based approach, a set of trade off solutions (or nondominated solutions) are identified by simultaneously considering all the objectives. Recent studies considering calibration of hydrological models have given more attention for Pareto-based multiobjective schemes. One of the effective methods based on Pareto schemes for solving multiobjective problem is Multi-Objective Differential Evolution (MODE), an extension of Differential Evolution (DE) algorithm, which is found to be very effective for various water resources system planning and operations (Reddy and Kumar 2007). In spite of several merits, it is hardly used in calibration problems of the hydrologic models. Nevertheless, the skillful performance of DE based algorithm relies on appropriate setting of its control parameters such as mutation factor and crossover probability, which are problem dependent and may require a lot of computational effort in tuning those parameters. Since DE itself is a self-learning scheme, DE can evolve the control parameters (mutation factor F and crossover rate CR). Therefore, this study incorporates a self-adaptation scheme for MODE which will dynamically adjust the control parameters during its execution. The present study aimed to instigate MOSaDE algorithm for parameter estimation of physically based Variable Infiltration Capacity (VIC) hydrological Model. The procedures of formulating the MOSaDE algorithm, specification of calibration criteria are presented and illustrated with a case study in the upper Krishna river sub-basin, India.

The remaining paper is presented as follows: Sect. 2 presents the various steps involved in MOSaDE algorithm, and brief description of the VIC hydrological model. Section 3 gives details about the study location and various types of data. Finally, the results of the study are presented in Sect. 4, and conclusion from the study is drawn in Sect. 5.

37.2 Methodology

37.2.1 MOSaDE Algorithm

Multiobjective self-adaptive differential evolution (MOSaDE) algorithm is an advancement of previously SaDE for optimization problem involving multiobjective using the concept of MODE presented by Reddy and Kumar (2007). Similar to SaDE, the MOSaDE automatically adapts the two control parameters (mutation factor and cross over probability) of the algorithm for generation of new trial vector (child population). The notion of Pareto dominance is incorporated within the SaDE to solve problems with multiple conflicting objectives. The notion of crowding distance and fast nondominated sorting is adopted from NSGA II (Deb et al. 2002) for maintaining diversity among the drawn solution and to sort the solutions in distinct groups or fronts. Various steps for the MOSaDE algorithm are given below.

i. Generate NP initial population $\left(V_{i,0} = \{v_{i,0}^1, v_{i,0}^2, \ldots, v_{i,0}^D\}\right)$ randomly from the possible ranges of the decision variables. Here $v_{i,0}^j$ denotes the initial value of jth decision variable in the ith individual, and D denotes the dimension of the problem. Create an empty external archive EAR $= \Theta$.

ii. Mutation factor F_i and Crossover rate CR_i for each individual are set according to:

$$\begin{aligned} F_i &= N_i(0.5, 0.3) \\ CR_i &= N_i(0.5, 0.1) \end{aligned} \tag{37.1}$$

where, $N_i(0.5,0.3)$ represents a Gaussian random number with mean as 0.5 and standard deviation of 0.3.

iii. First, evaluate each member of NP individuals, and then perform nondominating sorting in the current population to fill the external archive with nondominated solutions. Initialize the generation number G as 0.

iv. To generate NP child vectors from the NP parent vectors, first mutation for each individual is performed according to:

$$U_{i,G} = V_{rand_1,G} + F_i(V_{rand_2,G} - V_{rand_3,G}) \tag{37.2}$$

where $U_{i,G} = \left\{U_{i,G}^1, U_{i,G}^2, \ldots, U_{i,G}^D\right\}$ is the mutant vector, and $V_{rand1,G}$, $V_{rand2,G}$ and $V_{rand3,G}$ are arbitrarily selected individuals within the current generation. Next crossover is executed to produce trial vectors by the following equation:

$$x_{i,G}^j = \begin{cases} u_{i,G}^j & \text{if } \mathrm{rd}_j \le CR_{i,G} \text{ or } (j = rand_j) \\ v_{i,G}^j & \text{otherwise} \end{cases} \tag{37.3}$$

where $x_{i,G}^j$, $u_{i,G}^j$ and $v_{i,G}^j$ are the jth parameter from generation G for ith trial vector, ith mutant vector, and ith population vectors respectively; rd_j is uniformly distributed between [0,1]; and $rand_j$ is a random value in between [1,D].

v. Compute the fitness function values for each trial vector generated.

vi. Compare fitness function each trial vector with its corresponding parent vector. If a parent dominates, then discard the child vector. If a child dominates a parent vector, then replace the parent with child vector. If both of them are nondominated then store the child vector in a temporary population. Combine the temporary population with the current population and choose NP member from the current generation using a dynamic crowding operator with the nondominated sorting strategy. Empty the temporary population and update external archive *EAR* with nondominated solutions from the current generation and also make a check for dimension of *EAR*, so the size of *EAR* should not exceeds the predefined the maximum size (E_{max}) of *EAR*.

vii. The values of F and CR are revised by mean of evolution. First, note the F and CR values guiding the child vectors to the next generation for every 10 generations, and subsequently revised mean F and mean CR are computed by averaging those successful F and CR values, respectively. These revised mean F and mean CR are used to generate updated F and CR values for each individual as computed before with the previous standard deviation values.

viii. Increase the generation number by one, i.e., G += 1. Examine the stopping criterion (e.g., achieved maximum number of iterations), stop the algorithm in case stopping criteria is fulfilled or else proceed with step (iv) and repeat steps (iv) to (vii).

37.2.2 VIC Hydrological Model

The VIC model is formulated at University of Washington, which is a macroscale hydrologic model and has been widely used in various hydrological studies at global as well as regional scale (Liang et al. 1994). Being a physics based model, it can solve both the complete energy and water balances equations to simulate surface runoff, baseflow, canopy interception, snow pack, evapotranspiration, and many other output fluxes for each model grid at various temporal scale (i.e., monthly, daily, sub-daily). Another distinctive characteristic of this model is its consideration for multiple soil layers and multiple elevation bands within a single model grid, nonlinear baseflow estimation, representation of subgrid variability of different vegetation classes, and notion of variable infiltration curve.

The grid-based VIC model divides the study area into uniform grid cells, which are further divided into M types of land cover, where m = M (or the last tile) is the bare soil and m = 1, 2,…,$(M - 1)$ indicates $(M - 1)$ different vegetation tiles. The evapotranspiration, which is sum of bare soil evaporation, vegetation transpiration, and canopy evaporation, is calculated from each grid cells by means of Penman–Monteith equation. Among the three layers of soil in the VIC, the bottommost layer is only responsible for baseflow, which receives it, the soil moisture from the layer above it by means of gravitational flow. The top and middle layers characterize the bare soil evapotranspiration and soil dynamic response from the received precipitation, respectively. The VIC model uses the variable infiltration curve to characterize soil surface response, which can be formulated as:

$$c = c_m\left[1 - (1 - FA)^{\frac{1}{b_i}}\right] \tag{37.4}$$

where, c_m represents maximum infiltration capacity; c denotes the infiltration capacity, b_i is a parameter for controlling the shape of infiltration capacity curve; and FA represent fractional area where infiltration capacity is below the potential infiltration rate.

Water balance for each grid in the model was estimated by the principle of conservation of mass (i.e., the water entering a grid essentially leaves the grid and/or stored within the grid) and the dividing of incoming rainfall into runoff, evapotranspiration and water storage in the soil. Thus, the water balance can be expressed as:

$$\frac{dS}{dt} = R - ET - RO \tag{37.5}$$

where *R, ET*, *RO*, and *dS/dt* are amount of rainfall, evapotranspiration, total runoff and change of soil water respectively.

The VIC hydrological model produces various output fluxes at each individual model grids. As no routing system is incorporated in the model, a separate runoff model by Nijssen et al. (1997) is used along with the VIC hydrological model to transfer the runoff generated from each grid to the outlet of the watershed.

37.2.3 Design of VIC Model Calibration

This study aims to assess the suitability of the proposed MOSaDE-based parameters estimation approach and also to investigate the effect of consideration of various objectives on the overall calibration efficiency. One key element for any calibration framework is to characterize the distinguished hydrological processes by choosing suitable objective functions. Therefore, past studies regarding hydrological models have employed a wide range of objective functions for problem of parameter estimation. For instance, Madsen et al. (2002) have suggested that choice of objective functions should emphasize the high and low flows simultaneously. Therefore, Nash–Sutcliffe model efficiency (NSE) and logarithmic Nash–Sutcliffe model efficiency (LNSE) are chosen as the two conflicting objectives in this study for the current multiobjective parameter estimation problem. NSE and LNSE can be mathematically expressed as below:

$$NSE = 1 - \frac{\sum_{t=1}^{N}\left(Q_{sim,t} - Q_{obs,t}\right)^2}{\sum_{t=1}^{N}\left(Q_{obs,t} - \overline{Q}_{obs}\right)^2} \tag{37.6}$$

$$LNSE = 1 - \frac{\sum_{t=1}^{N}\left(\ln\left(Q_{sim,t}\right) - \ln\left(Q_{obs,t}\right)\right)^2}{\sum_{t=1}^{N}\left(\ln\left(Q_{obs,t}\right) - \ln\left(\overline{Q}_{obs}\right)\right)^2} \tag{37.7}$$

where, N is the number of time steps; $Q_{obs,t}$ and $Q_{sim,t}$ are the observed and simulated runoff values at time i; and $\overline{Q}_{obs}$ represents the average observed runoff.

The objective function NSE is used to put emphasis on the high flows, while the LNSE emphasizes the low flow values. Therefore, a trade-offs exist between these two criteria used for the calibration, which is helpful in many instances. For example,

Table 37.1 Details of the seven chosen VIC model parameters, units, description, and ranges

Parameter	Description	Units	Range
b_i	Infiltration parameter	dimensionless	0.0–0.4
D_s	Baseflow parameter[1]	fraction	0.0–1.0
Ds_{max} mm/day0–30	Maximum baseflow	mm/day	0–30
W_s	Baseflow parameter[2]	fraction	0.0–1.0
d_1	Thickness of first soil layer	meters	0–0.5
d_2	Thickness of second soil layer	meters	0.5–1.5
d_3	Thickness of third soil layer	meters	0.5–2.0

1 indicates fraction of the Ds_{max} where which non-linear baseflow occurs
2 indicates fraction of maximum soil moisture where non-linear baseflow occurs

one may look for parameter set suitable for flood simulation, while not much attention for low flows and vice versa. Generally, seven parameters of VIC hydrological model are considered for calibration, which are: i) b_{inflt} controls the shape of infiltration capacity curve which in turn adjusts the partitioning of incoming precipitation into infiltration and runoff. (a higher value of b_{inflt} suggests lower infiltration and more runoff), ii) W_s and D_s are responsible for controlling the baseflow and decide the departing speed of baseflow from the third soil layer, iii) Ds_{max} represents the maximum velocity of baseflow, and iv) d_1, d_2 and d_3, represent the depth of the soil layers and controls the proportion of transpiration, direct runoff and baseflow from the available water (thicker soil layers signifies slow runoff process) (Liang et al. 1994). The ranges of seven model parameters chosen in the present work are adopted from VIC documentation and user manual and details of the seven parameters are shown in Table 37.1.

37.3 Case Study

37.3.1 Study Area

The study area chosen in the current study is located in upper Krishna sub-basin, Maharashtra, India. The study area comprises Karad Tahsil of Satara district and some parts of Patan Tahsil. The study area falls in between Latitude 17°00′ N to 18°15′ N and Longitude 73°30 E to 74°30′ E. The study area and its location are shown in Fig. 37.1.The size of the study area is around 5462 km^2. The study area receives an annual average precipitation of nearly 1118 mm. The average elevation is 650 m from mean sea level.

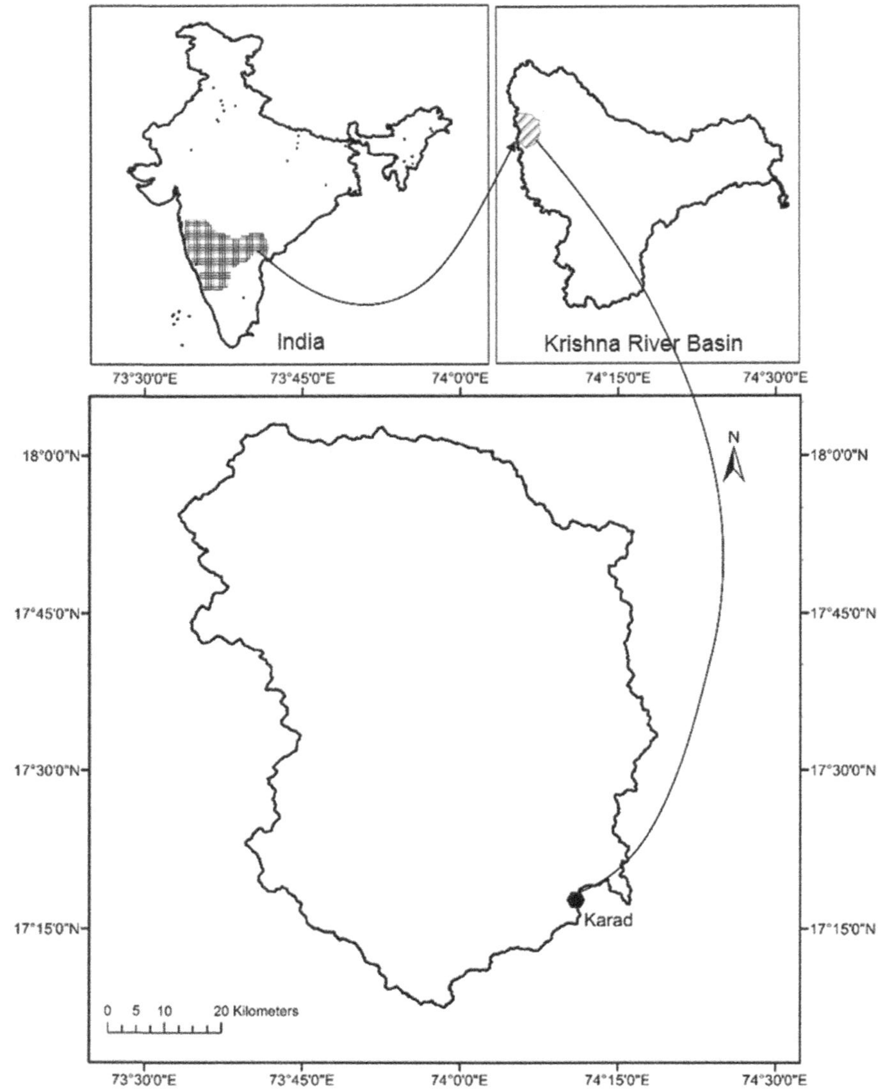

Fig. 37.1 Location map of the study area

37.3.2 Data Used

Daily precipitation (0.25° × 0.25°) and temperature data (1° × 1°) are taken from Indian Meteorological Department (IMD). Wind speed data are collected at 2.5° × 2.5° scale from National Centres for Environmental Prediction (NCEP). Soil Data were taken from Food and Agricultural Organization's (FAO) harmonized world soil database at 30 arc second (~1 km) spatial resolution. Global land cover map from Advanced Very High-Resolution Radiometer (AVHRR) was obtained from University of Maryland (1 km × 1 km). SRTM (90 m) DEM was collected from National Remote Sensing Centre of India. The data of daily observed streamflow at Karad station (basin outlet) were obtained from the WRIS-India.

37.4 Results and Discussion

In the study, the VIC model and the routing model were operated at daily time scale for the time period of 1995–1999. For the application of VIC model, the catchment is divided into grids of 0.25° spatial resolution. The VIC hydrological model produces various output fluxes (e.g., surface runoff, baseflow, evapotranspiration etc.) at each individual model grids. As no routing system is incorporated in the model, a separate runoff model by Nijssen et al. Therefore, to obtain the daily streamflows at the outlet (i.e., Karad Station) of the studied basin, a separate routing model was assigned. To calibrate the VIC model using MOSaDE algorithm, population size and external archive size of the MOSaDE were set to 50 and 100 respectively.

37.4.1 Objective Function Values and Approximated Pareto Front

Figure 37.2 shows the obtained Pareto front from the MOSaDE-based proposed parameter estimation scheme. It can be realized from the figure that the values of the two objective functions vary widely in the solutions from the Pareto front, indicating the fact that both the NSE and LNSE objective functions cannot be simultaneously improved by any single solution. The reason behind such trade-off is that improvement of NSE values will emphasize on matching the peak flows, which will thereby put less focus on the low flow simulation, resulting in a higher LNSE values, and vice versa. From the Fig. 37.2, it can also be seen that all the final solutions were in the first front or nondominated, and also uniformly distributed, which indicates the efficiency of the proposed methodology.

The daily NSE value in the final solutions ranged from 0.62 to 0.70, while the LNSE values varied from 0.1121 to 0.1361. This represents a satisfactory performance of the model considering NSE values, as various past studies have reported that a model can be considered to be calibrated well for streamflow simulation if daily NSE $\geq$ 0.60 (Krause et al. 2005). However, low flow simulation from the model was not quite skillful as it can be seen from the LNSE objective function values. In Fig. 37.2, the best single criterion solutions are represented by the points on the furthest ends of the Pareto fronts. Between these two extreme solutions, there exists other solutions with varying degree of trade-offs between NSE and LNSE Objective function values. Therefore, the users can choose a solution depending on their interest.

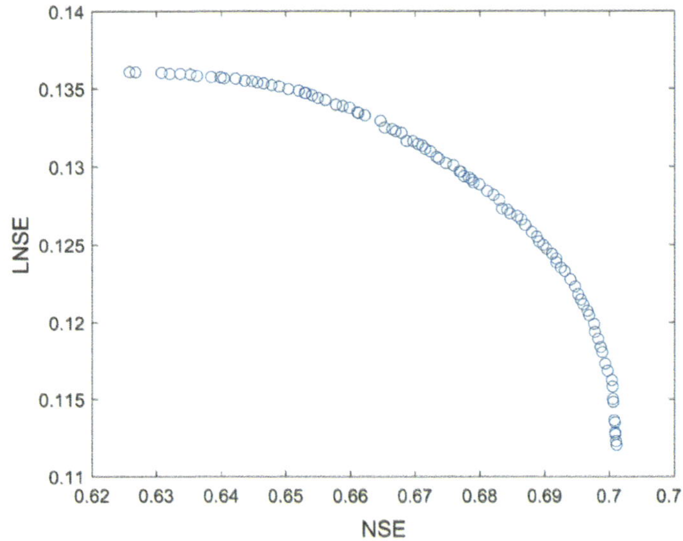

Fig. 37.2 Approximate Pareto front for VIC model using MOSaDE based on NSE and LNSE

37.4.2 Parameter Identifiability Analysis

The maximum and minimum values of the optimal parameter values from the 100 nondominated solutions generated by MOSaDE for the calibration of VIC model is displayed in Table 37.2.

The identifiability analysis in hydrological modeling refers to finding the optimal model structure and parameter sets that are most illustrative for a catchment of consideration. Therefore, estimation of appropriate parameter sets will support such model identification process. For identifiability analysis of VIC model parameters, all the parameter sets from Pareto front were considered. Then those parameter combinations are normalized between 0 and 1 using the given range of parameter values from Table 37.1, and shown in Fig. 37.3, where normalized parameter values are presented along the ordinate and the seven VIC model parameters are plotted

Table 37.2 Ranges of optimal VIC model parameters obtained from calibration using MOSaDE

Parameter	Initial Range	Minimum Value	Maximum Value
b_{infilt}	0.00001–0.4	0.01	0.23
D_s	0.0–1.0	0.003	0.004
Ds_{max}	0–30	27.8	30
W_s	0.0–1.0	0.72	0.87
d_1	0–0.5	0.10	0.10
d_2	0.5–1.5	0.89	0.93
d_3	0.5–2.0	0.55	1.22

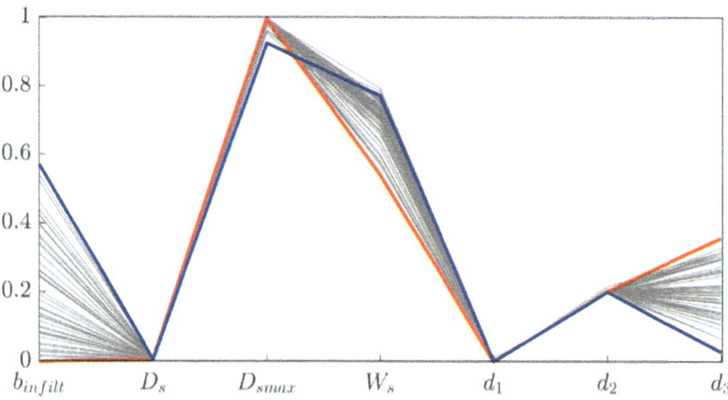

Fig. 37.3 Normalized parameter spread obtained from the MOSaDE-based calibration of VIC hydrological model

along the x-axis. Each grey line across the graph represents one member of final nondominated solutions. The two extreme solutions on the Pareto are shown by the blue and red lines in Fig. 37.3, and they represent the best parameter combination with respect to NSE (high flow) and LNSE (low flow) criterion.

It is promising to see the Pareto solutions have a tendency to stay closely (specially, D_s, D_{smax}, d_1 and d_2) in the parameter space. However, a large variability is also observed in certain parameters (e.g., b_{infilt}, d_3, and W_s), which are responsible for deciding the shape of the simulated hydrograph, indicating a sign of the high uncertainty associated with those parameters. By looking at red line and the blue in Fig. 37.3, it can be realized that concentrating on one between the two objectives results into different parameter combinations. A trend is also ostensible when going through the Pareto front solutions. For example, d_3 decreases for increasing NSE (high flow), while b_{infilt} and W_s increases with increase in NSE.

Correlation in the estimated model parameters indicates nonidentifiability of the parameters (Kuczera and Mroczkowski 1998). Therefore, the correlation of the seven VIC parameters from the Pareto optimal solution is given in Table 37.3. Few parameters have shown correlation value above 0.6 and one pair has correlation coefficient

Table 37.3 Correlation between the VIC parameters resulting from the Pareto optimal solutions of MOSaDE algorithm

Parameter	b_{infilt}	D_s	Ds_{max}	W_s	d_1	d_2	d_3
b_{infilt}	1	−0.47	−0.67	0.67	−0.03	−0.02	−0.97
D_s	–	1	0.22	−0.32	0.05	0.05	0.46
Ds_{max}	–	–	1	−0.36	0.12	−0.16	0.66
W_s	–	–	–	1	0.10	−0.19	−0.80
d_1	–	–	–	–	1	−0.21	0.01
d_2	–	–	–	–	–	1	0.01
d_3	–	–	–	–	–	–	1

more than 0.90. High correlations between d_3, b_{infilt}, and W_s can be seen from Table 37.3, which suggest nonidentifiability of these parameters. A wide spread of these three parameters can also be seen from Fig. 37.3, indicating their poor identifiability. However, correlation between other parameters is quite low, which refers that most of the VIC model parameters are well defined and suitable for optimization algorithms. Apparently, this analysis is carried out with two objective functions and therefore more efforts by using different optimization function combinations will certainly help to get more understanding of the parameter identifiability problem.

37.4.3 Runoff Simulation

A comparison of simulated flows (grey shaded area) from Pareto-optimal solutions obtained by MOSaDE algorithm and observed hydrograph (black dashes) is shown in Fig. 37.4. As the disagreement between the various parameter set in the obtained Pareto front was found to be smaller (almost equal for most of the parameters excluding b_{infilt}, d_3, and W_s), simulated hydrographs produced a very narrow range. Another reason for the narrow ranges in the simulated hydrograph might be associated to the idea of equifinality in hydrological model calibration (Beven, 1993), which states that different parameter sets can produce equally fit system output. Therefore, the narrow ranges in simulated hydrograph obtained by the MOSaDE can be due to the equifinality in VIC model calibration.

From Fig. 37.4, it can be seen that the larger uncertainty in the simulated flows is more during the low flows and recession periods as the hydrograph uncertainty ranges do not cover the observed flows during these periods. Therefore, the VIC model presented for the present study is more skillful to simulate the high flows than the low flows or recession flows. Nevertheless, it is apparent from Fig. 37.4 that the simulated streamflow's capture the trend of observation flows very well.

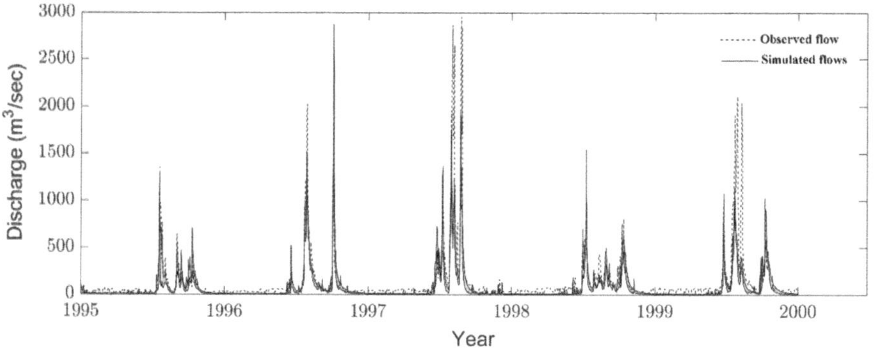

Fig. 37.4 Comparison of simulated flows from Pareto-optimal solutions with observed flow during 1995–1999 at Karad Hydro-observations Station

37.5 Conclusions

Parameter estimation of the hydrological models is considered as one of the most significant problems in the field of hydrology. Past studies have suggested that traditional single-objective calibration of hydrological models is often not sufficient to characterize the various features of hydrological system. Therefore, a multiobjective optimization method is employed in this study, which is an extension of SaDE, for performing the multiobjective calibration of VIC hydrological model in an efficient manner. The Nash–Sutcliffe efficiency and logarithmic Nash–Sutcliffe efficiency are chosen as the two objective functions to capture the characteristics of low and high flows simultaneously. Results from the study showed that the employed methodology performed well in the present case study. The uniform spread of Pareto front solutions provides more flexibility for stakeholders to select an appropriate parameter set based on their priorities. Furthermore, parameter identifiability analysis was carried to get more insight into the model structure and parameter interactions. Few parameters were found to be poorly identifiable, which might be caused due to reasons like equifinality of the calibration problem, multi-modality or even the choice of objective functions. Nevertheless, results from the study strongly support the viability of the proposed MOSaDE-based calibration methodology as an alternative for parameter estimation of complex hydrological models.

References

Beven KJ (1985) Hydrological forecasting. Wiley, Chichester, 405–435

Beven K (1993) Prophecy, reality and uncertainty in distributed hydrological modelling. Adv Water Resour 16(1):41–51

Deb K, Pratap A, Agarwal S, Meyarivan TAMT (2002) A fast and elitist multiobjective genetic algorithm: NSGA-II. IEEE Trans Evol Comput 6(2):182–197

Duan Q, Sorooshian S, Gupta V (1992) Effective and efficient global optimization for conceptual rainfall-runoff models. Water Resour Res 28(4):1015–1031

Jha MK, Kumar A, Nanda G, Bhatt G (2006) Evaluation of traditional and nontraditional optimization techniques for determining well parameters from step-drawdown test data. J Hydrol Eng 11(6):617–630

Jung BS, Karney BW (2006) Hydraulic optimization of transient protection devices using GA and PSO approaches. Journal of Water Resources Planning and Management 132(1):44–52

Krause P, Boyle DP, Bäse F (2005) Comparison of different efficiency criteria for hydrological model assessment. Advances in Geosciences 5:89–97

Kuczera G, Parent E (1998) Monte Carlo assessment of parameter uncertainty in conceptual catchment models: the Metropolis algorithm. J Hydrol 211(1):69–85

Liang X, Lettenmaier DP, Wood EF, Burges SJ (1994) A Simple hydrologically Based Model of Land Surface Water and Energy Fluxes for GSMs. J Geophys Res 99:415–428

Madsen H, Wilson G, Ammentorp HC (2002) Comparison of different automated strategies for calibration of rainfall-runoff models. J Hydrol 261(1):48–59

Nijssen B, Lettenmaier DP, Liang X, Wetzel SW, Wood EF (1997) Streamflow simulation for continental-scale river basins. Water Resour Res 33(4):711–724

Reddy MJ, Kumar DN (2007) Multiobjective differential evolution with application to reservoir system optimization. Journal of Computing in Civil Engineering 21(2):136–146
Ryu JH (2009) Application of HSPF to the distributed model intercomparison project: case study. J Hydrol Eng 14(8):847–857

Chapter 38
Adaptive Neuro-Fuzzy Inference System-Based Yield Forecast Using Climatic Variables

Kalpesh Borse and P. G. Agnihotri

Abstract Crop yield is affected by climate, prevailing during crop season and inputs applied. As such modeling, the cause and effect relationship between yield and these factors could provide an approach for pre-harvest yield forecast. Prediction of impacts of Climate Change (CC) on crop yields requires a model and its parameters, how crops respond to climate. Predictions from various models often disagree with the climatic variables and its impact. A common method is used to quantify impacts of CC is statistical models trained on historical yields and some simplified measurements of weather parameters, such as growing season average temperature and precipitation. CC is a really big apprehension to the entire world. Its direct impact on crop growth and yield is very important to understand. In the present study, the Fuzzy logic crop yield model was developed by considering different climatic variables. Temperature, rainfall, evaporation, humidity parameters were considered for the crop yield model. The model was developed by considering the 15-year crop yield data and the same period for the climatic variables. The triangular membership function is being adopted in the fuzzy model. In this study, a fuzzy rule-based system (FRBS) using the Takagi Sugeno-Kang approach has been used for developing the crop yield model. Model is validated by the coefficient of correlation and found that there is more than 0.9 coefficient of correlation between observed and evaluated yield.

Keywords Adaptive Neuro-Fuzzy inference system (ANFIS) · Crop yield · Climate change modeling · Fuzzy logic · Fuzzy rule-based system (FRBS)

38.1 Introduction

Farming is one of the main sectors to be impacted by different sources, such as climatic changes, soil attributes, seasonal changes, etc. Crop yield prediction is based on various kinds of data collected and extracted by using data mining techniques and different sources that are useful for the development of the crop. It is an art of

K. Borse (✉) · P. G. Agnihotri
Civil Engineering Department, Sardar Vallabhbhai National Institute of Technology, Surat, India
e-mail: kalpeshborse22@gmail.com

R. Jha et al. (eds.), *Climate Change Impacts on Water Resources*, Water Science and Technology Library 98, https://doi.org/10.1007/978-3-030-64202-0_38

forecasting crop and the quantity of yield in advance. Prediction of the crop yield can be extremely useful for the farmers and stakeholders. They can indenture their crop prior to harvest if they have an idea of the amount of yield they can expect, which gives often securing a more competitive price than if they were to wait until after harvest. The involvement of experts in the prediction of crop yield leads to issues such as lack of knowledge about natural events, the negation of personal perception and fatigue, etc.; such issues can be overcome by using the models and decision tools for crop yield prediction.

In a country such as India, it is possible to cultivate a large number of crops due to diverse climatic conditions. Among these crops such as rice (*Oryza sativa* L.) is an important food crop of the country. The total area under cultivation of Rice 431.96 (lakh hectors) (22.6% of the gross cropped area) and the total production of rice is 110.15 million tonnes during the year 2017–18 (Annual Report 2017–18, Department of Agriculture, GOI) constituting about 42.9% of total food production (Economic Survey 2017–2018).

The prediction of crop yield quantifies by using three major unbiased methods (i) biometrical characteristics (ii) weather variables and (iii) agricultural inputs (Agrawal et al. 2001). These methods can be used individually or in combination, to give a complex model. Prediction is a substantial aid in actual and efficient planning and is a more important aspect for a developing economy such as ours so that acceptable planning exercise is started for maintainable growth, overall development and poverty mitigation. Prediction of crop yield is of enormous utility to the government and planners in the formulation and implementation of various policies relating to food procurement, storage, distribution, price, import-export, etc. Some studies have been carried out to prediction of crop yield using weather parameters (Huda et al. 1975; Chowdhury and Sarkar 1981). However, such prediction studies based on statistical models need to be done on a constant basis and for different agro-climatic zones due to observable effects of changing environmental conditions and weather shifts at different locations and areas. Then, there is a need to developed area-specific forecasting models based on time series data to help the policymakers for taking effective decisions to counter adverse situations in food production.

In the recent period, soft computing techniques such as an artificial neural network (ANN), fuzzy logic, genetic algorithm and chaos theory have been extensively applied in the sphere of prediction and time series modeling. Adaptive neuro-fuzzy inference system (ANFIS), which is the integration of neural networks and fuzzy logic, has the potential to capture the benefits of both these fields in a single framework. ANFIS utilizes philological information from the fuzzy logic as well as the learning capability of an ANN.

This study was carried out to develop ANFIS based crop yield model based on the climatic constraints. ANFIS model was developed by considering the different combinations of climatic constraints. Detail explanation was given in the methodology section below.

38.2 Materials and Methods

38.2.1 Study Area

This study was carried out to develop prediction models for predicting the yield of Kharif rice at Nashik taluka of Nashik district of Maharashtra state in India. It is located at 20.33° N latitude and 73.25° E longitudes. It has a dry season from early October to mid-June and a wet season from June to early October. The annual average rainfall of this region is about 713.50 mm, which is subjected to large variation. Yearly yield data of rice (kg/ha) for 27 years (w.e.f. 1987–88 to 2014–15) was collected from the Department of Agriculture, Maharashtra State. Figure 38.1 shows the location map of the study area.

The time series daily climatic data of 27 years (from 1987–88 to 2014–15) were collected from WRDHP, Nashik (M.S.). Five climate parameters were studied; namely average daily temperature (T °C), average daily relative humidity (Rh%),

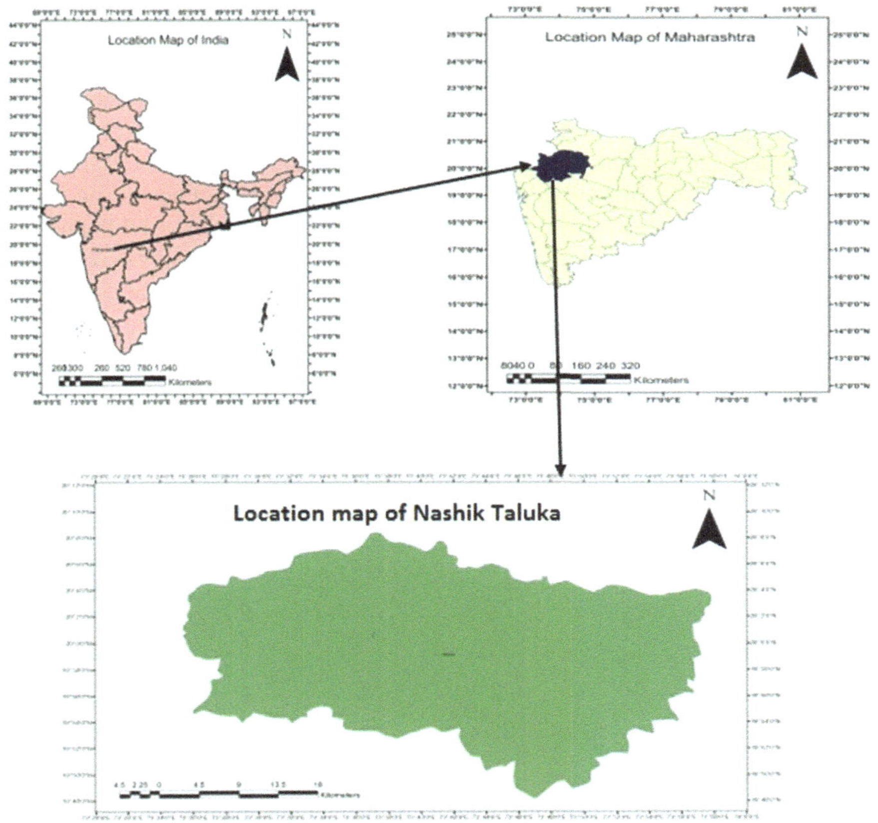

Fig. 38.1 Location map of study area

average daily total rainfall (P), the average weekly number of rainfall days (n) and average daily pan evaporation (E). Moreover, daily weather data related to Kharif (the autumn harvest is also known as the summer or monsoon crop in India) crop seasons starting from a fortnight before sowing up to one month before harvest were utilized for the development of models in the present study; therefore, the weather data for rice crop (Kharif season), from May 21 to October 22 in each year were employed.

38.3 Adaptive Neuro-Fuzzy Inference Systems (ANFIS)

ANFIS is a fuzzy algorithm based on Takagi Sugeno-Kang (TSK) fuzzy inference system (Jang et al. 1997; Loukas 2001). ANFIS is a powerful soft computing technique works based on the principle of two powerful computing techniques that is Artificial Neural Network (ANN) and Fuzzy logic. ANFIS utilizes philological information from the fuzzy logic as well as the learning capability of an ANN for automatic fuzzy if-then rule generation and parameter optimization. An ANFIS input interference panel in MATLAB is shown in Fig. 38.2.

ANFIS has a lot of advantages over individual computing tools such as the FUZZY system and ANN. In the FUZZY system, making rules is very important; whole predictions are based on how one makes the rules. If parameters involved more than one, it would become more completed to make all such rules. ANFIS solves this matter by taking the help of ANN. ANN optimizes the parameters and makes the rules. These rules are fetched into the FUZZY inference system. ANFIS presents some linearity with respect to some of its parameters; hence it increases the overhead of the computation process without increasing the efficiency. Optimization of fuzzy rules is also not so efficient as compared to manual rules making hence predictions are subjected to more uncertainty.

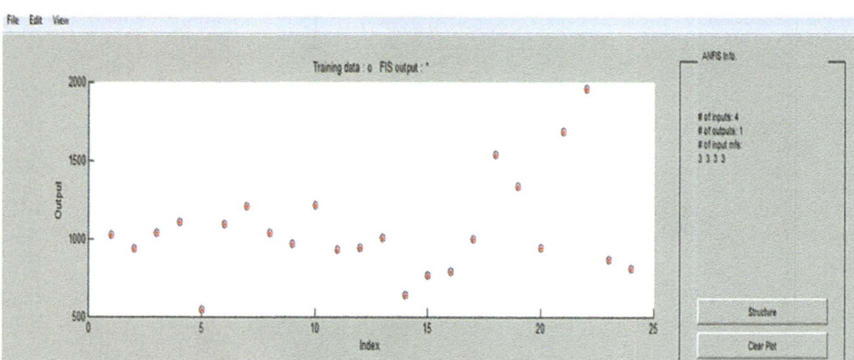

Fig. 38.2 ANFIS input panel

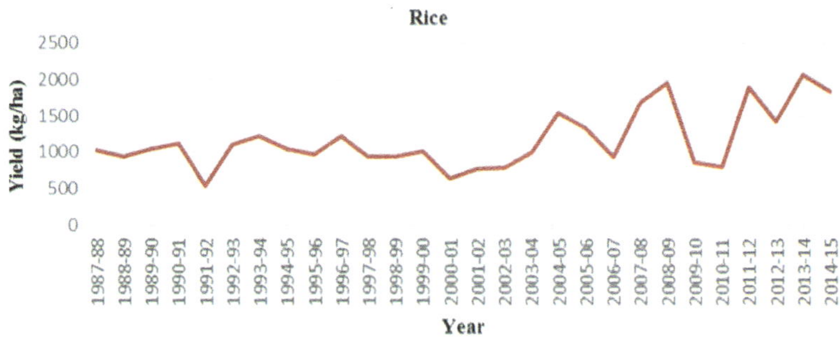

Fig. 38.3 Rice yield over time

38.4 Methodology

38.4.1 Model Input

Model parameters were selected based on the constraints which affect crop growth. In this particular study, parameters are purely climatic. Many experts have considered the parameters other than the climatic but the problem with the other parameters not easily available or very long experimentations are required. This study demonstrates the development of the crop yield model purely based on the climatic parameters such as Average rainfall (mm), Maximum Rainfall (mm), Total Rainfall (mm), Minimum Avg. Temperature (°C), Maximum Avg. Temperature (°C), Average Temperature (°C), Average Evaporation (mm), Relative humidity (%) are studied and a combination was tried to formulate the best model.

Among all the above-listed parameters, different combinations were tried to predict the crop yield and the most effective parameters were selected for the prediction of crop yield. Variation of crop yield over time is as shown in Fig. 38.3.

38.5 Result and Discussion

Among all the eight available Fuzzy membership functions, the product of two sigmoidal functions (psigmf) was used and hybrid optimization was applied while training the variables.

38.5.1 Model Training

The model was developed mainly base on the Maximum Rainfall (mm), Minimum Avg. Temperature (°C), Maximum Avg. Temperature (°C), Average Evaporation

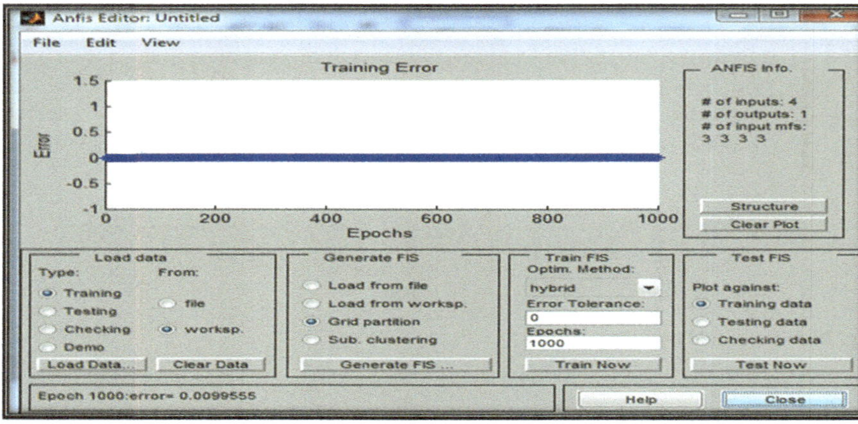

Fig. 38.4 Optimized results

(mm), Relative humidity (%). The selected error was zero while developing the model and reached zero at epoch 2 as shown in Fig. 38.4. From the model. Model characteristics were as shown in figure because it has the lesser error 0.00995 from the same number of epochs.

38.5.2 Model Validation

By selecting the independent set of data of the parameters mentioned in the model, predicted the crop yield and results are as shown in the Fig. 38.5.

Figure 38.6 shows the coefficient of correlation between observed and predicted yield it can be seen that R^2 is greater than 0.9.

Fig. 38.5 Model performance

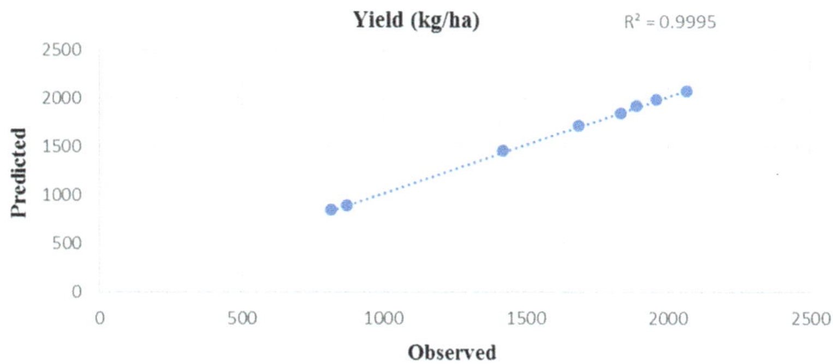

Fig. 38.6 Coefficient of correlation between observed and predicted yield

38.6 Conclusion

In this study, an ANFIS based model was developed for the crop yield prediction based on climatological parameters such as Maximum Rainfall (mm), Minimum Avg. Temperature (°C), Maximum Avg. Temperature (°C), Average Evaporation (mm), Relative humidity (%). With the help of listed parameters, the model was developed and the combination was tried to formulate the best model. All the parameter combination is performing to greater accuracy. The model was performing at an accuracy of 0.0099 kg/hectare accuracy. Coefficient of correlation between observed and predicted yield it can be seen that R^2 is greater than the 0.9. From the above results, one can conclude that the application of soft computing techniques to the prediction of crop yield has high accuracy and precise estimation.

References

Agrawal R, Jain RC, Mehta SC (2001) Yield forecast based on weather variables and agricultural inputs on agro-climatic zone basis. Ind J Agric Sci 71 (7), 487–490

Chowdhury A, Sarkar MB (1981) Estimation of rice yield through weather factors in a dry sub-humid region. Mausam 32(4):393–396

Huda AKS, Ghildyal BP, Tomar VS, Jain RC (1975) Contribution of climate variables in predicting rice yield. Agric Met 15:71–86

Jang J-SR (1993) ANFIS: adaptive network-based fuzzy inference systems. IEEE Trans Syst Man Cybern 23:665–685

Jang JSR, Sun CT, Mizutani E (1997) Neuro-fuzzy and soft computing, a computational approach to learning and machine intelligence. Prentice Hall, NJ, USA. ISBN 0-13-261066-3

Jones AJ (2004) New Tools in Non-linear Modeling and Prediction. CMS 1(2):109–149

Kim B, Park JH, Kim BS (2002) Fuzzy logic model of Langmuir probe discharge data. Comput Chem 26(6):573–581

Kokate KD, Sananse SL, Kadam JR (2000) Pre-harvest forecasting of rice yield in Konkan region of Maharashtra. J Maha Agric Univ 25(3):289–293

Loukas YL (2001) Adaptive neuro-fuzzy inference system: an instant and architecture-free predictor for improved QSAR studies. J Med Chem 44(17):2772–2783

Nayak PC, Sudheer KP, Rangan DM, Ramasastri KS (2004) A neuro-fuzzy computing technique for modelling hydrological time series. J Hydrol 291:52–66
Takagi T, Sugeno M (1985) Fuzzy identification of systems and its application to modeling and control. IEEE Trans Syst Man Cybern 15:116–132
Zadeh LA (1965) Fuzzy sets. Inf Control 8, 338–353

Chapter 39
Impact of Climate Change on Hydrological Parameters

Arunima Priyadarsini Patnaik and Bandita Naik

Abstract The increasing rate of global surface temperature is going to have a significant impact on local hydrological regimes and thus on water resources; this leads to the assessment of water resources potential resulting from the climate change impacts. The main parameters that are closely related to climate change are temperature, precipitation and runoff. Therefore, there is a growing need for an integrated analysis that can quantify the impacts of climate change on various aspects of water resources. Quantifying the impacts of land-use change and land cover practices on the hydrological response of a watershed have been an area of interest for hydrologists in recent years as this information could serve as a basis for developing sound watershed management interventions. The degree and type of land cover influence the rate of infiltration, runoff, and consequently the volumes of surface runoff and total sediment loads transported from a watershed. It often results in significant degradation of land resources such as loss of soil by erosion, nutrient leaching and organic matter depletion. However, very few studies in India have used the physically-based hydrological models along with the land use/land cover change conditions. Hence in this current work SWAT model has been used to assess the impact of LU/LC changes on daily and monthly streamflow of Mahanadi River Basin of Sambalpur region. The results of the study indicated that the though land-use patterns have changed, resulting in an increase in agricultural, barren and buildup land and decrease in forest cover leading to an increase in the runoff, but changes have not occurred as significantly as the changes in annual streamflow. However, the number of days of high-intensity rainfall has increased over a decade, which, along with the land-use changes, explains the increase in streamflow.

Keywords SWAT model · LU/LC changes · Daily and monthly runoff

A. P. Patnaik (✉) · B. Naik
Department of Civil Engineering, CAPGS, BPUT, Rourkela, India
e-mail: arunima.patnaik@gmail.com

R. Jha et al. (eds.), *Climate Change Impacts on Water Resources*, Water Science and Technology Library 98, https://doi.org/10.1007/978-3-030-64202-0_39

39.1 Introduction

General

Water is the foremost part of all living things and a major force constantly shaping the human lives on the earth. It is also a key factor in the air conditioning of the earth for human existence and in influence the progress of civilization. Among all natural resources, water is one of the most important and significant resources found on earth. The redistribution of water through the hydrological cycle is also responsible for the climate of any place, such as daily fluctuations of temperature, precipitation and wind speed and these type of changes in the ecosystem affect the hydrological cycle. Over the last 100 years, Odisha is facing extreme weather conditions in the form of natural disasters (flood, drought, heatwaves, earthquakes and cyclones). The natural calamities affected 25 of 30 districts of Odisha which results in damages, loss of properties and loss of human lives. Therefore is a necessity to study the impact of climate change on water resource in this region. An increase or decrease in precipitation pattern can result in an increase in the frequency of flood, droughts and change in water quality. Therefore, it is necessary to carry out analysis to find out calibration and validation of two climatic parameters, that is, temperature and precipitation.

The National Water Policy of India (2002) acknowledges that national perspectives are needed to regulate the improvement and management of water resources so that the scarce water resources can be developed and conserved on a balance and environmentally sound basis. Impact of land-use changes, watershed development to soil loss and growth of population, water quality and quality is among the most worthy topic in a watershed. The hydrological cycle can be distributed due to changes in land-use by alternating the base flow and annual mean discharge of the basin. The hydrological model plays an important role in simulating the process of rainfall-runoff, soil erosion under different situations. The impact of climate change is going to be most serving in developing countries due to poor capacity.

A hydrological model SWAT (Soil and Water Assessment Tool) model is used in the present analysis. This model is a physically-based, continuous-time model, developed by Dr. Arnold for United USDA-ARS (Agricultural Research Service). The SWAT model is used to simulate or predict runoff of different basins, sediment yield and pollution loading in watersheds. The model has the ability to use for small watersheds as well as the major river basin systems. It is distributed in a time interval hydrological model with an Arc GIS interface. Automated model calibration and validation check.

39.2 Study Area

Mahanadi River is the sixth-largest river in India and one of the major interstate east-flowing river in peninsular India. The Mahanadi basin lies between the latitude of 19°20′N to 23°35′N and longitude of 80°30′E to 86°50′E. Mahanadi basin is physically surrounded on the north by Central India hills, by the Eastern Ghats in the South and East and by Maikala hill range in the West. The total catchment area of the basin is 141600 km^2. The river enters Odisha through the Jharsuguda district subsequent to covering about a portion of its aggregate length. Before Sambalpur, it meets its tributary Ib. The Ib, which is the third biggest tributary of Mahanadi, ascends in town Pandrapt, Region Raighar (Chhattishgarh) and channels Raigarh region of Chhattisgarh and three districts of Odisha, to be specific Sundargarh, Jharsuguda and Sambalpur. After Sambalpur, the Mahanadi stream takes a southerly turn and it is joined by the Ong. The Ong flows towards Sartaipali, Padampur and Bijepur territory of Balangir and Bargarh area of Odisha. Physically, the basin is bounded in the north by the Central India Hills, the Eastern Ghats in the south and east, and the Maikala Hill Range in the west, lying within geographical co-ordinates of 80°30' E to 86°50' E and 19°20'N to 23°35'N. The location of the study area is shown in Fig. 39.1.

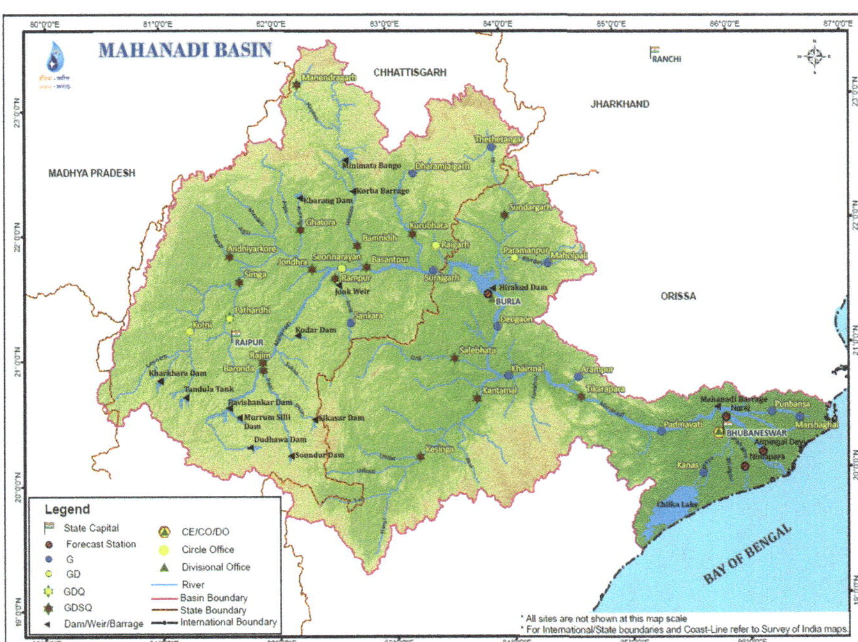

Fig. 39.1 Location of study area *Source* India WRIS website)

39.2.1 Rainfall

Based on the Indian Meteorological Department's (IMD) annual district rainfall figures from 1901 to 2000, the calculated average rainfall in the Mahanadi basin is 1406 mm. Overall, the Mahanadi basin is a high rainfall region, the lowest annual average being 1080 mm in the Kawardha district of Chhattisgarh, while the Jashpur district of Odisha has the highest annual average rainfall of 1653 mm. The western portion of the basin bordering Maharashtra receives the lowest rainfall. The central part of the basin receives moderate rainfall, while the northern, southern and delta regions experience the highest rainfall in the basin. Plots the seasonal distribution of rainfall in the basin. Most parts of the basin receive 80–90% of its annual rainfall from the southwest monsoon (i.e., from June to September). However, the amount of rainfall received is dependent on location in the basin. For example, as compared to the other parts of the basin, the districts located near the delta receive less rainfall (about 60–70% average annual rainfall) between June and September, but receive more rainfall (about 10–22% annual rainfall) from the northeast monsoon (i.e., from October to December). The districts with higher rainfall between October and December also show a marginal increase in the rainfall during March to May as compared to the basin average for this period. Figure 39.2 shows the Seasonal distribution of rainfall as a percentage of the annual average.

IMD's 100-year district-level data was also used to generate trends in the rainfall in the respective districts. Figure 39.3 shows a map of the linear trends observed on

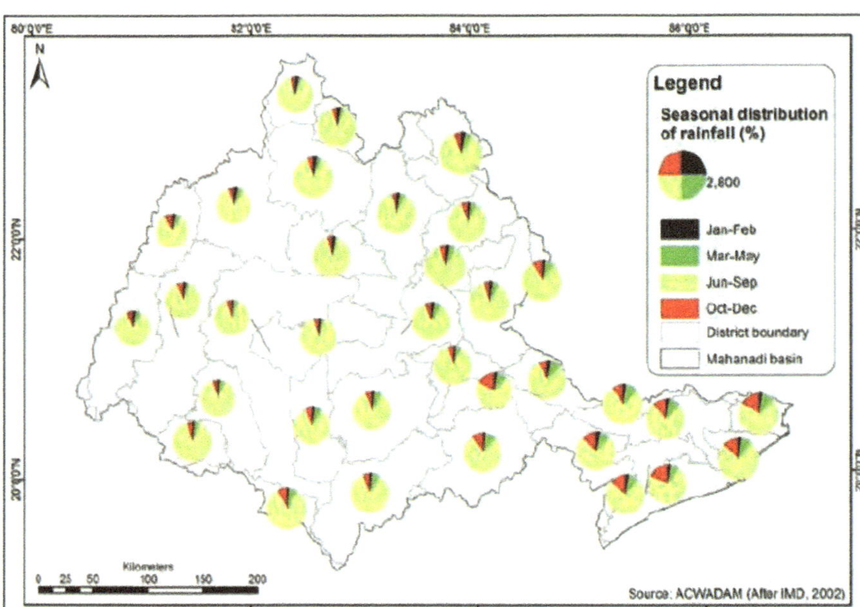

Fig. 39.2 Seasonal distribution of rainfall as a percentage of the annual average

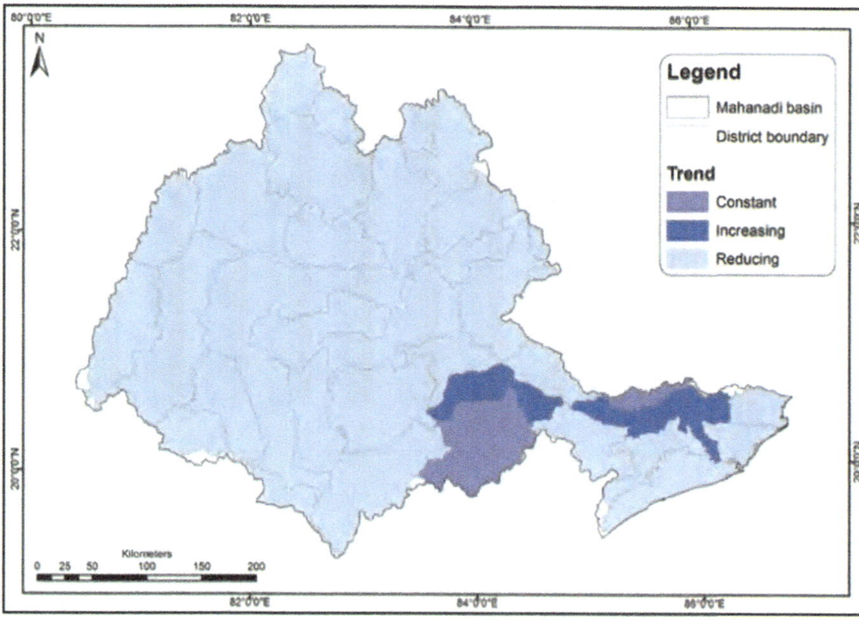

Fig. 39.3 Trends in annual rainfall in the Mahanadi basin. *Source* District level 100 years IMD data

plotting the 112 years (1901–2012) IMD district-level data set. The majority of the districts in the basin show a reducing trend computed from the long-term average rainfall. Only two districts each show a constant trend and an increasing trend.

39.2.2 Temperature

Daily temperature (maximum, minimum and mean) gridded data (1° x 1°) for 36 years (1969–2004) collected from IMD has been analyzed. Average monthly temperature variation for 36 years (1969–2004) is given in Fig. 39.4. Three parameters, namely minimum, maximum and mean temperature, indicate that December and January are the coldest months with a minimum temperature of 12°C. April and May are the hottest months in this region, where the maximum temperature ranges from 39°C to 40°C. As compared to the eastern portion and delta area, the western portion records the lowest and highest temperatures during winter and summer, respectively. The highest day temperature recorded in the basin is 50.3°C in June 2003.

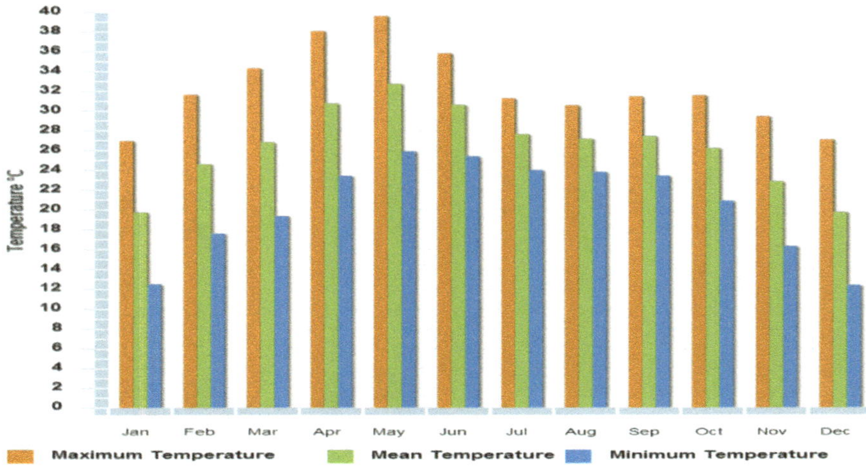

Fig. 39.4 Monthly average temperature (1969–2004) (CWC Report-2014)

39.2.3 Land-Use and Land Cover

Mahanadi valley is best known for its fertile soil and flourishing agriculture, which preliminary depends on a network of canals that arise from the river. Rice, oilseeds and sugarcane are the principal crops cultivated in Mahanadi valley. The basin has an area of about 79,900 km^{2,} which is about 57% of the basin area and 4% of the total area of the country. Except in the coastal plains of Odisha, the basin has an extensive area under forests. The sparse vegetation of the highlands contrasts with the moderately luxuriant vegetation of the river valleys. The coastal plains of Odisha, with a high incidence of rainfall, are predominantly rice growth areas. The land utilization pattern of the Mahanadi river basin comprises 37.275% forest area, 10.432% cultivated area, 91.137% area with other uncultivated land excluding fallow land, 4.967% fallow land and 38.187% net snow areas as shown in Fig. 39.5. The cultivated area is the total area used for sowing two or more crops in one calendar year. The net snow area is the area of snow for each crop but is counted only once. Out of the total annual irrigation water demand of 11 km^2 in the basin, the Kharif season utilizes 7 km^2 and Rabi season uses 4 km^2. Major land-use and associated water use changes that have taken place in the basin in the twentieth century are related to intensive irrigation of agricultural areas. In the last decade since 2004–05, land cropped in the Kharif season only (i.e., largely rain-fed land) has decreased marginally to 30% and land cropped twice or thrice (i.e., irrigated land) has increased substantially, from about 8–15%. The largest increases in irrigated land are in the plains of Chhattisgarh, with the development of major irrigation projects in the upper reaches of the Mahanadi and Seonath rivers. Fallow lands in the basin have decreased from about 17% to about 11% (15,507 km^2) in the last five years or so. This is in contrast to the MoA data, which shows that fallow and cultural wastelands have increased from 7% to about 9% (14,011 km^2).

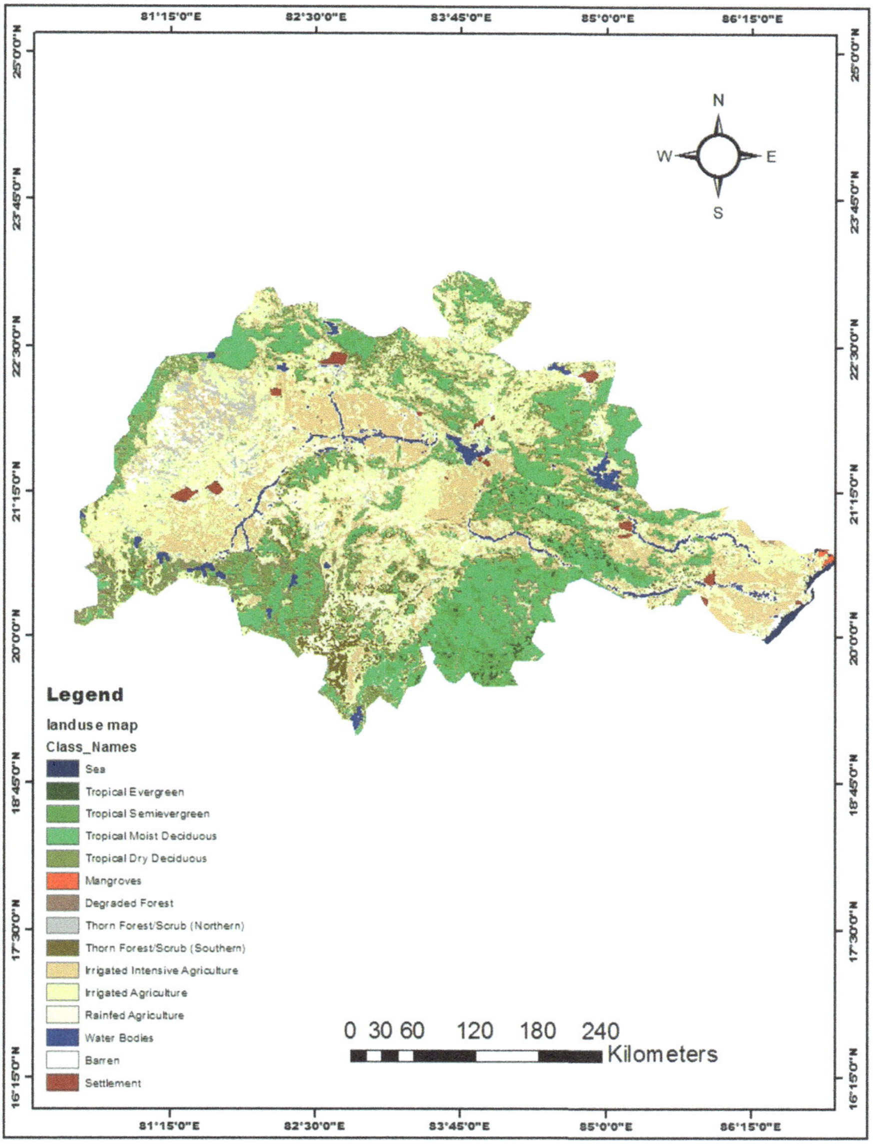

Fig. 39.5 Land-use and land cover data for the Mahanadi river basin

39.3 Methodology

39.3.1 Soil and Water Assessment Tool (SWAT)

The soil and water assessment tool (SWAT) is a continuous, long term, physically based conceptual model. This model operates at basin scale on daily time steps. SWAT model predicted for impacts of the land-use management, sediment and agricultural chemical yield for the development of the physically-based model. It stimulates the hydrologic cycle in two phases; land phase and routing phase. The land phase aims to control the amount of water, sediment, nutrient and pesticides loading. The routing phase aims to define the movement of water, sediments, etc., through the channel network of the watershed. It is a model with Arc GIS interface which has been developed by the USDA-ARS and the Blackland Research and Extension Center. In this SWAT model, the total catchment is firstly divided into sub-basins or sub-watersheds based on the topographic regions assumed to be lumping further divided into a series of HRUs (Hydrological response units) on the basis of soil, slope and land-use combinations. The Green-Ampt infiltration method is one of the options that this model offers to compute excess precipitation at the HRU level, the other one being the NCRS curve number method. Simulations are carried out for the components of the hydrological cycle, nutrient cycles, sediment yield and aggregate for the sub-basins. The SWAT model can be selected based on the data availability, provides the users with various options when the simulation is conducted for the hydrological parameters. The runoff model formed in SWAT is lumped at the sub-basin level because it computes an average value for spatially varied surface runoff. This one is the limitation of the SWAT model. A SWAT CUP (SWAT Calibration and Uncertainty procedure) model integrate various calibration and uncertainty analysis using the same interface. The SWAT CUP model can run SUFI2, GLUE and Parasol. The SWAT project contains input data and one calibration method to allow the user to run a calibration program until convergence is reached. User can save the calibration iteration for later use. Figure 39.6 shows the SWAT Model Flow Diagram.

39.4 Results and Discussion

Using Arc SWAT software DEM map, land-use map have been represented. By using SWAT-CUP sensitivity analysis of flow parameters, simulated, calibrated and validated results are observed. Monthly calibration and validation for streamflow were performed after conducting sensitivity analysis. The sensitivity analysis was performed to determine the optimal parameters best-fitted values based upon the observed data collected from the study area. Seventeen (2000–2016) year's meteorological and observed streamflow data were used for calibration and validation.

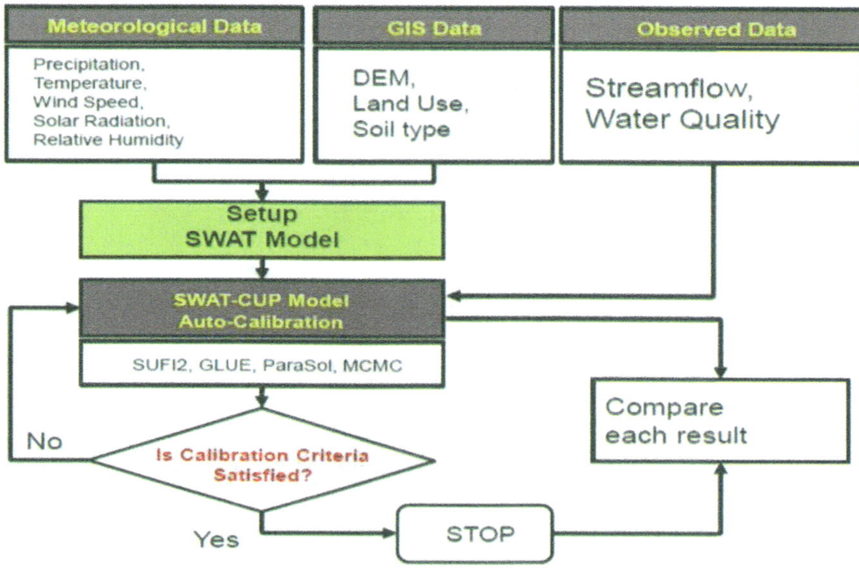

Fig. 39.6 SWAT model flow diagram

Also, analysis of land-use change and climate change and environmental impact assessment observed and modeled the discharge. Figures 39.7, 39.8 and 39.9 show the DEM map, Land-use the land cover map and Watershed delineation of study area, respectively.

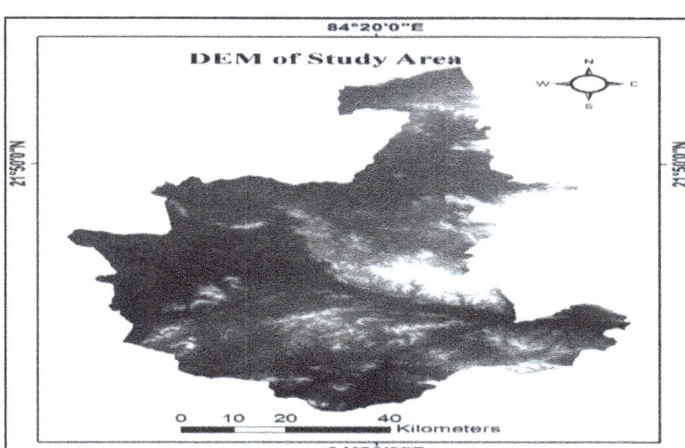

Fig. 39.7 DEM map

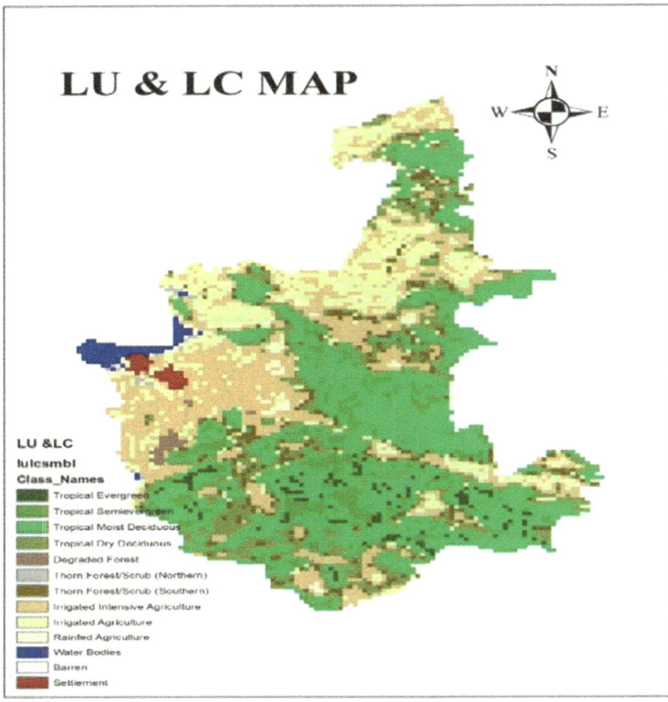

Fig. 39.8 Land-use land cover map

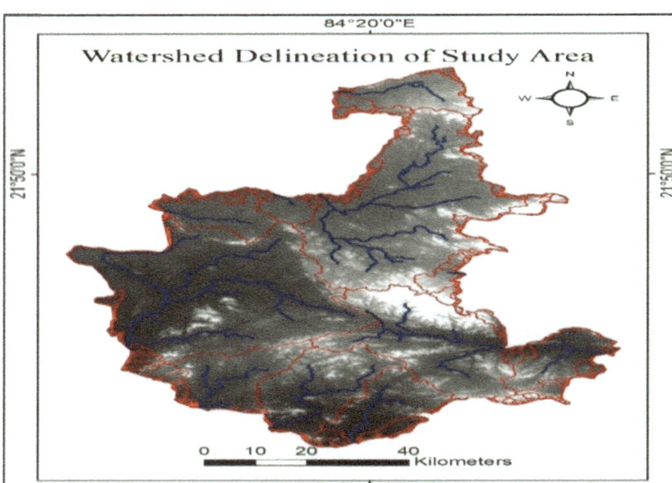

Fig. 39.9 Watershed delineation of the study area

Maps Obtain From Input Data

See Figs. 39.7, 39.8 and 39.9.

39.4.1 Sensitivity Analysis

Six parameters were considered for sensitivity analysis to identify the most sensitive parameters. The sensitive parameters which are obtained from the sensitivity were further carried out for calibration. The ranges of various flow calibration were referring to the SWAT CUP from the previously studied journals and user manual. The best-fitted values, which ranges of the parameter for the catchment, have been represented in Table 39.1. Figure 39.10 shown the Sensitivity analysis of flow calibration parameters.

Table 39.1 Ranges and best-fitted values of flow calibration parameters

Sl no.	Flow calibration parameters	Qualifier	Minimum	Maximum	Fitted value
1	Curve Number (CN2)	r_	−0.5	0.5	−0.3257
2	Base flow alpha factor (ALPHA_BF)	v_	0	1	0.121564
3	Groundwater delay(days) (GW_DELAY)	v_	30	350	200.01
4	Threshold depth of water(mm) (GWQMN)	v_	0	5000	4219.125
5	Groundwater revap coefficient (GW_REVAP)	v_	0.02	0.3	0.392711
6	Soil evaporation compensation factor (ESCO)	v_	0.01	1	0.512134

Fig. 39.10 Sensitivity analysis of flow calibration parameters

39.5 Conclusion

The study area for the present work is a catchment of the Mahanadi river basin. The present study has been conducted for the Mahanadi middle basin area of Ib tributary. SWAT model has been used, which runs under the Arc GIS interface and the model input in the form of runoff discharge for the basin. The simulation discharge value calibrates with the observed discharge for the time period of 2003–2011 and validation time period 2012–2016. Four sub-basin and five HRUs are found to exist in the region from the delineation result from Arc SWAT. For the SWAT-CUP model for six parameters taken for the calibration and validation of model analysis and the flow, calibration parameters are considered from the literature review and self-interpretation.

References

Abbaspour KC, Faramarzi M, Ghasemi SS, Yang H (2009) Assessing the impact of climate change on water resources in Iran. Water Resour Res 45(10)

Beven KJ, Kirkby MJ (1979) A physically based, variable contributing area model of basin hydrology/Un model a base physique de zone d'appel variable de l'hydrologie du bassin versant. Hydrol Sci J 24(1):43–69

Devak M, Dhanya CT (2014) Downscaling of precipitation in Mahanadi basin. India Int J Civil Eng Res 5:111–120

Gosain AK, Rao S, Arora A (2011) Climate change impact assessment of water resources of India. Current Sci, 356–371

Mohapatra M, Mohanty UC (2006) Spatio-temporal variability of summer monsoon rainfall over Orissa in relation to low pressure systems. J Earth Syst Sci 115(2):203–218

Mujumdar PP, Ghosh S (2008) Modeling GCM and scenario uncertainty using a possibilistic approach: application to the Mahanadi River, India. Water Resour Res 44(6)

Qiu LJ, Zheng FL, Yin RS (2012) SWAT-based runoff and sediment simulation in a small watershed, the loessial hilly-gullied region of China: capabilities and challenges. Int J Sedim Res 27(2):226–234

Chapter 40
Morphometric Analysis of Kosi River Basin, Bihar, India Using Remote Sensing and GIS Techniques

Niraj Kumar and Ramakar Jha

Abstract The present research work is aimed at studying morphological parameters of the Kosi basin, considering its importance in understanding hydrological behavior. For this purpose, radar data, that is, SRTM 1 Arc-Second Global elevation (30 m), is processed using ArcGIS 10.2 software for the analysis. The ArcSWAT Tool is used as an add-in in ArcGIS 10.2 to delineate the watershed and to generate stream networks. For analyzing the basin, several morphological parameters, namely linear aspects, areal aspects and relief aspects, are estimated. In this study, the drainage basin is 7th order with a lower mean bifurcation ratio of 2.21, indicating a structurally less disturbed basin. The form factor, elongation ratio and circulatory ratio show irregular and elongated basin shape. A low drainage density of 0.86 km/km^2 indicates coarse drainage texture and highly permeable subsoil materials. Lower stream frequency reveals gentle ground slope, less runoff and more infiltration. Lower relief aspect values justify the basin is in the valley region. This study is of greater importance in the geomorphological study of the basin and in other future investigations like prioritization of various sub-basins in the Kosi River basin.

Keywords Kosi river · SRTM · Morphological parameters · Linear aspects · Areal aspects · Relief aspects

40.1 Introduction

The river basin is the area of land drained by a river and its tributaries. It consists of all the surfaces of land drained by the mainstream channel and several tributaries. In order to understand the hydrological behavior of the basin, morphometric analysis of the drainage basin and its stream networks play an important role. The

N. Kumar (✉)
Department of Civil Engineering, Muzaffarpur Institute of Technology, Muzaffarpur, Bihar 842003, India
e-mail: nirajdsi10@gmail.com

R. Jha
Department of Civil Engineering, National Institute of Technology, Patna, Bihar 800004, India
e-mail: rjha43@gmail.com

R. Jha et al. (eds.), *Climate Change Impacts on Water Resources*, Water Science and Technology Library 98, https://doi.org/10.1007/978-3-030-64202-0_40

morphometric analysis of river basins is widely studied by researchers across the world since the origin of the concept in the year 1932. Horton (1932) was the first who introduced the morphometric analysis of river basin; the idea was later developed by many researchers. Horton (1932), Smith (1950) and Miller (1953) explained geomorphology as the science of landform origin and evolution and further recommended various parameters for the study of the morphology of the basin. Strahler (1964) explained that the morphometric analysis of a river basin provides a quantitative description of the drainage system, which is an important aspect of the characterization of basins. Clarke (1996) defined geomorphometry as the measurement and analysis of the surface of the earth and landform's dimensions. Jones (1999) studied morphometry and observed that the morphometric parameters affect catchment streamflow patterns through their influence on concentration time. Earlier, the researchers were using the conventional method for studying the morphometric characteristics of the basin; but with time, remote sensing and GIS technique gained the interest of researchers since it is a powerful tool for estimating the basin characteristics. The results obtained from remote sensing and GIS technique are more reliable and accurate; thus it is considered as the most effective tool. Avinash et al. (2011) have conducted the analysis of morphometric parameters for prioritization of sub-basins of Gurpur River. Yadav et al. (2014) have used remote sensing and GIS technique for morphometric analysis of Upper Tons basin of Northern Foreland of Peninsular India using the CARTOSAT satellite images. Choudhari et al. (2018) have conducted a study to identify the morphometric parameters of the Mula river basin, Maharashtra, to prioritize the sub-basins for groundwater potential. Many researchers (Singh and Singh 2011; Akbari et al. 2012; Soni et al. 2013; Choudhari et al. 2018) have used Shuttle Radar Topography Mission (SRTM) Digital Elevation Model (DEM) for estimating the morphometric parameters due to the high spatial resolution of the images.

In the present study, the hydrological behavior of the Kosi river basin has been conducted. Kosi River is known as "Sorrow of Bihar," as it is responsible for the highest number of flood incidents in the region in the last 30 years. The river has shifted nearly 150 km in the westward direction with extensive flooding in the basin region during the last 200 years (Gole and Chitale 1966). The main reason for this westward movement and frequent flooding in the region is due to sedimentation of the basin (Wells and Dorr 1987). The cause of the sedimentation in the basin region and shifting characteristics of the river is due to the hydrological behavior of the basin.

Keeping the above facts in mind, the present study aims to evaluate and analyze the various morphometric parameters of the Kosi river basin in Bihar using remote sensing and GIS technique. This study will be helpful in understanding the hydrological behavior of the basin.

40.2 Study Area

The Kosi River is one of the major tributaries of the Ganga River. It flows in the North Bihar plain, covering an area of 11175.68 km^2. The Kosi River originates in the Himalayan region in Nepal; after draining a large area in Tibet and Nepal, it enters into Bihar (India) region near Bhimnagar and joins the Ganga river near Kurshela, Katihar district, Bihar. The drainage basin is located between latitudes 25.35°–26.36° N and longitudes 86.53°–87.59° E (Fig. 40.1). The region is plain and the maximum elevation of the study area is 91 m. The Kosi basin has 29 sub-watersheds. The slope in the basin ranges from 0 to 35°. Ten districts of Bihar fall in the drainage basin.

40.2.1 Data Source

For the purpose of the morphometric analysis of the Kosi basin Shuttle Radar Topography Mission (SRTM) 1, Arc-Second Global elevation data with a spatial resolution of 30 m is processed using ArcGIS 10.2 software. These data have been extracted from the United States Geological Survey (USGS) EarthExplorer portal (http://srtm.usgs.gov/mission.php).

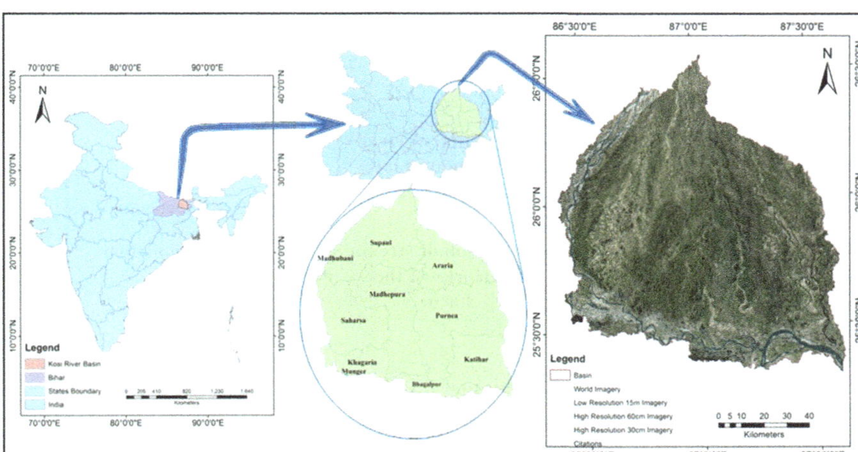

Fig. 40.1 Location of the study area with district boundaries

40.3 Methodology

The analysis consisted of several steps, which include study area selection; data collection (radar data from USGS EarthExplorer); watershed delineation and extraction of river basin boundary; extraction of stream networks of the basin; estimation of various morphometric parameters; interpretation of the results; and hydrological behavior of Kosi River basin (Fig. 40.2).

For this purpose, ArcGIS 10.2 software (ArcMap10.2) is used. The ArcSWAT Tool is used as an add-in in ArcGIS10.2 software to delineate the watershed and to generate stream networks. Spatial Analyst tool extensions such as Hydrology tool and Surface tool are used in ArcMap 10.2 for the analysis.

Since the basin region is large, four sets of data have been extracted, namely SRTM1N25E0876V3, SRTM1N25E087V3, SRTM1N26E086V3 and SRTM1N26E087V3 from the USGS EarthExplorer portal. These Digital Elevation

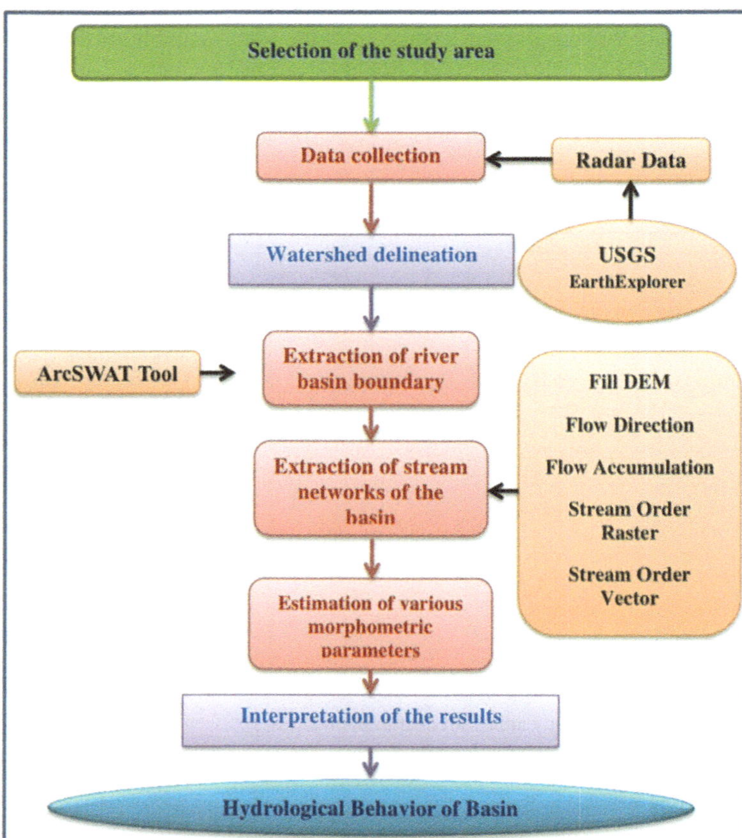

Fig. 40.2 Methodology

Models (DEM) are mosaicked using ArcGIS 10.2 software to get a single DEM representing the entire basin. The projection for the whole analysis of the Kosi river basin is WGS 1984 UTM Zone 44 N. The mosaicked DEM is processed using ArcGIS 10.2 software for the extraction of the basins and the stream networks. The Clip tool under raster processing in Data Management tool extension is used for clipping the projected mosaic DEM within the basin boundary to generate basin DEM. Further, the projected basin DEM is processed using various tools, say fill, flow direction, flow accumulation, stream order raster and stream order vector, stepwise under Hydrology tool in Spatial Analyst tool extension in ArcMap 10.2. The final step of the processing of the basin DEM is the stream network generation. The streams are obtained through the following steps like conditional (for flow accumulation), stream order and stream to feature under hydrology tool extension in ArcMap 10.2. The generated stream network shows the order of the streams.

The delineated basin and stream networks provide important results like length, area, perimeter, and stream number. These data are analyzed using standard formulas suggested by various scientists, namely Gravelius (1914), Horton (1932, 1945), Miller (1953), Schumm (1956), Strahler (1957, 1964), Hadley and Schumm (1961), Nookaratnam et al. (2005) as mentioned in Table 40.1 to compute the morphometric parameters. The study of the various morphometric parameters and comparing their values with standard values suggested by researchers helps in understanding the hydrological behavior of the Kosi river basin.

40.4 Results and Discussion

In the present study, various morphometric parameters are computed and these parameters are categorized under three heads, namely basin characteristics, linear aspects, areal aspects and relief aspects of the basin. These parameters are compared with standard values to understand the hydrological behavior of the Kosi river basin.

A. **Basin Characteristics**:

The **basin length, basin perimeter** and **basin area** are the most important parameters for the hydrologic design of a basin as these basin characteristics are required to compute the various morphological parameters. From the analysis conducted for basin characteristics, it is found that the basin length is high, having a value of 261.429 km, which indicates the elongated shape of the basin. Moreover, the basin area is 11175.679 km^2 and the basin perimeter of Kosi basin is 985.876 km (Table 40.2).

B. **Linear Aspects of the Basin**:

The linear aspects characterize the basin depending on the stream networks, drainage pattern and order of the streams. In this study, the parameters, namely stream order, stream length, mean stream length, stream length ratio and bifurcation ratio, are determined to understand the linear aspects of the basin.

Table 40.1 The formula used for computations of various morphometric parameters and their references

Morphometric parameters	Formula/definition	References
Basin characteristics		
Basin length, L_b (km)	$L_b = 1.312A^{0.568}$ where L_b = Basin length (km) A = Area of the basin (km^2)	Nookaratnam et al. (2005)
Linear aspects		
Stream order (u)	Hierarchical rank	Strahler (1964)
Stream Length (L_u)	Length of the stream	Horton (1945)
Mean stream length (L_{sm})	$L_{sm} = \frac{L_u}{N_u}$ where L_u = Mean stream length of a given order (km), N_u = Number of stream segment	Horton (1945)
Stream length ratio (RL)	$RL = \frac{L_u}{L_{u-1}}$ where L_u = Total stream length of order (u) L_{u-1} = The total stream length of its next lower order	Horton (1945)
Bifurcation Ratio (R_b)	$R_b = \frac{N_u}{N_{u+1}}$ where N_u = Number of stream segments present in the given order N_{u+1} = Number of segments of the next higher order	Schumm (1956)
Mean bifurcation ratio (R_{bm})	R_{bm} = Average of bifurcation ration of all orders	Schumm (1956)
Areal aspects		
Form factor (R_f)	$R_f = \frac{A}{L_b^2}$ where A = Area of the basin (km^2) L_b = Basin length (km)	Horton (1932, 1945)
Elongation ratio (R_e)	$R_e = \frac{\text{Dia of circle having area of basin}}{\text{Basin Length}}$ $R_e = \frac{2\sqrt{\frac{A}{\pi}}}{L_b}$ where A = Area of the basin (km^2) L_b = Basin length (km)	Schumm (1956)
Circularity ratio (R_c)	$R_c = \frac{\text{Basin area}}{\text{Circle area having perimeter of basin}}$ $R_c = \frac{4\pi A}{P^2}$ where A = Area of the basin (km^2) P = Perimeter (km)	Miller (1953), Strahler (1964)
Shape factor (Bs)	$B_s = \frac{1}{R_f} = \frac{L_b^2}{A}$ where L_b = Basin length (km) A = Area of the basin (km^2)	Horton (1932)

(continued)

Stream Order (u):

Strahler (1964), has proposed that the smallest fingertip tributaries are designated as 1st order; these two or more 1st order channels join to form 2nd order channel segments; these two or more 2nd order join to form 3rd order channel segments

Table 40.1 (continued)

Morphometric parameters	Formula/definition	References
Compactness co-efficient (C_c)	$C_c = \frac{\text{Basin Perimeter}}{\text{Perimeter of circle having basin area}}$ $C_c = \frac{0.2821\,P}{\sqrt{A}}$ where P = Perimeter (km) A = Area of the basin (km^2)	Gravelius (1914)
Drainage density (D_d) (km/km^2)	$D_d = \frac{L_u}{A}$ where L_u = Total length of stream (km) A = Area of the basin (km^2)	Horton (1932, 1945)
Length of overland flow (L_o) (km^2/km)	$L_o = \frac{1}{2D_d}$ where D_d = Drainage density (km/km^2)	Horton (1945)
Constant of channel maintenance (km^2/km)	$C = \frac{1}{D_d}$ where D_d = Drainage density (km/km^2)	Schumm (1956)
Stream frequency (F_s) ($Streams/km^2$)	$F_s = \frac{N_u}{A}$ where N_u = Total number of streams of all order A = Area of basin (km^2)	Horton (1932, 1945)
Relief aspects		
Basin relief (R) (m)	$R = H - h$ where H = Maximum elevation in meter h = Minimum elevation in meter	Hadley and Schumm (1961)
Relief ratio (R_r)	$R_f = \frac{R}{L_b}$ where R = Basin relief L_b = Basin length'(m)	Schumm (1956)
Ruggedness number (R_n)	$R_o = R * D_d$ where R = Basin relief D_d = Drainage density	Schumm (1956)

Table 40.2 Basin characteristics of Kosi River and their values

Sl. no.	Morphometric parameters	Symbol	
Basin characteristics			
1	Basin area	A	11175.679 km^2
2	Basin perimeter	P	985.876 km
3	Basin length	L_b	261.429 km

and so on. The highest order stream will be the one through which all discharge and sediment pass. In the present study, Strahler's Stream Ordering system, which is slightly modified by Horton's system, has been followed for ranking the stream segments because of its simplicity. From this study, it is observed that ten sub-basins are of 4th order; six sub-basins are of 5th order; eight sub-basins are of 6th order and the rest four sub-basins are of 7th order. Thus, the entire Kosi basin is of 7th order, as shown in Fig. 40.3, indicating the degree of streams branching within the basin region.

Stream Length (L_u):

Stream length (L_u) of a channel is a dimensional property that reveals the surface runoff characteristics of a drainage network. Stream length of channel usually decreases with increasing stream order as obtained in this study (Fig. 40.4). Short length streams are representative of areas with large slopes and finer texture, whereas longer length is generally indicative of low gradients. In this study, the plot of log stream Length w.r.t. stream order showed the linear pattern, which indicates the

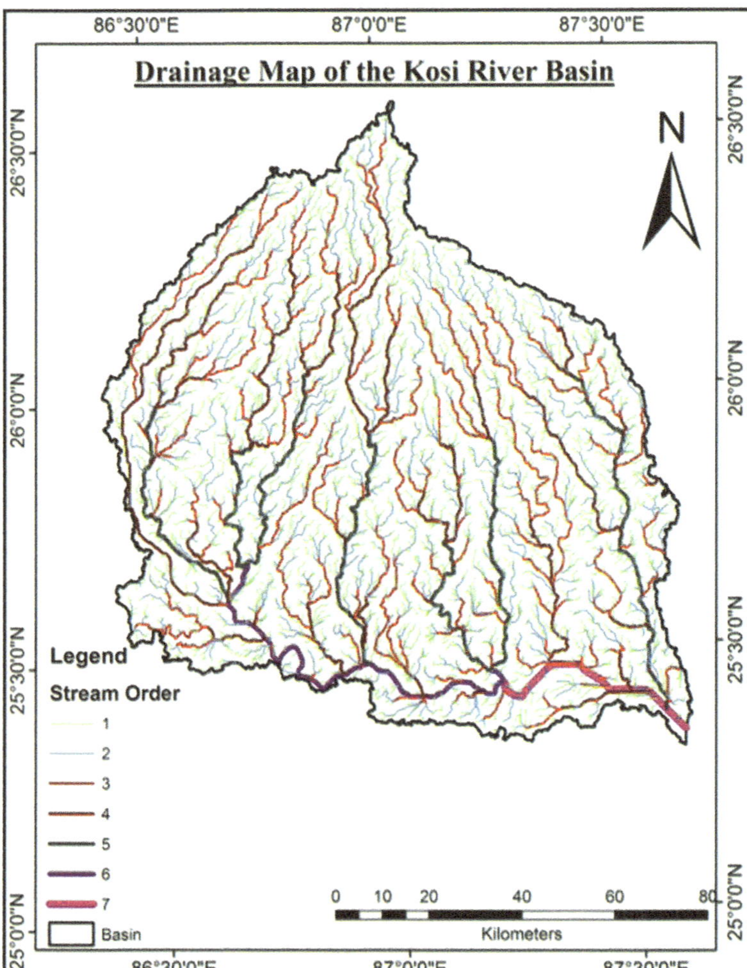

Fig. 40.3 Stream order map of Kosi River Basin showing basin is of 7th order

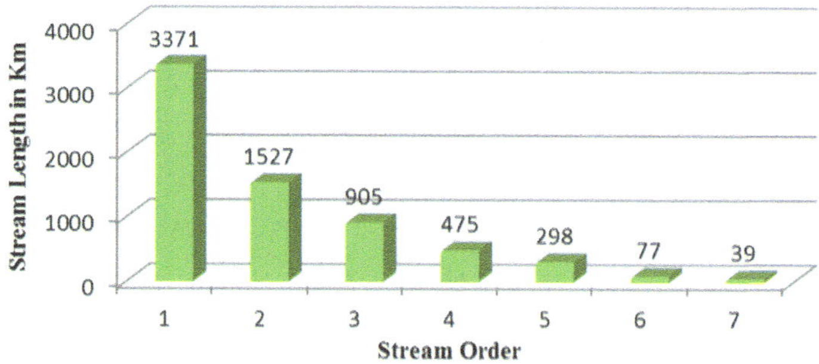

Fig. 40.4 The plot of stream length versus stream order

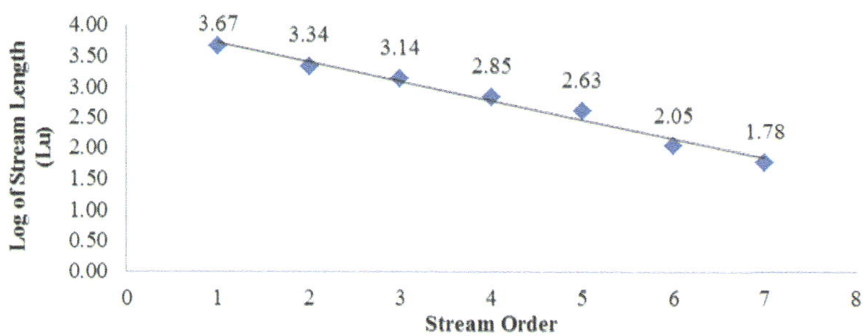

Fig. 40.5 Geometric relationship b/w stream orders and log of stream lengths

homogeneous rock material subjected to weathering erosion characteristics of the basin (Fig. 40.5).

Mean Stream Length (L_{sm}):

In the present study, Horton's law of stream length is used to compute mean stream length. For different basins, the mean stream length values are different as it is proportional to the size and topography of the basins. From this study, it is observed that mean stream length values for the basins have not much variation (Table 40.3).

Stream Length Ratio (RL):

In the present study, the mean stream length ratio value is observed to be 1.02, which is due to the slope and topography of these basins. The average stream length ratio value of the 6th order streams has a higher value compared to the other streams. This indicates that the area drained by 6th order streams is more permeable, gentler than streams of other orders.

Bifurcation Ratio (R_b):

The bifurcation ratio represents the branching pattern of the drainage network in a given watershed. In the present study, the basin has a lower mean bifurcation ratio of

Table 40.3 Linear aspects of Kosi River and their values

Sl. no.	Morphometric parameters	Symbol							
Linear aspects									
1	Stream order	u	1	2	3	4	5	6	7
2	Stream number	N_u	3371	1527	905	475	298	77	39
3	Bifurcation ratio	R_b	–	1/2	2/3	3/4	4/5	5/6	6/7
4	Mean bifurcation ratio	$(R_b)_b$mean	2.21						
5	Stream length	L_u	4681.9	2210.1	1394.8	705.0	424.3	111.0	59.9
6	Mean stream length	L_{sm}	1.39	1.45	1.54	1.48	1.42	1.44	1.54
			2/1	3/2	4/3	5/4	6/5	7/6	
7	Stream length ratio	RL	1.04	1.06	0.96	0.96	1.01	1.07	–
8	Mean stream length ratio	$(RL)_{mean}$	1.02						

2.21, indicating structurally less disturbed basins with high permeability and without any distortion in the drainage pattern

C. **Areal Aspects of the Basin**:

The analysis conducted for areal aspects consists of two parts, namely, **shape characteristics** and **drainage characteristics**. The shape characteristics include form factor, elongation ratio, circulatory ratio, basin shape factor, and compactness coefficient. **The form factor** outlines the form of the drainage basin. The smaller the value of form factor, the more elongated will be the basin, that is, the basin will be long narrow, having larger lengths. **Elongation ratio** values vary from 0 (in a highly elongated shape with high relief) to 1 (in a circular shape with low relief). The **circulatory ratio** is much influenced by the stream length, stream frequency and stream gradient of a different order. The circulatory ratio value varies from 0 (elongated shape) to 1 (circular shape), where value >0.5 suggests that the basin is more or less circular in shape. **Basin shape factor** is a measure of basin shape irregularity. **The compactness coefficient** is the relationship of the shape of the drainage basin to a circle. If the value of the basin shape factor and compactness coefficient is unity, then the basin would be a perfect circle. From the analysis conducted for shape characteristics, it is found that the form factor of 0.16, elongation ratio of 0.46, the circulatory ratio of 0.14, the basin shape factor of 6.12 and compactness coefficient of 2.63 show irregular and elongated basin shape with high permeability and moderate slope in the region (Table 40.4).

Table 40.4 Areal aspects of Kosi River and their values

Sl. no.	Morphometric parameters	Symbol	
Areal aspects			
1	Form factor	R_f	0.164
2	Elongation ratio	R_e	0.456
3	Circulatory ratio	R_c	0.144
4	Shape factor	B_s	6.116
5	Compactness co-efficient	C	2.631
6	Drainage density	D_d	0.858
7	Length of overland flow	L_o	0.583
8	Constant of channel maintenance	C_c	1.166
9	Stream frequency	F_s	0.599

The drainage characteristics include drainage density, length of overland flow, stream frequency and constant of channel maintenance. **Drainage density** is an important indicator of the linear scale of landform elements in stream eroded topography, which indicates the closeness of spacing of the channels. Drainage density varies with a wide dimension of geologic and climatic environments such that the low value occurs in the regions of highly permeable subsoil materials under dense vegetative cover with low relief, whereas the high value occurs in the region of weak impermeable sub-surface materials under sparse vegetative cover with high relief. In the present study, a low drainage density of 0.86 km/km^2 indicates coarse drainage texture and highly permeable subsoil under dense vegetative cover with low relief. **Length of overland flow** is used to describe the length of a flow path of water over the ground surface before it reaches the definite stream channels. Mathematically, it is half of the reciprocal of drainage density. Here, the basin has a higher value; thus the region has longer flow paths, gentle ground slope, leads to less runoff and high infiltration. **Constant of channel maintenance** depends on the permeability, rock type, vegetation cover, relief and duration of erosion; mathematically, it is the inverse of drainage density. The higher value of **Stream frequency** (>2.0 streams/km^2) shows steep slopes, less permeable rock materials, greater runoff, less infiltration, sparse vegetation and high relief conditions, whereas vice versa for low values (<1.0 streams/km^2). In the present study, a lower stream frequency of 0.60 streams/km^2 and a higher constant of channel maintenance of 1.17 km^2/km reveals gentle ground slope, less runoff and more infiltration.

D. **Relief Aspects of the Basin**:

The relief aspects of the basin include **basin relief**, **relief ratio** and **ruggedness number**. **Basin relief** controls the stream gradient and, therefore, influences the flood pattern and the amount of sediment that can be transported (Hadley and Schumm 1961). From the analysis conducted for relief aspects, it is found that the basin has a lower basin relief value of 91 m and a lower relief ratio value (Table 40.5). **Ruggedness number** indicates the structural complexity of the terrain. Its low value

Table 40.5 Relief aspects of Kosi River and their values

Sl. no.	Morphometric parameters	Symbol	
Relief aspects			
1	Basin relief	R	91
2	Relief ratio	R_r	0.00035
3	Ruggedness number	R_n	0.07806

justifies the low basin relief condition. This shows that the basin is in the valley region, which also validates the findings of the present study.

40.5 Conclusions

The present study has proved that the Remote Sensing and Geographic Information System technique is more efficient in watershed delineation and calculation of morphometric parameters as it provides reliable and accurate results. The Kosi river basin has stream order varying from 1 to 7, and the stream length is maximum in 1st order streams, showing a maximum number of streams in the lower order. The linear aspects of the basin, namely stream order, stream length ratio, bifurcation ratio, etc., indicate that the basin has a dendritic drainage pattern, high permeability and no structural control. The form factor, elongation ratio and circulatory ratio show more irregular, elongated and non-circular basin shapes. The drainage characteristics of the basin reveal that the basin has a coarse drainage texture and high permeable subsoil materials under dense vegetative cover. High length of overland flow and lower stream frequency is the indication of a long flow path of the basins with gentle ground slope, less runoff and more infiltration. The relief aspects justifying that the basin is in the valley region. Based on this study, prioritization of various sub-basins may be carried out, which will be very helpful in water, soil and natural resource management.

This study is of greater importance in the geomorphological study of the basin and further investigations in the prioritization of sub-basins of the Kosi River basin. The prioritization of sub-basins will be helpful in identifying the groundwater potential in the region. The results obtained in this study will be helpful in the generation of thematic maps of drainage, slope and elevation for flood risk mapping of the basin region. The results are also important for an extensive field survey.

References

Avinash K, Jayappa KS, Deepika B (2011) Prioritization of sub-basins based on geomorphology and morphometricanalysis using remote sensing and geographic informationsystem (GIS) techniques. Geocarto Int 26(7):569–592. https://doi.org/10.1080/10106049.2011.606925
Choudhari PP, Nigam GK, Singh SK, Thakur S (2018) Morphometric based prioritization of watershed for groundwater potential of Mula river basin, Maharashtra, India. Geol Ecol Landscapes 2(4):256–267. https://doi.org/10.1080/24749508.2018.1452482
Clarke JI (1996) Morphometry from maps: essay in geomorphology. Elsevier Publ. Co, New York, pp 235–274
Gole CV, Chitale SV (1966) Inland delta building activity of Kosi River. J Hydraulics Div. In:Proceedings of the American Society of Civil Engineers, 92 (HY2):111–126
Gravelius H (1914) Grundrifi der gesamten Gewcisserkunde. Band I: Flufikunde (Compendium of Hydrology, vol I. Rivers, in German). Goschen, Berlin
Hadley RF, Schumm SA (1961) Sediment sources and drainage basin characteristics in upper Cheyenne river basin. US Geological Survey, Water-Supply Paper, Washington, DC
Horton RE (1932) Drainage Basin characteristics. Trans Am Geophys Union 13:350–361
Horton RE (1945) Erosional development of streams and their drainage basin. Hydrophysical approach to quantitative morphology. Geol Soc Am Bull 56:275–370
Jones JAA (1999) Global hydrology: processes, resources and environmental management. Longman, 399
Miller VC (1953) A quantitative geomorphic study of drainage basin characteristics in the Clinch mountain area, New York. Department of Geology, ONR, Columbia University, Virginia and Tennessee, Project, NR 389-402, Technical Report 3
Nookaratnam K, Srinivastava YK, Venkateswara W, Amminedu E, Murthy KSR (2005) Check dam positioning by prioritization of micro watershed using SYI model and morphometric analysis remote sensing and GIS perspective. J Ind Soc Remote Sensing 33(1):25–28
Soni Sandeep Kr, Tripathi Shashikant, Maurya Abhishek Kr (2013) GIS based morphometric characterization of Mini watershed—Rachhar nala of Anuppur district Madhya Pradesh. Int J Adv Technol Eng Res (IJATER) 3:32–38
Akbari S, Karthik A, Venkatesh G, Ratu ST, Penjor K (2012) Estimation of geomorphology parameters for small catchment using GIS. ISSN 0974-5904, 05(04)(01):976–981
Smith KG (1950) Standards for grading textures of erosional topography. Am J Sci 248:655–688
Strahler AN (1957) Quantitative analysis of watershed geomorphology. Trans Am Geophys Union 38:913–920
Strahler AN (1964) Quantitative geomorphology of drainage basins and channel networks. In: ByVenTe Chow (ed) Handbook of applied hydrology. McGraw Hill Book Company, New York
Schumm SA (1956) Evolution of drainage systems and slopes in Badlands at Perth Amboy, New Jersey. Geological Soc Am Bull 67:597–646
Singh V, Singh UC (2011) Basin morphometry of Maingra River, district Gwalior, Madhya Pradesh, India. Int J Geomatics Geosci 1(4):891–902
Yadav SK, Singh SK, Gupta M, Srivastava PK (2014) Morphometric analysis of Upper Tons basin from Northern Foreland of Peninsular India using CARTOSAT satellite and GIS. Geocarto Int 29(8):895–914
Wells NA, Dorr JA (1987) Shifting of the Kosi river, northern India. Geol 15:204–207

Chapter 41
Simulation of Impact of Climate Change on the Performance of a Reservoir System in Eastern India

Satabdi Saha, Debasri Roy, and Rajib Das

Abstract The work deals with the impact of changing climate on performance of DVC (Damodar Valley Corporation), reservoir system (Fig. 41.1) (comprising Tilaiya, Maithon, Konar and Panchet dams (multipurpose dams), Tenughat dam (single purpose) and Durgapur barrage on river Damodar of India using HEC-5, a simulation model for reservoir operation (developed by HEC, USA). The projected climate has been taken from projected climate output (A1B scenario) of Regional Climate Model PRECIS of Hadley Centre, UK prepared under IPCC (Intergovernmental Panel on Climate Change). Integrated operation and reservoir guide/ rule curve of DVC Authority was used from 1st of June (water year starting date in India) for a future period (2018–2024). Simulations were executed for 1985–1990 (baseline period) and for 2018–2024 (future period) with changed climate data (with current demand and projected demand). The system performance (on a seasonal basis) using "Performance Indices" (viz. volume reliability, time reliability, flood control index) for the future scenario (with current demand) became better than that for baseline condition in four seasons for Panchet reservoir. Reliability of meeting demands (M&I) in November and December in 2020 (with 58% increased projected inflow in comparison to baseline) with projected demand showed improvement (by 14%) over those in October and in months of January to May in 2020 at Maithon reservoir. It was noted that remarkably increased projected inflow (unseasonal) into the Panchet (by 1286.8%) and Maithon (by 1014.4%) reservoirs with reference to a baseline condition in March and May of 2022, respectively, caused flooding (unseasonal and

S. Saha (✉) · D. Roy · R. Das
School of Water Resources Engineering, Jadavpur University, 188, Raja S.C. Mallick Road, Kolkata, West Bengal 700032, India
e-mail: satabdi.biet@gmail.com

D. Roy
e-mail: debasri_roy1@yahoo.co.in

R. Das
e-mail: rajibdas79@gmail.com

S. Saha
Regent Education and Research Foundation, Bara Kanthalia, Barrackpore, Kolkata, West Bengal 700121, India

R. Jha et al. (eds.), *Climate Change Impacts on Water Resources*, Water Science and Technology Library 98, https://doi.org/10.1007/978-3-030-64202-0_41

high) downstream Panchet reservoir and bank full flow (unseasonal) downstream Maithon dam in March and May. The results from the study would guide concerned authorities for operating reservoirs for anticipated climate change.

Keywords Climate change · Reservoir operation · Performance indices · HEC-5 model · Guide curve

41.1 Introduction

Changing climate is currently an issue of great concern. Expected consequences of changed climate are the modification of the hydrological cycle, leading to change in rainfall pattern, increase in quantum and frequency of floods [6] and intensification of drought and associated water issues [5]. Thus, climate change can impact reservoir operation reasonably (and consequently on its performance) as the operation which caters to various water demands as also control flood, is dependent on climate and weather. South Asia (including India), which is houses poor people, emerges to be vulnerable to changed climate and associated adverse socio-economic effects.

In this background, the work reports impact of changed climate on DVC (Damodar Valley Corporation) reservoir system performance (Fig. 41.1) located on Damodar river in India using simulation models (HEC-5 1998 & HEC-HMS 3.4) (developed by HEC, USA). HEC-HMS (Hydrologic Modeling System) model was adopted for the generation of inflow to the reservoir system. Simulation model HEC-5 was employed to analyze DVC system performance in fulfilling demands along with controlling flood under changing climate (with both existing and projected demands) using volume reliability, time reliability, flood indices.

Some studies are discussed here.

Ahmadi et al. (2015) studied the performance of Karoon-4 dam in Iran for future periods (using the output of HadCM3 climate model (A2 emission scenario) and adopting optimization algorithm, NSGA-II for optimization of operation rules of the reservoir in conjunction with hydrologic model IHACRES. The result indicated increased reliability and reduced vulnerability for the generation of hydropower with adaptive management strategy (under climate change).

François et al. (2015) estimated the changed performance of multiobjective reservoirs (French Alps) using Dynamic Programming with clear, short, far-sighted management policies for multimodel climatic simulations for SRES-A1B scenario. Changes in hydro-meteorological variables rather than the changes in management strategy affected the system performance appreciably.

Li et al. (2010) investigated the impact of changing climate (based on Coupled Global Climate Model data) on flow and reservoir performance in, Northern American Prairie watershed. The analysis unveiled an increase of occurrence of high peak flow, an increase of volume of available water and also high reliability of existing operation rules for drought protection, controlling of the flood.

Minville et al. (2010) studied the impact of changing climate on reservoir operations (medium-term) for Peribonka water system Quebec, Canada. The projections indicated an increase of inflow, early flow peaks, greater flow volumes for spring flood and sensitiveness of reservoir to operating rules with the conclusion that rules be reviewed for considering modified seasonal hydrological regime.

41.2 Study Area

India's first river development project, DVC, was developed on the river Damodar. Damodar, with three tributaries—Bokaro, Konar and Barakar—is an important tributary system of Bhagirathi- Hugli (distributory of Ganga). The four dams namely Tilaiya (on river Barakar), Konar (on river Konar), Maithon (on river Barakar, downstream Tilaiya dam) and Panchet (on river Damodar) and Durgapur barrage (on river Damodar, downstream confluence of Barakar river and Damodar river) (Figs. 41.1, 41.2) were commissioned in 1953–1959. Tenughat dam (a single purpose reservoir) was commissioned in 1978 on Damodar river, upstream Panchet dam.

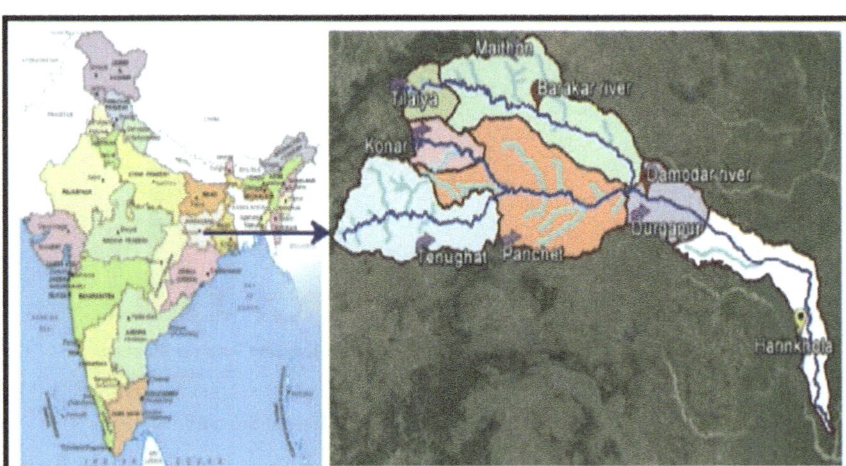

Fig. 41.1 DVC system (Topographical view)

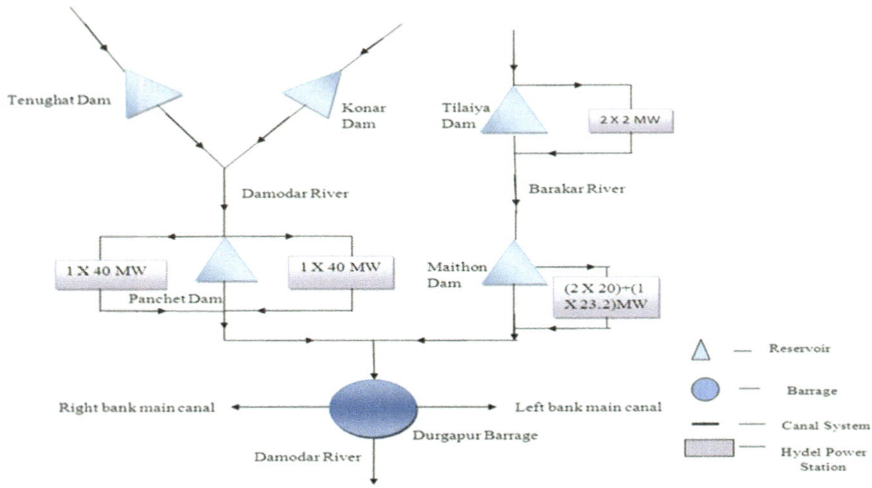

Fig. 41.2 Outline Plan of DVC system

41.3 Data Acquisition

Reservoir operation data, flow, release from the reservoir (1996–2013), Muskingum routing parameters, channel capacity, characteristic curves of the reservoir, guide (or rule) curves of the reservoir, the demand of water (Fig. 41.3) were obtained from DVC Authority. (0.5 × 0.5°) grid wise rainfall for 1978–2007 and grid wise (1 × 1°) daily mean temperature for 1985–2007 were obtained from National Climatic Centre, IMD, Pune. Daily rainfall and temperature, A1B scenario of PRECIS model (0.44° × 0.44° lat./long.—grid wise) was obtained for 2018–2024 from IITM, GoI, Pune. Other meteorological data (daily-1985–2007), namely relative humidity,

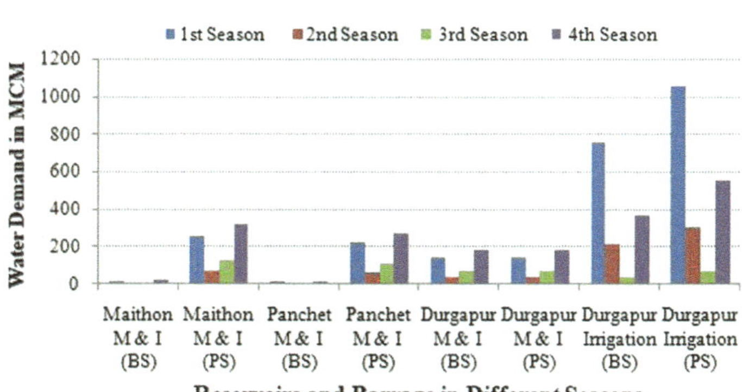

Fig. 41.3 Baseline and projected water demands in MCM at reservoirs and barrage (BS denotes baseline scenario and PS denotes projected scenario)

actual sunshine hours, wind speed, were obtained from IMD, GoI, Kolkata and those for 2014–2025 were obtained from IITM, GoI, Pune. Information on the current, future demand for water was taken from the report prepared by WAPCOS (Water and Power Consultancy Services 2010).

41.4 Methodology

41.4.1 HEC-HMS Model

HEC-HMS model simulates the precipitation-runoff process of the watershed by disintegrating the hydrologic cycle in pieces. Many common methods of hydrology have been considered in the model. This model was used for generating inflow to the reservoir system (HEC-HMS 2000).

41.4.2 HEC-5 Model

The HEC-5 model simulates reservoir operation (river network) for controlling flood, supplying water, generation of hydropower and preserving water quality.

41.4.3 Performance Indices

The impact assessment was done using 'Performance Indices' for 1 June to 30 September (1st season), October (2nd season), 1 November to 31 December (3rd season), 1 January to 31 May (4th season) separately to consider variability in inflow, the pattern of water demand, rule curve operation for four seasons.

The performance indices are:

1. Volume reliability (RI_0)—Ratio of released water volume and target demand.
2. Time reliability (RI_1)—Ratio of stages for which demands could be met and the total stages in the simulation.
3. Flood Index (RI_2)—Ratio between stages for which river overflows bank and stages for which pool level lies in the flood storage zone.

41.5 Result and Discussions

The system performance (on a seasonal basis) for the future scenario (with current demand) became better than that for baseline conditions in four seasons for the

Panchet reservoir. Reasonably good local catchment flow and releases from upstream reservoirs viz., Tenughat and Konar reservoirs (due to sufficient inflow in respective catchments) helped to maintain reliability level as that of baseline scenario for Panchet reservoir.

However, a decrease of volume reliability (by 49 and 48%) and time reliability (by 51 and 55%) (Fig. 41.4a–d) with reference to baseline period was noted at Panchet reservoir in meeting at site projected M&I demands in 4th season of 2019 and 2020 respectively. Poor performance in both years is driven by corresponding high water requirements projected for the future.

Also, the study indicated decreased volume and time reliabilities with current demand (by 31 and 65%) and future demand (by 35 and 61%) (Fig. 41.4a–d) over the corresponding baseline condition for Maithon reservoir in 2nd season of 2020. The analysis unveiled that decrease of projected inflow (by 80%) in 2020 and rule curve bounded operation in 2nd season (for both existing and future conditions) resulted in poor performance for the Maithon reservoir. Further, rule curve bounded operation, projected demand and decrease of projected inflow (by 75%) in 2023 in 2nd season led to decrease of volume (by 33%) and time (by 31%) reliabilities for Maithon reservoir in 2023.

Again, reliability of meeting M&I demands in 3rd season of 2020 (with 58% increased projected inflow with reference to that of baseline) with projected demand showed improvement (by 14%) (not shown) over that of 2nd season and 4th season in 2020 at Maithon reservoir. But still, the performance was unable to be restored to that of the baseline scenario. This is due to the detrimental effect of the future (increased) demands as reliability is limited by available inflow and high demand.

Decreased of projected inflow at Durgapur Barrage in 4th season, 2019 (by 15.84%) in and in 2nd season, 2020 (by 17.42%) resulted in poor performance with projected demand—the volume reliability was found to decrease by 56 and 62% in 4th season and by 19 and 41% (Fig. 41.4a–d) in 2nd season of 2019 and 2020 respectively. Guide curve bounded operation in Maithon and Panchet reservoirs in 2nd season and enhanced demand resulted in poor performance (decrease of reliability) for at site demand of Durgapur Barrage. However, with current demand and projected flow, reliability remained unchanged for Durgapur Barrage.

It was noted that remarkably increased projected inflow (4th season) into the Panchet (by 1286.8%) and Maithon (by 1014.4%) (Fig. 41.6a–d) reservoirs with reference to a baseline condition in March and May of 2022 respectively caused flooding (4th season) downstream Panchet reservoir and bank full flow (4th season) downstream Maithon reservoir in March and May.

An increase of volume reliabilities in 4th season (by 7–35% and by 26–27%) was noted at Maithon and Panchet reservoirs due to the aforementioned increased projected inflow. Remarkably good performance (increased by 4–56% compared to baseline period) was noted in the case of Konar reservoir in 2nd season and 3rd season of future years (not shown).

High flow spanning over four days in 4th season of 2022 from Konar, Tenughat and Panchet catchments caused flooding downstream Panchet reservoir for one day only due to elegant flood control operation executed by reservoirs during the full duration

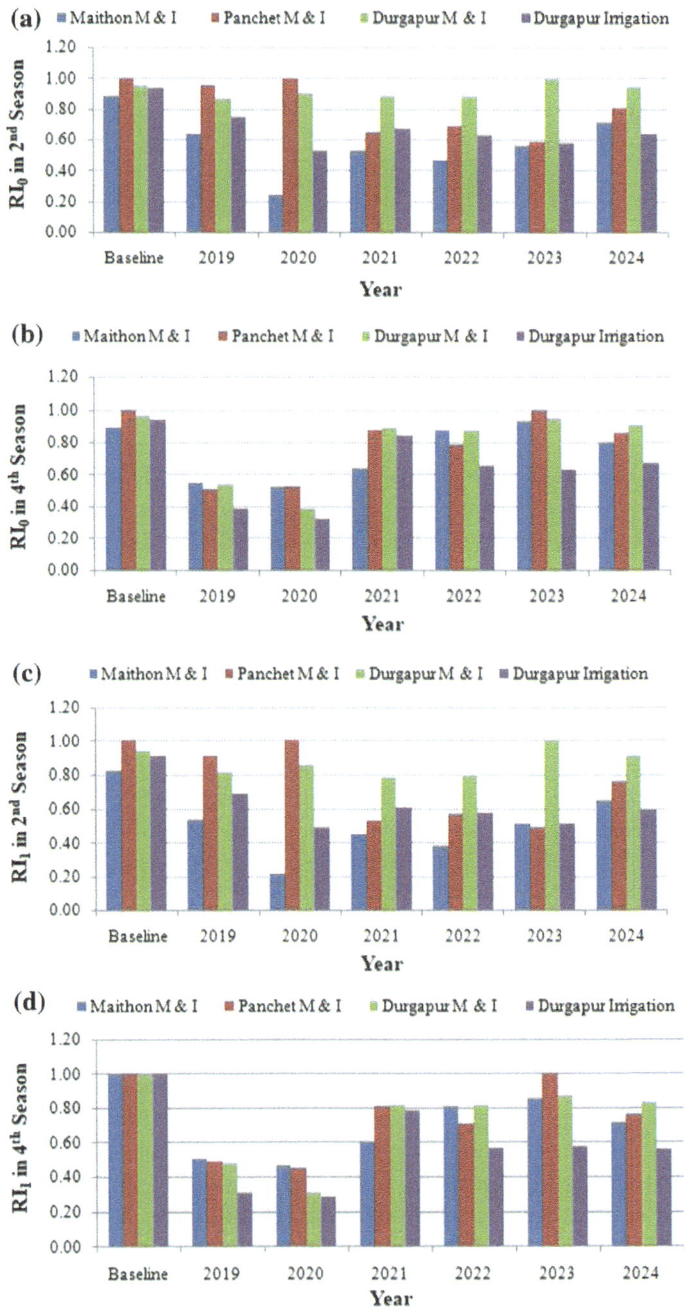

Fig. 41.4 (a–d) Performance indices in baseline and projected years for reservoir system

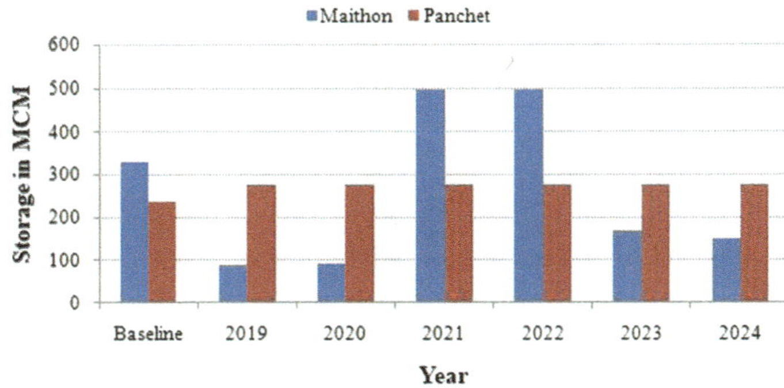

Fig. 41.5 End of period storage on 31 May in baseline and projected years at Maithon and Panchet reservoirs

of high flow event. High flow for three consecutive days in March 2023 in the upper stretch of river Damodar system caused flooding downstream Panchet reservoir for one day and kept flood storage in Panchet reservoir full for six consecutive days.

A decrease of volume reliability (by 14% with reference to baseline period) was noted in 1st season of 2023 (not shown) with projected demand at Panchet reservoir. Comparatively low inflow (by 44% with reference to the baseline period) was noted during this period.

End of period storage at Maithon reservoir at the end of 4th season was found to be increased (by 49% with reference to baseline period) in 2021 and 2022 (Fig. 41.5) due to increased projected inflow in 4th season (Fig. 41.6a–d) (by 407% and 1014.35% with reference to baseline inflow) in 2021 and 2022 respectively. This led to comparatively better performance (volume reliabilities increased by 12% and by 35% with reference to volume reliabilities in 4th season of other projected years) in 2021 and 2022, respectively, at Maithon reservoir.

41.6 Conclusions

Simulations of the impact of changing climate on the performance of the DVC reservoir system were executed for the baseline period (1985–1990) and future period (2018–2024) (with current demand and projected demand) with the simulation models HEC-HMS and HEC-5. The impact assessment was executed with performance indices (volume reliability, time reliability and flood index) for four seasons. The analysis unveiled the following aspects.

The system performance (on a seasonal basis) for the future scenario (with current demand) became better than that for the baseline period in four seasons for the Panchet reservoir. Again, reliability of meeting M&I projected demands in 3rd season in 2020 (with 58% increased projected inflow with reference to that of baseline) with projected demand showed improvement (by 14%) over that of 2nd season and 4th

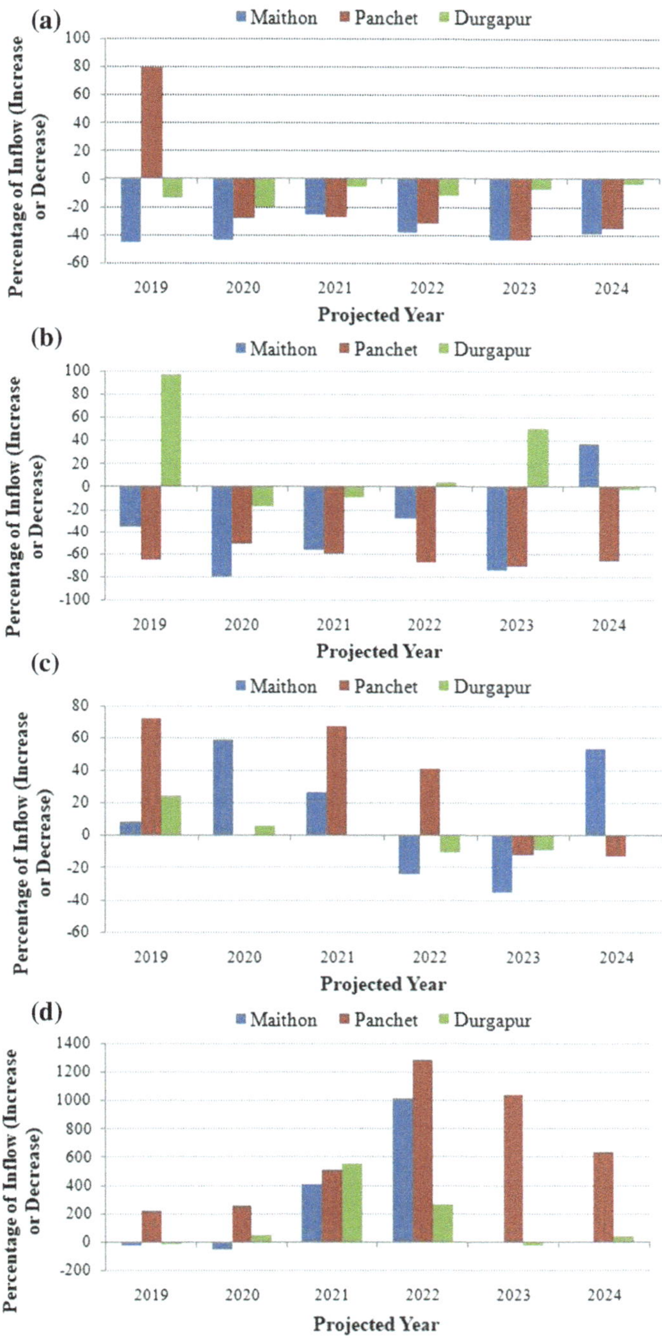

Fig. 41.6 (a–d) Percentage of projected inflow with reference to a baseline condition in 1st, 2nd, 3rd and 4th season respectively

season in 2020 at Maithon reservoir. It was noted that remarkably increased projected inflow (4th season) into the Panchet (by 1286.8%) and Maithon (by 1014.4%) reservoirs with reference to baseline period in March and May of 2022 respectively caused flooding (4th season) downstream Panchet reservoir and bank full flow (4th season) downstream Maithon reservoir in March and May.

The results from the study would guide concerned authorities for operating reservoirs for expected climate change.

Acknowledgements The authors convey sincere gratitude for kind cooperation extended (supplying data and related matter) by officials and personnel of DVC, Maithon, India; IITM (Indian Institute of Tropical Meteorology), GoI, Pune and IMD (India Meteorological Department), GoI, Kolkata.

References

Ahmadi M, Haddad OB, Loáiciga HA (2015) Adaptive reservoir operation rules under climatic change. Water Resour Manage 29:1247–1266

François B, Hingray B, Creutin JD, Hendrickx F (2015) Estimating water system performance under climate change: influence of the management strategy modeling. Water Resour Manage 29:4903–4918

HEC-5 (1998) Simulation of flood control and conservation systems. User's manual, Version 8.0, US Army Corps of Engineers, Hydrologic Engineering Centre, Davis, California, USA

HEC (2000) Hydrologic modelling system HEC-HMS. Technical Reference Manual, US Army Corps of Engineers, Hydrologic Engineering Centre, CA, USA

IPCC (1990) Climate change: the IPCC scientific assessment. Report prepared for intergovernmental panel on climate change by working group I. Cambridge University Press, UK

IPCC (2007) Climate change synthesis report, contribution of working group I, II and III to the fourth assessment report (AR4) of the intergovernmental panel on climate change. Cambridge University Press, UK

Li L, Xu H, Chen X, Simonovic SP (2010) Streamflow forecast and reservoir operation performance assessment under climate change. Water Resour Manage 24:83–104

Minville M, Brissette F, Leconte R (2010) Impacts and uncertainty of climate change on water resource management of the Peribonka River system (Canada). J Water Resour Plan Manage 136:376–385

WAPCOS Limited (2010) Master plan for Damodar River valley system, vol 1

Chapter 42
Assessing the Impact of Spatial Resolution on Land Surface Model Based on Hydrologic Simulations

Aiendrila Dey and Renji Remesan

Abstract Though land surface models (LSMs) are originally developed for representing water fluxes, carbon fluxes and energy fluxes between land and atmosphere, recently LSMs are being used for hydrological simulation because it has some positive traits in comparison to conventional hydrological models. In this study, the Joint UK Land Environment Simulator (JULES), a land surface scheme of the Met Office Unified Model (UM), is implemented to study the effect of different spatial resolutions on streamflow simulation at the Krishna River basin (catchment area 2,60,000 km^2), India. The meteorological datasets used here are WFDEI (WATCH-Forcing-Data-ERA-Interim) global data at the resolution of $1^\circ \times 1^\circ$ and $2^\circ \times 2^\circ$. The simulation is run for 2001–2008 period with 3 years (2001–2003) of spin-up with 50 spin-up cycle and further simulation of years from 2004–2008. To assess the performance of the stream flow simulation, appropriate statistical parameters such as mean error (ME), root-mean-square-error (RMSE), percentage BIAS (PBIAS) are used to see error statistics. The study results indicate that spatial resolution of routing, driving and ancillary data has a significant effect on model output at the river basin scale. Though we have implemented the simulations in coarser resolutions, such studies on hydrological fluxes with a change of spatial resolution are important to know associated uncertainties with driving data and routing data resolution selection, this, in turn, would also help researchers to make meaningful management decisions to deal with future water security issues.

Keywords JULES · Input resolution · Krishna River basin · India

42.1 Introduction

Water is the most vital element for humans and all other living beings, and its availability is getting diminished drastically due to various factors, especially in developing countries. Owing to monsoon rainfall and glacial snowmelt, India is well

A. Dey (✉) · R. Remesan
School of Water Resources, IIT Kharagpur, Kharagpur 721302, India
e-mail: aiendrila.dey@iitkgp.ac.in

R. Jha et al. (eds.), *Climate Change Impacts on Water Resources*, Water Science and Technology Library 98, https://doi.org/10.1007/978-3-030-64202-0_42

nourished with freshwater, but the temporal and spatial variations of freshwater availability are highly dynamic; Existing water use pattern in the country is also getting distorted by the combined effect of climate change and other anthropogenic aspects (Mall et al. 2006). Some studies have shown strong evidences of the impact of land use and climate changes on water resources in various parts of the Indian subcontinent with compounded effects of changes in cropping pattern, land use and land cover, over exploitation of groundwater etc. which have influenced hydrological processes and hydrological behavior of river basins (Mall et al. 2006; Rahman and Rosolem 2017). Most of the Indian River basins are densely populated and intensively cultivated regions, so even a small change in climate and land use land cover can make an enormous effect on the hydrology of the basin (Dwarakish and Ganasri 2015). Thus, reliable prediction and understanding of climate change impact on hydrological fluxes and comprehensive assessment of future water availability through integrated models are a matter of greater importance in Indian scenarios.

To account for multifaceted complex hydrological issues in river basins, policy-driven outputs from hydrological and land surface models are encouraged and employed (George et al. 2009). Different hydrological models are available in the literature for determining future climate change impact on hydrological fluxes in Indian River basins. But the hydrological parameters in models are very sensitive to the spatial resolution of input data. Vema et al. (2018) observed the impact of different spatial resolution gridded precipitation data and varying sub-basin sizes on river basin discharge and sediment yield on the SWAT model. They noticed that sediment yield and peak flow of hydrograph decreased with finer resolution precipitation data. Haddeland et al. (2002) applied the VIC model to Arkansas-Red and Columbia River basin and observed that total runoff decreases with the increase in spatial resolution. Fu et al. (2011) discussed how recharge, stream discharge, runoff and groundwater head have changed with six different spatial resolution precipitation input at Alergaarde catchment in Denmark with the help of the MIKE SHE model. So, from past studies, it can be said that it is very important to choose a correct spatial resolution for simulating hydrological variables, especially in the case of large scale hydrological models and land surface models. In recent years, land surface models are getting popularized for estimating hydrological fluxes; thus, this study focusing on hydrological simulations with JULES LSM, which has the capability to simulate energy, water and carbon fluxes at different spatial and temporal scales.

In this study, we have applied the JULES land surface model to study the hydrological behaviour in the Krishna river basin at default settings. This study further explored the influence of scaling aspects on river flow prediction when the JULES model is driven with two different spatial resolutions ($1^\circ \times 1^\circ$, $2^\circ \times 2^\circ$) input data.

42.2 Study Area

Krishna is the fourth largest river basin in India, lies in the Southern Indian peninsula, located between 73°17′ to 81°9′ east longitudes and 13°10′ to 19°22′ north latitudes covers three states of Karnataka, Andhra Pradesh and Maharashtra and serves a population of approximately 70 million (George et al. 2009). The length of the Krishna River is 1300 km and the drainage area of this basin is 258,948 km^2. It originates from the Western Ghats near Jor village of Satara district of Maharashtra at an altitude of 1337 m just north of Mahabaleshwar. The main tributaries of this river are Ghataprabha River, Malaprabha River, Bhima River, Tungabhadra River and Musi River. The cultivable area of this basin is about 203,000 km, which is 75.86% of the total basin area.

42.3 Materials and Methods

42.3.1 JULES Model

JULES, a third-generation LSM represents energy, water and carbon flux exchanges between the land surface (Slevin et al. 2015) and atmosphere and also represents some other gasses such as ozone and methane. In JULES total of nine surface types, that is, five plant functional types (Needle leaf trees, Broadleaf trees, C4 (tropical) grass, shrubs), C3 (temperature) grass, and four non-vegetation types (Inland Water, Urban, Land-ice, Bare soil) are normally used. For representing each surface type, the tile approach is used, and for each tile, separate energy balance is calculated (Rahman and Rosolem 2017). JULES consider four layers of soil column in the subsurface. The thickness of the soil column is 0.1, 0.25, 0.65 and 2 m, respectively, varying from the top layer to bottom layer (Tsarouchi et al. 2014). Darcy–Richards's equation is used for the calculation of water exchange between soil layers (Tsarouchi et al. 2014). In this model, precipitation is partitioned into canopy interception and through fall (Tsarouchi et al. 2014). The potential evaporation is estimated by the Penman-Monteith equation and it is assumed that canopy evaporation will occur at the potential rate (Tsarouchi et al. 2014). So, it is very advantageous to use JULES model because it has a flexible and modular structure where one can introduce new elements of science as well as new modules into the model (Best et al. 2011). The input driving data required by JULES are incoming short wave radiation, longwave radiation, precipitation, temperature, specific humidity, wind speed and surface pressure (Tsarouchi et al. 2014). Spatial datasets, that is, soil properties (hydraulic and thermal parameter values), land cover and DEM-derived gridded flow directions and accumulated areas, are the ancillary data required by JULES.

Two schemes (TOPMODEL and PDM) are introduced in JULES to represent sub-grid scale heterogeneity in soil moisture. TOPMODEL is a more complicated scheme that depends on the topographic information to calculate saturation excess runoff and represents heterogeneity throughout the soil column (Best et al. 2011). PDM is much simpler than TOPMODEL scheme and represents heterogeneity only on the top soil layer depends upon soil water content, soil water content at saturation and shape parameter. TOPMODEL is suitable for calculating water table and wetland fractions and PDM is suitable for representing surface runoff. Two river routing model, Total Runoff Integrating Pathway (TRIP) and Rapid Flow Model (RFM) is included in JULES 2011, which follows 1-D kinematic wave equation described by Bell et al. (2007) and Dadson et al. (2011), (MacKellar et al. 2013).

42.3.2 Data Sets

In this paper, we use driving data from WATCH-Forcing-Data-ERA-Interim (WFDEI) to drive JULES LSM. WFDEI is the updated version of WFD datasets based on ERA-Interim reanalysis data available for the period of 1979–2012 (Weedon et al. 2014) at 3 hourly time steps. WFDEI datasets have two precipitation products-GPCC version 6 and 7 (Global Precipitation Climatology Centre) and CRU TS3.1/TS3.21 (Climate Research Unit at the University of East Anglia). These two precipitation data are referred as WFDEI-GPCC and WFDEI-CRU. The driving datasets available in WFDEI are- 10 m wind speed (ms^{-1}), 2 m temperature, surface pressure, 2 m specific humidity, Downward shortwave radiation flux, Downward longwave radiation flux, rainfall rate and snowfall rate (Weedon et al. 2014).

Soil properties, land cover classification data, vegetation fraction, TOPMODEL parameters are used in this model at two horizontal spatial resolution ($2° \times 2°$, $1° \times 1°$) provided by Slevin et al. (2017). In this dataset, Harmonized World Soil Database version 1.2 was used for calculating different soil parameters such as soil hydraulic properties, soil thermal properties, etc. (Slevin et al. 2017).

42.3.3 Model Experiments

In this study, JULES LSM is simulated for the period of 2001–2008 with 3-year spin-up period using different resolution meteorological datasets Two different resolution ($1°$ longitude $\times$ $1°$ latitude and $2°$ longitude $\times$ $2°$ latitude) used in this study for evaluating the impact of spatial resolution variability on hydrological fluxes. A complete input dataset (land use and soil map, model parameters, time series of the meteorological dataset) has been allotted for each grid point. It is mandatory to reach the equilibrium state for the model before each run. For initialization of the model internal state and achieving its equilibrium spin-up period plays an important role (Tsarouchi et al. 2014). In this paper, 3 years (2001–2003) of the spin-up period is used with 50 spin-up cycles. After completion of each spin-up cycle, the model ensures its equilibrium by comparing the value of soil moisture and soil temperature at the beginning of the cycle. If the value of simulated soil moisture and soil temperature is less than 1 kg/m^2 and 1%, respectively, then the model is considered to be fully spun-up and then it continues to do the main simulation (Zulkafli et al. 2013).

42.4 Results and Discussion

In this study, the Total Runoff Integrating Pathways (TRIP) river flow routing scheme has been applied to the Krishna river basin for streamflow simulation. Model simulated daily streamflow hydrograph is compared with observed data at two gauging stations (Karad and Wadenapally) of the Krishna basin. To study the effect of spatial resolution on the JULES model derived streamflow output, the observed data is compared with modelled data by means of statistical parameters like mean error (ME), root-mean-square-error (RMSE), and percentage BIAS (PBIAS). While Karad lies on the upstream side of the Krishna basin, Wadenapally is at the downstream location. Figure 42.1 shows the model simulated daily streamflow values with observed data at Karad gauging station at two different spatial resolutions.

The figure indicates that the temporal pattern of the streamflow hydrograph is well evaluated by the model in the upstream Karad gauging station at 1° resolution. Though at both spatial resolutions, the model estimates the timing of peaks and recession limb reasonably well, the peak value is underestimated at 1° resolution and the streamflow response is overestimated at 2° spatial resolution. The value of co-relation co-efficient is 0.448 and 0.432 at 1° and 2° resolution, respectively. JULES model showed a different behavior at the downstream location of the basin. The comparison between daily observed and simulated streamflow at the Wadenapally gauging station is shown in Fig. 42.2. This figure indicates that model shows very poor performance at the Wadenapally gauging station compared to the Karad gauging station at both resolutions as the model is not able to take the effect of storage structure on streamflow hydrograph. As shown in Fig. 42.2 JULES underestimate the streamflow response at 1° resolution with large negative bias in summer season and post-monsoon season and overestimate at 2° resolution with large positive bias in monsoon and winter season.

The temporal pattern is also not well captured by the model at 1° resolution and the simulated peak is formed just before the observed peak. Flow routing at large spatial scales can result in an overestimation of high flows (Arora et al. 2001); this could be the reason for larger positive bias in 2° resolution. The co-relation co-efficient is 0.296 and 0.339 at 1° and 2° spatial resolution, respectively. The root mean square error (RMSE) and mean error (ME) for both river basins at 1° resolution and 2° resolution are shown in Table 42.1. If we perform the flood routing at 5–25 km spatial resolution just like in traditional hydrological applications, JULES performance may improve considerably. Another important thing to note that this study has not performed calibration of flood routing parameters separately for the Krishna basin; instead, we are using default values.

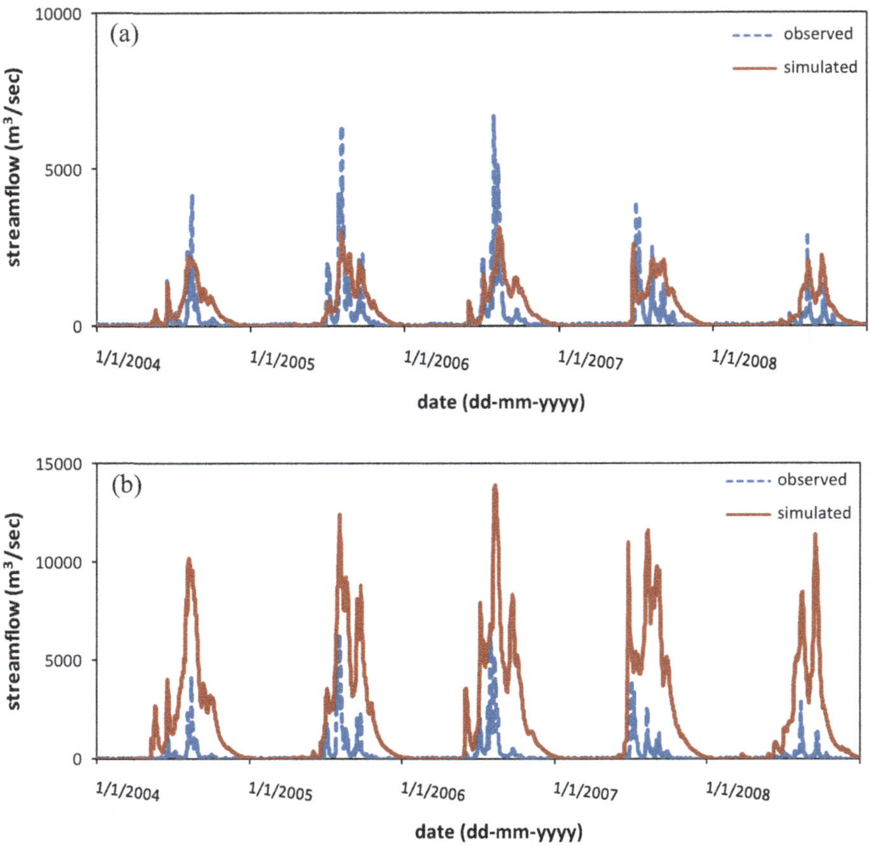

Fig. 42.1 Observed and simulated streamflow hydrograph at Karad gauging station during 2004–2008 periods **a** at 1° resolution **b** at 2° resolution

The values indicate that the model calculates the streamflow better at winter and summer months compared to monsoon and post-monsoon periods. This itself indicates that more refinement is needed on parameters that decide base flow. The RMSE and ME values are quite high for both resolutions at the Wadenapally gauging station compared to the Karad gauging station. The model settings with default parameters indicate a clear connection with the hydrological performance of JULES and model/input grid resolutions. The differences in results from upstream and downstream gauging stations indicate the scale influence of river flow routing scheme in the JULES. A detailed diagnostic approach of model evaluation needs to be done to see overall water balance predicted by JULES model focusing on other components like soil moisture, evapotranspiration and base flow. However, the model simulated values presented in this paper are very preliminary results and future improvement by parameterization will be done for better prediction of hydrological fluxes.

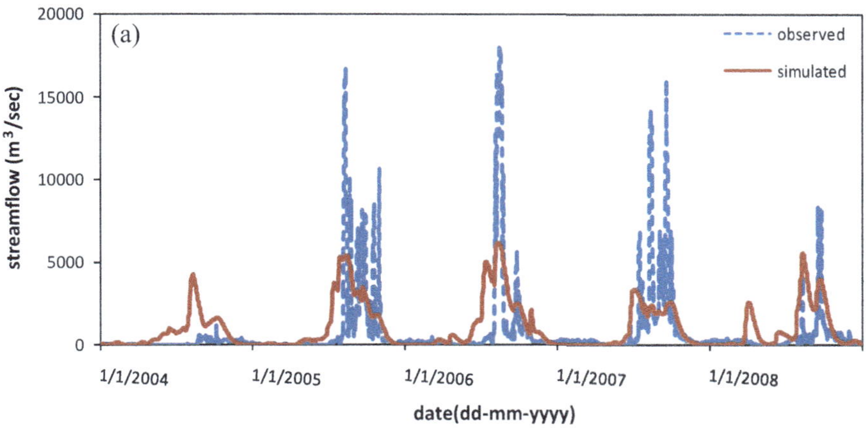

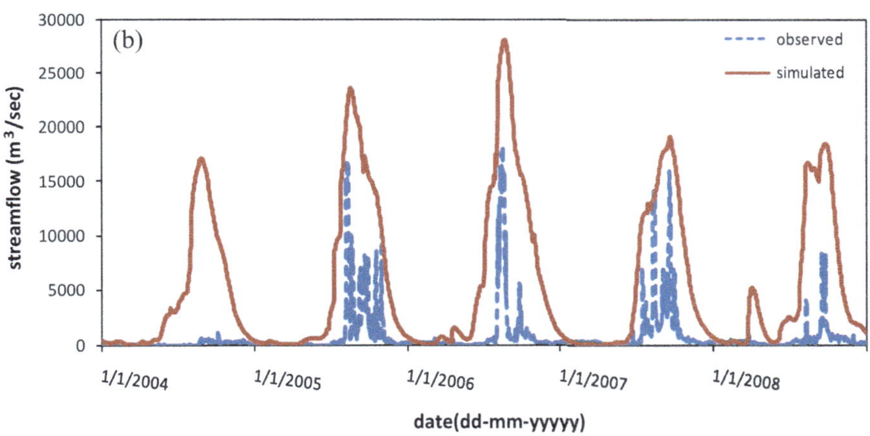

Fig. 42.2 Observed and simulated streamflow hydrograph at Wadenapally gauging station during 2004–2008 periods **a** at 1° resolution **b** at 2° resolution

42.5 Conclusion

This study has applied JULES in the Krishna River basin to investigates how effectively this model represents the hydrological fluxes and water balance at different spatial scales (1° × 1°, 2° × 2°) with default settings. Our preliminary results indicate that though, our current preliminary settings are failing to reach peak values properly, the JULES model can be considered as a promising alternative for large scale simulation in base flow dominated Indian rivers like Krishna if driven in proper spatial-temporal scales. When we have driven the model for river runoff simulation with 1° routing data, the model performance was better with 1° than 2° resolutions, which would get better if further simulated with a high resolution routing grid. It

Table 42.1 Streamflow statistics for different seasons when the JULES model simulated at 1° and 2° resolution during 2004–2008 periods

Gauging station	Resolution	Performance score	DJF	MAM	JJA	SON
Karad	1°	ME	−22.05	34.79	399.9	239.45
		RMSE	26	195.31	765.28	332
		PBIAS	−81.61	106.85	56.54	174.5
	2°	ME	3.94	294.2	4058.95	1425.24
		RMSE	32.11	916.63	4864.2	1,986
		PBIAS	14.58	903.55	573.89	1,039
Wadenapally	1°	ME	−141.09	286.86	914.95	−198.75
		RMSE	178.01	603.42	2669.06	1151.27
		PBIAS	−79.92	289.25	61.11	−13.26
	2°	ME	163.21	804.62	10471.33	7146.305
		RMSE	335.65	1454.02	11828.8	8457.49
		PBIAS	92.45	811.32	699.39	476.86

DJF = December to February, MAM = March to May, JJA = June to August, SON = September to November

shows that the resolution of river routing data is crucial in better hydrology representation in JULES and the modeler needs to ensure that the model has an appropriate river routing geometry representation before actual simulation. In short, our study indicates that hydrologic predictions may become unreliable with large uncertainties at river basins if the model is not properly selecting river routing, driving and ancillary data resolution. It directs to the need for further research on JULES to identify the correct resolution of the model grid and routing grid for the Krishna basin to represent the hydrological functions effectively.

References

Arora V, Seglenieks F, Kouwen N (2001) Soulis E Scaling aspects of river flow routing. Hydrol Process 15(3):461–477. https://doi.org/10.1002/hyp.161

Bell VA, Kay AL, Jones RG, Moore RJ (2007) Development of a high resolution grid-based river flow model for use with regional climate model output. Hydrol Earth Syst Sci 11(1):532–549. https://doi.org/10.5194/hess-11-532-2007

Best MJ, Pryor M, Clark DB, Rooney GG, Essery RLH, Ménard CB, Edwards JM, Hendry MA, Porson A, Gedney N, Mercado LM, Sitch S, Blyth E, Boucher O, Cox PM, Grimmond CSB, Harding RJ (2011) The Joint UK Land Environment Simulator (JULES), model description—part 1: energy and water fluxes. Geosci Model Dev 4(3):677–699. https://doi.org/10.5194/gmd-4-677-2011

Dadson SJ, Bell VA, Jones RG (2011) Evaluation of a grid-based river flow model configured for use in a regional climate model. J Hydrol 411(3–4):238–250. https://doi.org/10.1016/j.jhydrol.2011.10.002

Dwarakish GS, Ganasri BP (2015) Impact of land use change on hydrological systems: a review of current modeling approaches. Cogent Geosci 1(1):1–18. https://doi.org/10.1080/23312041.2015.1115691

Fu S, Sonnenborg TO, Jensen KH, He X (2011) Impact of Precipitation Spatial Resolution on the Hydrological Response of an Integrated Distributed Water Resources Model. Vadose Zo. J. 10(1):25. https://doi.org/10.2136/vzj2009.0186

George BA, Nawarathna B, Malano HM, Parthasaradhi G (2009) Assessing water security across the Krishna River Basin, Most, (July), 2009

Haddeland I, Matheussen BV, Lettenmaier DP (2002) Influence of spatial resolution on simulated streamflow in a macroscale hydrologic model. Water Resour Res. 38(7): 29-1-29–10. https://doi.org/10.1029/2001wr000854

MacKellar NC, Dadson SJ, New M, Wolski P (2013) Evaluation of the JULES land surface model in simulating catchment hydrology in Southern Africa. Hydrol Earth Syst Sci Discuss 10(8):11093–11128. https://doi.org/10.5194/hessd-10-11093-2013

Mall RK, Singh R, Gupta A, Srinivasan G, Rathore LS (2006) Impact of climate change on Indian agriculture: a review. Clim Change 78(2–4):445–478. https://doi.org/10.1007/s10584-005-9042-x

Rahman M, Rosolem R (2017) Towards a simple representation of chalk hydrology in land surface modelling. Hydrol Earth Syst Sci 21(1):459–471. https://doi.org/10.5194/hess-21-459-2017

Slevin D, Tett SFB, Williams M (2015) Multi-site evaluation of the JULES land surface model using global and local data. Geosci Model Dev 8(2):295–316. https://doi.org/10.5194/gmd-8-295-2015

Slevin D, Tett SF, Exbrayat B, Bloom J-F, Williams M (2017) Global evaluation of gross primary productivity in the JULES land surface model v3.4. Gosci Model Dev 10(7), 2651–2670. https://doi.org/10.5194/gmd-10-2651-2017

Tsarouchi GM, Buytaert W, Mijic A (2014) Coupling a land-surface model with a crop growth model to improve et flux estimations in the Upper Ganges basin, India. Hydrol Earth Syst Sci 18(10):4223–4238. https://doi.org/10.5194/hess-18-4223-2014

Vema VK, Thomas J, Athira P, Kurian C, Sudheer KP (2018) Uncertainty in the SWAT Model Simulations due to Different Spatial Resolution of Gridded Precipitation Data, January 2018

Weedon GP, Balsamo G, Bellouin N, Gomes S, Best MJ, Viterbo P (2014) The WFDEI meteorological forcing data set: WATCH Forcing Data methodology applied to ERA-Interim reanalysis data. Water Resour Res 50(9):7505–7514. https://doi.org/10.1002/2014WR015638

Zulkafli Z, Buytaert W, Onof C, Lavado W, Guyot JL (2013) A critical assessment of the JULES land surface model hydrology for humid tropical environments. Hydrol Earth Syst Sci 17(3):1113–1132. https://doi.org/10.5194/hess-17-1113-2013

Chapter 43
Infilling Missing Monthly Maximum and Minimum Temperature Dataset by EM Algorithm Followed by Distribution Based Statistical Assessment Using Eight Absolute Homogeneity Tests

P. Kabbilawsh, D. Sathish Kumar, and N. R. Chithra

Abstract In the current research, missing value analysis and tests for the presence of homogeneity were applied to the temperature records obtained from seven meteorological observatories spread throughout the state of Kerala. The monthly mean maximum and mean minimum temperature datasets of observatories managed by the Indian Meteorological Department (IMD) for the period 1969–2015 were considered. During analysis, every observatory was studied independently and those observatories which are having missing values for five continuous years, and more were rejected. The missing records were estimated using the Expectation Maximization Algorithm (EMA). The infilled dataset needs to be hydrologically as well as statistically stable for later hydrological and meteorological assessments. The reliability of datasets was tested using eight statistical absolute homogeneity tests. Before applying the homogeneity tests on the datasets, their assumptions must be fulfilled; one predominant assumption is regarding the normal distribution of the dataset. Thereby, the datasets are checked for normality behaviour by four statistical tests, namely Skewness z-ratio, Kurtosis z-ratio, Kolmogorov–Smirnov (KS) and Shapiro–Wilk (SW) test. Out of 14 datasets consisting of seven mean monthly maximum and seven mean monthly minimum temperature, ten datasets were found to be normal at 95% level of confidence (LOC) and the rest four were highly skewed and kurtotic. Following the normality tests, eight statistical absolute homogeneity tests were applied individually on the annual scale by which non-homogeneous stations were detected and eliminated. The datasets which resembled normal behaviour were tested using parametric homogeneity tests such as Linear Regression, Student's t-test, Cumulative Deviation and Worsley Likelihood Ratio test. Out of normally distributed datasets, only three datasets were statistically homogenous at 99% LOC, and seven datasets failed to clear the homogeneity test. Remaining four non-normally distributed datasets was checked for homogeneity by applying non-parametric tests,

P. Kabbilawsh (✉) · D. Sathish Kumar · N. R. Chithra
Department of Civil Engineering, National Institute of Technology, Calicut 673601, India
e-mail: kabbilawsh_p160006ce@nitc.ac.in

R. Jha et al. (eds.), *Climate Change Impacts on Water Resources*, Water Science and Technology Library 98, https://doi.org/10.1007/978-3-030-64202-0_43

such as Distribution-Free CUSUM, Rank-Sum, Median Crossing and Turning Points test. Out of four datasets, three were homogenous at 99% LOC and one failed homogeneity test.

Keywords Expectation Maximization Algorithm (EMA) · Test for normality · Skewness and kurtosis z-ratio · Parametric and non-parametric homogeneity test · Kolmogorov–Smirnov (KS) and Shapiro–Wilk (SW) test

43.1 Introduction

A large number of studies have been carried out in the last decade on the global, continental and regional level for long-term temperature variations (Hänsel et al. 2016); following it several attempts have been made to develop a model for understanding and extrapolating the temperature variations (Mohanty et al. 1997). The four deciding factors in the development of a precise model are the length of data, amount of missing values in the observatories used for regional level analysis, adequacy of the number of stations occupying the study area and the quality of the dataset. The quality of temperature datasets decides the accuracy and reliability of model outcomes in studies related to climate change, drought modelling, flood modelling and modelling associated with hydrologic and meteorological studies. Long-term datasets are termed as 'homogenous' when the values present in the dataset undergo changes in their respective indigenous pattern by variations caused only due to climate variability or climate change or climate resilience. Inhomogeneities are non-climatic induced changes or alterations in the indigenous model of long-term time series. Inhomogeneities may be caused due to change in traditional methodology followed for measuring the parameter (rainfall, temperature, humidity, the percentage of particulate matter), sudden changes in conditions in the area surrounding the station or a decrease in the infallibility of the measuring device.

In the current study, the Expectation-Maximization algorithm (EMA) is applied as a point estimator for filling data gaps. Maximum Likelihood Estimation (MLE) is a statistical method for finding the parameters of any particular probability model. EMA is an iterative MLE method that can infill unobserved/missing or incomplete data. The literature indicates that EMA is generally applied for two specific cases. The first case is when long-term time-series datasets have gaps because of shortfalls or errors while observations are recorded. The second case is when likelihood estimates of the parameter of the function are determined by taking into consideration that function has an inherent latent behaviour.

Techniques used for infilling data gaps can be classified into two approaches. First one is the distribution-based approach (DIBA), and latter one is the data-based approach (DABA). The statistical algorithm is rewritten to enhance the estimate of missing values and estimates of particular distribution, which are all determined on one stage in DIBA, Whereas in DABA, filling the data gaps is the first step, on completion of which it proceeds to the second stage (which is the estimation of the distribution parameter). EMA behaves both as DIBA and DABA depending upon the application. If EMA is applied to obtain parameter estimation in relation to a particular incident, then EMA can be termed as DIBA. EMA is used to obtain standard results such as variance, covariance matrix and vector of means, which is then used as input data for analysis in proceeding the step; this makes EMA more of a DABA.

Application of EMA as a data filling technique is not much explored in hydrological and meteorological studies. A few examples of EMA in the area of hydrology is listed here. A comparative analysis of artificial neural schemes and EMA was performed (Nelwamondo et al. 2007) and proved that EMA is better suited when there is little or no interdependency between the input variables. Simulated surface temperature, a typical climate data set was used to demonstrate the applicability of EMA over traditional non-iterative methods and proved it leads to better estimates (Schneider 2001). EMA was used to fill missing monthly temperature values of stations using adjacent stations as reference stations, and subsequently, the performance of infilled data was analysed by several statistical criteria such as correlation coefficient, efficiency and normalised root mean square error (Firat et al. 2010). EMA was used to develop a new daily rainfall model that used concurrent values of the surrounding station to increase the accuracy of estimation, which was completely different from the conventional daily rainfall generation scheme (Ahn 2009). EMA along with singular value decomposition performed well for patching monthly rainfall, which was concluded based on their performance test (Makhuvha et al. 1997). From all these recommendations, EMA was applied in the current study for effectively handling data gaps.

43.2 Expectation Maximization Algorithm (EMA)

Inverse dependence among the model's parameter and the missing data values forms the basis of formulation in EMA. The inherent assumption of EMA is missing values are missing completely at random (MCAR), which implies the probability of unobserved/missing data is independent of the missing values itself (Rubin 1976). The EMA is comprised of two stages: The Expectation stage (E-stage) and the Maximization stage (M-stage). Considering the available observed/present data and along with distribution's parameter estimates, E-stage finds the predicted expectation of the missing value. The M-stage determines the model's parameter estimates in order to maximise the value of the full data log-likelihood function from the E-stage. The two stages are continuously applied again and again consecutively until the iteration converges. For each cycle, EMA updates the measure θ_n, of the latent parameter θ for n number of iterations.

43.3 Check for Assumption of Normality

A normal distribution is a type of distribution that is always symmetrical about the mean of the population and has the shape of a bell-curved in nature. Many parametric statistical tests require the dependent variable to be at least approximately normally distributed to some extent. Normality tests are those which are applied to the given sample data to check whether the given sample data belongs to a normally distributed population and also measure the degree of normality behaviour. The following are the basic requirements for a dataset to be termed as normally distributed.

- The dataset should be continuous, and the measure of the dataset should be scale type and not ordinal or nominal in nature.
- The Skewness and Kurtosis value for a particular dataset should be of a smaller magnitude and preferably closer to zero. A minor deviation from zero is acceptable when the overall values are not large enough in comparison to their respective standard error.
- The Skewness and Kurtosis are divided by standard errors, and they are termed as Skewness and Kurtosis *z*-scores, respectively (Cramer and Duncan 1998). The *z*-score should not be significantly different from the required LOC (such as 95% or 99%).
- The null hypothesis of the Shapiro–Wilk (SW) and Kolmogorov–Smirnov (KS) normality tests is that data are normally distributed (Razali and Wah 2011). In order to reject the null hypothesis *p*-value should be greater than the α-value. If so, then we fail to reject the null hypothesis and the data are approximately normal.

43.4 Homogeneity Test

The homogeneity tests are broadly classified as 'Absolute techniques' which makes use of series data itself and 'Relative techniques' which makes use of information obtained from surrounding observatory, commonly known as reference observatories. Literature survey reveals that there are many relative and absolute techniques. However, in comparison to relative techniques, absolute techniques have been frequently used due to the non-availability of an observatory in the surrounding area and mathematical complexity invovled in applying relative methods (Wijngaard et al. 2003). Absolute techniques are preferred over relative techniques as they are developed to reduce the effect of autocorrelation in meteorological datasets (precipitation and temperature). The absolute techniques are further classified as the distribution based and non-distribution-based techniques, depending on their normality behaviour. Eight well-known homogeneity tests have been selected to test the datasets. Out of eight tests four were applied to those datasets that cleared the normality test and rest four were applied to those datasets that failed to clear the normality test.

Distribution-based absolute homogeneity tests applied are Cumulative Deviation test, Worsley's likelihood ratio test, Student's t-test and Linear Regression test. Non-distribution based absolute homogeneity tests used are Distribution Free CUSUM Test, Rank sum test, Turning points test and Median crossing test.

43.4.1 Distribution-Based Absolute Homogeneity Test

The following test needs the dataset to be normally distributed and it's termed as parametric tests.

43.4.1.1 Cumulative Deviation Test

This test checks whether the average of two parts of a dataset is significantly different (for an unknown time of change). This test's main function is to find a change in the average value of a time series dataset after q observations.

$$E(x_i) = \mu \; \mathrm{i} = 1, 2, 3 \ldots \mathrm{q} \tag{1}$$

$$E(x_i) = \mu + \Delta \; \mathrm{i} = \mathrm{q} + 1, \mathrm{q} + 2, \ldots \mathrm{n}$$

where μ is the average of the dataset belonging to the series before the change point and Δ is the change in the average value. The cumulative deviations from the average values are calculated as:

$$S_0^* = 0 \; S_k^* = \sum_{i=1}^{k} (x_i - \bar{x}) \; k = 1, 2, 3 \ldots n \tag{2}$$

and the rescaled adjusted partial sums are obtained by dividing the S_k^* values by the standard deviation:

$$S_k^{**} = \left\{ \frac{S_k^*}{D_k} \right\}$$

$$D_x^2 = \sum_{i=1}^{n} \frac{(x_i - \bar{x})^2}{n} \tag{3}$$

The test statistic Q is: $Q = \max|S_k^{**}|$ and is determined for every single year, with the largest score expressing the most probable change point. A negative score of S_k^* expresses that the second part of the dataset has a larger average value than the earlier part and vice versa.

43.4.1.2 Worsley Likelihood Ratio Test

This test checks if the average of values in two sections of a dataset is different (for an unknown time of change). The function of this test is to find a change point in series, the average value after n observations and to calculate the value of $S_{k.}^{*}$ It also weighs the value of S_k^* by depending on their relative position in the time series. If the outcome of the series is negative, it means the second part of the series has a higher average value in comparison to its earlier part and vice versa.

$$Z_k^* = \frac{S_k^*}{[K(n-k)]^{0.5}}$$
$$Z_k^{**} = \left\{\frac{Z_k^*}{D_x}\right\} \tag{4}$$

The test statistic is given by $W = \frac{(n-2)^{0.5}V}{(1-V^2)^{0.5}}$ where $V = \max|Z_k^{**}|$

Several comprehensive materials on the Student's t-test, Distribution-free CUSUM test and Linear regression test are available which address the definition, derivation, graphical plots and criterion with clarity (Shao and Zhang 2010).

43.4.2 *Non-distribution Based Absolute Homogeneity Test*

The following test does not need the dataset to be normally distributed and it is termed as non-parametric tests.

43.4.2.1 Rank-Sum Test

The method breaks the series into two parts and checks whether the median in two different periods is different. The data is ranked from smallest to largest. In the case of ties, average values are used. The number of data values in the smaller group is given as n and while the larger group is given by m. A test statistic s is defined as the sum of ranks of observation in the smaller group, the theoretical mean and standard deviation of s under H_o is given by

$$\mu = \left\{\frac{n(N+1)}{2}\right\}$$
$$\sigma = \left[\frac{nm(N+1)}{12}\right]^{0.5} \tag{5}$$

The standardised form of the test statistic Z_{rs} is computed as:

$$Z_{rs} = \left\{\frac{s - 0.5 - \mu}{\sigma}\right\} \text{ if } s > 0$$
$$Z_{rs} = 0 \text{ if } s = 0$$
$$Z_{rs} = \left\{\frac{s + 0.5 - \mu}{\sigma}\right\} \text{ if } s < 0 \quad (6)$$

43.4.2.2 Median Crossing Test

The n time series values are replaced by 0 if $x_i < x_{median}$ and by 1 if $x_i > x_{median}$. If the time series data come from a random process, then m (the number of times 0 is followed by 1 or 1 is followed by 0) is approximately normally distributed with:

$$\mu = (n - 1)/2$$
$$\sigma = (n - 1)/4 \quad (7)$$

The z-statistic is, therefore, given by (critical test statistic values for various significance levels can be obtained from the normal probability tables):

$$Z = |m - \mu|\sigma^{0.5} \quad (8)$$

43.4.2.3 Turning Points Test

The dataset consisting of n values are assigned a score of one if $x_{i-1} < x_i > x_{i+1}$ or if $x_{i-1} > x_i < x_{i+1}$., else they are given a value as 0. The number of times 1 appears (m) is approximately normally distributed with:

$$\mu = 2(n - 1)/3$$
$$\sigma = 16(n - 29)/3 \quad (9)$$

The z-statistic is, therefore (critical test statistic values for various significance levels can be obtained from normal probability tables):

$$Z = |m - \mu|\sigma^{0.5} \quad (10)$$

43.5 Study Area and Available Data

The state of Kerala is located approximately between 8°15'N and 12°50'N latitudes and between 74°50'E and 77°30'E longitudes. As the monsoon enters the Indian sub-continent through Kerala, the state is widely recognized as "Gateway of summer monsoon" over India. The mean maximum monthly temperature (MMAX), highest maximum monthly temperature (HMAX), mean maximum monthly temperature (MMIN), lowest minimum monthly temperature (LMIN), date of occurrence of HMAX and LMIN, number of occasions of temperature value more than MMAX and lower than MMIN, respectively, for seven stations for the period 1969–2015 covering the entire state of Kerala was provided by Indian Meteorological Department (IMD). The spatial locations of the observation sites are shown in Fig. 43.1.

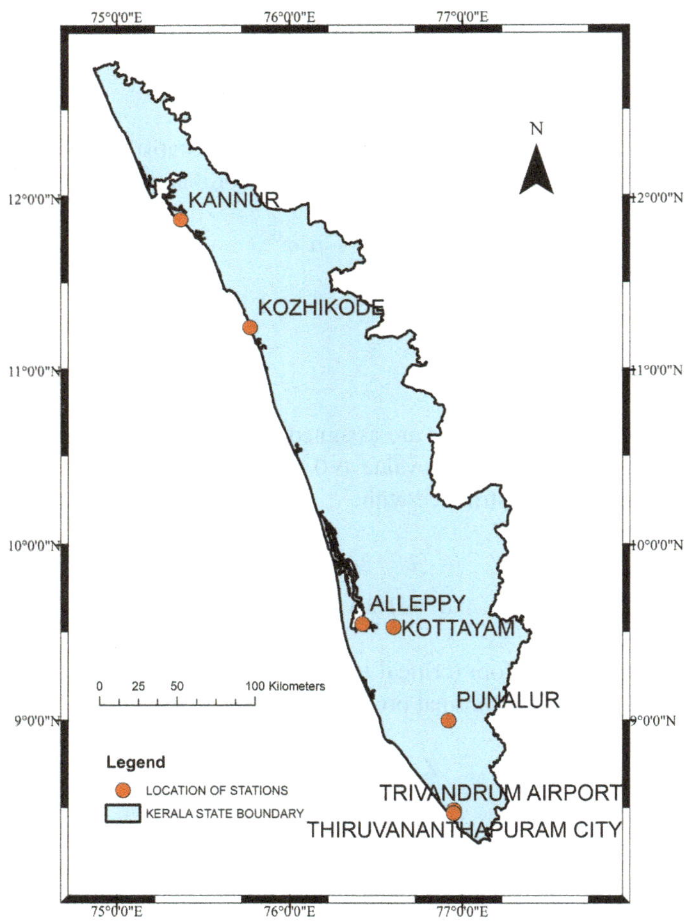

Fig. 43.1 Location of seven meteorological observatories

The time gap (1969–2015) was decided by assessing the records of the observatories for using as many gauging sites as possible. The mean maximum monthly temperature (MMAX) and lowest minimum monthly temperature (LMIN) records are calculated by averaging the daily maximum and minimum temperature data at the stations. Following the same procedure, annual maximum and minimum temperature are averaged values of MMAX and LMIN. The number of missing values in MMAX, HMAX, MMIN, LMIN are 98, 98, 132 and 121, respectively. The total number of monthly values available was 14575. The total number of values estimated by EMA was 449 monthly values, which accounted for 2.98% of overall data (Table 43.1).

43.6 Result and Discussion

The annual and seasonal mean values are used for climate change and climate variability analysis. Therefore all four test statistics namely—Skewness z-ratio, Kurtosis z-ratio, Kolmogorov-Smirnov(KS) and Shapiro-Wilk(SW) test statistic must be satisfied for a series to be fit and termed as normally distributed. The methodology is depicted in the flow chart shown in Fig. 43.2.

43.6.1 Normality Test Result

The test procedure for checking the normality is explained using the mean minimum monthly temperature (MMIN) belonging to Trivandrum airport station. From Table 43.2, it is possible to conclude that the Kurtosis value of south-west monsoon (−0.938), pre-monsoon (0.312) and winter season (0.960) and its corresponding Kurtosis z-ratio also lie within permissible limits (−1.96, 1.96). Whereas, for the post-monsoon season (−2.078), the value is less than −1.96 (95% LOC). The other three statistics, obtained from, skewness z-ratio (0.118), Kolmogorov–Smirnov (0.153), Shapiro–Wilk (0.135) tests indicate that the series is normally distributed. In the current study, only annual mean maximum and maximum temperature are used. Therefore, the test statistics of annual minimum and maximum temperatures are only needed to be taken into consideration for deciding the behaviour of distribution. All the four statistics are within limits for MMIN variable belonging to Trivandrum airport station, and thereby the time-series can be considered to be normally distributed. Therefore, parametric homogeneity test can be applied to it. A similar analysis was carried out for all seven stations, and the results are presented in Table 43.3. For the Trivandrum airport station (MMAX), the Skewness *z*-ratio (0.032) and Kurtosis z-ratio (−0.405) is close to zero. The *p*-values of KS (0.2) and SW (0.346) tests are greater than the α-value (0.05) by a big margin. Only post-monsoon Kurtosis *z*-ratio (−2.922) is lesser than 95% (1.96) and 99% (2.34) LOC. The distribution is said to be a normal distribution. For Thiruvananthapuram station, MMIN dataset, the kurtosis *z*-ratio (−1.924) passed the test by a small margin. Other

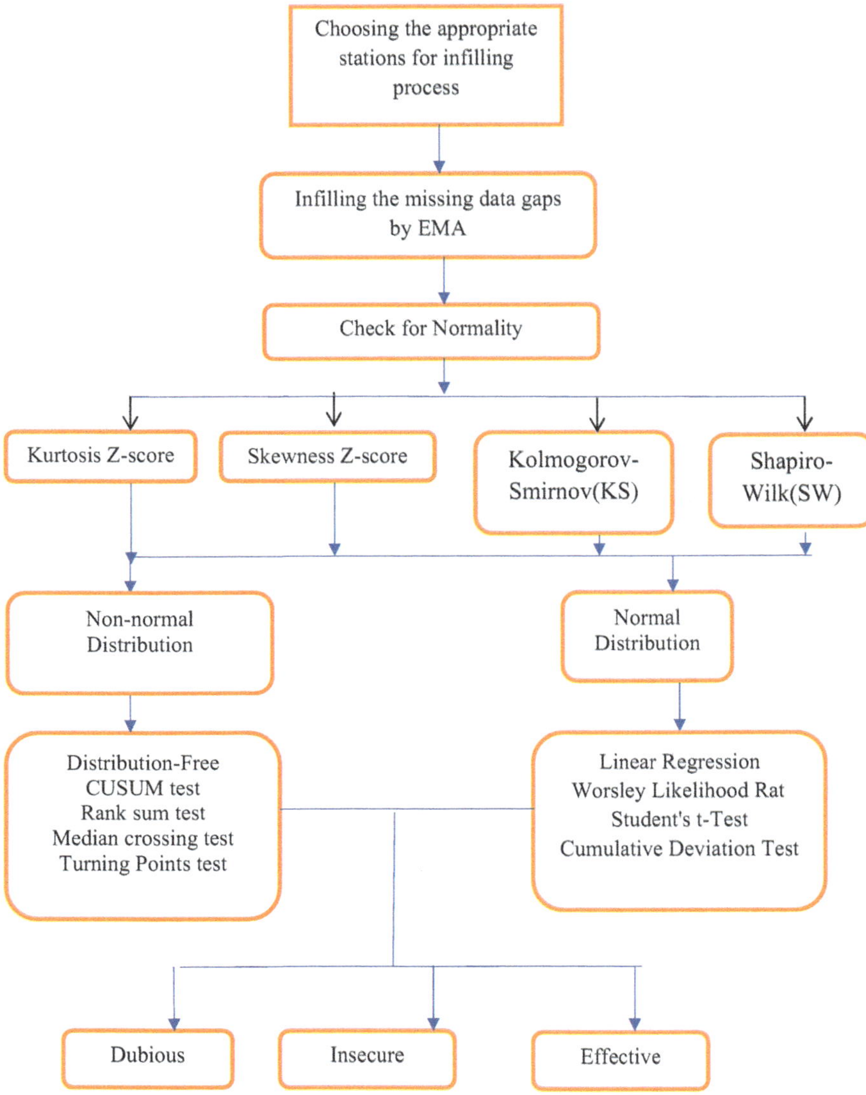

Fig. 43.2 The methodology followed in the current study

test statistics obtained from KS (0.2), SW (0.611) and skewness z-ratio (−0.046) are within limits. For the post-monsoon datatset, Skewness z-ratio (−2.248) is less than 1.96 (95% LOC). Although the distribution is slightly skewed, it is generally considered as the normal distribution (Table 43.3).

The four test statistics-Skewness z-ratio, Kurtosis z-ratio, Kolmogorov–Smirnov, Shapiro–Wilk obtained for the MMAX dataset, for annual as well as for all four seasons are within acceptable limits and thereby, it is considered as normally

Table 43.1 Amount of missing data present in various meteorological observatories

	Station name	Start year	End year	N*	Number of values							
					MMAX		HMAX		MMIN		LMIN	
					P*	M*	P*	M*	P*	M*	P*	M*
1	Kozhikode	1969	2015	47	564	0	564	0	564	0	564	0
2	Cannanore	1981	2015	34	404	16	404	16	404	16	404	16
3	Alleppy	1969	2015	47	549	15	548	16	548	16	549	15
4	Punalur	1969	2015	47	532	32	534	30	503	61	510	54
5	Kottayam	1973	2011	43	500	16	498	18	496	20	498	18
6	Thiruvananthapuram	1969	2015	47	564	0	564	0	564	0	564	0
7	Trivandrum airport	1969	2015	47	545	19	546	18	545	19	546	18
Total number of missing values						98		98		132		121

N*, Total number of years
P*, Number of monthly temperature values present
M*, Number of monthly temperature values missing

distributed. It is completely fit to be tested by all parametric tests. For Kottayam MMIN dataset Skewness z-ratio (−3.482), Kurtosis z-ratio (2.268) does lie within 95% LOC (−1.96 to 1.96). The p-value of the KS test (0.014) and SW test (0.001) is less than the α-value (0.05). The distribution is highly skewed as well as kurtotic to be termed as a normal distribution. Similar tests were carried out for all other stations, and the results are tabulated (Table 43.3). Out of fourteen datasets, ten datasets were found to be normally distributed and the rest four were found to be non-normally distributed.

43.6.2 Classification of the Results Based on Homogeneity Tests

In all eight absolute homogeneity tests, the null hypothesis assumes the dataset to be homogenous in nature, and the alternative hypothesis is claimed to be non-homogeneous in nature. We aim to fail the test statistic so that we can fail to reject the null hypothesis for the series to homogenous. After applying eight homogeneity tests on seven selected stations for the temperature dataset, the results were classified into three categories. The subjective explanation of the three categories is as follows, and it is benchmarked with reference to the inter-annual standard deviation of testing variable series.

Category 1: 'Effective'—three and more out of four tests fail to reject the null hypothesis. The amount of inhomogeneities and its effect on series is negligible and adequately less in comparison to the inter-annual standard deviation of testing

Table 43.2 Results of the normality test for monthly minimum temperature belonging to Trivandrum airport station

	Skewness value	Skewness z-ratio	Kurtosis value	Kurtosis z-ratio	Kolmogorov–Smirnov	Shapiro–Wilk
January	0.231	0.666	1.603	2.354	0.200	0.093
February	0.155	0.447	0.465	0.683	0.027	0.220
March	−0.799	−2.303	1.991	2.924	0.010	0.031
April	0.379	1.092	0.999	1.467	0.200	0.207
May	−0.211	−0.608	0.057	0.084	0.040	0.249
June	−0.401	−1.156	−0.619	−0.909	0.200	0.125
July	0.150	0.432	−0.011	−0.016	0.200	0.906
August	0.267	0.769	0.256	0.376	0.200	0.279
September	0.018	0.052	0.669	0.982	0.153	0.324
October	−0.278	−0.801	0.030	0.044	0.200	0.680
November	−0.765	−2.205	2.250	3.304	0.200	0.058
December	−0.833	−2.401	0.828	1.216	0.128	0.075
Annual	−0.563	−1.622	1.136	1.668	0.200	0.145
Post_monsoon	0.118	0.340	1.415	2.078	0.153	0.135
Southwest_monsoon	−0.350	−1.009	−0.938	−1.377	0.184	0.049
Pre_monsoon	0.108	0.311	0.312	0.458	0.200	0.993
Winter	−0.728	−2.098	0.960	1.410	0.200	0.160

Table 43.3 Results of normality test for the annual minimum and maximum temperature of all the seven stations

Station name	Type of variable	Skewness z-ratio	Kurtosis z-ratio	KS	SW	Type of distribution
Trivandrum airport	MMAX	0.032	−0.405	0.200	0.346	Normal
Thiruvananthapuram	MMIN	−0.046	−1.924	0.200	0.611	Normal
Thiruvananthapuram	MMAX	−0.925	−1.079	0.157	0.118	Normal
Kottayam	MMIN	−3.482	2.268	0.014	0.001	Non-Normal
Kottayam	MMAX	−3.850	4.106	0.003	0.001	Non-Normal
Punalur	MMIN	−1.628	1.611	0.200	0.179	Normal
Punalur	MMAX	−1.628	1.611	0.200	0.179	Normal
Alleppy	MMIN	−4.291	4.452	0.001	0.000	Non-Normal
Alleppy	MMAX	1.706	0.937	0.153	0.113	Normal
Cannanore	MMIN	0.925	0.508	0.200	0.340	Normal
Cannanore	MMAX	0.925	0.508	0.200	0.340	Normal
Kozhikode	MMIN	−0.813	−0.232	0.200	0.675	Normal
Kozhikode	MMAX	2.859	1.360	0.019	0.008	Non-Normal

variable series that they will largely escape detection. We can conclude the series is statistically significant for climate change analysis.

Category 2: 'Insecure'—only two tests out of four tests reject the null hypothesis and the rest two fail to reject the null hypothesis. The magnitude of inhomogeneity is more than the acceptable level given by the inter-annual standard deviation of the testing variable series. Further trend or variability analysis should be done after applying some correction factors, and results should be evaluated more critically from the aspect of the presence of possible inhomogeneities.

Category 3: 'Dubious'—none of the tests fails to reject the null hypothesis. The magnitude of inhomogeneity present outstrips the level given by the inter-annual standard deviation of the testing variable series. Further trend or forecasting analysis carried on would give borderline outcomes and should be considered to be counterfeit in nature. Thereby, it would be better to ignore these datasets for further statistical analysis.

43.6.3 Homogeneity Test for Non-Normally Distributed Datasets

The distribution-free CUSUM test for the Kozhikode station had a break in the year 1996, and the value of deviation was 20, which is more than an acceptable limit of 11.175. The dataset rejects the null hypothesis and the data shows a statistically significant step at 99% LOC. For the rank-sum test, the year 1992 was chosen as

Table 43.4 Results of non-parametric homogeneity tests for non-normally distributed dataset

Station name	Type of data	Distribution-free CUSUM test		Rank sum test (year of break = 1992	Median crossing test	Turning points test	Result
		Y*	D*				
Kozhikode	MMAX	1996	20	−5.097	5.013	−1.411	Dubious
Alleppy	MMAX	1981	13	−1.224	2.654	−2.117	Effective
Kottayam	MMIN	2003	11	−1.173	2.469	−2.341	Effective
Kottayam	MMAX	1994	8	1.883	2.160	−2.341	Effective

Y*—Year of break and D*—Deviation

the break year since it is in the middle of the gap period. The value of rank-sum z-statistic for the Kozhikode station is −5.097, which is less than −2.567, which is unacceptable for α-value of 99% LOC. It means the median of 1969–1991 and 1992–2015 is significantly different. The *z*-score of median crossing test for the Kozhikode station is 5.013, which is more than 2.576 at 99% LOC, which indicates that the datasets are not obtained by a random process. Turning test z-score is −1.411, which lies within the limit. Since the time-series of Kozhikode failed to clear at least two tests, it is not considered as homogeneous. In the current classification, it would be called dubious in nature (Table 43.4). For the rest three stations, at least three test statistics were passed by the station. For example, Alleppy had a CUSUM z-score of 13 and failed to fall within the limit of 11.175 by a small margin. However, the other three test statistics were within limits and, thereby, the station was termed as homogenous and effective as per classification followed here (Table 43.4).

43.6.4 Homogeneity Tests for Normally Distributed Datasets

The linear regression test statistic value for the Kozhikode MMAX variable is 8.912, which is more than the acceptable value of 2.692 (99% LOC). It means data shows statistically significant increasing change at α-value less than 0.01, and it rejects the null hypothesis (Table 43.5). The student t-test gave a value of 3.56, which is more than 2.69 (99% LOC). The inference is that the mean value of the dataset in the period 1969–1991, and 1992–2015 is statistically significant at α-value less than 0.01, and thereby it rejects the null hypothesis.

For cumulative deviation test statistic value is 2.526, which is more than the acceptable value of 1.514 (99% LOC). It means data shows a statistically significant step jump at the year 1994 at α-value less than 0.01, and thereby it rejects the null hypothesis. A positive value of the result means that the average value of the time period 1969–1994 is smaller than that of 1994–2015. Similarly, for the Worsley Likelihood Ratio test statistic value was 7.040, which is more than an acceptable limit of 3.29 (99% LOC). Overall, Kozhikode MMAX datasets reject the null hypothesis in all

Table 43.5 Results of parametric homogeneity tests for normally distributed dataset

Station name	Type of data	Linear regression	Worsley likelihood ratio		Student's t-test	Cumulative deviation test		Result
			Y*	V*		Y*	V*	
Kozhikode	MMIN	8.912	1994	7.0406	3.560	1994	2.526	Dubious
Cannanore	MMAX	7.544	1985	5.934	-2.828	1994	2.074	Dubious
Cannanore	MMIN	2.582	2012	4.998	1.411	2011	1.194	Effective
Alleppy	MMIN	-2.234	2000	4.36	1.395	1999	1.752	Insecure
Punalur	MMAX	0.774	1985	1.412	0.056	1986	0.681	Effective
Punalur	MMIN	0.803	1984	4.44	0.372	1976	1.269	Effective
Thiruvananthapuram	MMAX	12.168	1994	8.318	-2.63	1993	2.656	Dubious
Thiruvananthapuram	MMIN	4.205	2008	4.94	0.067	2004	1.627	Dubious
Trivandrum airport	MMAX	5.09	1978	6.376	0.059	1986	1.961	Dubious
Trivandrum airport	MMIN	6.261	1993	5.915	0.043	1993	2.263	Dubious

Y*—Year of break and V*—Values

four statistics, and therefore the series is termed as non-homogenous falling under the dubious category. The datasets of Cannanore (MMAX), Thiruvananthapuram (MMAX and MMIN) and Trivandrum airport (MMAX and MMIN) rejected the null hypothesis for four test statistics. The datasets are non-homogeneous and fall under the dubious category as per classification followed. In the case of Alleppy (MMIN) dataset, both linear regression and the Student's t-test-statistic are within the acceptable limit and fail to reject the null hypothesis. However, the dataset rejected the null hypothesis for cumulative deviation and Worsley likelihood ratio tests. Overall, out of four statistical tests, the dataset failed to clear two tests. Therefore, the series is categorized as non-homogenous and falling under insure category.

Cannanore (MMIN) and Punalur (MMAX and MMIN) have at least three test statistic results within acceptable limits. Thereby, the datasets are considered as homogenous and fall under the category 'effective'. Out of ten datasets, six datasets fall under the category 'dubious', one falls under the category 'secure', and the remaining three datasets are under the category 'effective'.

43.7 Conclusion

Investigating meteorological and hydrological variables in order to understand and find inherent latent behaviour helps to predict, forecast, and model the climatic system and its variation. To obtain decisive and trustworthy outcomes, input data needs to be homogenous in nature. In the current study undertaken, right from infilling data, narrowing down the distribution type and the performances of homogeneity tests were questioned.

As a further study in continuation of this research, performances of homogeneity test under high autocorrelation could be analysed. Simulation tests for deciding the right homogeneity test for structural breaks in a series in relation to a reference series can be carried out. Once the irregularity is detected based on the outcomes of the test, the series needs to be homogenised.

References

Kim TW, Ahn H (2009) Spatial rainfall model using a pattern classifier for estimating missing daily rainfall data, pp 367–376
Duncan D (1998) Fundamental statistics for social research: step-by-step calculations and computer techniques using SPSS for Windows. New York
Firat M, Dikbas F, Koç AC, Gungor M (2010) Missing data analysis and homogeneity test for Turkish precipitation series. Sadhana-Acad Proc Eng Sci 35(6): 707–720
Hänsel S, Medeiros DM, Matschullat J, Petta RA, de Mendonça Silva I (2016) Assessing homogeneity and climate variability of temperature and precipitation series in the capitals of North-Eastern Brazil. Front Earth Sci 4(March): 1–21
Makhuvha T, Pegram G, Sparks R, Zucchini W (1997) Patching rainfall data using regression methods. 1 Best subset selection, EM and pseudo-EM methods: Theory J Hydrol 198(1–4):289–307
Mohanty UC, Ravi N, Madan OP, Paliwal RK (1997) Forecasting minimum temperature during winter and maximum temperature during summer at Delhi. Meteorol Appl 4(1):37–48
Nelwamondo FV, Mohamed S, Marwala T (2007) Missing data: a comparison of neural network and expectation maximization techniques. Curr Sci 93(11):1514–1521
Razali NM, Wah YB (2011) Power comparisons of Shapiro-Wilk, Kolmogorov-Smirnov, Lilliefors and Anderson-Darling tests. J Stat Model Anal 2(1):21–33
Rubin DB (1976) Inferences and missing data. Biometrika 63(3):581–592
Schneider T (2001) Analysis of incomplete climate data: estimation of Mean Values and covariance matrices and imputation of missing values. J Clim 14(5):853–871
Shao X, Zhang X (2010) Testing for change points in time series. 105(491):1228–1240
Wijngaard JB, Klein Tank AMG, Können GP (2003) Homogeneity of 20th century European daily temperature and precipitation series. Int J Climatol 23(6):679–692

Chapter 44
Multisite Monthly to Daily Naturalised Streamflow Disaggregation Using Daily Flow Pattern Hydrograph

Vivek Verma

Abstract Conservation and preservation of environmental wildlife and fisheries became an important criterion in this era. Therefore, the environmental protection flow standards for major river basins and estuaries have been established in the state of Texas enacted by federal laws. The environmental flow issues can be addressed effectively and efficiently on a daily time step, and thus, a new technique, daily flow pattern hydrograph, is developed and implemented for disaggregating monthly naturalised streamflows to daily while preserving monthly volumes. Constitution of flow pattern hydrograph emphasises numerous issues such as the impact of water resources development at upstream, comparative analyses of available flows, geology, hydrology, the geographical location of gaging stations, etc. The previous parametric and nonparametric studies dealing with streamflow disaggregation methods mostly preserve spatial and temporal statistical characteristics such as variance, skewness and maximum and minimum values. This study has been conducted from the perspective of capturing low flows, and therefore a statistical parameter, duration curve, also known as flow frequency metric, is employed to evaluate and analyse the result. This study is performed for the Trinity River basin located at the state of Texas. However, this technique can be employed in any other river basin irrespective of its shape, and size.

44.1 Introductions and Background

The Texas Commission on Environmental Quality (TCEQ), earlier known as Texas Natural Resource Conservation Commission (TNRCC), and its ten contractors developed Water Availability Modelling (WAM) system during 1997–2004 adjourned by the Texas Legislature in 1997 (Wurbs 2005; Wurbs 2006). WAM system consists of a generalised Water Right Analysis Package (WRAP) simulation model, input datasets for all river basins of Texas and other supporting databases. The software,

---

V. Verma (✉)
Civil and Environmental Engineering, Florida International University, Miami, FL, USA
e-mail: vivekverma10@gmail.com

R. Jha et al. (eds.), *Climate Change Impacts on Water Resources*, Water Science and Technology Library 98, https://doi.org/10.1007/978-3-030-64202-0_44

input files and other relevant information are publicly available at https://ceprofs.civil.tamu.edu/rwurbs/wrap.htm.

The Water Right permit system maintained by TCEQ administers new permits or revises old, existing permits using the WAM system. WAM system is routinely applied by the water community (private consulting firms, water districts, river authorities and others) on a monthly time step to evaluate and analyse available water quantity. This is also useful for assessing local and regional planning. Texas Water Development Board (TWDB) utilises the WAM system for statewide planning activities.

Senate Bill 2 (SB2) enacted by the 77th Texas Legislature in the year 2001 with the prime focus on balancing human water requirements as well as preserving, conserving and maintaining aquatic flora and fauna, wildlife and fisheries. This led to systematic studies, also known as Texas Instream Flow Program (TIFP), related to flow regimes incorporating numerous environmental factors such as water quality, habitat, hydrology, biology, basically encompassing a wide range of interdisciplinary studies to support ecological environment.

The detailed studies conducted for TIFP in SB2 was consuming lots of time as it was incorporating various above-mentioned environmental variables. Thus, Senate Bill 3 (SB3) enacted by the 80th Texas Legislature was passed in the year 2007 to quickly establish environmental instream flow standards and recommendations for all the river basins and estuaries of Texas according to the best available science. The refinements of these adopted studies are to be conducted every 10 years.

Assessment of aforementioned established environmental streamflow standards requires monitoring of low flows, which can be addressed adequately on a daily time step. Previously, the WAM system was routinely applied by the water community on a monthly time step. Thus, this recent inclusion to support and analyse environmental instream flows was the prime motivation to expand WAM monthly system from the perspective of capturing low flows, which leads to systematic study of numerous approaches adopted in the literature for disaggregating monthly streamflows to daily. Most of them are cited below with their adopted strategies and applications.

Earlier hydrological models preserved statistical parameters at the annual time scale by linking the past only through statistics at the annual level (Valencia and Schaake 1972, 1973). This model has been modified and improved by linking past through current values obtained after disaggregation (Mejia and Rousselle 1976). A new algorithm is presented for simultaneously disaggregating flows at number of locations along the drainage network. The probability distribution of disaggregated flows is approximated with the help of monthly flow patterns using k nearest neighbour nonparametrically (Kumar et al. 2000).

A nonparametric stochastic technique was employed for streamflow disaggregation. Annual or monthly flows are generated for a long period of analysis based on historical flows, which are then disaggregated to daily preserving statistical characteristics (Prairie et al. 2007; Nowak 2010). The elementary method was adopted for disaggregating both regulated and unregulated streamflow by preserving both statistical parameters as well as mass balance (Acharya and Ryu 2014).

Parametric and nonparametric models, are the two classical aforementioned approaches widely used in the literature. These techniques preserve statistical characteristics such as mean, median, skew, variance and other properties resulting from disaggregating coarser (monthly/annually) to finer (daily/monthly/weekly) streamflows. However, little work has been done by researchers for disaggregating monthly flows to daily from the perspective of capturing low flows. Therefore, this research work explores a new avenue for developing and implementing a new technique, with the point of view of capturing low flows, which utilises daily flow pattern hydrograph for disaggregating monthly naturalised flows to daily while preserving monthly volumes.

Daily flow pattern hydrograph is a pattern representing actual existing conditions that disaggregates monthly naturalised flows to daily in the ratio of its constituted flows, which requires comparative analyses of distinct available flows at gaging stations. The results of which are evaluated with the help of the duration curve, discussed in the next paragraph. Those flows were adopted in the formulation of the daily flow pattern hydrograph, which represents actual river basin flow characteristics. The composition of daily flow pattern hydrograph requires significant time and expertise in the field of hydrology. It is applicable for disaggregating both regulated and unregulated streamflows in a watershed.

The results obtained by disaggregating monthly flows to daily are analysed with the help of the flow duration curve, also known as flow frequency metrics. It is widely used by water resources engineers for comparing flow characteristics across the river basin and watersheds (Vogel and Fennessey 1995). It is a good indicator and particularly used for capturing low flows and thus adopted for evaluating the results (Vogel and Fennessey 1994; Smakhtin 2001). Additionally, it provides geologic characteristics of drainage basins (Searcy 1959).

A flow duration curve exhibits the relationship between the exceedance frequencies and the magnitude of mean flows. The exceedance frequencies in percentage represent a number of days average discharge equalled or exceeded the indicated magnitudes during the day (Singh 1971). It is computed on the basis of Weibull formula

$$P = \frac{m}{N+1} * 100\% \tag{1}$$

where P is Exceedance frequencies, m is rank and N is the sample size.

44.2 Study Area

Trinity river basin is situated in the state of Texas. It originates near Texas–Oklahoma border at north and confluences at Galveston Bay east of Houston. It is the largest river basin with its watershed (18,000 square miles) entirely enclosed within the State of Texas. Mean annual precipitation ranges from 36 inches at headwater to 52 inches

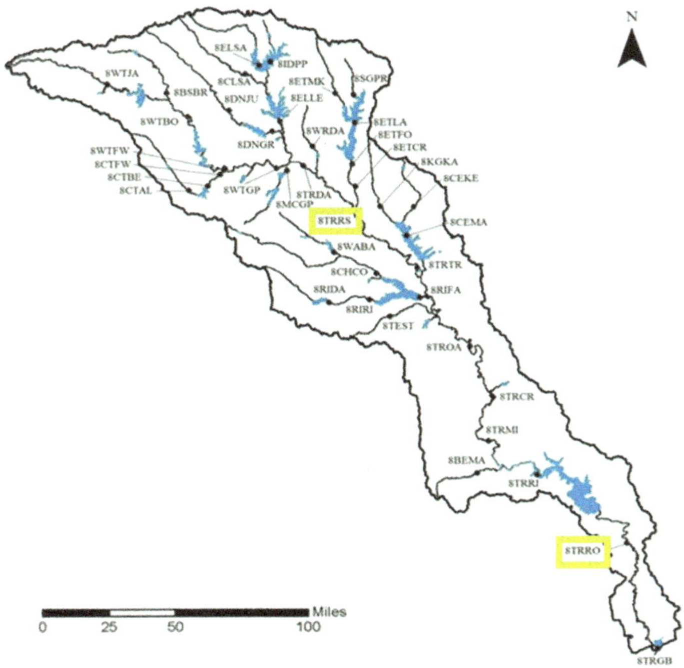

Fig. 44.1 Selected gaging stations at Trinity River basins

near its confluence point. The control points (gaging stations) 8TRRO (Trinity River at Romayor) and 8TRRS (Trinity River near Rosser), as illustrated in Fig. 44.1, are selected in this technical paper for two reasons. First, it has a relatively long period of observed streamflow records and second establishment of environmental instream flow standard as per SB3.

44.3 Methodology

Systematics study of the geographical location of gaging stations, the impact of water resources development at upstream, comparative analyses of available flows, geology, hydrology, regulation of upstream flows, etc., are performed for each gaging station for formulating daily flow pattern hydrographs. However, these selected gaging stations do not have any missing flows. However, many stations that have missing flows were addressed with the help of available streamflows from other nearby gaging stations, based on the criteria of reflecting actual river flow characteristics. Gaging stations having a long period of missing flow records were discarded. The impact of upstream development is particularly important and illustrated in Fig. 44.2, which represents streamflow hydrograph at gaging station Evadale located in Neches

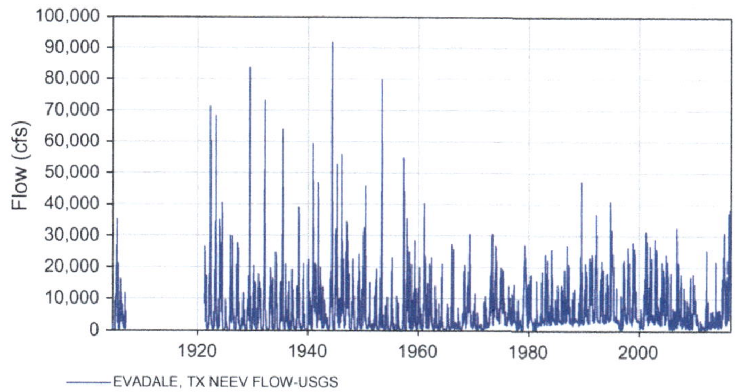

Fig. 44.2 USGS Daily gage flows at Evadale on Neches River, TX

river basin. Frequent high flows and simultaneous low flows are observed during 1920–1960s. The frequency of high flows reduces and low flows become slightly higher after the 1960s. It seems from the streamflow vs time graph that upstream development may take place around the 1960s which regulated the streamflows.

Once-daily flow pattern hydrograph is constituted, they are inputted to the WRAP/WAM modelling system, which disaggregates monthly naturalised flows to daily while preserving monthly volumes.

44.4 Results and Discussion

The frequency analysis has been done for USGS gage, USACE and WAM daily flows and presented through Tables 44.1 and 44.2, developed within HEC-DSSVue. The first column in these tables represents daily flows obtained from USGS gaging stations. The second column illustrates the US Army Corps of Engineers (USACE) daily unregulated flows obtained from USGS gage flow by removing the effects of their multi-purpose flood operating reservoirs. Trinity river basins have eight such reservoirs in total.

Flow pattern hydrograph is constituted with the help of these flows, which becomes the input for the WRAP/WAM model. The third column illustrates the daily naturalised flows obtained by disaggregating monthly naturalised flows to daily. Similarly, the fourth and fifth column represents the same thing as that of column 1 and 3, respectively, but for a different hydrological period of analysis, that is, from 2010 to 2015.

Flow frequency metrics are divided into two subperiods, that is, from 1940–2009 to 2010–2015. This division depends upon the period of analysis of available data of streamflows. The following statistical metrics are presented in Tables 44.1 and 44.2:

Table 44.1 Flow frequency metrics for control point 8TRRO

Statistical parameters	8TRRO Trinity River at Romayor				
	Frequency Metrics 1940–2009			Frequency Metrics 2010–2015	
	USGS	USACE	WAM	USGS	WAM
	Daily	Daily	Daily	Daily	Daily
	(cfs)	(cfs)	(cfs)	(cfs)	(cfs)
Mean	8167	9731	9128	7775	8890
SD	11798	14176	13558	13845	18261
Min	104	129	0.0	697	0.0
Max	117000	154419	175475	71800	264175
Percentage exceedance					
0.2%	78290	109882	99461	70965	140858
0.5%	62716	87852	83443	69488	111753
1%	55300	70670	66102	66350	89944
2%	45300	53962	51836	59550	77804
5%	32800	36428	35096	43350	47015
10%	23400	25001	23933	24200	21960
15%	18100	18667	18055	16850	14652
20%	13500	14821	14160	8880	11162
30%	7680	9575	8992	4210	5975
40%	4570	6253	5859	2500	3611
50%	2800	4315	3915	1780	2374
60%	2000	3025	2648	1460	1809
70%	1450	2132	1763	1240	1319
80%	1060	1449	1076	1120	483
85%	910	1138	768	1050	287
90%	726	853	525	1030	66
95%	495	594	275	968	0
98%	339	406	102	907	0
99%	255	320	26	876	0
99.5%	197	267	0	833	0
99.8%	149	229	0	821	0

- The average magnitude of mean daily flows.
- Standard deviation (SD) of daily flows.
- The minimum value of daily flow.
- The maximum value of daily flow.
- Daily streamflows are analogous to the exceedance frequencies, calculated on the basis of the Weibull formula.

Table 44.2 Flow frequency metrics for control point 8TRRS

Statistical parameters	8TRRS Trinity River near Rosser				
	Frequency metrics 1940–2009			Frequency metrics 2010–2015	
	USGS	USACE	WAM	USGS	WAM
	Daily	Daily	Daily	Daily	Daily
	(cfs)	(cfs)	(cfs)	(cfs)	(cfs)
Mean	3239	4012	3740	3847	3973
SD	5806	9359	9079	7244	11856
Min	64	4	0.0	481	0.0
Max	133000	171289	189464	67600	139642
Percentage exceedance					
0.2%	48945	93773	84070	58788	115674
0.5%	34800	63226	59151	47508	92842
1%	27000	46290	45289	37954	61992
2%	19262	31598	30258	27908	44195
5%	13150	16547	15903	17585	20641
10%	9060	9486	9005	11400	6698
15%	6570	6296	5894	7141	4442
20%	4700	4460	4152	4304	2991
30%	2450	2604	2394	1890	1741
40%	1450	1722	1529	1230	1196
50%	1020	1180	1022	976	759
60%	818	849	710	849	503
70%	639	614	457	766	209
80%	471	410	247	710	27
85%	393	326	170	683	0
90%	300	260	96	650	0
95%	182	186	7	620	0
98%	140	132	0	589	0
99%	126	101	0	567	0
99.5%	116	75	0	552	0
99.8%	107	52	0	540	0

It is evident from Table 44.1 (Frequency Metrics 1940–2009) that maximum flow increases respectively in observed, unregulated and naturalised flow (USGS, USACE and WAM). Whereas the mean flow is minimum (8,167 cfs) for observed flow and maximum (9,731 cfs) for unregulated flow. The mean flow of naturalised flow is 9,128 cfs. It is clear that the naturalised and unregulated flows are very near to the observed flow. The standard deviation (SD) of these flows are very close. This statistical analysis indicates that naturalised flow generated by WRAP/WAM model is very close to the observed flows. However, it encounters some zeroes while capturing very low flows, particularly while capturing exceedance frequencies of 99.5 and 99.8%. This is due to the reason that monthly flows generated from the WRAP/WAM simulation model produce zero flows, especially during drought periods (1960s, 1990s and 2010s). Therefore, disaggregating nil monthly flows will result in zero daily naturalised flows, which is apparent from Frequency Metrics of 1940–2009 and 2010–2015.

An almost similar trend is observed in Trinity River near Rosser, as illustrated in Table 44.2. Mean flow increases in observed, naturalised and unregulated flow (USGS, USACE and WAM), respectively. Standard deviation is similar for unregulated and naturalised flows but varies substantially when compared with observed flows. SD of these flows is almost two times SD of observed flows. As already discussed above, daily naturalised flows obtained from disaggregating WRAP/WAM monthly naturalised were not able to capture very low flows because monthly naturalised flows generated from WRAP/WAM simulation model are zero, and thus disaggregating zero monthly flows will result in zero daily flows, irrespective of any flow pattern hydrograph, which is evident from the Frequency Metrics 1940–2009 and 2010–2015.

44.5 Conclusion

Texas streamflow is highly variable temporally and spatially, including exceptional floods and droughts with seasonal variations. Thus, the general compilation strategy for daily flow pattern hydrograph is discussed in this article, which is applicable for any other basins in the world. It disaggregates monthly flows to daily from the perspective of capturing low flows. Previous studies related to disaggregation focused on replicating statistical characteristics of watersheds. This is a generalised technique that has been applied for disaggregating regulated and unregulated streamflows. Daily flow pattern hydrograph has been formulated depending upon the different sites, location, availability of data and other variable factors. The sources of daily flows may vary between sites depending upon data availability and hence suitable strategy has been adopted for different control points. Different pragmatic issues such as flow alterations, filling in gaps of missing data, etc., have been substituted with other available flows (with similar river basin characteristics) for that specific time period on the basis of replicating actual river basin flow characteristics. This paper demonstrates an effective and efficient technique for generating daily naturalised flows using flow pattern hydrograph, which addresses the environmental instream flow issues effectively. Daily flow pattern hydrograph has got its own utilities in water resources engineering.

References

Acharya A, Ryu JH (2014) Simple method for streamflow disaggregation. J Hydrol Eng 19(3):509–519
Kumar DN et al (2000) Multisite disaggregation of monthly to daily streamflow. Water Resourc Res 36(7):1823–1833
Mejia JM, Rousselle J (1976) Disaggregation models in hydrology revisited. Water Resourc Res 12(2):185–186
Nowak K et al. (2010) A nonparametric stochastic approach for multisite disaggregation of annual to daily streamflow. Water Resourc Res 46(8):n/a–n/a
Prairie J et al (2007) A stochastic nonparametric technique for space-time disaggregation of streamflows. Water Resourc Res 43(3):n/a–n/a
Searcy JK (1959) Flow-duration curves, US Government Printing Office Washington
Singh KP (1971) Model flow duration and streamflow variability. Water Resourc Res 7(4):1031–1036
Smakhtin VU (2001) Low flow hydrology: a review. J Hydrol 240(3):147–186
Valencia DR, Schaake JC (1972) A disaggregation model for time series analysis and synthesis. School of Engineering, Massachusetts Institute of Technology
Valencia DR, Schaake JC (1973) Disaggregation processes in stochastic hydrology. Water Resourc Res 9(3):580–585
Vogel RM, Fennessey NM (1994) Flow-duration curves. I: new interpretation and confidence intervals. J Water Resourc Plann Manag 120(4):485–504
Vogel RM, Fennessey NM (1995) Flow duration curves II: a review of applications in water resources planning. JAWRA J Am Water Resourc Assoc 31(6):1029–1039
Wurbs RA (2005) Texas water availability modeling system. J Water Resourc Plann Manag 131(4):270–279
Wurbs RA (2006) Methods for developing naturalized monthly flows at gaged and ungaged sites. J Hydrol Eng 11(1):55–64

Chapter 45
Error Analysis of TMPA Near Real-Time Precipitation Estimates for an Indian Monsoon Region

Ashish Kumar and RAAJ Ramsankaran

Abstract Timely measurement of precipitation across a large area is essential for flood/drought/landslide forecasting. Satellite-based precipitation estimates (SPEs) are one of the promising sources of precipitation data for the above mentioned near real-time (NRT) applications. However, these estimates have inherent errors and uncertainty. Hence, it is crucial to assess and quantify them to support effective NRT applications. Therefore, as a pilot study, the Tropical Rainfall Measuring Mission (TRMM) Multi-satellite Precipitation Analysis (TMPA) Real Time (RT) version-7 (V7) product is evaluated during the monsoon of 2002–2013 across a large river basin called Krishna River Basin, located in South-central part of India. It shall be noted that this is the first of its kind study in India. Here, the Total Error in TMPA 3B42RT is decomposed into individual error components such as Hit Error, Miss Precipitation and False Precipitation. The obtained error components are then analyzed spatially and intensity wise. Our results indicate that the leading source of error in TMPA 3B42RT is Hit Error, which shows significant spatial differences between the orographic region and non-orographic regions of the basin. Intensity wise, the TMPART-V7 overestimates very light to moderate precipitation and vice versa for heavy to very heavy precipitation. However, TMPA-RT V7 has excellent detection ability for heavy to very heavy precipitation.

Keywords Satellite Precipitation Estimates (SPEs) · TMPA · TRMM3B42RT · India · Error components · Precipitation intensity · Topography · Indian monsoon region

---

A. Kumar · R. Ramsankaran
Hydro-Remote Sensing Applications (H-RSA) Group, Department of Civil Engineering, Indian Institute of Technology Bombay, Mumbai 400076, India
e-mail: ashishsnh647@gmail.com

R. Ramsankaran (✉)
Interdisciplinary Program in Climate Studies, Indian Institute of Technology Bombay, Mumbai 400 076, India
e-mail: ramsankaran@civil.iitb.ac.in

R. Jha et al. (eds.), *Climate Change Impacts on Water Resources*, Water Science and Technology Library 98, https://doi.org/10.1007/978-3-030-64202-0_45

45.1 Introduction

'Precipitation' is one of the most important components of the hydrological cycle. Particularly, Real Time (RT) precipitation in finer spatial and temporal scale is indispensable for applications such as flood forecasting, drought monitoring, etc. Rain gauge data is considered as a reliable source for such type of applications. However, its distribution is scarce and uneven in most of the developing nations like India, Bangladesh, etc. Furthermore, most often, it is not available in near real-time to be utilized in disaster preparedness. Due to the emergence of remote sensing technology, several sources of RT global precipitation data (PERSIANN; Sorooshian et al. 2000, CMORPH; Joyce et al. 2004, TMPA; Huffman et al. 2007, IMERG; Huffman et al. 2014) are available in the public domain. Hence, RT Satellite-based Precipitation Estimates (SPEs) can be an alternate choice for real-time applications.

As satellites do not measure precipitation directly, instead of retrieving it from remote sensing observations (brightness temperatures), there is a need for a comprehensive evaluation of SPEs with observed precipitation datasets to know their accuracy. Plenty of evaluation studies have been carried out at different spatial and temporal scales across the world (Dinku et al. 2007; Bharti et al. 2016, etc.). Most of these studies use conventional measures (e.g., linear correlation coefficient, mean error, mean absolute error, bias, root mean square error, probability of detection, false alarm ratio, critical success index, equitable threat score, etc.) to assess the error. However, the mentioned measures are unable to capture entire error characteristics in SPEs due to the complex nature of the errors (Tian et al. 2009; Tang et al. 2015).

Practically, two-stages are involved in the retrieval process of SPEs. The first stage is screening, i.e., detection of raining and non-raining areas and the second stage is to establish a relationship between the observed brightness temperatures and rain rates. The errors can be introduced in either of the processes. Two types of errors are present in the screening process. If raining areas are not detected, it will result in Miss Precipitation. At the same time, if no-raining areas are mistakenly considered as raining ones, it could lead to False Precipitation. On the other hand, even, when the raining areas are correctly detected, there still exist retrieval errors in converting the brightness temperature to rain rate (i.e., Hit Error). The nature of errors in both the stages are different and thus, it is imperative to separate the errors into various components like Miss Precipitation, False Precipitation and Hit Error. It would be extremely helpful for end-users as well as algorithm developers to get a better understanding of the entire error characteristics of SPEs. Tian et al. (2009) have given a promising scheme for decomposition and analysis of errors (Miss Precipitation, False Precipitation and Hit Error) of the SPEs. Recently, some of the researchers (Gebregiorgis et al. 2012; Yong et al. 2016) have included this scheme in their study to get in-depth information regarding the error characteristics of SPEs.

Some of the studies (Yong et al. 2014; Himanshu et al. 2017) mentioned that SPEs are underestimating heavy precipitation intensity and overestimating light to moderate precipitation intensity. However, most of these studies have investigated the dependency of the precipitation intensity on conventional statistical measures such as bias and RMSE only but not on different individual error components. To the best of our knowledge, such type of detailed studies is not available for any of the Indian river basins. In view of the above, the main goal of this study is to carry out a detailed error analysis of a RT satellite-based precipitation Estimates over a larger river basin in India. Accordingly, the present study focuses on two aspects (i) Decomposition of the Total Error into independent error components and (ii) Intensity analysis of the independent error components.

This article is organised into six sections. Following this introduction section, a description of the study area and the data used are presented in Sects. 45.2 and 45.3, respectively. The methodology used in this research is introduced in Sect. 45.4. Then, in Sect. 45.5, the obtained results of the error analysis are presented. Finally, the summary and major conclusions arrived by the obtained results as well as the recommendations suggested for future studies, are given in Sect. 45.6.

45.2 Study Area

The test site for this study is the Krishna River Basin (KRB), one of the largest river basins in India, which is about 250,000 sq-km. The extent of KRB lies between 73°17′ to 81°9′ East longitudes and 13°10′ to 19°22′ North latitudes (Fig. 45.1). The mean annual precipitation of KRB is around 859 mm (Ministry of Water Resources 2014). However, it varies from less than 400 mm in the north-west to a maximum of more

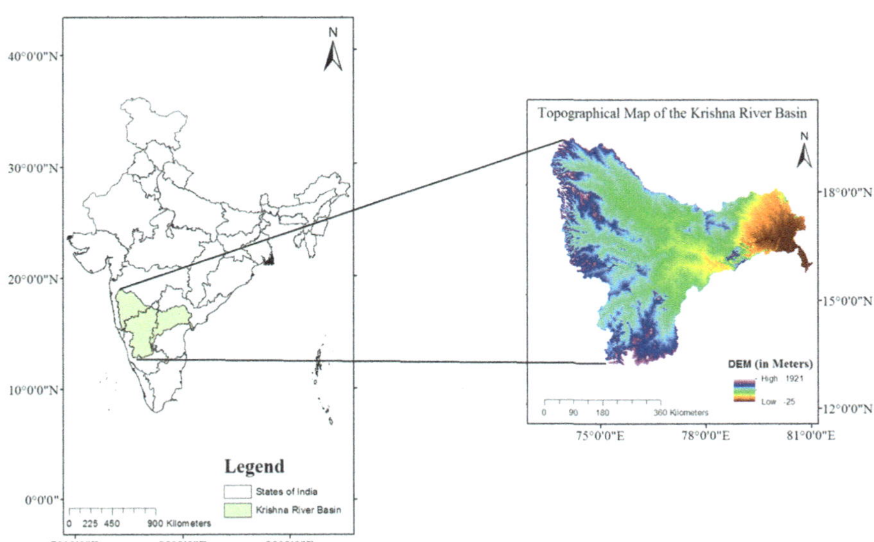

Fig. 45.1 Geographical location (left) and elevation profile (right) of the Krishna River Basin, India

than 3000 mm on the western periphery and 1000 mm in the deltaic region. KRB receives approximately 80–85% of the total annual precipitation during southwest Monsoon (~15th June to 14th October). KRB is extremely vulnerable to floods every year (National Rainfall Area Authority 2011). In addition, the terrain of KRB is very complex due to the presence of different types of landform zones, that is, hilly terrain (orographic region) on the western periphery, a plateau region in the interior basin and plains on the deltaic portion. KRB also has a noticeable variation in topography, as shown in Fig. 45.1. This basin is mostly dominated by a semi-arid region. However, a dry sub-humid area in the deltaic region, arid in the rain-shadow eastern parts of the western periphery and a little humid band (North to South) along the western periphery are also present. Significant spatial variability of precipitation, complexity in terrain and high vulnerability of floods make the KRB a suitable study site for the assessment of RT SPEs. Another important reason for selecting KRB is that it is a snow-free basin because RT SPEs have significant limitations over snow-fed basins.

45.3 Datasets

The observed precipitation for assessment is obtained from the Indian Meteorological Department (IMD) for the monsoon of 2002–2013. It is a fine resolution (0.25° × 0.25°) gridded daily precipitation (at 3 GMT) dataset over India developed by Pie et al. (2014). This gridded daily precipitation data is prepared by comparatively well spread rain gauge stations over the Indian land region following thorough, rigorous quality control. Several studies (Prakash et al. 2015; Shah and Mishra 2015, etc.) have been conducted and used this IMD Gridded Precipitation as reference data for assessment of satellite-based precipitation estimates. IMD gridded precipitation is the certified officially released product to use in hydrological and meteorological applications over the Indian land region at the present time. Although IMD Gridded precipitation has inherently interpolation error, for larger-scale analysis, this gridded precipitation shall be considered as one of the reliable products across India.

The satellite-based real-time precipitation TMPA3B42-RT Version 7 (Hereafter referred as 3B42RT) is used in the present study. Although, Global Precipitation Measurement (GPM) era multi-satellite product IMERG at very finer resolution is available, but not used in the present study due to its shorter span of data availability. Availability of 3B42RT in near real-time (within 6–9 h of the observation) to the end-users make it suitable for real-time applications such flood monitoring, drought monitoring, landslide monitoring. 3B42RT relies on microwave observations from the low orbiting satellites. The spatial and temporal gaps in the microwave observations are filled with infrared (IR) data in it. The spatial and temporal resolution of 3B42RT are 0.25° × 0.25° and 3-hourly.

45.4 Methodology

The framework of the study is shown in Fig. 45.2. Both the precipitation datasets such as IMD and 3B42RT are acquired for the monsoon of 2002–2013. IMD accumulates daily precipitation ending at 0300 UTC. Hence, we estimated daily precipitation from three-hourly 3B42RT precipitation products ending at the same time for the sake of homogeneity in the analysis. Estimation of the different error components is carried out in the subsequent step following Tian et al. (2009). Under this scheme, Total Error is decomposed into three independent error components, that is, Hit Error, Miss Precipitation and False Precipitation. According to this scheme, if S (unit: mm/day) and O (unit: mm/day) are the satellite-based precipitation estimates (3B42RT) and observed precipitation (IMD) respectively, then the different independent error components for each grid are obtained by the following equations (Eqs. 45.1, 45.2 and 45.3)

$$\text{False Precipitation (mm/day)} = S - O, \text{ when } S > O \text{ and } O = 0 \tag{45.1}$$

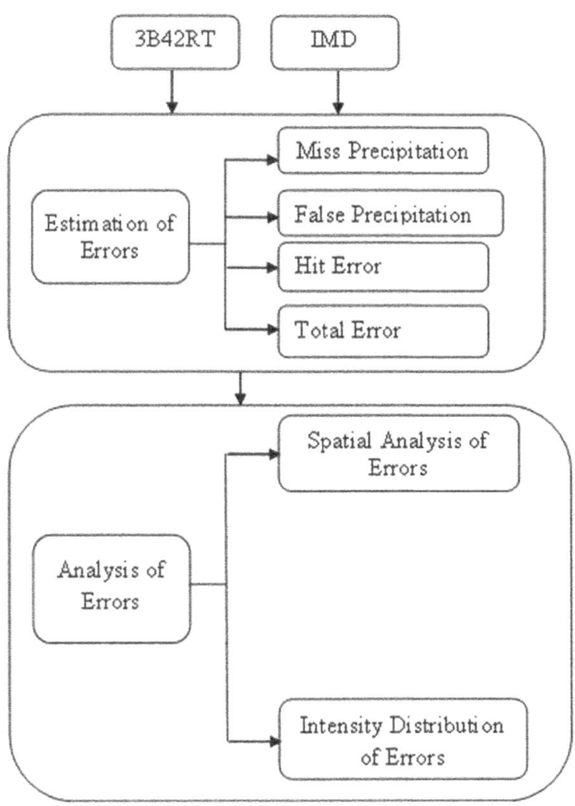

Fig. 45.2 Framework of the methodology

$$\text{Miss Precipitation (mm/day)} = \text{S} - \text{O}, \text{ when S} = \text{O and O} > 0 \tag{45.2}$$

$$\text{Hit Error (mm/day)} = \text{S} - \text{O}, \text{ when S} > \text{O and O} > 0 \tag{45.3}$$

$$\text{Total Error (mm/day)} = \text{Hit Error} + \text{Miss Precipitation} + \text{False Precipitation} \tag{45.4}$$

Also, Total Error is linked with independent error components by the following equation (Eq. 45.4)

Once the independent error components are obtained, the errors have been analysed spatially, and intensity wise as described in the following sub-sections.

45.4.1 Spatial Analysis of Errors

Spatial analysis has been carried out to understand the spatial variation of independent error components, including the Total Error in different parts of the basin. The error components (Hit Error, Miss Precipitation and False Precipitation) and Total Error for each day in each grids are quantified by Eqs. 45.1, 45.2, 45.3 and 45.4, respectively. Then, average error components, including the Total Error are quantified at each of the grid.

45.4.2 Intensity Distribution of Errors

A clear understanding of the variation of errors (Hit Error, Miss Precipitation, False Precipitation and Total Error) with precipitation intensity is needed to find out the precipitation intensity for which under/overestimation, Miss/False Precipitation exist in 3B42RT. For this analysis, we computed the distribution of Miss, False, Hit and Total Precipitation amount as functions of precipitation intensity (mm/day). Classification of Precipitation intensity bin is based on the Indian Meteorological Department 2017 (Appendix 1). In this analysis, for each precipitation intensity classes, the accumulated IMD observed precipitation and the corresponding accumulated 3B42RT precipitation are estimated. Finally, the obtained accumulated values are rescaled by the number of months (12) and grids (355) for clear visualization.

45.5 Results and Discussion

Results of the error analysis are presented in this section. The analysis is categorized as mentioned in Sect. 45.4, that is, spatial analysis of errors and intensity wise analysis of errors.

45.5.1 Spatial Analysis of Errors

The spatial pattern of errors is shown in Fig. 45.3a–d. Each figure represents the monsoonal average of the Total Error and its three independent error components: Hit Error, Miss Precipitation and False Precipitation. 3B42RT overestimates (Fig. 45.3a) the observed precipitation in most parts of the basin except the western

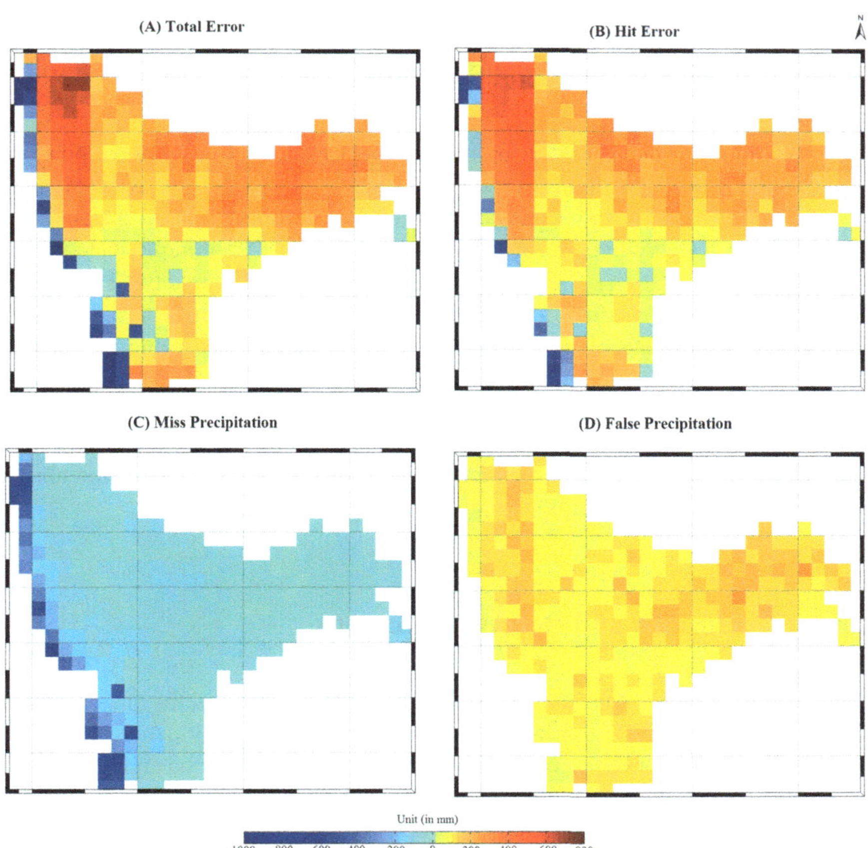

Fig. 45.3 Error components for monsoon seasonal-average: **a** Total error; **b** Hit error; **c** Missed precipitation; and **d** False precipitation

periphery/orographic region, where it shows considerable underestimation (represented by blue colour). Similar spatial characteristics (i.e., overestimation and underestimation) as Total Error (Fig. 45.3a) observed in Hit Error (Fig. 45.3b) also in most parts of the basin indicates that the major contribution to the Total Error is from Hit Error. A substantial difference in Hit Error (Fig. 45.3b) is noticed in the basin, that is, an underestimation on the western periphery/orographic region and overestimation in other parts of the basin. These underestimations during hit events in the western periphery/orographic region may be caused due to the problem in 3B42RT sensors for accurate quantification of precipitation intensity in orographic regions. Moreover, the amount of Miss Precipitation (Fig. 45.3c) is considerably high in the western periphery/orographic region as compared with other parts of the basin. This is probably due to the inability of current satellite precipitation retrievals in detecting the warm rain processes or short-lived convective storms over orographic regions. Unlike the other error components, the spatial pattern of False Precipitation (Fig. 45.3d) exhibits an almost uniform pattern across the basin.

45.5.2 Intensity Distribution of Errors

In this section, the intensity distribution of Total Precipitation, Hit Precipitation, Miss Precipitation and False Precipitation are plotted in Fig. 45.4a–c. Total Precipitation (Fig. 45.4a) and Hit Precipitation (Fig. 45.4b) of 3B42RT shows a significant overestimation for very light to moderate precipitation (< 35.5 mm/day) and underestimation for heavy to very heavy precipitation (64.5–244.5 mm/day). The above-obtained results of overestimation/underestimation are consistent with past studies (Tang et al. 2014; Yong et al. 2016) carried out in other parts of the world. The problem of overestimation and underestimation mainly arise due to the non-unique relation between brightness temperature (T_B) and precipitation intensity. Mostly, Detection errors (Miss and False Precipitation) are observed for the intensities up to rather heavy precipitation (< 64.5 mm/day) indicates the excellent detection ability for heavy to very heavy precipitation.

45.6 Conclusions and Recommendations

Spatial analysis shows that the Estimation Error (Hit Error) is the major source of total error in 3B42RT and its magnitude varies considerably across the KRB. Severe underestimations present in the orographic regions of the basin indicate that the TRMM sensors and/or 3B42RT algorithm are unable to quantify the orographic precipitation. Along with estimation error, detection error particularly, Miss Precipitation is also found to be a serious problem in the orographic region. Likewise, intensity analysis indicates that the overestimation issue in the estimation stage (hit event) exists in light to moderate precipitation and vice versa for heavy to very heavy

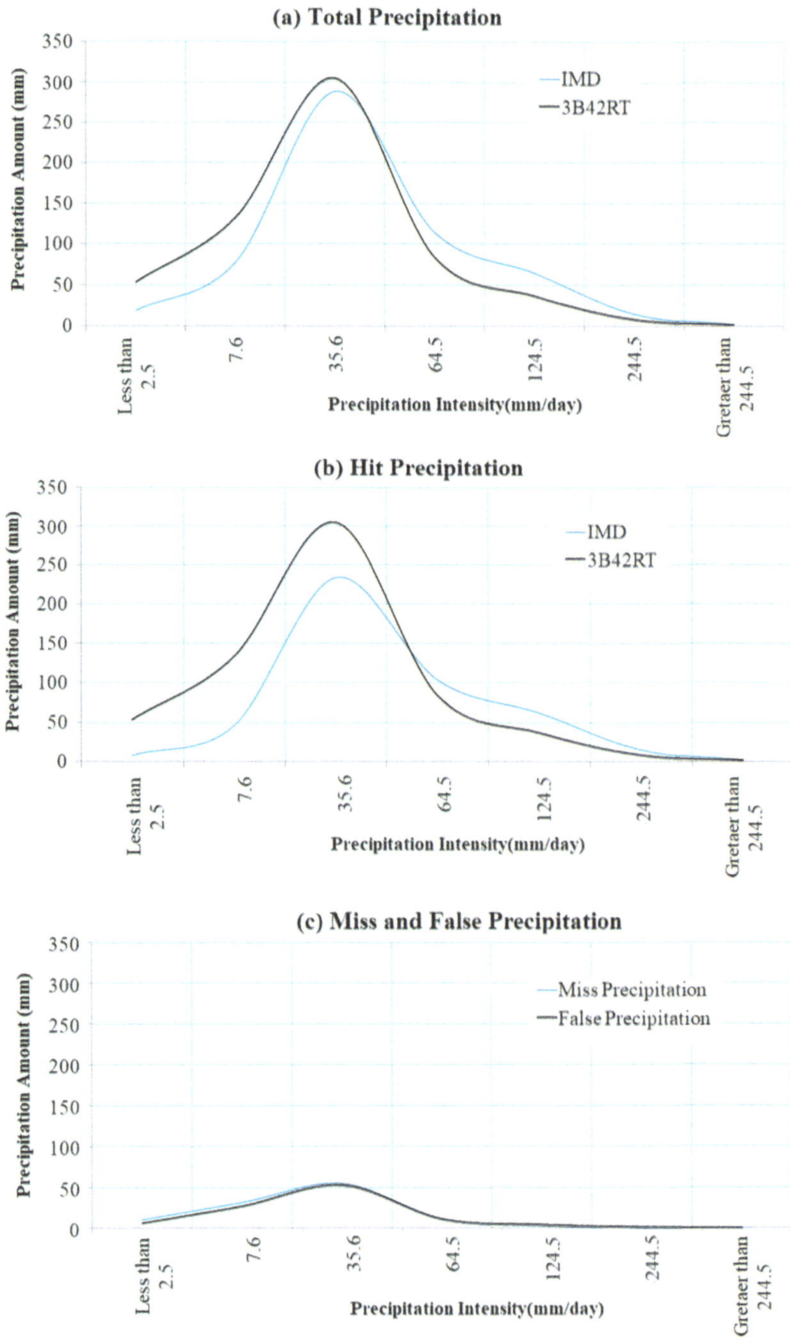

Fig. 45.4 Intensity distribution (horizontal axis represents the precipitation intensity and vertical axis represents the rescaled accumulated daily precipitation in each bin)

Table 45.1 Classification of rainfall according to India Meterological Department, India

Classes	Rainfall in a day (mm)
Very light rain	1–2.4
Light rain	2.5–7.5
Moderate	7.6–35.5
Rather heavy	35.6–64.4
Heavy	64.5–124.4
Very heavy	124.5–244.5
Extremely heavy	>244.5

Source India Meteorological Department (2018) (http://imd.gov.in/section/nhac/wxfaq.pdf) (accessed on 13th July 2018)

precipitation events. However, 3B42RT has excellent detection ability in heavy to very heavy precipitation.

Considering these research findings, it is found that the 3B42RT error components vary spatially and precipitation intensity-wise in KRB during monsoon season. It shall be noted that this study alone is not sufficient to make changes in the 3B42RT algorithm; nevertheless, it endorses that similar studies have to be carried out for different SPEs in different basins before using them for any applications such as flood modelling, drought monitoring, etc. Such analysis shall help us to better understand the behavior of that satellite rainfall product's error components in a given region so that one can make efficient use of available SPEs. It shall be noted that though this study is performed on 3B42RT products, it will be useful for GPM products as well because the GPM products are based on the 3B42RT algorithms. It is also recommended that further studies should include the influence of other factors, such as topography, climate and season on error components of SPE products.

Supportive Information

See Table 45.1

Acknowledgments The authors are sincerely thankful to the TMPA science team for providing the TMPA-RT V7 precipitation data at free of cost through their data repository. The authors are also grateful to the India Meteorology Department (IMD) for preparing the precipitation data set and making it available for the research community. The authors also acknowledge the United State Geological Survey (https://earthexplorer.usgs.gov/) for providing SRTM DEM at free of cost for research purpose.

References

Bharti V, Singh C, Ettema J, Turkington TA (2016) Spatiotemporal characteristics of extreme rainfall events over the Northwest Himalaya using satellite data. Int J Climatol 36(12):3949–3962

Dinku T, Ceccato P, Grover-Kopec E, Lemma M, Connor SJ, Ropelewski CF (2007) Validation of satellite rainfall products over East Africa's complex topography. Int J Remote Sens 28(7):1503–1526

Gebregiorgis AS, Tian Y, Peters-Lidard CD, Hossain F (2012) Tracing hydrologic model simulation error as a function of satellite rainfall estimation bias components and land use and land cover conditions. Water Resourc Res 48:W11509

Himanshu SK, Pandey A, Yadav B (2017) Assessing the applicability of TMPA-3B42V7 precipitation dataset in wavelet-support vector machine approach for suspended sediment load prediction. J Hydrol 550:103–117

Huffman GJ, Bolvin DT, Nelkin EJ, Wolff DB, Adler RF, Gu G, Hong Y, Bowman KP, Stocker EF (2007) The TRMM multisatellite precipitation analysis (TMPA): Quasi-global, multiyear, combined-sensor precipitation estimates at fine scales. J Hydrometeorol 8(1):38–55

Huffman GJ, Bolvin DT, Braithwaite D, Hsu K, Joyce R, Xie P (2014) NASA Global Precipitation Measurement (GPM) Integrated Multi-satellitE Retrievals for GPM (IMERG). Algorithm Theoretical Basis Document (ATBD), Version 4.4, NASA, 30 p

India Meteorological Department (2018) Met. Terminologies and Glossary. http://www.imdgoa.gov.in/index.php?option=com_content&view=article&id=271. Accessed 13 July 2018

Joyce RJ, Janowiak JE, Arkin PA, Xie P (2004) CMORPH: A method that produces global precipitation estimates from passive microwave and infrared data at high spatial and temporal resolution. J Hydrometeorol 5(3):487–503

Ministry of Water Resources (2014) Krishna Basin. http://www.india-wris.nrsc.gov.in/Publications/BasinReports/Krishna%20Basin.pdf. Accessed 18 July 2018

National Rainfed Area Authority (2011) Impact of high rainfall/floods on ground water resources in the Krishna River basin (During 1999–2009). http://nraa.gov.in/pdf/krbasin_2009.pdf. Accessed 8 July 2018

Pai DS, Sridhar L, Rajeevan M, Sreejith OP, Satbhai NS, Mukhopadhyay B (2014) Development of a new high spatial resolution ($0.25° \times 0.25°$) long period (1901–2010) daily gridded rainfall data set over India and its comparison with existing data sets over the region. Mausam 65(1):1–18

Prakash S, Mitra AK, AghaKouchak A, Pai DS (2015) Error characterization of TRMM Multi-satellite Precipitation Analysis (TMPA-3B42) products over India for different seasons. J Hydrol 529:1302–1312

Shah RD, Mishra V (2015) Development of an experimental near-real-time drought monitor for India. J Hydrometeorol 16(1):327–345

Sorooshian S, Hsu KL, Gao X, Gupta HV, Imam B, Braithwaite D (2000) Evaluation of PERSIANN system satellite–based estimates of tropical rainfall. Bull Am Meteorol Soc 81(9):2035–2046

Tang L, Tian Y, Lin X (2014) Validation of precipitation retrievals over land from satellite-based passive microwave sensors. J Geophys Res Atmos 119(8):4546–4567

Tang L, Tian Y, Yan F, Habib E (2015) An improved procedure for the validation of satellite-based precipitation estimates. Atmos Res 163:61–73.

Tian, Y., Peters-Lidard, CD., Eylander, JB., Joyce, RJ., Huffman, GJ., Adler, RF., Hsu, KL., Turk, FJ., Garcia, M., Zeng, J (2009) Component analysis of errors in satellite-based precipitation estimates. J Geophys Res Atmos 114:D24101

Yong B, Chen B, Gourley JJ, Ren L, Hong Y, Chen X, Wang W, Chen S, Gong L (2014) Intercomparison of the Version-6 and Version-7 TMPA precipitation products over high and low latitudes basins with independent gauge networks: Is the newer version better in both real-time and post-real-time analysis for water resources and hydrologic extremes? J Hydrol 508:77–87
Yong B, Chen B, Tian Y, Yu Z, Hong Y (2016) Error-component analysis of trmm-based multisatellite precipitation estimates over mainland China. Remote Sens 8(5):440

Chapter 46
Comparison of Selection of Predictors for Statistical Downscaling of Precipitation Using Different Statistical Techniques

Kumar Keshav, Vivekanand Singh, and Roshni Thendiyath

Abstract Selection of predictors for statistical downscaling is crucial for establishing the best relationship between predictors (such as Relative humidity, Geopotential height, u-wind, specific humidity, etc.) and predictand (such as precipitation, temperature, etc.). In this paper, the statistical method (factor analysis, correlation coefficient, etc.) are used for selecting the best predictors for downscaling predictand. The selection of potential predictors enhances the performance, computational times and model outputs. The Bagmati river basin, Bihar, India, is a highly flood-prone area. During rains, the Bagmati river carries a discharge much in excess of the channel capacity and brings a huge quantity of debris and sediment from the erodible slopes of the Himalayan. Due to this, the Bagmati river basin is selected for the study to compare different statistical techniques. In this study, 33 years (i.e., from 1981 to 2014) National Centers for Environmental Prediction/National Center for Atmospheric Research (NCEP/NCAR) reanalysis outputs data and 33 years observed rainfall data from Indian Meteorological Department (IMD) Pune are used for different predictors. After comparing different statistical methods, factor analysis gave a better result than other statistical methods.

Keywords Downscaling · Factor analysis · Predictand · Potential predictor

46.1 Introduction

There are many factors that influence precipitation, which is unpredictable and very difficult to select variables/factors that affect precipitation. Various factor generates a number of data. The massive data generation, which is experienced in many

K. Keshav (✉) · V. Singh · R. Thendiyath
Department of Civil Engineering, National Institute of Technology Patna, Bihar 800005, India
e-mail: keshavpotters@gmail.com

V. Singh
e-mail: vsingh@nitp.ac.in

R. Thendiyath
e-mail: roshni@nitp.ac.in

R. Jha et al. (eds.), *Climate Change Impacts on Water Resources*, Water Science and Technology Library 98, https://doi.org/10.1007/978-3-030-64202-0_46

real-world applications nowadays, calls for multivariate methods for data analysis. Consideration of various factor increases the computational time, reduces the efficiency, performance, and output of the model. Implementation of statistical techniques reduces the computational time and increases the efficiency of the model. In order to minimize the influence of such noisy variables, some data reduction is usually necessary, variable selection method.

General Circulation Models (GCMs) provide the future estimations of different emission impacts, including the changes in precipitation and temperature of the earth's system, but for the regional studies, their resolution is too coarse for any analysis because in this resolution significant amount of information which includes the regional topography, etc., is ignored. GCM is a coarse resolution (Sylwia 2014) model up to 3.5° * 3.5° or 300 * 300 km grid resolution. It is composed of many homogeneous grid cells that represent horizontal and vertical areas on the globe, but the output from the GCM model cannot be directly fed into the hydrological model as these models perform well on local climate variables like precipitation, temperature.

Downscaling is the method of establishing a relationship between coarse-resolution GCM variables such as air temperature, geopotential heights, relative humidity, specific humidity, u wind, v wind, etc., and local climate variable such as precipitation, temperature, dew point, etc. This can be done by dynamical or statistical downscaling (Harpham and Wilby 2005). Dynamical downscaling uses RCM that takes the output from GCM directly and gives information about the more complex topography of the area at a higher resolution. Since error associated with dynamical downscaling and the complexity involved in it is higher than statistical downscaling, this method limited its use. While statistical downscaling though involves huge data requirements, but it is easier to operate and the error associated with it is also less. Statistical downscaling can be done by using transfer function, weather generator or by weather typing. The statistical downscaling using transfer function is the most efficient method of downscaling as it establishes a non-linear relationship between predictors (NCEP data) and the predictand (observed data) and its high potential for complex, non-linear, time-varying input and output.

Selection of predictors for statistical downscaling is crucial for establishing the best relationship between predictors (such as Relative humidity, Geopotential height, u-wind, specific humidity, etc.) and predictand (such as precipitation, temperature, etc.). In this paper, a statistical method (factor analysis, correlation coefficient, etc.) are used for selecting the best predictors for downscaling predictand. Factor analysis can help us establish that sets of observed variables are, in fact, all measuring the same underlying factor (perhaps with varying reliability) and so can be combined to form a more reliable measure of that factor. Factor analysis is a collection of methods used to examine how underlying constructs influence the responses on a number of measured variables. There are basically two types of factor analysis: exploratory and confirmatory. Exploratory factor analysis (EFA) attempts to discover the nature of the constructs influencing a set of responses. Confirmatory factor analysis (CFA) tests whether a specified set of constructs is influencing responses in a predicted way.

Chitra and Thampi (2013) used the ANN technique to study the GCM performance in two rain gauge stations, one in the Chaliyar river basin in Kerala and other stations close to it for two monsoon period southwest monsoon and south eat monsoon. Potential predictors for these two seasons was determined based on the correlation coefficient between NCEP data and observed data and also between NCEP and GCM data and separate model were developed for each season. GCM model used for simulation in this paper was third-generation CGCM3 for the twentieth century, 20 C3M. Simulated GCM data were corrected for biases and errors. Fiseha et al. (2012) presented his paper on statistical downscaling of precipitation and temperature in the Upper Tiber basin. This method utilized regression-based technique and stochastic weather generator Long Ashton Research Station Weather Generator (LARS-WG), compared their performance and estimated the strength of the model. HadCM3, GCM predictions of A2 and B2 scenarios were used to simulate the future predictions. Regression-based technique produced quite a fair result for precipitation and maximum and minimum temperature except for a few extremities. Cheng (2012) downscaled eight GCM's with regression-based downscale technique to project changes in the frequency of future daily streamflow events for a selected weather station for various meteorological sites in Canada in four selected river basins. This study thus added an extra application to determine a return period for daily streamflow.

Flood in north Bihar is a continuous and recurrent unabated phenomenon since the origin of the Himalayan Rivers, which transverse in these areas. The floods in Bihar are caused due to the southwest monsoon. North Bihar comprises of the alluvial plain of the Gangetic basin and is drained by a number of rivers originating from Nepal and outfalls into the Ganges. Bagmati River is one of these rivers. During rains, these rivers carry discharge much in excess of the channel capacity and bring a huge quantity of debris and sediment from the erodible slopes of Himalayan. After emerging from the hills, their slope remains fairly steep, which gradually becomes flattered towards the plains; thereby the velocity decreases, resulting in deposition of silt in river beds, particularly in the flatter reaches. This reduces the capacity of the river to carry the flood discharge and the floodwater overtops the bank causing inundation of the adjoining areas. It also happens sometimes that the river erodes its bank and changes the course abruptly, thus opening up new channels and causing damage to neighboring areas. There are some pockets in Bihar which are perennially waterlogged. This river is very vulnerable to high current and much devastation during a flood in the downstream reached for which hydrological studies have to be carried out. Flood is mostly happened due to precipitation. So, the Downscaling of precipitation depends on the selection of potential predictors.

46.2 Study Area and Data Used

The Bagmati River is a perennial river of Nepal and India, especially of North-Bihar. The Bagmati River originates from Shivpuri ranges of hills in Nepal at latitude 27° 47′ N and longitude 85° 17′E, and 16 km North-East of Kathmandu at an altitude of 1500 m above Mean Sea Level. The Bagmati River flows south-westerly 10-kilometer along the Kathmandu Valley, which is mainly rice-pattie in terraces up the slope. A number of obstructive rock surface obstruct the flow down in the valley; among these, the Pashupatinath Temple is constructed upon. After crossing the temple, the river flows south across the plain, where it is linked by the Monahara River and bends west direction. In Kathmandu, the river flows into many old important locations. Bagmati River joins river Kosi on the downstream side at Badlaghat and ultimately gets in into the Ganges. Its total length is 589 km and the total catchment area is 14384 km^2. From this area, 6500 km^2 falls in the Bihar region. It passes nearly 195 km in Nepal and rests 394 km in Bihar. It is one of the main rivers of north Bihar, known for large streams and much destruction during the flood. Major tributaries are Lakhandei, Lalbakeya and Adhwara group of rivers. The average annual rainfall of Bagmati Basin, including Adhwara, is 1255 mm.

For this study, monthly rainfall data of 34 years, that is, from 1981 to 2014, at five rain gauge stations, namely Dheng, Kamtaul, Muzzafarpur, Benibag and Hayaghat in Bagmati river basin have been used. Another set of mean monthly data of 32 years, that is, from 1981 to 2014, is obtained from NCEP/NCAR REANALYSIS project for the area under consideration at different pressure levels (1000 mb to 100 mb). A relationship is established between observed IMD data, which form the predictant (Wetterhall et al. 2005) and NCEP data by product-moment correlation formula (Chitra and Thampi 2013) and factor analysis method to obtain the potential predictor. These potential predictors are the atmospheric variables at a different pressure level, which is best correlated with the observed data taken.

46.3 Methodology

46.3.1 Selection of Potential Predictor

There are many factors that influence precipitation, which is unpredictable and very difficult to select variables/factors that affect precipitation. Various factor generates a number of data. The massive data generation, which is experienced in many real-world applications nowadays, calls for multivariate methods for data analysis. Consideration of various factor increases the computational time, reduces the efficiency, performance, and output of the model. Implementation of statistical techniques reduces the computational time and increases the efficiency of the model. In order to minimize the influence of such noisy variables, some data reduction is usually necessary, variable selection method. The selection of predictors for statistical downscaling is crucial for establishing the best relationship between predictors (such as Relative humidity, Geopotential height, u-wind, specific humidity, etc.) and predictant (such as precipitation, temperature, etc.). Two statistical techniques are used to find out the best predictors for downscaling the precipitation. The first technique is the Pearson product-moment correlation and another technique is factor analysis.

46.3.2 Pearson Product-Moment Correlation

The correlation coefficient between these set of data is calculated based on the product-moment correlation coefficient formula as given below:

$$\rho_{xy} = \frac{\sum_{n=1}^{N} (x_t - \bar{x})(y_t - \bar{y})}{N\sigma_x\sigma_{\bar{y}}} \tag{46.1}$$

where,

x_t =predictant, y_t = predictor, ρ_{xy} = correlation coefficient
σ_y =standard deviation of predictor, N = total number of data

In statistical problem, Pearson product-moment correlation evaluates the linear correlation between two variables in a population X and Y and gives a value between −1 to +1 developed by Karl Pearson from a related idea by Francis Galton in the 1880s. Here + signifies total positive correlation, that is, all data points lying on the same side of the line for which X increases when Y increases, −1 signifies total negative correlation, that is, total data points lying on the line and for which Y increases if X decreases and 0 signifies non-linear correlation between the variables as shown in the Fig. 46.1.

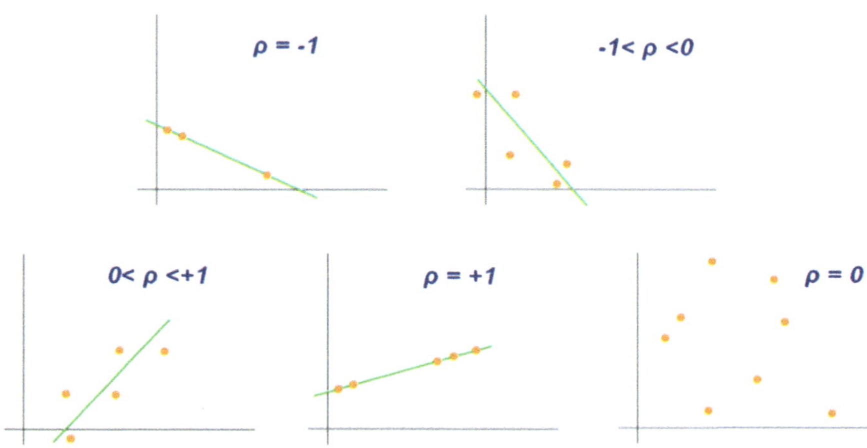

Fig. 46.1 Example of scatter diagram with different value of correlation

46.4 Factor Analysis

Factor analysis is a technique that is used to reduce a large number of variables into fewer numbers of factors. This technique extracts maximum common variance from all variables and puts them into a common score. As an index of all variables, we can use this score for further analysis. Factor analysis is part of the general linear model (GLM) and this method also assumes several assumptions: there is a linear relationship, there is no multicollinearity, it includes relevant variables into the analysis, and there is a true correlation between variables and factors.

Factor loading: Factor loading is basically the correlation coefficient for the variable and factor. Factor loading shows the variance explained by the variable on that particular factor. In the SEM approach, as a rule of thumb, 0.7 or higher factor loading represents that the factor extracts sufficient variance from that variable.

Eigenvalues: Eigenvalues is also called characteristic roots. Eigenvalues show variance explained by that particular factor out of the total variance. From the commonality column, we can know how much variance is explained by the first factor out of the total variance. For example, if our first factor explains 68% variance out of the total, this means that 32% variance will be explained by the other factor.

Factor score: The factor score is also called the component score. This score is all row and columns, which can be used as an index of all variables and can be used for further analysis. We can standardize this score by multiplying a common term. With this factor score, whatever analysis we will do, we will assume that all variables will behave as factor scores and will move.

Criteria for determining the number of factors: According to the Kaiser Criterion, Eigenvalues is a good criterion for determining a factor. If Eigenvalues is greater than one, we should consider that a factor and if Eigenvalues is less than one, then we should not consider that a factor. According to the variance extraction rule, it should be more than 0.7. If the variance is less than 0.7, then we should not consider that a factor.

Rotation method: The rotation method makes it more reliable to understand the output. Eigenvalues do not affect the rotation method, but the rotation method affects the Eigenvalues or percentage of variance extracted. There is a number of rotation methods available: (1) No rotation method, (2) Varimax rotation method, (3) Quartimax rotation method, (4) Direct oblimin rotation method, and (5) Promax rotation method. Each of these can be easily selected in SPSS, and we can compare our variance explained by those particular methods.

46.5 Result and Discussion

46.5.1 Selection of Potential Predictor by Pearson Product Moment Correlation

Potential predictors like geopotential heights, relative humidity, specific humidity, air temperature at different pressure level available in the NCEP/NCAR REANALYSIS model is correlated with observed mean monthly precipitation obtained from IMD, Pune. A value between 0.5 to 1.0 is considered a good correlation, and therefore those variables are selected as potential predictors for precipitation. The calculation for Potential predictor obtained for precipitation is shown in Table 46.1.

In Table 46.1, the correlation matrix is generated between rainfall and air temperature at different pressure levels (i.e., 250–1000 mb). Correlation of air temperature at different pressure levels is a good correlation with rainfall, but air temperature at 250 and 300 mb shows the best correlation, that is, 0.6998 and 0.6987 among other pressure level air temperature and air temperature at 500 mb pressure shows the least correlation with rainfall whose value is 0.5534. In Table 46.2 show the correlation value of all the predictors at different pressure level (i.e., 100–1000 mb) with rainfall. In Table 46.2, Geopotential height at pressure level at 100, 150, 200, 250, 300 and 400 mb shows the good correlation with precipitation and geopotential height at 1000, 925, 850 and 700 mb shows a poor correlation with precipitation.

46.5.2 Selection of Potential Predictor by Factor Analysis

Potential predictors like geopotential heights, relative humidity, specific humidity, air temperature at different pressure level available in NCEP/NCAR REANALYSIS model is factor analysed with observed mean monthly precipitation obtained from IMD, Pune and select the number of factors that influence/affect the precipitation. The following table shows the Eigenvalue and Factor pattern in Tables 46.3 and 46.4

In Table 46.3, among three factors f1, f2 and f3. f1 has eigenvalue greater than 1 and other factor f2 and f3 have eigenvalue less than 1. Variability of factor, f1 is **91.0269%** and other factors shows the variability only 8.98%. So, The Factor, whose eigenvalue is greater than 1 and Variability is nearest to 100% has been selected. In Table 46.4, by considering factor f1, the factor pattern of all the predictors are shown. From the above Table 46.4, specific humidity at pressure level 600 and 700mb has the value **0.9963** and **0.9918,** which is very close to 1. it shows that these two factors affect precipitation more.

Similarly, factor analysis can be done on other predictors and results are shown below in Table 46.5.

From Table 46.5, Geopotential height (Geop) at pressure level 50 mb and 70 mb is selected as a potential predictor, which shows value, that is, **0.9934** and **0.9968** very close to 1. Specific humidity (SH) at pressure levels 600 mb and 700 mb is

Table 46.1 Correlation matrix of rainfall versus air temperature at the different pressure level

Variables	rainfall	airtemp250mb	airtemp300mb	airtemp 400mb	airtemp500mb	airtemp600mb	airtemp700mb	airtemp850mb	airtemp925mb	airtemp1000mb
rainfall	**1**	0.6998	0.6987	0.6898	0.6890	0.6794	0.6482	0.5991	0.5747	0.5534
airtemp250mb	0.6998	**1**	0.9934	0.9684	0.9448	0.9225	0.8845	0.8267	0.7998	0.7713
airtemp300mb	0.6987	0.9934	**1**	0.9889	0.9695	0.9458	0.9012	0.8346	0.8056	0.7747
airtemp400mb	0.6898	0.9684	0.9889	**1**	0.9925	0.9725	0.9249	0.8502	0.8195	0.7876
airtemp500mb	0.6890	0.9448	0.9695	0.9925	**1**	0.9916	0.9526	0.8801	0.8503	0.8204
airtemp600mb	0.6794	0.9225	0.9458	0.9725	0.9916	**1**	0.9814	0.9237	0.8985	0.8730
airtemp700mb	0.6482	0.8845	0.9012	0.9249	0.9526	0.9814	**1**	0.9779	0.9628	0.9466
airtemp850mb	0.5991	0.8267	0.8346	0.8502	0.8801	0.9237	0.9779	**1**	0.9977	0.9920
airtemp925mb	0.5747	0.7998	0.8056	0.8195	0.8503	0.8985	0.9628	0.9977	**1**	0.9978
airtemp1000mb	0.5534	0.7713	0.7747	0.7876	0.8204	0.8730	0.9466	0.9920	0.9978	**1**

Table 46.2 Potential predictor for precipitation

Pressure level (mb)	Geopotential heights	Air temperature	Relative humidity	Specific humidity
1000	−0.66	0.55	0.66	0.73
925	−0.67	0.57	0.66	0.73
850	−0.68	0.6	0.69	0.74
700	−0.42	0.65	0.73	0.75
600	0.28	0.68	0.74	0.76
500	0.54	0.69	0.7	0.75
400	0.62	0.7	0.64	0.74
300	0.66	0.7	0.44	0.73
250	0.67	0.69		
200	0.68	0.14		
150	0.69	−0.29		
100	0.7	−0.29		

Table 46.3 Eigenvalues

	f1	f2	f3
Eigenvalue	8.1924	0.2100	0.0044
Variability (%)	91.0269	2.3334	0.0486
Cumulative (%)	91.0269	93.3603	93.4089

Table 46.4 Factor pattern of rainfall and specific humidity at different pr. level

	f1
RAINFALL	0.7581
SH 300mb	0.9477
SH 400mb	0.9589
SH 500mb	0.9773
SH 600mb	0.9961
SH 700mb	0.9913
SH 850mb	0.9818
SH 925mb	0.9777
SH 1000mb	0.9749

selected as a potential predictor, which shows value, that is, **0.9961** and **0.9913,** very close to 1. Relative humidity (rh) at pressure levels 600 mb and 700 mb is selected as the potential predictor, which shows value, that is, **0.9550** and **0.9675** close to 1. Air temperature (air temp) at pressure level 600 mb and 700 mb is selected as a potential predictor, which shows value, that is **0.9901** and **0.9892** close to 1.

Table 46.5 Potential Predictor for Precipitation by factor analysis

	f1		f1		f1
RAINFALL	**0.6914**	RAINFALL	**0.6846**	RAINFALL	**0.7581**
Geop250mb	**0.9724**	Geop10mb	**0.9392**	SH 300mb	**0.9477**
Geop300mb	**0.9700**	Geop20mb	**0.9715**	SH 400mb	**0.9589**
Geop400mb	**0.9549**	Geop30mb	**0.9799**	SH 500mb	**0.9773**
Geop500mb	**0.8937**	Geop50mb	**0.9934**	SH 600mb	**0.9961**
Geop600mb	0.6281	Geop70mb	**0.9968**	SH 700mb	**0.9913**
Geop700mb	−0.3298	Geop100mb	**0.9864**	SH 850mb	**0.9818**
Geop850mb	**−0.8840**	Geop150mb	**0.9718**	SH 925mb	**0.9777**
Geop925mb	**−0.9110**	Geop200mb	**0.9626**	SH 1000mb	**0.9749**
Geop1000mb	**−0.9141**				
	f1		f1		f1
RAINFALL	**0.7320**	Rainfall	**0.6802**	rainfall	**0.5198**
rh300	**0.6516**	airtemp250mb	**0.9369**	airtemp10mb	0.3883
rh400	**0.8967**	airtemp300mb	**0.9510**	airtemp20mb	**0.9115**
rh500	**0.9595**	airtemp400mb	**0.9630**	airtemp30mb	**0.8341**
rh600	**0.9550**	airtemp500mb	**0.9764**	airtemp50mb	0.4484
rh700	**0.9675**	airtemp600mb	**0.9901**	airtemp70mb	−0.0663
rh850	**0.9518**	airtemp700mb	**0.9892**	airtemp100mb	−0.2064
rh925	**0.9203**	airtemp850mb	**0.9485**	airtemp150mb	0.2151
rh1000	**0.9162**	airtemp925mb	**0.9253**	airtemp200mb	**0.7087**
		airtemp1000mb	**0.9005**		

46.6 Conclusions

Pearson product-moment correlation approach between rainfall and predictors (i.e., geopotential height, air temperature, relative humidity, specific humidity) at different pressure level, that is, from 100 to 1000mb gives positive correlation value between 0.5 and 0.75. Many predictors have the same correlation value. It is difficult to choose the best predictor for predictand (precipitation). So, factor analysis can be done to refine the Pearson product-moment correlation result in a better way. Using factor analysis, among three factors f1, f2 and f3. f1 has eigenvalue greater than 1 and other factor f2 and f3 have eigenvalue less than 1. The variability of factor, f1 is **91.0269%** and other factors show the variability only 8.98%. So, The Factor, whose eigenvalue is greater than 1 and variability is nearest to 100% has been selected. Factor pattern of rainfall with other predictors give a better value, which is nearer to 1. Geopotential height (Geop) at pressure level 50 mb and 70 mb is selected as a potential predictor, which shows value, that is, **0.9934** and **0.9968** very close to 1. Specific humidity (SH) at pressure levels 600 mb and 700 mb is selected as a potential predictor, which shows value, that is, **0.9961** and **0.9913** very close to 1. Relative humidity (rh) at pressure levels 600 mb and 700 mb is selected as a potential predictor, which shows value, that is, **0.9550** and **0.9675** close to 1.

Finally, after a comparison of the two statistical methods, factor analysis gave a better and clear figure than the Pearson product-moment correlation approach.

References

Cheng CS, Auld H, Li Q, Li G (2012) Possible impacts of climate change on extreme weather events at local scale in south–central Canada. Climatic Change 112(3–4):963–979

Chitra NR, Thampi SG (2013) International Journal of Innovative Research in Science, Engineering and Technology 2:6–11

Fiseha BM, Melesse AM, Romano E, Volpi E, Fiori A et al. (2012) Statistical downscaling of precipitation and temperature for the upper tiber basin in central Italy. Int J Water Sci 1

Ghosh S, Mujumdar PP (2008) Statistical downscaling of GCM simulations to streamflow using relevance vector machine. Adv Water Resourc 31(1):132–146

Harpham C, Wilby RL (2005) Multi-site downscaling of heavy daily precipitation occurrence and amounts. J Hydrol 312:235–255

Hunter D, Yu H, Pukish III MS, Kolbusz J, Wilamowski BM (2003) Selection of proper neural network sizes and architectures—a comparative study. IEEE Trans Indus Inform 8:228–240

Raje D, Mujumdar PP (2009) A conditional random field– based downscaling method for assessment of climate change

Skelly WC, Henderson-Sellers ANN (1996) Grid box or grid point: what type of data do GCMs deliver to climate impacts researchers? Int J Climatol 079–1086

Trzaska S, Schnarr E (2014) A review of downscaling methods for climate change projections. African And Latin American Resilience To Climate Change Center for International Earth Science Information Network (CIESIN), p 56

Tisseuil C, Vrac M, Lek S, Wade AJ (2010) Statistical downscaling of river flows. J Hydrol 385:279–291

Veitch D (2005) Wavelet neural networks and their application in the study of dynamical systems. Dissertation submitted for the MSc in Data Analysis, Networks And Non-linear Dynamics

Wetterhall F, Halldin S, Xu CY (2005) Statistical precipitation downscaling in central Sweden with the analogue method. J Hydrol 306:136–174